SCHAUM'S OUTLINE OF

THEORY AND PROBLEMS

of

STRENGTH of MATERIALS
Second Edition

•

BY

WILLIAM A. NASH, Ph.D.

Professor of Civil Engineering
University of Massachusetts

SCHAUM'S OUTLINE SERIES
McGRAW-HILL, INC.

New York   St. Louis   San Francisco   Auckland   Bogotá
Caracas   Lisbon   London   Madrid   Mexico   Milan
Montreal   New Delhi   Paris   San Juan   Singapore
Sydney   Tokyo   Toronto

07-045894-4

18 19 20 21 22 23 24 25 26 27 28 29 30 SH SH 9 8 7 6 5 4 3 2 1

Cover design by Amy E. Becker.

Preface

This second edition of *Theory and Problems of Strength of Materials* adheres to the basic plan of the first edition, but with a considerable broadening of scope. As in the earlier edition, the contents are divided into chapters covering duly-recognized areas of theory and study. Each chapter begins with a summary of the pertinent definitions, principles, and theorems, followed by graded sets of solved and supplementary problems. Derivations of formulas and proofs of theorems are included among the solved problems. The problems have been chosen and solutions arranged so that the principles are clearly established. They serve to illustrate and amplify the theory, provide the repetition of basic principles so vital to effective teaching, and bring into sharp focus those fine points which are essential to a complete understanding.

Since publication of the first edition, course offerings in strength of materials have become more sophisticated and frequently tend to include topics in plastic analysis and design, treatments of shear centers, curved beams, and use of singularity functions to describe beam behavior. Also, it has become rather common to introduce the student to strain energy methods of analysis and the elements of the theory of elasticity. The present edition includes solved problems in all these areas, as well as in others not previously represented.

The author is deeply indebted to his wife, Verna B. Nash, and his children Rebecca and Phillip, for their patience and understanding during the preparation of the manuscript.

WILLIAM A. NASH

Amherst, Massachusetts
March 1972

CONTENTS

CONTENTS

CONTENTS

Chapter 1

Tension and Compression

INTERNAL EFFECTS OF FORCES

In this book we shall be concerned with what might be called the *internal effects* of forces acting on a body. The bodies themselves will no longer be considered to be perfectly rigid as was assumed in statics; instead, the calculation of the deformations of various bodies under a variety of loads will be one of our primary concerns in the study of strength of materials.

AXIALLY LOADED BAR

The simplest case to consider at the start is that of an initially straight metal bar of constant cross-section, loaded at its ends by a pair of oppositely directed collinear forces coinciding with the longitudinal axis of the bar and acting through the centroid of each cross-section. For static equilibrium the magnitudes of the forces must be equal. If the forces are directed away from the bar, the bar is said to be in *tension*; if they are directed toward the bar, a state of *compression* exists. These two conditions are illustrated in Fig. 1-1.

Under the action of this pair of applied forces, internal resisting forces are set up within the bar and their characteristics may be studied by imagining a plane to be passed through the bar anywhere along its length and oriented perpendicular to the longitudinal axis of the bar. Such a plane is designated as *a-a* in Fig. 1-2(a). For reasons to be discussed later, this plane should not be "too close" to either end of the bar. If for purposes of analysis the portion of the bar to the right of this plane is considered to be removed, as in Fig. 1-2(b), then it must be replaced by whatever effect it exerts upon the left portion. By this technique of introducing a cutting plane, the originally internal forces now become external with respect to the remaining portion of the body. For equilibrium of the portion to the left this "effect" must be a horizontal force of magnitude P. However, this force P acting normal to the cross-section *a-a* is actually the resultant of distributed forces acting over this cross-section in a direction normal to it.

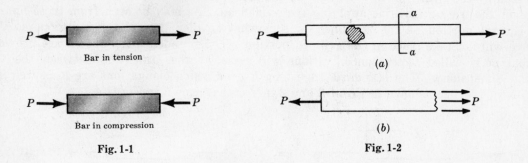

Bar in tension

(a)

Bar in compression

(b)

Fig. 1-1

Fig. 1-2

1

DISTRIBUTION OF RESISTING FORCES

At this point it is necessary to make some assumption regarding the manner of variation of these distributed forces, and since the applied force P acts through the centroid it is commonly assumed that they are uniform across the cross-section. Such a distribution is probably never realized exactly because of the random orientation of the crystalline grains of which the bar is composed. The exact value of the force acting on some very small element of area of the cross-section is a function of the nature and orientation of the crystalline structure at that point. However, over the entire cross-section the variation is described with reasonable engineering accuracy by the assumption of a uniform distribution.

NORMAL STRESS

Instead of speaking of the internal force acting on some small element of area, it is better for comparative purposes to treat the normal force acting over a *unit* area of the cross-section. The intensity of normal force per unit area is termed the normal *stress* and is expressed in units of force per unit area, e.g. lb/in². The phrase *total stress* is sometimes used to denote the resultant axial force in pounds. If the forces applied to the ends of the bar are such that the bar is in tension, then *tensile stresses* are set up in the bar; if the bar is in compression we have *compressive stresses*. It is essential that the line of action of the applied end forces pass through the centroid of each cross-section of the bar.

TEST SPECIMENS

The axial loading shown in Fig. 1-2(a) occurs frequently in structural and machine design problems. To simulate this loading in the laboratory, a test specimen is held in the grips of either an electrically driven gear-type testing machine or a hydraulic machine. Both of these machines are commonly used in materials testing laboratories for applying axial tension.

In an effort to standardize materials testing techniques the American Society for Testing Materials, commonly abbreviated A.S.T.M., has issued specifications that are in common use throughout this country. More than a score of different type specimens are prescribed for various metallic and nonmetallic materials for both axial tension and axial compression tests. For the present only two of these will be mentioned here, one for metal plates thicker than 3/16 in. and appearing as in Fig. 1-3, the other for metals over 1.5 in. thick and having the appearance shown in Fig. 1-4. The dimensions shown are those specified by the A.S.T.M. but the ends of the test specimens may be of any shape to fit the grips of the testing machine applying the axial load. As may be seen from these figures, the central portion of the specimen is somewhat smaller than the end regions so that failure will not take place in the gripped portion. The rounded fillets shown are provided so that no so-called stress concentrations will arise at the transition between the two lateral dimensions. The standard gage length over which elongations are measured is 8 in. for the specimen shown in Fig. 1-3 and 2 in. for that shown in Fig. 1-4.

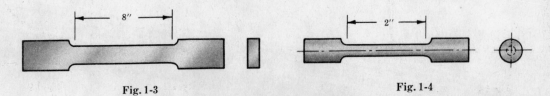

Fig. 1-3   Fig. 1-4

The elongations are measured by either mechanical or optical extensometers or by cementing an electric resistance-type strain gage to the surface of the material. This resistance strain gage consists of a number of very fine wires oriented in the axial direction of the bar. As the bar elongates, the electrical resistance of the wires changes and this change of resistance is detected on a Wheatstone bridge and interpreted as elongation.

NORMAL STRAIN

Let us suppose that one of these tension specimens has been placed in a tension-compression testing machine and tensile forces gradually applied to the ends. The elongation over the gage length may be measured as indicated above for any predetermined increments of the axial load. From these values the elongation per unit length, which is termed *normal strain* and denoted by ϵ, may be found by dividing the total elongation Δ by the gage length L, i.e. $\epsilon = \Delta/L$. The strain is usually expressed in units of inches per inch and consequently is dimensionless. *Total strain* is sometimes used to denote the elongation in inches.

STRESS-STRAIN CURVE

As the axial load is gradually increased in increments, the total elongation over the gage length is measured at each increment of load and this is continued until fracture of the specimen takes place. Knowing the original cross-sectional area of the test specimen the *normal stress*, denoted by σ, may be obtained for any value of the axial load merely by the use of the relation

$$\sigma = \frac{P}{A}$$

where P denotes the axial load in pounds, and A the original cross-sectional area. Having obtained numerous pairs of values of normal stress σ and normal strain ϵ, the experimental data may be plotted with these quantities considered as ordinate and abscissa respectively. This is the *stress-strain curve* or *diagram* of the material for this type of loading. Stress-strain diagrams assume widely differing forms for various materials. Figure 1-5 is the stress-strain diagram for a medium-carbon structural steel, Fig. 1-6 is for an alloy steel, and Fig. 1-7 is for hard steels and certain nonferrous alloys. For nonferrous alloys and cast iron the diagram has the form indicated in Fig. 1-8, while for rubber the plot of Fig. 1-9 is typical.

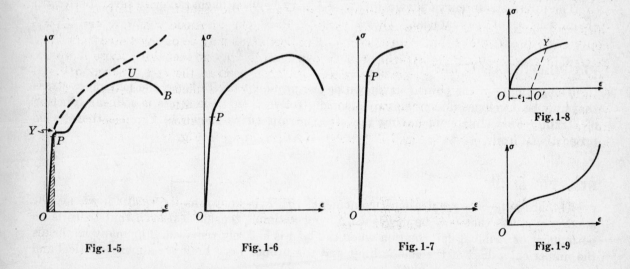

Fig. 1-5   Fig. 1-6   Fig. 1-7   Fig. 1-8

Fig. 1-9

DUCTILE AND BRITTLE MATERIALS

Metallic engineering materials are commonly classed as either *ductile* or *brittle* materials. A *ductile material* is one having a relatively large tensile strain up to the point of rupture (for example, structural steel or aluminum) whereas a *brittle material* has a relatively small strain up to this same point. An arbitrary strain of 0.05 in/in is frequently taken as the dividing line between these two classes of materials. Cast iron and concrete are examples of brittle materials.

HOOKE'S LAW

For any material having a stress-strain curve of the form shown in Fig. 1-5, 1-6, or 1-7, it is evident that the relation between stress and strain is linear for comparatively small values of the strain. This linear relation between elongation and the axial force causing it (since these quantities respectively differ from the strain or the stress only by a constant factor) was first noticed by Sir Robert Hooke in 1678 and is called *Hooke's law*. To describe this initial linear range of action of the material we may consequently write

$$\sigma = E\epsilon$$

where E denotes the slope of the straight-line portion OP of each of the curves in Figs. 1-5, 1-6, and 1-7.

MODULUS OF ELASTICITY

The quantity E, i.e. the ratio of the unit stress to the unit strain, is the *modulus of elasticity* of the material in tension, or, as it is often called, *Young's modulus*. Values of E for various engineering materials are tabulated in handbooks. Since the unit strain ϵ is a pure number (being a ratio of two lengths) it is evident that E has the same units as does the stress, for example lb/in². For many common engineering materials the modulus of elasticity in compression is very nearly equal to that found in tension. *It is to be carefully noted that the behavior of materials under load as discussed in this book is restricted (unless otherwise stated) to the linear region of the stress-strain curve.*

MECHANICAL PROPERTIES OF MATERIALS

The stress-strain curve shown in Fig. 1-5 may be used to characterize several strength characteristics of the material. They are:

PROPORTIONAL LIMIT

The ordinate of the point P is known as the *proportional limit*, i.e. the maximum stress that may be developed during a simple tension test such that the stress is a linear function of strain. For a material having the stress-strain curve shown in Fig. 1-8 there is no proportional limit.

ELASTIC LIMIT

The ordinate of a point almost coincident with P is known as the *elastic limit*, i.e. the maximum stress that may be developed during a simple tension test such that there is no permanent or residual deformation when the load is entirely removed. For many materials the numerical values of the elastic limit and the proportional limit are almost identical and the terms are sometimes used synonymously. In those cases where the distinction between the two values is evident the elastic limit is almost always greater than the proportional limit.

ELASTIC AND PLASTIC RANGES

That region of the stress-strain curve extending from the origin to the proportional limit is called the *elastic range*; that region of the stress-strain curve extending from the proportional limit to the point of rupture is called the *plastic range*.

YIELD POINT

The ordinate of the point Y, denoted by σ_{yp}, at which there is an increase in strain with no increase in stress is known as the *yield point* of the material. After loading has progressed to the point Y, yielding is said to take place. Some materials exhibit two points on the stress-strain curve at which there is an increase of strain without an increase of stress. These are called *upper* and *lower yield points*.

ULTIMATE STRENGTH OR TENSILE STRENGTH

The ordinate of the point U, the maximum ordinate to the curve, is known either as the *ultimate strength* or the *tensile strength* of the material.

BREAKING STRENGTH

The ordinate of the point B is called the *breaking strength* of the material.

MODULUS OF RESILIENCE

The work done on a unit volume of material, as a simple tensile force is gradually increased from zero to such a value that the proportional limit of the material is reached, is defined as the *modulus of resilience*. This may be calculated as the area under the stress-strain curve from the origin up to the proportional limit and is represented as the shaded area in Fig. 1-5. The units of this quantity are in-lb/in³. Thus, resilience of a material is its ability to absorb energy in the elastic range.

MODULUS OF TOUGHNESS

The work done on a unit volume of material as a simple tensile force is gradually increased from zero to the value causing rupture is defined as the *modulus of toughness*. This may be calculated as the entire area under the stress-strain curve from the origin to rupture. Toughness of a material is its ability to absorb energy in the plastic range of the material.

PERCENTAGE REDUCTION IN AREA

The decrease in cross-sectional area from the original area upon fracture divided by the *original* area and multiplied by 100 is termed *percentage reduction in area*. It is to be noted that when tensile forces act upon a bar, the cross-sectional area decreases, but calculations for the normal stress are usually made upon the basis of the original area. This is the case for the curve shown in Fig. 1-5. As the strains become increasingly larger it is more important to consider the instantaneous values of the cross-sectional area (which are decreasing), and if this is done the *true* stress-strain curve is obtained. Such a curve has the appearance shown by the dashed line in Fig. 1-5.

PERCENTAGE ELONGATION

The increase in length (of the gage length) after fracture divided by the initial length and multiplied by 100 is the *percentage elongation*. Both the percentage reduction in area and the percentage elongation are considered to be measures of the *ductility* of a material.

WORKING STRESS

The above-mentioned strength characteristics may be used to select a so-called *working stress*. Throughout this book all working stresses will be within the elastic range of the material. Frequently such a stress is determined merely by dividing either the stress at yield or the ultimate stress by a number termed the *safety factor*. Selection of the safety factor is based upon the designer's judgment and experience. Specific safety factors are sometimes specified in building codes. See Problems 1.4, 1.12, 1.13.

STRAIN HARDENING

If a ductile material can be stressed considerably beyond the yield point without failure, it is said to *strain harden*. This is true of many structural metals.

The nonlinear stress-strain curve of a brittle material, shown in Fig. 1-8, characterizes several other strength measures that cannot be introduced if the stress-strain curve has a linear region. They are:

YIELD STRENGTH

The ordinate to the stress-strain curve such that the material has a predetermined permanent deformation or "set" when the load is removed is called the *yield strength* of the material. The permanent set is often taken to be either 0.002 or 0.0035 in. per in. These values are of course arbitrary. In Fig. 1-8 a set ϵ_1 is denoted on the strain axis and the line $O'Y$ is drawn parallel to the initial tangent to the curve. The ordinate of Y represents the yield strength of the material, sometimes called the *proof stress*.

TANGENT MODULUS

The rate of change of stress with respect to strain is known as the *tangent modulus* of the material. It is essentially an instantaneous modulus given by $E_t = d\sigma/d\epsilon$.

There are other characteristics of a material that are useful in design considerations. They are:

COEFFICIENT OF LINEAR EXPANSION

This is defined as the change of length per unit length of a straight bar subject to a temperature change of one degree. The value of this coefficient is independent of the unit of length but does depend upon the temperature scale used. Usually we will consider the Fahrenheit scale, in which case the coefficient denoted by α is given for steel, for instance, as 6.5×10^{-6} per F°. Temperature changes in a structure give rise to internal stresses just as do applied loads. See Problem 1.7.

POISSON'S RATIO

When a bar is subject to a simple tensile loading there is an increase in length of the bar in the direction of the load, but a decrease in the lateral dimensions perpendicular to the load. The ratio of the strain in the lateral direction to that in the axial direction is defined as *Poisson's ratio*. It is denoted in this book by the Greek letter μ. For most metals it lies in the range 0.25 to 0.35. See Problems 1.16–1.20.

GENERAL FORM OF HOOKE'S LAW

The simple form of Hooke's law has been given for axial tension when the loading is entirely along one straight line, i.e. uniaxial. Only the deformation in the direction of the load was considered and it was given by

$$\epsilon = \frac{\sigma}{E}$$

In the more general case an element of material is subject to three mutually perpendicular normal stresses $\sigma_x, \sigma_y, \sigma_z$, which are accompanied by the strains $\epsilon_x, \epsilon_y, \epsilon_z$ respectively. By superposing the strain components arising from lateral contraction due to Poisson's effect upon the direct strains we obtain the general statement of Hooke's law:

$$\epsilon_x = \frac{1}{E}[\sigma_x - \mu(\sigma_y + \sigma_z)] \qquad \epsilon_y = \frac{1}{E}[\sigma_y - \mu(\sigma_x + \sigma_z)] \qquad \epsilon_z = \frac{1}{E}[\sigma_z - \mu(\sigma_x + \sigma_y)]$$

See Problems 1.17 and 1.20.

ELASTIC VERSUS PLASTIC ANALYSIS

Stresses and deformations in the plastic range of action of a material are frequently permitted in certain structures. Some building codes allow particular structural members to undergo plastic deformation, and certain components of aircraft and missile structures are deliberately designed to act in the plastic range so as to achieve weight savings. Furthermore, many metal-forming processes involve plastic action of the material. For small plastic strains of low- and medium-carbon structural steels the stress-strain curve of Fig. 1-5 is usually idealized by two straight lines, one with a slope of E, representing the elastic range, the other with zero slope representing the plastic range. This plot, shown in Fig. 1-10, represents a so-called *elastic, perfectly-plastic material*. It takes no account of still larger plastic strains occurring in the strain hardening region shown as the right portion of the stress-strain curve of Fig. 1-5. See Problem 1.21.

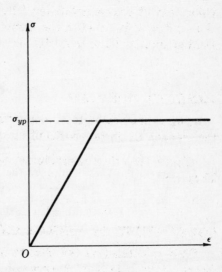

Fig. 1-10

CLASSIFICATION OF MATERIALS

This entire discussion has been based upon the assumptions that two characteristics prevail in the material. They are that we have a

HOMOGENOUS MATERIAL, one with the same elastic properties (E, μ) at all points in the body, and an

ISOTROPIC MATERIAL, one having the same elastic properties in all directions at any one point of the body. Not all materials are isotropic. If a material does not possess any kind of elastic symmetry it is called *anisotropic,* or sometimes *aeolotropic.* Instead of having two independent elastic constants (E, μ) as an isotropic material does, such a substance has 21 elastic constants. If the material has three mutually perpendicular planes of elastic symmetry it is said to be *orthotropic.* The number of independent constants is 9 in this case. This book considers only the analysis of isotropic materials.

DYNAMIC EFFECTS

In determination of mechanical properties of a material through a tension or compression test, the rate at which loading is applied sometimes has a significant influence upon the results. In general, ductile materials exhibit the greatest sensitivity to variations in loading rate, whereas the effect of testing speed on brittle materials, such as cast iron, has been found to be negligible. In the case of mild steel, a ductile material, it has been found that the yield point may be increased as much as 170 percent by extremely rapid application of axial force. It is of interest to note, however, that for this case the total elongation remains unchanged from that found for slower loadings.

Solved Problems

1.1. Determine the total elongation of an initially straight bar of length L, cross-sectional area A, and modulus of elasticity E if a tensile load P acts on the ends of the bar.

The unit stress in the direction of the force P is merely the load divided by the cross-sectional area, i.e. $\sigma = P/A$. Also the unit strain ϵ is given by the total elongation Δ divided by the original length, i.e. $\epsilon = \Delta/L$. By definition the modulus of elasticity E is the ratio of σ to ϵ, i.e.

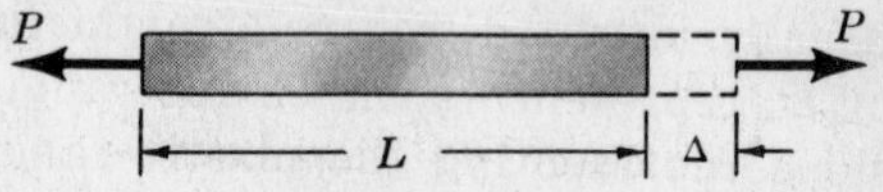

Fig. 1-11

$$E = \frac{\sigma}{\epsilon} = \frac{P/A}{\Delta/L} = \frac{PL}{A\Delta} \quad \text{or} \quad \Delta = \frac{PL}{AE}$$

Note that Δ has the units of length, perhaps in. or ft.

1.2. A surveyors' steel tape 100 ft long has a cross-section of 0.250 in. by 0.03 in. Determine the elongation when the entire tape is stretched and held taut by a force of 12 lb. The modulus of elasticity is 30×10^6 lb/in^2.

$$\text{elongation } \Delta = \frac{PL}{AE} = \frac{(12)(100 \times 12)}{(0.250)(0.03)(30 \times 10^6)} = 0.0640 \text{ in.}$$

1.3. A steel bar of cross-section 1 in^2 is acted upon by the forces shown in Fig. 1-12(a). Determine the total elongation of the bar. For steel, $E = 30 \times 10^6$ lb/in^2.

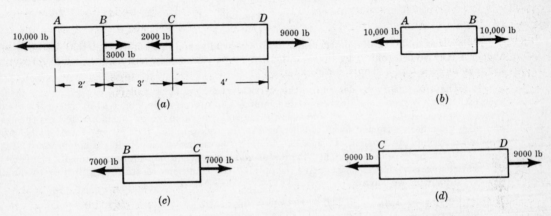

Fig. 1-12

The entire bar is in equilibrium, hence all portions of it are also. The portion of the bar between A and B has a resultant force of 10,000 lb acting over every cross-section, hence a free-body diagram of this 2 ft length appears as in Fig. 1-12(b) above. The force at the right end of this segment must be 10,000 lb to maintain equilibrium with the applied force at the left end. The elongation of this portion is

$$\Delta_1 = \frac{PL}{AE} = \frac{10,000(24)}{(1)(30 \times 10^6)} = 0.0080 \text{ in.}$$

The force acting in the segment between B and C is found by considering the algebraic sum of the forces to the left of a section between B and C. This indicates that a resultant force of 7000 lb acts to the left, i.e. the section has a tensile force acting upon it. This same result could of course have been obtained by considering the algebraic sum of the forces to the right of this section. Consequently the free-body diagram of the segment BC appears as in Fig. 1-12(c). The elongation of this portion is

$$\Delta_2 = \frac{7000(36)}{(1)(30 \times 10^6)} = 0.0084 \text{ in.}$$

Similarly, the force acting over any cross-section between C and D must be 9000 lb to maintain equilibrium with the applied load at D. The free-body diagram of the segment CD appears as in Fig. 1-12(d). The elongation of this portion is

$$\Delta_3 = \frac{9000(48)}{(1)(30 \times 10^6)} = 0.0144 \text{ in.}$$

The total elongation is consequently $\Delta = 0.0080 + 0.0084 + 0.0144 = 0.0308$ in.

1.4. The Howe truss shown in Fig. 1-13(a) supports the single load of 120,000 lb. If the working stress of the material in tension is taken to be 20,000 lb/in^2, determine the required cross-sectional area of bars DE and AC. Find the elongation of bar DE over its 20 ft length. Assume that the limiting value of the working stress in tension is the only factor to be considered in determining the required area. Take the modulus of elasticity of the bar to be 30×10^6 lb/in^2.

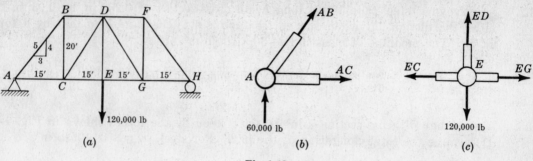

Fig. 1-13

This truss is statically determinate both externally and internally, i.e. the reactions at the supports may be determined by the equations of static equilibrium and also the axial force in each bar may be found by a simple statics analysis.

It is first necessary to determine the vertical reactions at A and H. By symmetry these are each 60,000 lb. A free-body diagram of the joint at A appears as in Fig. 1-13(b) where the unknown forces in the bars have been denoted as AB and AC, the same designations as the bars themselves, and they have been assumed to be tensile forces. In this manner if they are found to be positive they actually indicate tension. If they are found to be negative they indicate compression, and the signs thus obtained are in agreement with the usual sign convention designating tensile forces as positive and compressive forces as negative. Applying the equations of static equilibrium to the above free-body diagram we have

$$\Sigma F_v = 60,000 + \tfrac{4}{5}(AB) = 0 \qquad \text{or} \qquad AB = -75,000 \text{ lb}$$
$$\Sigma F_h = \tfrac{3}{5}(-75,000) + AC = 0 \qquad \text{or} \qquad AC = 45,000 \text{ lb}$$

Likewise, a free-body diagram of the point at E appears as in Fig. 1-13(c) above. From statics,
$$\Sigma F_v = ED - 120,000 = 0 \qquad \text{or} \qquad ED = 120,000 \text{ lb}$$

The simple consideration of trusses used here assumes all bars are so-called two-force members, i.e. subject to either axial tension or compression and no other loadings.

For axial loading the stress is given by $\sigma = P/A$, where P is the axial force and A the cross-sectional area of the bar. Here, the stress is given as 20,000 lb/in^2 in each bar and the areas are thus given by

$$A_{DE} = 120,000/20,000 = 6 \text{ in}^2 \qquad \text{and} \qquad A_{AC} = 45,000/20,000 = 2.25 \text{ in}^2$$

The elongation of a bar under axial tension is given by $\Delta = PL/AE$. For bar DE we have

$$\Delta = \frac{(120,000)(240)}{(6)(30 \times 10^6)} = 0.160 \text{ in.}$$

1.5. Determine the total increase of length of a bar of constant cross-section hanging vertically and subject to its own weight as the only load. The bar is initially straight.

The normal stress (tensile) over any horizontal cross-section is caused by the weight of the material below that section. The elongation of the element of thickness dy shown is

$$d\Delta = (Ay\gamma/AE)\,dy$$

where A denotes the cross-sectional area of the bar and γ its specific weight (weight/unit volume). Integrating, the total elongation of the bar is

$$\Delta = \int_0^L \frac{Ay\gamma\,dy}{AE} = \frac{A\gamma}{AE}\frac{L^2}{2} = \frac{(A\gamma L)L}{2AE} = \frac{WL}{2AE}$$

where W denotes the total weight of the bar. Note that the total elongation produced by the weight of the bar is equal to that produced by a load of half its weight applied at the end.

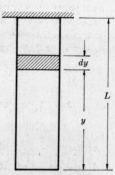

Fig. 1-14

1.6. A steel wire 1/4 in. in diameter is used for hoisting purposes in building construction. If 500 ft of the wire is hanging vertically, and a load of 300 lb is being lifted at the lower end of the wire, determine the total elongation of the wire. The specific weight of the steel is 0.283 lb/in³ and $E = 30 \times 10^6$ lb/in².

The total elongation is caused partially by the applied force of 300 lb and partially by the weight of the wire. The elongation due to the 300 lb load is

$$\Delta_1 = \frac{PL}{AE} = \frac{(300)(500 \times 12)}{\frac{1}{4}\pi(\frac{1}{4})^2(30 \times 10^6)} = 1.27 \text{ in.}$$

From Problem 1.5 the elongation due to the weight of the wire is

$$\Delta_2 = \frac{WL}{2AE} = \frac{\frac{1}{4}\pi(\frac{1}{4})^2(500 \times 12)(0.283)(500 \times 12)}{2(\frac{1}{4}\pi)(\frac{1}{4})^2(30 \times 10^6)} = 0.170 \text{ in.}$$

Thus the total elongation is $\Delta = 1.27 + 0.17 = 1.44$ in.

1.7. A straight aluminum wire 100 ft long is subject to a tensile stress of 10,000 lb/in². Determine the total elongation of the wire. What temperature change would produce this same elongation? Take $E = 10 \times 10^6$ lb/in² and α (the coefficient of linear expansion) $= 12.8 \times 10^{-6}$/F°.

The total elongation is given by $\Delta = \dfrac{PL}{AE} = \dfrac{(10,000)(100 \times 12)}{10 \times 10^6} = 1.20$ in.

A rise in temperature of ΔT would cause this same expansion if

$$1.20 = (12.8 \times 10^{-6})(100 \times 12)(\Delta T) \quad \text{and} \quad \Delta T = 78.2 \text{ F}°$$

1.8. Two prismatic bars are rigidly fastened together and support a vertical load of 10,000 lb as shown. The upper bar is steel having specific weight 0.283 lb/in³, length 35 ft, and cross-sectional area 10 in². The lower bar is brass having specific weight 0.300 lb/in³, length 20 ft and cross-sectional area 8 in². For steel $E = 30 \times 10^6$ lb/in², for brass $E = 13 \times 10^6$ lb/in². Determine the maximum stress in each material.

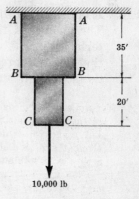

Fig. 1-15

The maximum stress in the brass bar occurs just below the junction at section *B-B*. There, the vertical normal stress is caused by the combined effect of the load of 10,000 lb together with the weight of the entire brass bar below *B-B*.

The weight of the brass bar is $W_b = (20 \times 12)(8)(0.300) = 576$ lb.

The stress at this section is $\sigma = \dfrac{P}{A} = \dfrac{10,000 + 576}{8} = 1320$ lb/in².

The maximum stress in the steel bar occurs at section *A-A*, the point of suspension, because there the entire weight of the steel and brass bars gives rise to normal stress, whereas at any lower section only a portion of the weight of the steel would be effective in causing stress.

The weight of the steel bar is $W_s = (35 \times 12)(10)(0.283) = 1185$ lb.

The stress across section *A-A* is $\sigma = \dfrac{P}{A} = \dfrac{10,000 + 576 + 1185}{10} = 1180$ lb/in².

1.9. A solid truncated conical bar of circular cross-section tapers uniformly from a diameter d at its small end to D at the large end. The length of the bar is L. Determine the elongation due to an axial force P applied at each end. See Fig. 1-16.

The coordinate x describes the distance from the small end of a disc-like element of thickness dx. The radius of this small element is readily found by similar triangles:

$$r = \frac{d}{2} + \frac{x}{L}\left(\frac{D-d}{2}\right)$$

The elongation of this disc-like element may be found by applying the formula for extension due to axial loading, $\Delta = PL/AE$. For the element, this expression becomes

$$d\Delta = \frac{P\,dx}{\pi\left[\dfrac{d}{2} + \dfrac{x}{L}\left(\dfrac{D-d}{2}\right)\right]^2 E}$$

The extension of the entire bar is obtained by summing the elongations of all such elements over the bar. This is of course done by integrating. If Δ denotes the elongation of the entire bar,

$$\Delta = \int_0^L d\Delta = \int_0^L \frac{4P\,dx}{\pi[d + (x/L)(D-d)]^2 E} = \frac{4PL}{\pi Dd\,E}$$

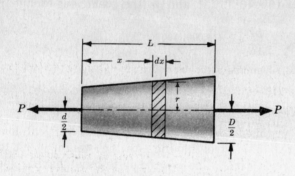

Fig. 1-16

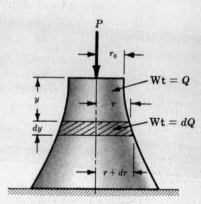

Fig. 1-17

1.10. A body having the form of a solid of revolution supports a load P as shown in Fig. 1-17. The radius of the upper base of the body is r_0 and the specific weight of the material is γ lb/ft³. Determine how the radius should vary with the altitude in order that the compressive stress at all cross-sections should be constant. The weight of the solid is not negligible.

Let y be measured from the upper base as shown and let Q denote the weight of that portion of the body of altitude y. Then dQ represents the increment to Q in the increment of altitude dy. Let r and $r + dr$ denote the radii of the upper and lower surfaces respectively of this horizontal element and A and $A + dA$ the corresponding areas. Considering the normal compressive stresses acting over both surfaces of this element, we have

$$\frac{P+Q}{A} = \frac{P+Q+dQ}{A+dA} = \sigma = \text{constant}$$

from which

$$\frac{dA}{dQ} = \frac{A}{P+Q} = \frac{1}{\sigma} \tag{1}$$

The increment of area between the upper and lower faces of the element is $dA = \pi(r+dr)^2 - \pi r^2 = 2\pi r\,dr$; the increment of weight is $dQ = \pi r^2\gamma(dy)$. Then from (1), $\dfrac{2\pi r(dr)}{\pi r^2\gamma(dy)} = \dfrac{1}{\sigma}$. Integrating,

$$2\ln r = (\gamma/\sigma)y + C_1$$

Applying the boundary condition that $r = r_0$ when $y = 0$, we find $C_1 = 2 \ln r_0$. Also from the conditions at the upper base, $\sigma = P/\pi r_0^2$. Finally, $r = r_0 e^{\gamma \pi r_0^2 y/2P}$.

1.11. Two identical steel bars are pin-connected and support a load of 100,000 lb as shown in Fig. 1-18(a). Find the required cross-sectional area of the bars so that the normal stress in them is no greater than 30,000 lb/in². Also find the vertical displacement of the point B. Take $E = 30 \times 10^6$ lb/in².

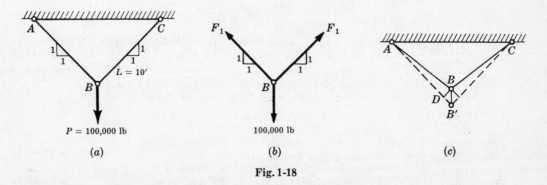

Fig. 1-18

A free-body diagram of the pin at B is shown in Fig. 1-18(b), where F_1 represents the force (lb) in each bar.

From statics: $\Sigma F_v = 2(1/\sqrt{2})F_1 - 100,000 = 0$ or $F_1 = 70,700$ lb

Hence the required area is $A = 70,700/30,000 = 2.35$ in².

Because our study of strength of materials is restricted to the case of *small* deformations, the basic geometry of the structure is essentially unchanged. Thus we can denote the position of the deformed bars by the dashed lines shown in Fig. 1-18(c), and the angle $DB'B$ is very nearly 45°. The elongation of the left bar is represented by DB' and is found from the expression for axial extension (Problem 1.1) to be

$$DB' = \frac{(70,700)(120)}{(2.35)(30 \times 10^6)} = 0.120 \text{ in.}   \text{ and thus }   BB' = \frac{0.120}{\cos 45°} = 0.170 \text{ in.}$$

1.12. The two steel bars AB and BC are pinned at each end and support the load of 60,000 lb shown in Fig. 1-19(a). The metal is annealed cast steel, having a yield point of 60,000 lb/in². Safety factors of 2 for tensile members and 3.5 for compressive members are adequate. Determine the required cross-sectional areas of these bars and also the horizontal and vertical components of displacement of point B. Take $E = 30 \times 10^6$ lb/in².

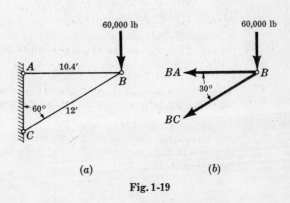

Fig. 1-19

A free-body diagram of the joint at B appears as in Fig. 1-19(b) if the unknown forces are assumed to be tensile.

From statics: $\Sigma F_v = -60,000 - BC \sin 30° = 0$ or $BC = -120,000$ lb

$\Sigma F_h = -BA - BC \cos 30° = 0$ or $BA = 104,000$ lb

The working stresses are given by $60,000/2 = 30,000$ lb/in² for tension and $60,000/3.5 = 17,100$ lb/in² for compression.

The required areas are found by dividing the axial force in each bar by the allowable working stress. Consequently

$$A_{AB} = \frac{104,000}{30,000} = 3.47 \text{ in}^2 \quad \text{and} \quad A_{BC} = \frac{120,000}{17,100} = 7.04 \text{ in}^2$$

To investigate the displacement of point B it is first necessary to calculate the axial deformation of each of the bars. From the expression derived in Problem 1.1 we find the extension of AB to be

$$\Delta_{AB} = \frac{(104,000)(10.4 \times 12)}{(3.47)(30 \times 10^6)} = 0.124 \text{ in.}$$

and the compression of BC to be $\quad \Delta_{BC} = \dfrac{(120,000)(12 \times 12)}{(7.04)(30 \times 10^6)} = 0.082 \text{ in.}$

The location of the point B after deformation has occurred may be determined by realizing that the bar AB actually lengthens 0.124 in. and also rotates as a rigid body about the pin at A. Further, the bar BC shortens 0.082 in. and also rotates about the pin at C.

Figure 1-19(c) illustrates the movement of point B to its deflected position designated by B'. It is to be observed that the deformations of the structure are small, hence the displacement due to rotation about A of the elongated bar AB may be represented by the straight line B_1B' rather than a circular arc with center at A. The same reasoning applies to the rotation of bar BC. From the geometry of the given sketch we immediately have for the displacement components of point B:

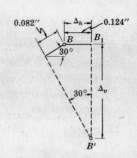

$$\Delta_h = 0.124 \text{ in.}$$

$$\Delta_v = \frac{0.124 + 0.082 \cos 30°}{\tan 30°} + 0.082 \sin 30° = 0.379 \text{ in.}$$

Fig. 1-19(c)

1.13. The pin-connected frame shown in Fig. 1-20 is supported at A in such a manner that vertical displacement is not permitted, but horizontal displacement is possible. Both bars are steel and of cross-sectional area 1 in². Bar AB is heated 80 F° above the reference temperature when the system is stress-free, but bar AC is kept at the reference temperature. Assume that both bars remain straight and determine the stresses in each. Take $\alpha = 6.5 \times 10^{-6}/\text{F}°$.

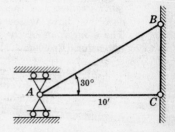

Fig. 1-20

For the purpose of analysis let us first assume that there is no connection between the bars at A. Then, due to the 80° temperature rise, bar AB will elongate an amount

$$80(6.5 \times 10^{-6})(10 \times 12)/\cos 30° = 0.072 \text{ in.}$$

Following this elongation, bar AB may be considered to undergo a rigid body rotation about point B, in which case the left end of the bar now lies at some point A' which is to the left of A.

However, the bars are actually connected at the left ends, hence there is a tensile force F_2 in AC stretching point A on the horizontal bar to the left, and for horizontal equilibrium at point A there is thus a compressive force F_1 in AB. The free-body diagram of the pin A thus appears as in Fig. 1-21.

Fig. 1-21

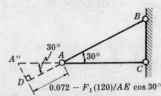

Fig. 1-22

For equilibrium: $F_2 - F_1 \cos 30° = 0$

or $F_2 = F_1 \cos 30°$

Figure 1-22 indicates how bar AB under the combined action of the thermal elongation and the axial compressive force may be regarded as rotating as a rigid body about B with the left end coming to its final location at point A'', which is the final position of point A. From the geometry of this figure

$$\overline{AD} = \overline{AA''} \cos 30°$$

Using the results of Problem 1.1 we may write

$$0.072 - \frac{F_1(120)}{AE \cos 30°} = \frac{F_2(120)}{AE} \cos 30° = \frac{F_1(\cos 30°)^2(120)}{AE}$$

where A represents the cross-sectional area of each bar. Solving:

$$F_1 = 9420A \qquad F_2 = 8150A$$

from which

$$\sigma_{AB} = 9420 \text{ lb/in}^2 \quad \text{compression} \qquad \sigma_{AC} = 8150 \text{ lb/in}^2 \quad \text{tension}$$

1.14. Consider two thin rods or wires as shown in Fig. 1-23(a) below, which are pinned at $A, B,$ and C and are initially horizontal and of length L when no load is applied. The weight of each wire is negligible. A force Q is then applied (gradually) at the point B. Determine the magnitude of Q so as to produce a prescribed vertical deflection δ of the point B.

This is an extremely interesting example of a system in which the elongations of all the individual members satisfy Hooke's law and yet for geometric reasons deflection is *not* proportional to force.

Each bar obeys the relation $\Delta = PL/AE$ where P is the axial force in each bar and Δ the axial elongation. Initially each bar is of length L and after the entire load Q has been applied the length is L'. Thus

$$L' - L = \frac{PL}{AE} \tag{1}$$

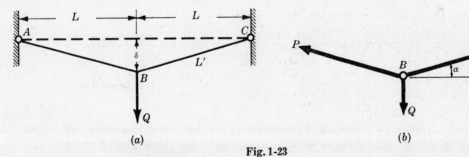

(a)   (b)

Fig. 1-23

The free-body diagram of the pin at B is shown in Fig. 1-23(b) above. From statics,

$$\Sigma F_v = 2P \sin \alpha - Q = 0 \quad \text{or} \quad Q = 2P\left(\frac{\delta}{L'}\right)$$

Using (1), $$Q = 2\frac{(L'-L)AE}{L}\frac{\delta}{L'} = \frac{2\delta AE}{L}\left(1 - \frac{L}{L'}\right) \tag{2}$$

But $$(L')^2 = L^2 + \delta^2 \tag{3}$$

Consequently $$Q = \frac{2\delta AE}{L}\left(1 - \frac{L}{\sqrt{L^2 + \delta^2}}\right) \tag{4}$$

Also, from the binomial theorem we have

$$\sqrt{L^2 + \delta^2} = L\left(1 + \frac{\delta^2}{L^2}\right)^{1/2} = L\left(1 + \frac{1}{2}\frac{\delta^2}{L^2} + \cdots\right) \tag{5}$$

and thus

$$1 - \frac{L}{L\left(1 + \frac{1}{2}\frac{\delta^2}{L^2}\right)} \approx 1 - \left(1 - \frac{1}{2}\frac{\delta^2}{L^2}\right) = \frac{1}{2}\frac{\delta^2}{L^2} \tag{6}$$

From this we have the approximate relation between force and displacement,

$$Q \approx \frac{2AE\delta}{L}\frac{\delta^2}{2L^2} = \frac{AE\delta^3}{L^3} \tag{7}$$

which corresponds to (4).

Thus the displacement δ is *not* proportional to the force Q even though Hooke's law holds for each bar individually. It is to be noted that Q becomes more nearly proportional to δ as δ becomes larger, assuming that Hooke's law still holds for the elongations of the bars. In this example superposition does *not* hold. The characteristic of this system is that the action of the external forces is *appreciably* affected by the small deformations which take place. In this event the stresses and displacements are not linear functions of the applied loads and superposition does not apply.

Summary: A material must follow Hooke's law if superposition is to apply. But this requirement alone is not sufficient. We must see whether or not the action of the applied loads is affected by small deformations of the structure. If the effect is substantial, superposition does not hold.

1.15. For the system discussed in Problem 1.14 let us consider wires each of initial length 5 ft, cross-sectional area 0.1 in² and with $E = 30 \times 16^6$ lb/in². For a load Q of 20 lb determine the central deflection δ by both the exact and the approximate relations given there.

The exact expression relating force and deflection is $Q = \frac{2\delta AE}{L}\left(1 - \frac{L}{\sqrt{L^2 + \delta^2}}\right)$. Substituting the given numerical values, $20 = \frac{2\delta(0.1)(30 \times 10^6)}{(60)}\left(1 - \frac{60}{\sqrt{(60)^2 + \delta^2}}\right)$. Solving by trial and error we find $\delta = 1.131$ in.

The approximate relation between force and deflection is $Q \approx \frac{AE\delta^3}{L^3}$. Substituting,

$$20 \approx \frac{(0.1)(30 \times 10^6)\delta^3}{(60)^3} \quad \text{from which} \quad \delta \approx 1.129 \text{ in.}$$

1.16. A square steel bar two inches on a side and 4 feet long is subject to an axial tensile force of 64,000 lb. Determine the decrease in the lateral dimension due to this load. Consider $E = 30 \times 10^6$ lb/in² and $\mu = 0.3$.

The loading is axial, hence the stress in the direction of the load is given by

$$\sigma = \frac{P}{A} = \frac{64,000}{4} = 16,000 \text{ lb/in}^2$$

The simple form of Hooke's law for uniaxial loading states that $E = \sigma/\epsilon$. The strain ϵ in the direction of the load is thus $16,000/30,000,000 = 0.000533$.

The ratio of the lateral strain to the axial strain is denoted as Poisson's ratio, i.e.

$$\mu = \frac{\text{lateral strain}}{\text{axial strain}}$$

The axial strain has been found to be 0.000533. Consequently, the lateral strain is μ times that value, or $(0.3)(0.000533) = 0.000166$. Since the lateral strain is 0.000166, the change in a two inch length is 0.000332 in. which represents the decrease in the lateral dimension of the bar.

It is to be noted that the definition of Poisson's ratio as the ratio of two strains presumes that only a single uniaxial load acts on the member.

1.17. Consider a state of stress of an element such that a stress σ_x is exerted in one direction, lateral contraction is free to occur in a second (z) direction, but is completely restrained in the third (y) direction. Find the ratio of the stress in the x-direction to the strain in that direction. Also, find the ratio of the strain in the z-direction to that in the x-direction.

Let us examine the general statement of Hooke's law discussed earlier. If in those equations we set $\sigma_z = 0$, $\epsilon_y = 0$ so as to satisfy the conditions of the problem, then Hooke's law becomes

$$\epsilon_x = \frac{1}{E}[\sigma_x - \mu(\sigma_y + 0)] \tag{a}$$

$$\epsilon_y = \frac{1}{E}[\sigma_y - \mu(\sigma_x + 0)] = 0 \tag{b}$$

$$\epsilon_z = \frac{1}{E}[0 - \mu(\sigma_x + \sigma_y)] \tag{c}$$

From (b),

$$\sigma_y = \mu\sigma_x$$

Consequently, from (a)

$$\epsilon_x = \frac{1}{E}(\sigma_x - \mu^2\sigma_x) = \frac{1 - \mu^2}{E}\sigma_x$$

Solving this equation for σ_x as a function of ϵ_x and substituting in (c) we have

$$\epsilon_z = -\frac{\mu}{E}(\sigma_x + \mu\sigma_x) = -\frac{\mu(1 + \mu)}{E}\frac{\epsilon_x E}{1 - \mu^2} = -\frac{\mu\epsilon_x}{1 - \mu}$$

We may now form the ratios

$$\frac{\sigma_x}{\epsilon_x} = \frac{E}{1 - \mu^2} \quad \text{and} \quad -\frac{\epsilon_z}{\epsilon_x} = \frac{\mu}{1 - \mu}$$

The first quantity, $E/(1 - \mu^2)$, is usually denoted as the *effective modulus of elasticity* and is useful in the theory of thin plates and shells. The second ratio, $\mu/(1 - \mu)$, is called the *effective value of Poisson's ratio*.

1.18. Consider an elemental block subject to uniaxial tension. Derive approximate expressions for the change of volume per unit volume due to this loading.

The strain in the direction of the forces may be denoted by ϵ_x. The strains in the other two orthogonal directions are then each $-\mu\epsilon_x$. Consequently, if the initial dimensions of the element are dx, dy, and dz then the final dimensions are

$$(1 + \epsilon_x)\, dx, \quad (1 - \mu\epsilon_x)\, dy, \quad (1 - \mu\epsilon_x)\, dz$$

and the volume after deformation is

$$V' = [(1 + \epsilon_x)\, dx][(1 - \mu\epsilon_x)\, dy][(1 - \mu\epsilon_x)\, dz]$$

$$= (1 + \epsilon_x)(1 - 2\mu\epsilon_x)\, dx\, dy\, dz$$

$$= (1 - 2\mu\epsilon_x + \epsilon_x)\, dx\, dy\, dz$$

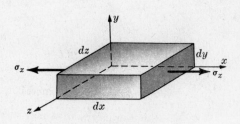

Fig. 1-24

since the deformations are so small that the *squares* and *products* of strains may be neglected.

Since the initial volume was $dx\,dy\,dz$, the change of volume per unit volume is

$$\frac{\Delta V}{V} = (1 - 2\mu)\epsilon_x$$

Hence, for a tensile force the volume increases slightly, for a compressive force it decreases.

Also, the cross-sectional area of the element in a plane normal to the direction of the applied force is given approximately by $A = (1 - \mu\epsilon_x)^2\,dy\,dz = (1 - 2\mu\epsilon_x)\,dy\,dz$.

1.19. A square bar of aluminum 2 in. on a side and 10 in. long is loaded by axial tensile forces at the ends. Experimentally, it is found that the strain in the direction of the load is 0.001 in/in. Determine the volume of the bar when the load is acting. Consider $\mu = 0.33$.

From Problem 1.18 the change of volume per unit volume is given by

$$\Delta V/V = \epsilon(1 - 2\mu) = 0.001(1 - 0.66) = 0.00034$$

Consequently, the change of volume of the entire bar is given by

$$\Delta V = 2(2)(10)(0.00034) = 0.0136 \text{ in}^3$$

The original volume of the bar in the unstrained state is 40 in^3. Since a tensile force increases the volume, the final volume under load is 40.0136 in^3. It is to be noted that ordinary methods of physical measurement would not lead to accuracy of six significant figures.

1.20. The general three-dimensional form of Hooke's law in which strain components are expressed as functions of stress components has already been presented. Occasionally it is necessary to express the stress components as functions of the strain components. Derive these expressions.

Given the previous expressions

$$\epsilon_x = \frac{1}{E}\left[\sigma_x - \mu(\sigma_y + \sigma_z)\right] \tag{1}$$

$$\epsilon_y = \frac{1}{E}\left[\sigma_y - \mu(\sigma_x + \sigma_z)\right] \tag{2}$$

$$\epsilon_z = \frac{1}{E}\left[\sigma_z - \mu(\sigma_x + \sigma_y)\right] \tag{3}$$

let us introduce the notation

$$e = \epsilon_x + \epsilon_y + \epsilon_z \tag{4}$$

$$\theta = \sigma_x + \sigma_y + \sigma_z \tag{5}$$

With this notation, (1), (2), and (3) may be readily solved by determinants for the unknowns $\sigma_x, \sigma_y, \sigma_z$ to yield

$$\sigma_x = \frac{\mu E}{(1 + \mu)(1 - 2\mu)}e + \frac{E}{1 + \mu}\epsilon_x \tag{6}$$

$$\sigma_y = \frac{\mu E}{(1 + \mu)(1 - 2\mu)}e + \frac{E}{1 + \mu}\epsilon_y \tag{7}$$

$$\sigma_z = \frac{\mu E}{(1 + \mu)(1 - 2\mu)}e + \frac{E}{1 + \mu}\epsilon_z \tag{8}$$

These are the desired expressions.

Further information may also be obtained from (1) through (5). If (1), (2), and (3) are added and the symbols e and θ introduced, we have

$$e = \frac{1}{E}(1 - 2\mu)\theta \tag{9}$$

For the special case of a solid subjected to uniform hydrostatic pressure p, $\sigma_x = \sigma_y = \sigma_z = -p$. Hence

$$e = \frac{-3(1 - 2\mu)p}{E} \quad \text{or} \quad \frac{p}{e} = -\frac{E}{3(1 - 2\mu)}$$

The quantity $E/3(1 - 2\mu)$ is often denoted by K and is called the *bulk modulus* or *modulus of volume expansion* of the material. Physically, the bulk modulus K is a measure of the resistance of a material to change of volume without change of shape or form.

We see that the final volume of an element having sides dx, dy, dz prior to loading and subject to strains ϵ_x, ϵ_y, ϵ_z is $(1 + \epsilon_x)\,dx\,(1 + \epsilon_y)\,dy\,(1 + \epsilon_z)\,dz = (1 + \epsilon_x + \epsilon_y + \epsilon_z)\,dx\,dy\,dz$.

Thus the ratio of the increase in volume to the original volume is given approximately by

$$e = \epsilon_x + \epsilon_y + \epsilon_z$$

This change of volume per unit volume, e, is defined as the *dilatation*.

1.21. A straight bar of uniform circular cross-section 2 in. in diameter is loaded by the axial force of 120,000 lb as shown in Fig. 1-25(a). The bar is rigidly fastened so that the ends are unable to approach one another. Use the stress-strain diagram shown in Fig. 1-10 with $\sigma_{yp} = 30,000$ lb/in^2 and determine the stresses in the two regions of the bar.

Let us denote the axial stress in the upper region of the bar by σ_1 and that in the lower region by σ_2. The free-body diagram of that portion of the bar surrounding the point of application of the load appears in Fig. 1-25(b). Thus, for vertical equilibrium we have, since area $= \pi(1)^2 = \pi$ in^2,

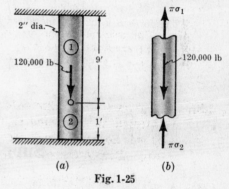

$$\pi\sigma_1 + \pi\sigma_2 - 120,000 = 0$$

or $\qquad\qquad \sigma_1 + \sigma_2 = 38,200 \qquad\qquad (a)$

Obviously the point of application of the 120,000 lb force moves downward. Thus, the extension of the upper region 1 must be numerically equal to the compression of the lower region 2.

Fig. 1-25

Let us first assume that all stresses are below the yield point. Using the result of Problem 1.1, we have

$$\frac{(\pi\sigma_1)(9)(12)}{\pi E} = \frac{(\pi\sigma_2)(1)(12)}{\pi E} \quad \text{or} \quad 9\sigma_1 = \sigma_2$$

Substituting this value in the equilibrium equation (a) one finds

$$\sigma_1 = 3820 \text{ lb/in}^2 \quad \text{and} \quad \sigma_2 = 34,400 \text{ lb/in}^2$$

These would be the correct stresses if the material were elastic up to a stress of 34,400 lb/in^2. But the maximum stress that can be attained in region 2 is 30,000 lb/in^2, i.e. the yield-point stress. Thus, the correct stress in that region is $\sigma_2 = 30,000$ lb/in^2, and from (a) we then find $\sigma_1 = 8,200$ lb/in^2 as the correct stress in region 1.

1.22. In wire-drawing a round wire is slowly drawn through a die in the shape of a truncated cone. In this manner the diameter is reduced from D_1 to D_2 as indicated. The die is split along a diametral plane but restrained from separating. Given a coefficient of friction ν between the wire and the die, determine the normal force q_n exerted on the wire by the die.

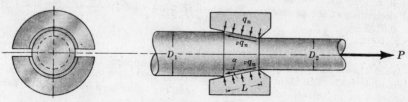

Fig. 1-26

The forces exerted by the die on the wire are shown in Fig. 1-26. The area of wire in contact with the die is given by $(A_1 - A_2)/\sin \alpha$ where A_1 is the cross-sectional area corresponding to D_1 and A_2 is that corresponding to D_2. Summing forces horizontally we have for equilibrium:

$$P - \left(\frac{A_1 - A_2}{\sin \alpha}\right)(q_n \sin \alpha + \nu q_n \cos \alpha) = 0 \qquad (a)$$

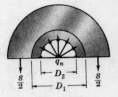

Fig. 1-27

The free-body diagram of the upper half of the die appears in Fig. 1-27, where S is the splitting force between the two halves of the die. Vertical equilibrium of this upper half of the die may be computed by projecting the radial forces q_n exerted by the wire on the die onto the horizontal plane, i.e. by considering them to act vertically over the horizontal area which is trapezoidal in shape and formed by the projection of the conical surface of contact of wire and die on the horizontal plane. Thus:

$$-\frac{S}{2} - \frac{S}{2} + (q_n \cos \alpha)\left(\frac{D_1 + D_2}{2}\right)L - q_n\left(\frac{D_1 + D_2}{2}\right)L\nu \sin \alpha = 0 \qquad (b)$$

Solving (a) and (b) simultaneously we find

$$q_n = \frac{\pi S}{(A_1 - A_2)(\cot \alpha - \nu)}$$

The force S is usually measured by mounting electric resistance strain gages on the outer surface of the die.

Supplementary Problems

1.23. A straight bar of uniform cross-section is subject to axial tension. The cross-sectional area of the bar is 1 in² and its length is 12 ft. If the total elongation is 0.0910 in. under a load of 19,000 lb, find the modulus of elasticity of the material.   *Ans.* $E = 30 \times 10^6$ lb/in²

1.24. Compute the height to which a vertical concrete wall may be built given an ultimate compressive strength of 2400 lb/in² and a safety factor of 4. The specific weight of concrete is 150 lb/ft³. *Ans.* $h = 576$ ft

1.25. A hollow right-circular cylinder is made of cast iron and has an outside diameter of 3 in. and an inside diameter of 2.5 in. If the cylinder is loaded by an axial compressive force of 10,000 lb, determine the total shortening in a 2 ft length. Also determine the normal stress under this load. Take the modulus of elasticity to be 15×10^6 lb/in² and neglect any possibility of lateral buckling of the cylinder.   *Ans.* $\Delta = 0.00738$ in., $\sigma = 4620$ lb/in²

1.26. A solid circular steel rod 1/4 in. in diameter and 15 in. long is rigidly fastened to the end of a square brass bar 1 in. on a side and 12 in. long, the geometric axes of the bars lying along the same line. An axial tensile force of 1200 lb is applied at each of the extreme ends. Determine the total elongation of the assembly. For steel, $E = 30 \times 10^6$ lb/in² and for brass $E = 13.5 \times 10^6$ lb/in². *Ans.* 0.0133 in.

1.27. The truss shown in Fig. 1-28 is pin-connected and supports the single force of 30,000 lb. All bars are made of SAE 1020 steel having a yield point of 35,000 lb/in². For tension members a safety factor of 2 is sufficient. Determine the required cross-sectional areas of bars CD and AB. *Ans.* area $CD = 0.86$ in², area $AB = 1.07$ in²

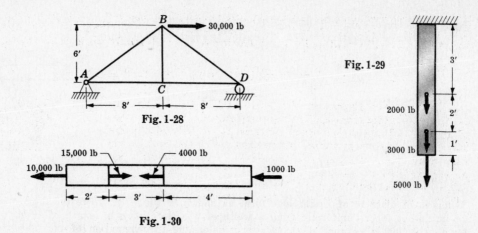

Fig. 1-29

Fig. 1-28

Fig. 1-30

1.28. A steel bar of uniform cross-section is suspended vertically and carries a load of 5000 lb at its lower extremity, as shown in Fig. 1-29. One foot above this a vertical force of 3000 lb is applied and 2 ft above this last point a load of 2000 lb is applied. The total length of the bar is 6 ft and its cross-sectional area is 1 in². The modulus of elasticity is 30×10^6 lb/in². Determine the total elongation of the bar.   *Ans.* 0.0204 in.

1.29. A brass bar of cross-sectional area 1.5 in² is subject to the axial forces shown in Fig. 1-30. Determine the total elongation of the bar. For brass, $E = 13 \times 10^6$ lb/in².   *Ans.* 0.000616 in.

1.30. Steel railroad rails are laid with their adjacent ends 1/8 in. apart when the temperature is 60° F. The length of each rail is 39 ft. The material is steel with $E = 30 \times 10^6$ lb/in² and $\alpha = 6.5 \times 10^{-6}/$F°. (*a*) Compute the gap between adjacent ends when the temperature is −10° F. (*b*) At what temperature will adjacent ends just be in contact? (*c*) Find the compressive stress in the rails when the temperature is 110° F. Neglect any possibility of buckling of the rails.
Ans.   gap = 0.338 in., $T = 101.3°$ F, $\sigma = 1690$ lb/in²

1.31. The following data were obtained during the tensile test of a circular cold-rolled steel specimen of diameter 0.507 in.

Axial load (lb)	Elongation in 2″ gage length	Axial load (lb)	Elongation in 2″ gage length
0	0	6950	0.0120
1250	0.0004	6950	0.0160
1850	0.0006	6900	0.0200
2400	0.0008	6950	0.0240
3050	0.0010	7000	0.0500
3640	0.0012	7750	0.1000
4250	0.0014	9350	0.2000
4850	0.0016	9900	0.3000
5450	0.0018	10100	0.4000
6050	0.0020	10100	0.5000
6700	0.0022	9900	0.6000
7250	0.0024	9500	0.7000
6900	0.0040	8900	0.7500
6950	0.0080		

At the time of rupture the final diameter of the bar at the section where failure occurred was 0.295 in. The original 2 in. gage length had increased to 2.750 in. From the given data determine the proportional limit of the material, the modulus of elasticity, the percentage reduction of area, the percentage elongation, and the breaking strength.

Ans. proportional limit = 36,000 lb/in², $E = 30 \times 10^6$ lb/in², percentage reduction of area = 66.3, percentage elongation = 37.5, breaking strength = 44,000 lb/in²

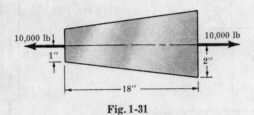

Fig. 1-31	Fig. 1-32

1.32. A flat steel plate is of trapezoidal form as shown in Fig. 1-31. The thickness of the plate is 0.5 in. and it tapers uniformly from a width of 2 in. to 4 in. in a length of 18 in. If an axial force of 10,000 lb is applied at each end, determine the elongation of the plate. Take $E = 30 \times 10^6$ lb/in².   *Ans.* 0.00416 in.

1.33. A solid conical bar of circular cross-section is suspended vertically as shown in Fig. 1-32. The length of the bar is L, the diameter of its base D, the modulus of elasticity is E, and the weight per unit volume γ. Determine the elongation of the bar due to its own weight.
Ans. $\Delta = \gamma L^2/6E$

1.34. A tripod with equal legs carries a vertical load W at its apex at a height h above the horizontal ground. The feet are prevented from sliding on the ground by light cords, each of length L, joining the midpoints of the legs. The cords are identical, each of cross-sectional area A and Young's modulus E. Determine the elongation of each cord.   *Ans.* $2WL^2/9hAE$

1.35. In Fig. 1-33, $AB, AC, BC, CD,$ and BD are pin-jointed rods. Point B is attached to point E by a spring whose unstretched length is 1 ft and whose spring constant is 25 lb/in. Neglecting the weight of all bars and the spring, determine the magnitude of the load W applied at D that makes CD horizontal.   *Ans.* 44 lb

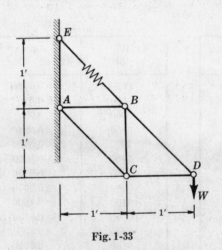
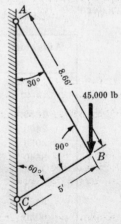

Fig. 1-33	Fig. 1-34

1.36. The steel bars AB and BC are pinned at each end and support the load of 45,000 lb, as shown in Fig. 1-34. The material is structural steel having a yield point of 35,000 lb/in² and safety factors of 2 and 3.5 are satisfactory for tension and compression respectively. Determine the size of each bar and also the horizontal and vertical components of displacement of point B. Take $E = 30 \times 10^6$ lb/in². Neglect any possibility of lateral buckling of bar BC.

Ans. area AB = 2.23 in², area BC = 2.25 in², Δ_h = 0.0130 in. (to right),
Δ_v = 0.0625 in. (downward)

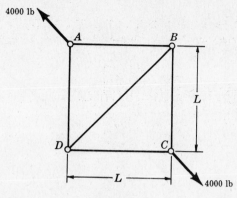

Fig. 1-35

1.37. The five-bar, pin-jointed structure shown in Fig. 1-35 is loaded by two forces of 4000 lb each, acting along a diagonal. All bars are steel having $E = 30 \times 10^6$ lb/in^2 and a cross-sectional area of 1 in^2. Determine the increase in length of AC.   *Ans.* 0.000455L

1.38. A solid circular brass bar 1 in. in diameter is subject to an axial tensile force of 10,000 lb. Determine the decrease in the diameter of the bar due to this load. For brass $E = 13.5 \times 10^6$ lb/in^2 and $\mu = 0.28$.   *Ans.* 0.000264 in.

1.39. A square steel bar is 2 in. on a side and 10 in. long. It is loaded by an axial tensile force of 40,000 lb. If $E = 30 \times 10^6$ lb/in^2 and $\mu = 0.3$, determine the change of volume per unit volume. *Ans.* 0.000133

1.40. Consider the square aluminum bar described in Problem 1.19 but with the axial loading reversed so as to cause compression. The compressive strain is considered to be 0.001 in/in. Determine the volume of the bar when the compressive force has been applied.   *Ans.* 39.9864 in^3

1.41. Consider a state of stress of an element in which a stress σ_x is exerted in one direction and lateral contraction is completely restrained in each of the other two directions. Find the effective modulus of elasticity and also the effective value of Poisson's ratio.

Ans.   eff. mod. $= \dfrac{E(1-\mu)}{(1-2\mu)(1+\mu)}$,   eff. Poisson's ratio $= 0$

1.42. Consider the state of stress in a bar subject to compression in the axial direction. Lateral expansion is restrained to half the amount it would ordinarily be if the lateral faces were load-free. Find the effective modulus of elasticity.   *Ans.* $\dfrac{E(1-\mu)}{1-\mu-\mu^2}$

1.43. A bar of uniform cross-section is subject to uniaxial tension and develops a strain in the direction of the force of 1/800. Calculate the change of volume per unit volume. Assume $\mu = 1/3$. *Ans.* 1/2400 (increase)

1.44. A straight aluminum rod of diameter 1.25 in. is subjected to an axial tensile force of 10,000 lb. If $E = 10 \times 10^6$ lb/in^2, $\mu = 1/4$, determine:

(a) the unit stress	*Ans.* (a)	8150 lb/in^2
(b) the unit strain	(b)	0.000815 in/in
(c) the elongation in an 8 in. gage length	(c)	0.00653 in.
(d) the change in diameter	(d)	−0.000255 in.
(e) the change in cross-sectional area	(e)	−0.00050 in^2
(f) the change in volume in an 8 in. gage length	(f)	0.00400 in^3

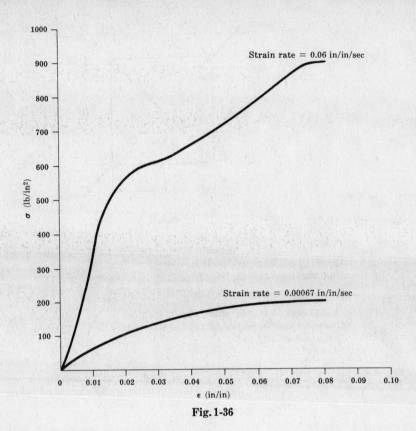

Fig. 1-36

1.45. A plastic formed by mixing equal parts by weight of Shell Epon Resin 815 and Shell Curing Agent V-40 has the compressive stress-strain curves shown for the two loading rates indicated in Fig. 1-36. To show the importance of loading rate on this substance, determine the ratio of stress at the strain rate of 0.06 in/in/sec to that at 0.00067 in/in/sec when the strain is 0.08. In a third test, for which the stress-strain curve is not shown but is of the form of the upper of the two curves indicated, the strain rate was 720 in/in/sec. For this case a yield point of 4500 lb/in² was attained. Determine the ratio of this yield point to the peak stress at a strain rate of 0.00067 in/in/sec.   *Ans.* 4.5; 22.5

Chapter 2

Statically Indeterminate Force Systems
Tension and Compression

DEFINITION OF A DETERMINATE FORCE SYSTEM

If the values of all the external forces which act on a body can be determined by the equations of static equilibrium alone, then the force system is *statically determinate*. The problems in Chapter 1 were all of this type.

Example 1.

The bar shown in Fig. 2-1 is loaded by the force P. The reactions are R_1, R_2, and R_3. The system is statically determinate because there are three equations of static equilibrium available for the system and these are sufficient to determine the three unknowns.

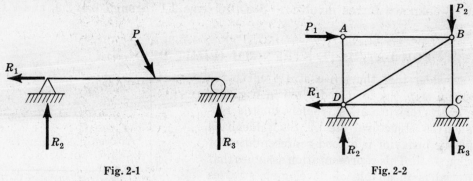

Fig. 2-1   Fig. 2-2

Example 2.

The truss $ABCD$ shown in Fig. 2-2 is loaded by the forces P_1 and P_2. The reactions are R_1, R_2, and R_3. Again, since there are three equations of static equilibrium available, all three unknown reactions may be determined and consequently the external force system is statically determinate.

The above two illustrations refer only to external reactions and the force systems may be defined as statically determinate *externally*.

DEFINITION OF AN INDETERMINATE FORCE SYSTEM

In many cases the forces acting on a body cannot be determined by the equations of statics alone because there are more unknown forces than there are equations of equilibrium. In such a case the force system is said to be *statically indeterminate*.

Example 3.

The bar shown in Fig. 2-3 is loaded by the force P. The reactions are R_1, R_2, R_3, and R_4. The force system is statically indeterminate because there are four unknown reactions but only three equations of static equilibrium. Such a force system is said to be indeterminate to the first degree.

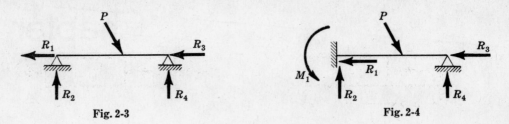

Fig. 2-3   Fig. 2-4

Example 4.

The bar shown in Fig. 2-4 above is statically indeterminate to the second degree because there are five unknown reactions R_1, R_2, R_3, R_4, and M_1 but only three equations of static equilibrium. Consequently the values of all reactions cannot be determined by use of statics equations alone.

METHOD OF ELASTIC ANALYSIS

The approach that we will consider here is called the *deformation method* because it considers the deformations in the system. Briefly, the procedure to be followed in analyzing an indeterminate system is first to write all equations of static equilibrium that pertain to the system and then *supplement* these equations with additional equations based upon the deformations of the structure. Enough equations involving deformations must be written so that the total number of equations from both statics and deformations is equal to the number of unknown forces involved. See Problems 2.1 through 2.4, 2.9, and 2.10.

ANALYSIS FOR ULTIMATE STRENGTH (LIMIT DESIGN)

We consider that the stress-strain curve for the material is of the form indicated in Fig. 2-5, i.e. one characterizing an extremely ductile material such as structural steel. Such idealized elasto-plastic behavior is a good representation of low-carbon steel. This representation assumes that the material is incapable of developing stresses greater than the yield point.

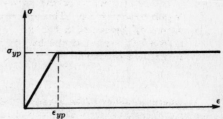

Fig. 2-5

In a statically indeterminate system any inelastic action changes the conditions of constraint. Under these altered conditions the loading that the system can carry usually increases over that predicted on the basis of completely elastic action everywhere in the system. Design of a statically indeterminate structure for that load under which some or all of the regions of the structure reach the yield point and cause "collapse" of the system is termed *limit design*. The *ultimate load* corresponding to such design is of course divided by some factor of safety to determine a *working load*. The term limit design, when used in this manner, applies only to statically indeterminate structures. For applications, see Problems 2.11 through 2.14.

Solved Problems

ELASTIC ANALYSIS

In Problems 2.1 through 2.10 it is assumed that the system is acting within the linear elastic range of action of the material.

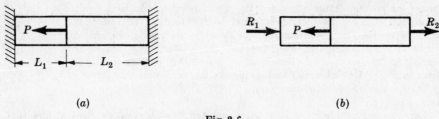

(a)　　　　　　　　　　　　　　　　　　　　(b)

Fig. 2-6

2.1. The bar shown in Fig. 2-6(a) is of constant cross-section and is held rigidly between the walls. An axial load P is applied to the bar at a distance L_1 from the left end. Determine the reactions of the walls upon the bar.

We must first draw the free-body diagram of the bar, showing the applied load P together with the reactions of the walls upon the bar. These reactions are denoted by R_1 and R_2, as shown in Fig. 2-6(b).

There is only one equation of static equilibrium, namely

$$\Sigma F_h = R_1 - P + R_2 = 0$$

Since this equation contains two unknowns (R_1 and R_2) the problem is statically indeterminate. Consequently this statics equation must be supplemented by an additional equation based upon the deformations of the bar.

The shortening of the portion of the bar of length L_1 must be equal to the elongation of the region of length L_2. This fact furnishes the basis for the equation concerning deformations. The change of length of a bar due to axial loading was given in Problem 1.1. The axial force acting in the left region of the bar is R_1(lb) and in the right region R_2(lb). The equation relating deformations becomes

$$\frac{R_1 L_1}{AE} = \frac{R_2 L_2}{AE}$$

where A denotes the cross-sectional area of the bar and E the modulus of elasticity. From this equation we have $R_1 L_1 = R_2 L_2$, and solving this simultaneously with the statics equation we find

$$R_1 = \frac{PL_2}{L_1 + L_2} \quad \text{and} \quad R_2 = \frac{PL_1}{L_1 + L_2}$$

Knowing these reactions, it is evident that the elongation of the right portion (L_2) of the bar is

$$\Delta_e = \frac{R_2 L_2}{AE} = \frac{PL_1 L_2}{(L_1 + L_2)AE}$$

and the shortening of the left portion (L_1) of the bar is

$$\Delta_c = -\frac{R_1 L_1}{AE} = -\frac{PL_1 L_2}{(L_1 + L_2)AE}$$

Thus 　　　　　　　　　　　　　　　　$$\Delta_e = -\Delta_c$$

2.2. Consider a steel tube surrounding a solid aluminum cylinder, the assembly being compressed between infinitely rigid cover plates by centrally applied forces as shown in Fig 2-7(a). The aluminum cylinder is 3 in. in diameter and the outside diameter of the steel tube is 3.5 in. If $P = 48,000$ lb, find the stress in the steel and also in the aluminum. For steel, $E = 30 \times 10^6$ lb/in² and for aluminum $E = 4 \times 10^6$ lb/in².

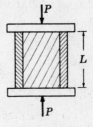

Fig. 2-7(a)

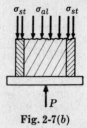

Let us pass a horizontal plane through the assembly at any elevation except in the immediate vicinity of the cover plates and then
remove one portion or the other, say the upper portion. In that event
the portion that we have removed must be replaced by the effect it
exerted upon the remaining portion and that effect consists of vertical
normal stresses distributed over the two materials. The free-body
diagram of the portion of the assembly below this cutting plane is
shown in Fig. 2-7(b) where σ_{st} and σ_{al} denote the normal stresses existing in the steel and aluminum respectively.

Fig. 2-7(b)

Let us denote the resultant force carried by the steel by P_{st} (lb) and that carried by the
aluminum by P_{al}. Then $P_{st} = A_{st}\sigma_{st}$ and $P_{al} = A_{al}\sigma_{al}$ where A_{st} and A_{al} denote the cross-sectional areas of the steel tube and the aluminum cylinder respectively. There is only one equa-
tion of static equilibrium available for such a force system and it takes the form

$$\Sigma F_v = P - P_{st} - P_{al} = 0$$

Thus, we have one equation in two unknowns, P_{st} and P_{al}, and hence the problem is statically
indeterminate. In that event we must supplement the available statics equation by an equation
derived from the deformations of the structure. Such an equation is readily obtained because the
infinitely rigid cover plates force the axial deformations of the two metals to be identical.

The deformation due to axial loading is given by $\Delta = PL/AE$. Equating axial deformations
of the steel and the aluminum we have

$$\frac{P_{st}L}{A_{st}E_{st}} = \frac{P_{al}L}{A_{al}E_{al}}$$

or $\dfrac{P_{st}L}{\frac{\pi}{4}[(3.5)^2 - (3)^2](30 \times 10^6)} = \dfrac{P_{al}L}{\frac{\pi}{4}(3)^2(4 \times 10^6)}$ from which $P_{st} = 2.71P_{al}$

This equation is now solved simultaneously with the statics equation, $P - P_{st} - P_{al} = 0$, and we
find $P_{al} = 0.27P$, $P_{st} = 0.73P$.

For a load of $P = 48,000$ lb this becomes $P_{al} = 12,900$ lb and $P_{st} = 35,100$ lb. The desired
stresses are found by dividing the resultant force in each material by its cross-sectional area:

$$\sigma_{al} = \frac{12,900}{\frac{\pi}{4}(3)^2} = 1820 \text{ lb/in}^2, \qquad \sigma_{st} = \frac{35,100}{\frac{\pi}{4}[(3.5)^2 - (3)^2]} = 13,700 \text{ lb/in}^2$$

2.3.   The bar AB is absolutely rigid and is supported by three rods as shown in Fig.
2-8(a). The two outer rods are steel, and each has a cross-sectional area of 0.50 in².
The central rod is copper and of area 1.5 in². For steel, $E = 30 \times 10^6$ lb/in² and for
copper $E = 17 \times 10^6$ lb/in². All rods are 7 ft long. The three rods are equally spaced
and the applied loads of 12,000 lb are each applied midway between the rods. Neglecting the weight of the bar AB, determine the force in each of the vertical bars.
The bar AB remains horizontal after the loads have been applied.

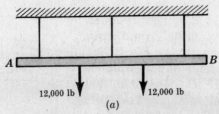

(a)

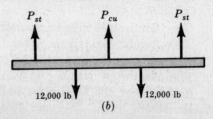

(b)

Fig. 2-8

First draw a free-body diagram of the bar AB showing all the forces acting on it. These forces include the two applied loads and the reactions of the vertical rods upon bar AB. If the force in each of the steel rods is denoted by P_{st} (lb) and that in the copper by P_{cu} (lb), then this diagram appears as in Fig. 2-8(b) above. The condition of symmetry has been used in stating that the forces in the steel rods are equal; hence there remains only one equation of static equilibrium, namely $\Sigma F_v = 2P_{st} + P_{cu} - 24{,}000 = 0$. Thus we have one equation containing two unknowns and the problem is statically indeterminate. This statics equation must be supplemented by an additional equation coming from the deformations of the structure.

Such an equation is readily determined because the elongations of the steel and copper rods are equal. The elongation due to axial loading is $\Delta = PL/AE$, and applying that expression to the steel and copper rods we have

$$\frac{P_{st}(84)}{0.5(30 \times 10^6)} = \frac{P_{cu}(84)}{1.5(17 \times 10^6)} \quad \text{or} \quad P_{st} = 0.588 P_{cu}$$

This equation may now be solved simultaneously with the statics equation to yield

$$2(0.588 P_{cu}) + P_{cu} - 24{,}000 = 0$$

Solving, $P_{cu} = 11{,}000$ lb and $P_{st} = 6500$ lb.

2.4. The bar AB is considered to be absolutely rigid and is horizontal before the load of 40,000 lb is applied as shown in Fig. 2-9(a). The connection at A is a pin, and AB is supported by the steel rod EB and the copper rod CD. The length of CD is 3 ft, of EB is 5 ft. The cross-sectional area of CD is 0.8 in², the area of EB is 0.5 in². Determine the stress in each of the vertical rods and the elongation of the steel rod. Neglect the weight of AB. For copper $E = 17 \times 10^6$ lb/in², for steel $E = 30 \times 10^6$ lb/in².

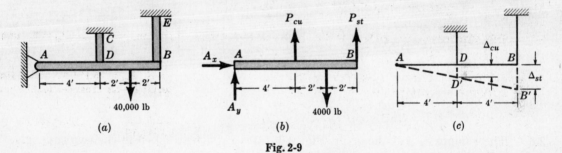

Fig. 2-9

The first step in the solution is to draw a free-body diagram of the bar AB, showing all forces acting on it. This appears as in Fig. 2-9(b). From statics we have

$$\Sigma F_h = A_x = 0 \tag{1}$$

$$\Sigma M_a = 4P_{cu} + 8P_{st} - 40{,}000(6) = 0 \tag{2}$$

$$\Sigma F_v = A_y + P_{cu} + P_{st} - 40{,}000 = 0 \tag{3}$$

Since the last two statics equations contain three unknowns the problem is statically indeterminate. Hence we must look to the deformations of the system for another equation. Because the bar AB is rigid the only movement it can undergo is a rigid body rotation about the pin at A as center. The dashed line in Fig. 2-9(c) indicates the final position of the bar AB after the load of 40,000 lb had been applied. Initially the bar was horizontal, as shown by the solid line.

The lower ends of the rods are originally at D and B, and they move to D' and B' after application of the 40,000 lb load. Because the bar AB is rigid, the similar triangles ADD' and ABB' furnish a simple relation between the deformations of the two vertical bars, namely $\Delta_{cu}/4 = \Delta_{st}/8$, where Δ_{cu} and Δ_{st} denote the elongations of the copper and steel rods respectively. Thus, the additional equation based upon deformations is $\Delta_{st} = 2\Delta_{cu}$.

But the elongation under axial loading is given by $\Delta = PL/AE$. Using this equation in the last deformation relation we get

$$\frac{P_{st}(60)}{0.5(30 \times 10^6)} = \frac{2P_{cu}(36)}{0.8(17 \times 10^6)} \quad \text{or} \quad P_{st} = 1.33P_{cu}$$

Solving this equation simultaneously with the statics equation (2) we find

$$4P_{cu} + 8(1.33P_{cu}) = 240,000, \quad P_{cu} = 16,400 \text{ lb} \quad \text{and} \quad P_{st} = 21,800 \text{ lb}$$

The stresses are given by the simple relation $\sigma = P/A$. Thus $\sigma_{cu} = 16,400/0.8 = 20,500 \text{ lb/in}^2$ and $\sigma_{st} = 21,800/0.5 = 43,600 \text{ lb/in}^2$.

2.5. The copper bar of Fig. 2-10 is of uniform cross-section and is rigidly attached to the walls as shown. The rod is 5 ft long and has a cross-sectional area of 2.5 in². At a temperature of 80°F the rod is stress-free in the configuration shown. Determine the stress in the rod when the temperature has dropped to 50°F. Assume that the supports do not yield. For copper, $E = 16 \times 10^6 \text{ lb/in}^2$ and $\alpha = 9.3 \times 10^{-6}/\text{F}°$.

One approach to this problem is to assume that the bar is cut free from the wall at the right end. In that event it is free to contract when the temperature falls and the bar shortens an amount $\Delta = (9.3 \times 10^{-6})(60)(30) = 0.0167$ in., by definition of the coefficient of linear expansion (Chapter 1).

It is next necessary to find the axial P that must be applied to the bar to stretch it 0.0167 in., i.e. to restore the right end to its true position, because we know that actually that end does not move at all when the temperature drops. To determine this force P, use the equation

$$\Delta = \frac{PL}{AE} \quad \text{which gives} \quad 0.0167 = \frac{P(60)}{2.5(16 \times 10^6)} \quad \text{or} \quad P = 11,200 \text{ lb}$$

The axial stress set up by this force $\sigma = P/A = 11,200/2.5 = 4500 \text{ lb/in}^2$.

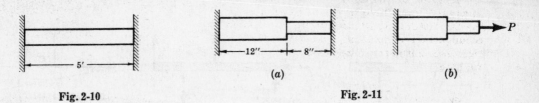

Fig. 2-10   (a)   (b)   Fig. 2-11

2.6. The composite bar shown in Fig. 2-11(a) is rigidly attached to the two supports. The left portion of the bar is copper, of uniform cross-sectional area 12 in² and length 12 in. The right portion is aluminum, of uniform cross-sectional area 3 in² and length 8 in. At a temperature of 80°F the entire assembly is stress-free. The temperature of the structure drops and during this process the right support yields 0.001 in. in the direction of the contracting metal. Determine the minimum temperature to which the assembly may be subjected in order that the stress in the aluminum does not exceed 24,000 lb/in². For copper $E = 16 \times 10^6 \text{ lb/in}^2$, $\alpha = 9.3 \times 10^{-6}/\text{F}°$ and for aluminum $E = 10 \times 10^6 \text{ lb/in}^2$, $\alpha = 12.8 \times 10^{-6}/\text{F}°$.

As in the previous problem it is perhaps simplest to consider that the bar is cut just to the left of the supporting wall at the right and is then free to contract due to the temperature drop ΔT. The total shortening of the composite bar is given by

$$(9.3 \times 10^{-6})(12)\Delta T + (12.8 \times 10^{-6})(8)\Delta T$$

according to the definition of the coefficient of linear expansion. It is to be noted that the shape of the cross-section has no influence upon the change in length of the bar due to a temperature change.

Even though the bar has contracted this amount it is still stress-free. However, this is not the complete analysis because the reaction of the wall at the right has been neglected by cutting the bar there. Consequently, we must represent the action of the wall by an axial force P applied to the bar, as shown in Fig. 2-11(b). For equilibrium, the resultant force acting over any cross-section of either the copper or the aluminum must be equal to P. The application of the force P stretches the composite bar by an amount

$$\frac{P(12)}{12(16 \times 10^6)} + \frac{P(8)}{3(10 \times 10^6)}$$

If the right support were unyielding we would equate the last expression to the expression giving the total shortening due to the temperature drop. Actually the right support yields 0.001 in. and consequently we may write

$$\frac{P(12)}{12(16 \times 10^6)} + \frac{P(8)}{3(10 \times 10^6)} \; = \; (9.3 \times 10^{-6})(12)\Delta T + (12.8 \times 10^{-6})(8)\Delta T \; - \; 0.001$$

The stress in the aluminum is not to exceed 24,000 lb/in^2 and since it is given by the formula $\sigma = P/A$, the maximum force P becomes $P = A\sigma = 3(24,000) = 72,000$ lb. Substituting this value of P in the above equation relating deformations, we find $\Delta T = 115$ F$^\circ$. Therefore the temperature may drop 115 F$^\circ$ from the original 80°F. The final temperature would be -35°F.

2.7. Consider the tapered (conical) steel bar of Fig. 2-12, which has both ends attached to unyielding supports. The bar is initially stress-free. If the temperature of the entire bar drops 40 F$^\circ$ determine the maximum normal stress in the bar. Take $E = 30 \times 10^6$ lb/in^2 and $\alpha = 6.5 \times 10^{-6}/F^\circ$.

Perhaps the simplest technique for solving this problem is to imagine that one end of the bar, say the right end, is temporarily cut free from its support. In this case the bar contracts an amount $40(36)(6.5 \times 10^{-6}) = 0.00935$ in. due to the temperature drop.

Next, let us find the axial force P which must be applied to the "free" right end so as to elongate the bar 0.00935 in., i.e. so that the true boundary condition of complete fixity will be satisfied at the right end. Setting up the coordinate system shown, we have

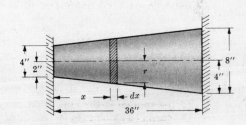

Fig. 2-12

$$r \; = \; 2 + 2x/36 \; = \; 2 + x/18$$

Since the angle of taper is comparatively small, the tensile force may be assumed to be uniformly distributed over any one cross-section. Also, since there are no abrupt changes of cross-section we may determine the elongation of the shaded disc-like element of thickness dx by applying $\Delta = PL/AE$, where $L = dx$, to the disc and then integrate over the entire bar:

$$0.00935 \; = \; \int_0^{36} \frac{P\,dx}{\pi(2 + x/18)^2 E} \; = \; \int_0^{36} \frac{324P\,dx}{E\pi(36 + x)^2} \; = \; \frac{324P}{72E\pi}$$

Solving, $P = 196,000$ lb, where P is the resultant axial force acting over any cross-section, i.e. the force necessary to restore the bar to its original length.

It is to be noted that the resultant force over every vertical cross-section is P (lb) for equilibrium of any portion of the bar. However, since the cross-sectional area of the bar varies from one end to the other the stress varies from a maximum value at the left end, where the cross-sectional area is a minimum, to a minimum at the right end, where the area is a maximum.

The maximum stress at the left end is given by $\sigma_{max} = \dfrac{196,000}{\pi(2)^2} = 15,600$ lb/in^2.

2.8. A hollow steel cylinder surrounds a solid copper cylinder and the assembly is subject to an axial loading of 50,000 lb as shown in Fig. 2-13(a). The cross-sectional area of the steel is 3 in^2, while that of the copper is 10 in^2. Both cylinders are the same length before the load is applied. Determine the temperature rise of the entire system required to place all of the load on the copper cylinder. The cover plate at the top of the assembly is rigid. For copper $E = 16 \times 10^6$ lb/in^2, $\alpha = 9.3 \times 10^{-6}/F^\circ$, while for steel $E = 30 \times 10^6$ lb/in^2, $\alpha = 6.5 \times 10^{-6}/F^\circ$.

One method of analyzing this problem is to assume that the load as well as the upper cover plate is removed and that the system is allowed to freely expand vertically because of a temperature rise ΔT. In that event the upper ends of the cylinders assume the positions shown by the dashed lines in Fig. 2-13(b).

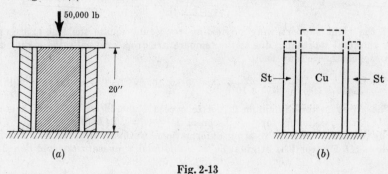

Fig. 2-13

The copper cylinder naturally expands upward more than the steel one because the coefficient of linear expansion of copper is greater than that of steel. The upward expansion of the steel cylinder is $(6.5 \times 10^{-6})(20)\Delta T$, while that of the copper is $(9.3 \times 10^{-6})(20)\Delta T$.

This is not of course the true situation because the load of 50,000 lb has not as yet been considered. If all of this axial load is carried by the copper then only the copper will be compressed and the compression of the copper is given by

$$\Delta_{cu} = \frac{PL}{AE} = \frac{50,000(20)}{10(16 \times 10^6)}$$

The condition of the problem states that the temperature rise ΔT is just sufficient so that all of the load is carried by the copper. Thus, the expanded length of the copper indicated by the dashed lines in the above sketch will be decreased by the action of the force. The net expansion of the copper is the expansion caused by the rise of temperature minus the compression due to the load. The change of length of the steel is due only to the temperature rise. Consequently we may write

$$(9.3 \times 10^{-6})(20)\Delta T - \frac{50,000(20)}{10(16 \times 10^6)} = (6.5 \times 10^{-6})(20)\Delta T \quad \text{or} \quad \Delta T = 111 \text{ F}^\circ$$

2.9. The rigid bar AD is pinned at A and attached to the bars BC and ED, as shown in Fig. 2-14(a). The entire system is initially stress-free and the weights of all bars are negligible. The temperature of bar BC is lowered 50 F° and that of the bar ED is raised 50 F°. Neglecting any possibility of lateral buckling, find the normal stresses in bars BC and ED. For BC, which is brass, assume $E = 14 \times 10^6$ lb/in², $\alpha = 10.4 \times 10^{-6}$/F° and for ED, which is steel, take $E = 30 \times 10^6$ lb/in², and $\alpha = 6.5 \times 10^{-6}$/F°. The cross-sectional area of BC is 1 in², of ED is 0.5 in².

Let us denote the forces acting on AD by P_{st} and P_{br} acting in the assumed directions shown in the free-body diagram, Fig. 2-14(b). Since AD rotates as a rigid body about A (as shown by the dashed line) we have $\Delta_{br}/10 = \Delta_{st}/25$, where Δ_{br} and Δ_{st} denote the axial compression of BC and the axial elongation of DE respectively.

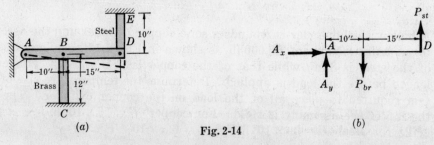

Fig. 2-14

The total change of length of BC is composed of a shortening due to the temperature drop as well as a lengthening due to the axial force P_{br}. The total change of length of DE is composed of a lengthening due to the temperature rise as well as a lengthening due to the force P_{st}. Hence we have

$$\frac{2}{5}\left[(6.5 \times 10^{-6})(10)(50) + \frac{P_{st}(10)}{0.5(30 \times 10^6)}\right] \;=\; -(10.4 \times 10^{-6})(12)(50) + \frac{P_{br}(12)}{1(14 \times 10^6)}$$

or
$$0.856P_{br} - 0.267P_{st} \;=\; 7750$$

From statics,
$$\Sigma M_A \;=\; 10P_{br} - 25P_{st} \;=\; 0$$

Solving these equations simultaneously, $P_{st} = 4030$ lb and $P_{br} = 10{,}100$ lb.

Using $\sigma = P/A$ for each bar, we obtain $\sigma_{st} = 8060$ lb/in² and $\sigma_{br} = 10{,}100$ lb/in².

2.10. Consider the statically indeterminate pin-connected framework shown in Fig. 2-15(a). Before the load P is applied the entire system is stress-free. Find the axial force in each bar caused by the vertical load P. The two outer bars are identical and have cross-sectional area A_i, the middle bar has area A_v. All bars have the same modulus of elasticity, E.

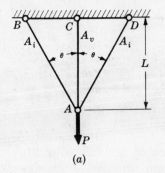
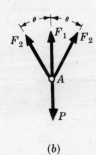
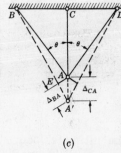

(a)   (b)   (c)

Fig. 2-15

The free-body diagram of the pin at A appears as in Fig. 2-15(b), where F_1 and F_2 denote axial forces (lb) in the vertical and inclined bars. From statics we have

$$\Sigma F_v \;=\; F_1 + 2F_2 \cos \theta - P \;=\; 0$$

This is the only statics equation available since we have made use of symmetry in stating that the forces in the inclined bars are equal. Since it contains two unknowns, F_1 and F_2, the force system is statically indeterminate. Hence we must examine the deformations of the system to obtain another equation. Under the action of the load P the bars assume the positions shown by the dashed lines in Fig. 2-15(c).

Because the deformations of the system are *small* the basic geometry is essentially unchanged and the angle $BA'A$ may be taken to be θ. AEA' is a right triangle and AE, which is actually an arc having a radius equal in length to the length of the inclined bars, is perpendicular to BA'. The elongation of the vertical bar is thus represented by AA' and that of the inclined bars by EA'. From this small triangle we have the relation

$$\Delta_{BA} \;=\; \Delta_{CA} \cos \theta$$

where Δ_{BA} and Δ_{CA} denote elongations of the inclined and vertical bars respectively.

Since these bars are subject to axial loading their elongations are given by $\Delta = PL/AE$. From that expression we have

$$\Delta_{BA} = \frac{F_2(L/\cos\theta)}{A_i E} \quad \text{and} \quad \Delta_{CA} = \frac{F_1 L}{A_v E}$$

Substituting these in the above equation relating Δ_{BA} and Δ_{CA} we have

$$\frac{F_2 L}{A_i E \cos\theta} = \frac{F_1 L}{A_v E}\cos\theta \quad \text{or} \quad F_2 = F_1\frac{A_i}{A_v}\cos^2\theta$$

Substituting this in the statics equation we find $F_1 + 2F_1(A_i/A_v)\cos^3\theta = P$, or

$$F_1 = \frac{P}{1 + 2(A_i/A_v)\cos^3\theta} \quad \text{and} \quad F_2 = \frac{P\cos^2\theta}{(A_v/A_i) + 2\cos^3\theta} \tag{1}$$

ULTIMATE STRENGTH (LIMIT DESIGN)

In each of the following problems the elasto-plastic behavior of the material is assumed to follow the idealized stress-strain curve of Fig. 2-16.

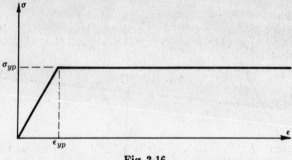

Fig. 2-16

The ultimate load, or limit load, determined in each of the following problems is the maximum possible load that can be applied to each system provided the stress-strain curve is of the type indicated and the material has infinite ductility, i.e. the flat region of the curve extends indefinitely to the right.

2.11. Consider the system composed of three vertical bars as indicated in Fig. 2-17(a). The outer bars of length L are equally spaced from the central bar and a load P is applied to the rigid horizontal member. Using limit design, determine the ultimate load P. The values of A and E are identical in all three bars.

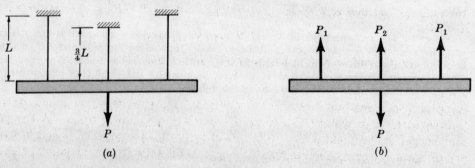

Fig. 2-17

Let us analyze the action as the load P increases from an initial value of zero, i.e. as it is slowly applied. For equilibrium we have:

$$2P_1 + P_2 = P \tag{1}$$

where P_1 represents the force in each of the outer bars and P_2 is the force in the inner bar [see Fig. 2-17(b)]. Since the horizontal member is rigid, the vertical elongation of each of the outer bars must equal that of the central bar. Thus

$$\frac{P_1 L}{AE} = \frac{P_2(3L/4)}{AE} \tag{2}$$

or

$$P_1 = \tfrac{3}{4}P_2 \tag{3}$$

Substituting this value in (1) we find

$$P_2 = \tfrac{2}{5}P \qquad P_1 = \tfrac{3}{10}P \tag{4}$$

The system thus begins to yield when $P_2 = \sigma_{yp}A$. Thus

$$P_{yp} = \tfrac{5}{2}\sigma_{yp}A$$

From the time of yielding of the central bar, the system deforms as if supported by only the two outside bars (which still act elastically) together with a constant force $\sigma_{yp}A$ supplied by the central bar. The value of P increases until yielding begins in each of the outer bars, i.e. when $P_1 = \sigma_{yp}A$. The ultimate load is thus

$$P_u = 2P_1 + P_2 = 2\sigma_{yp}A + \sigma_{yp}A = 3\sigma_{yp}A$$

It is to be noted that the deformation equation (2) is not employed to determine the ultimate load.

2.12. Reconsider Problem 2.10 for the case of three bars of equal cross-sectional area. Determine the ultimate load-carrying capability of the system.

For $A_i = A_v = A$ the force in the vertical bar exceeds that in either inclined bar as indicated by (1) of Problem 2.10. Thus, as P increases, the central vertical bar is the first to enter the inelastic range of action and its stiffness (effective value of AE) decreases. Any additional increase in the load P will cause no further increase in F_1 which will remain at the limit value $F_1^* = \sigma_{yp}A$. The central bar can now be replaced by a constant upward vertical force F_1^* and the system is now reduced to a statically determinate system consisting of the two outer bars subject to an applied load $P - F_1^*$. The load P can now be increased until the outer bars also develop the yield stress. It is not necessary to consider deformations of the system; we need look only at the equilibrium relation

$$P = F_1^* + 2F_2 \cos\theta \tag{1}$$

As the load P increases still more the outer bars also reach the yield point and the force in each of them becomes

$$F_2^* = \sigma_{yp}A \tag{2}$$

The ultimate load thus corresponds to the situation when $F_1^* = F_2^* = \sigma_{yp}A$ and this load is found from (1) as

$$P_u = \sigma_{yp}A(1 + 2\cos\theta) \tag{3}$$

This *limit load* should be divided by some safety factor to obtain a *working load*.

2.13. Suppose the three-bar system of Problem 2.10 is to withstand a load $P = 50,000$ lb. Compare the bar weights required if the design is based upon (a) the peak stress just reaching the yield point, and (b) ultimate load analysis. Assume that all bars are of identical cross-section, that $\theta = 45°$, and take the yield point of the material to be 36,000 lb/in².

(*a*) According to the elastic theory of Problem 2.10 the force in the vertical bar becomes

$$F_1 = \frac{2P}{2+\sqrt{2}} = 29,400 \text{ lb}$$

If the stress in that bar is equal to the yield point, we have a required cross-sectional area of $F_1 = A_1\sigma_{yp}$.   Hence

$$29,400 = A_1(36,000) \quad \text{or} \quad A_1 = 0.816 \text{ in}^2$$

(*b*) If the ultimate load analysis of Problem 2.12 is employed the stresses in all three bars are equal to the yield point and from (*3*) of Problem 2.12 we find a cross-sectional area of

$$50,000 = 36,000 A_2[1+2(0.707)] \quad \text{or} \quad A_2 = 0.578 \text{ in}^2$$

Ultimate load analysis thus implies a 29 percent saving in cross-sectional area and the same weight saving.

2.14. A system composed of a rigid horizontal member *AB* supported by four bars is indicated in Fig. 2-18(*a*). The bars have identical cross-sections and are made of the same material. Determine the ultimate load *P* that may be applied to the system.

Since the member *AB* is rigid it is evident that, upon application of a sufficiently large load *P*, *AB* may rotate as a rigid body about either point *A* or point *B*. (The ultimate load implies plastic deformation in bar 2, hence it is not necessary to consider rotation about *C*.) It is necessary to determine the ultimate loads corresponding to these two possibilities and then to select the smaller.

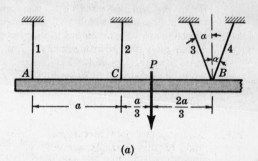

(*b*)

Let us first assume that yielding first begins in bars 1 and 2, in which case their effect can be represented by the two constant forces $\sigma_{yp}A$ as indicated in Fig. 2-18(*b*). The bars 3 and 4 are still in the elastic range of action and the forces in them are unknown. However, it is not necessary to determine the forces since the ultimate load P'_u may be determined by summing moments about point *B*:

$$P'_u\left(\frac{2a}{3}\right) - \sigma_{yp}A(a) - \sigma_{yp}A(2a) = 0$$

Solving:
$$P'_u = 4.5\sigma_{yp}A$$

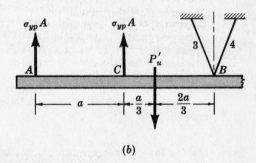

Next, let us consider that yielding begins in bars 2, 3, and 4 as indicated in Fig. 2-18(*c*). Bar 1 is still in the elastic range of action. Taking moments about point *A*:

$$(\sigma_{yp}A\cos\alpha)4a + \sigma_{yp}Aa - P''_u\frac{4a}{3} = 0$$

Solving:
$$P''_u = \tfrac{3}{4}\sigma_{yp}A(1+4\cos\alpha)$$

It is evident from inspection of P'_u and P''_u that for all values of the angle α, the value of P''_u is the smaller of the two and thus P''_u represents the ultimate load. When the applied load reaches this value the system is essentially converted into a mechanism and the rigid bar rotates about point *A*. Even in this condition bar 1 is not working to its full capacity.

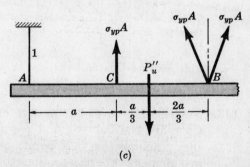

(*c*)

Fig. 2-18

Supplementary Problems

2.15. A square bar 2 in. on a side is held rigidly between the walls and loaded by an axial force of 40,000 lb as shown in Fig. 2-19. Determine the reactions at the ends of the bar and the extension of the right portion. Take $E = 30 \times 10^6$ lb/in^2.

Ans. left reaction = 24,000 lb, right reaction = 16,000 lb, extension = 0.00080 in.

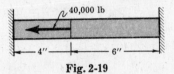

Fig. 2-19

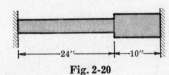

Fig. 2-20

2.16. Two initially straight bars are joined together and attached to supports as in Fig. 2-20. The left bar is brass for which $E = 14 \times 10^6$ lb/in^2, $\alpha = 10.4 \times 10^{-6}$/F$^\circ$ and the right bar is aluminum for which $E = 10 \times 10^6$ lb/in^2, $\alpha = 12.8 \times 10^{-6}$/F$^\circ$. The cross-sectional area of the brass bar is 1 in^2, that of the aluminum bar is 1.5 in^2. Let us suppose that the system is initially stress-free and that the temperature then drops 40F$^\circ$.

(a) If the supports are unyielding, find the normal stress in each bar.

(b) If the right support yields 0.005 in., find the normal stress in each bar. The weight of the bars is negligible.

Ans. (a) $\sigma_{br} = 6350$ lb/in^2, $\sigma_{al} = 4230$ lb/in^2; (b) $\sigma_{br} = 4250$ lb/in^2, $\sigma_{al} = 2840$ lb/in^2

2.17. A steel tube of 2 in. outside diameter and 1.75 in. inside diameter surrounds a solid brass cylinder 1.5 in. in diameter. Both are joined to a rigid cover plate at each end. The assembly is stress-free at a temperature of 80°F. If the temperature is then raised to 250°F, determine the stresses in each material. For brass $E = 14 \times 10^6$ lb/in^2, $\alpha = 10.4 \times 10^{-6}$/F$^\circ$; for steel $E = 30 \times 10^6$ lb/in^2, $\alpha = 6.5 \times 10^{-6}$/F$^\circ$.

Ans. $\sigma_{st} = 10,500$ lb/in^2, $\sigma_{br} = -4400$ lb/in^2

2.18. A compound bar is composed of a strip of copper between two cold-rolled steel plates. The ends of the assembly are covered with infinitely rigid cover plates and an axial tensile load P is applied to the bar by means of a force acting on each rigid plate as shown in Fig. 2-21. The width of all bars is 4 in., the steel plates are each 1/4 in. thick and the copper is 3/4 in. thick. Determine the maximum load P that may be applied. The ultimate strength of the steel is 80,000 lb/in^2 and that of the copper is 30,000 lb/in^2. A safety factor of 3 based upon the ultimate strength of each material is satisfactory. For steel $E = 30 \times 10^6$ lb/in^2 and for copper $E = 13 \times 10^6$ lb/in^2.   *Ans.* $P = 76,200$ lb

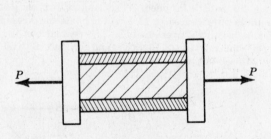

Fig. 2-21

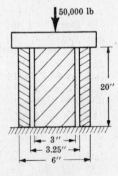

Fig. 2-22

2.19. An aluminum right-circular cylinder surrounds a steel cylinder as shown in Fig. 2-22. The axial compressive load of 50,000 lb is applied through the infinitely rigid cover plate shown. If the aluminum cylinder is originally 0.010 in. longer than the steel before any load is applied, find the normal stress in each when the temperature has dropped 50F$^\circ$ and the entire load is acting. For steel take $E = 30 \times 10^6$ lb/in^2, $\alpha = 6.5 \times 10^{-6}$/F$^\circ$, and for aluminum assume $E = 10 \times 10^6$ lb/in^2, $\alpha = 12.8 \times 10^{-6}$/F$^\circ$.   *Ans.* $\sigma_{st} = 945$ lb/in^2, $\sigma_{al} = 2170$ lb/in^2

2.20. The rigid horizontal bar AB is supported by three vertical wires as shown in Fig. 2-23 and carries a load of 24,000 lb. The weight of AB is negligible and the system is stress-free before the 24,000 lb load is applied. After the load is applied, the temperature of all three wires is raised 25 F°. Find the stress in each wire as well as the location of the applied load in order that AB remain horizontal. For the steel wire take $E = 30 \times 10^6$ lb/in², $\alpha = 6.5 \times 10^{-6}$/F°, for the brass wire $E = 14 \times 10^6$ lb/in², $\alpha = 10.4 \times 10^{-6}$/F°, and for copper $E = 17 \times 10^6$ lb/in², $\alpha = 9.3 \times 10^{-6}$/F°. Neglect any possibility of lateral buckling of any of the wires.

     Ans. $\sigma_{st} = 32{,}300$ lb/in², $\sigma_{br} = 22{,}400$ lb/in², $\sigma_{cu} = 21{,}400$ lb/in², $x = 0.273$ ft

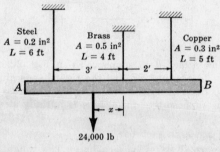

Fig. 2-23

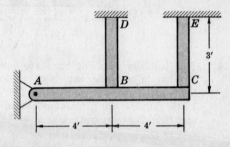

Fig. 2-24

2.21. The bar AC is absolutely rigid and is pinned at A and attached to bars DB and CE as shown in Fig. 2-24. The weight of AC is 10,000 lb and the weights of the other two bars are negligible. Consider the temperature of both bars DB and CE to be raised 70 F°. Find the resulting normal stresses in these two bars. DB is copper for which $E = 15 \times 10^6$ lb/in², $\alpha = 9.3 \times 10^{-6}$/F° and the cross-sectional area is 2 in², while CE is steel for which $E = 30 \times 10^6$ lb/in², $\alpha = 6.5 \times 10^{-6}$/F° and the cross-section is 1 in². Neglect any possibility of lateral buckling of the bars.

     Ans. $\sigma_{st} = 9100$ lb/in², $\sigma_{cu} = -4100$ lb/in²

2.22. Consider the rigid bar BD which is supported by the two wires shown in Fig. 2-25. The wires are initially stress-free and the weights of all members are to be neglected. Find the tension in each wire after the load P has been applied to the extreme end of the bar. The two wires have the same modulus of elasticity.

     Ans.   force in $AD = \dfrac{2P}{(A_1L_2^2H/2A_2L_1^3) + 2H/L_2}$,   force in $AC = \dfrac{2P}{(4HA_2L_1^2/A_1L_2^3) + H/L_1}$

2.23. The three bars shown in Fig. 2-26 support the vertical load of 5000 lb. The bars are all stress-free and joined by the pin at A before the load is applied. The load is put on gradually and simultaneously the temperature of all three bars decreases 15 F°. Calculate the stress in each bar. The outer bars are each brass and of cross-sectional area 0.4 in². The central bar is steel and of area 0.3 in². For brass $E = 13 \times 10^6$ lb/in² and $\alpha = 10.4 \times 10^{-6}$/F° and for steel $E = 30 \times 10^6$ lb/in² and $\alpha = 6.3 \times 10^{-6}$/F°.   *Ans.* $\sigma_{br} = 3550$ lb/in², $\sigma_{st} = 10{,}000$ lb/in²

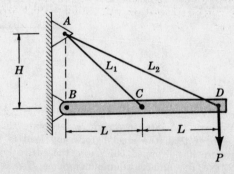

Fig. 2-25

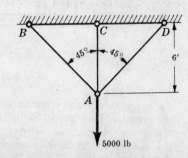

Fig. 2-26

2.24. The five-bar assembly of Fig. 2-27 was found to be slightly defective, i.e. points A and C which ought to have coincided failed to coincide by a distance Δ. After these points had been forced to coincide, the joint at that point was pinned. Determine the forces existing in each bar. All bars have the same cross-sectional area.

Ans. $F_1 = F_2 = F_3 = \left(\dfrac{\sqrt{3}}{2 + 3\sqrt{3}}\right)\dfrac{\Delta AE}{L}$ $F_4 = F_5 = -\left(\dfrac{1}{2 + 3\sqrt{3}}\right)\dfrac{\Delta AE}{L}$

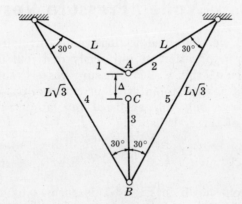

Fig. 2-27

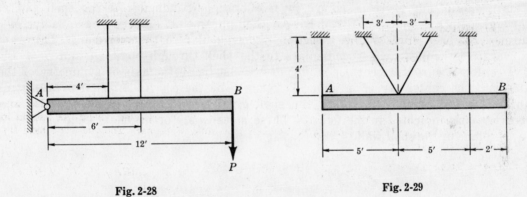

Fig. 2-28   **Fig. 2-29**

2.25. The rigid bar AB in Fig. 2-28 is supported by two vertical rods and is pinned at A. If the yield point of the steel in the rods is 36,000 lb/in^2 and the cross-sectional area of each rod is 0.2 in^2, determine the ultimate load P that may be applied.   *Ans.* 6000 lb

2.26. The rigid bar AB is supported by the four rods shown in Fig. 2-29. The rods are each circular in cross-section and of 2 in. diameter. They have $E = 30 \times 10^6$ lb/in^2 and a yield point of 40,000 lb/in^2. Using limit design determine the maximum weight of the bar AB. Assume that the weight is uniformly distributed along the length.   *Ans.* 378,000 lb

Chapter 3

Thin-Walled Pressure Vessels

In Chapters 1 and 2 we examined various cases involving uniform normal stresses acting in bars. Another application of uniformly distributed normal stresses occurs in the approximate analysis of thin-walled pressure vessels, such as cylindrical, spherical, conical, or toroidal shells subject to internal or external pressure from a gas or a liquid. In this chapter we will treat only thin shells of revolution and restrict ourselves to axisymmetric deformations of these shells.

NATURE OF STRESSES

The shell of revolution shown in Fig. 3-1 is formed by rotating a plane curve (the meridian) about an axis lying in the plane of the curve. The radius of curvature of the meridian is denoted by r_1 and this of course varies along the length of the meridian. This radius of curvature is defined by two lines perpendicular to the shell and passing through points B and C of Fig. 3-1. Another parameter, r_2, denotes the radius of curvature of the shell surface in a direction perpendicular to the meridian. This radius of curvature is defined by perpendiculars to the shell through points A and B of Fig. 3-1. The center of curvature corresponding to r_2 must lie on the axis of symmetry of the shell although the center for r_1 in general does not lie there. An internal pressure p acting normal to the curved surface of the shell gives rise to *meridional stresses* σ_ϕ and *hoop stresses* σ_θ as indicated in the figure. These stresses are orthogonal to one another and act in the plane of the shell wall.

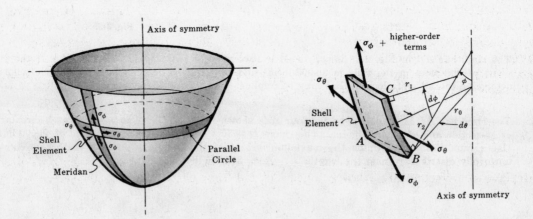

Fig. 3-1

In Problem 3.12 it is shown that

$$\frac{\sigma_\phi}{r_1} + \frac{\sigma_\theta}{r_2} = \frac{p}{h}$$

where h denotes the shell thickness.   A second equation may be obtained by consideration of the vertical equilibrium of the entire shell above some convenient parallel circle, as indicated in Problem 3.12.   The derivation of the above equation assumes that the stresses σ_ϕ and σ_θ are uniformly distributed over the wall thickness.

Applications of this analysis to cylindrical shells are to be found in Problems 3.1 through 3.5; to spherical shells in Problems 3.6 through 3.8; to conical shells in Problem 3.11; and to toroidal shells in Problem 3.13.

LIMITATIONS

The ratio of the wall thickness to either radius of curvature should not exceed approximately 0.10.   Also there must be no discontinuities in the structure.   The simplified treatment presented here does not permit consideration of reinforcing rings on a cylindrical shell as shown in Fig. 3-2, nor does it give an accurate indication of the stresses and deformations in the vicinity of end closure plates on cylindrical pressure vessels. Even so, the treatment is satisfactory in many design problems.

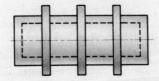

Fig. 3-2

The problems which follow are concerned with stresses arising from a uniform *internal* pressure acting on a thin shell of revolution.   The formulas for the various stresses will be correct if the sense of the pressure is reversed, i.e. if external pressure acts on the container.   However, it is to be noted that an additional consideration, beyond the scope of this book, must then be taken into account.   Not only must the stress distribution be investigated but another study of an entirely different nature must be carried out to determine the load at which the shell will *buckle* due to the compression.   A buckling or instability failure may take place even though the peak stress is far below the maximum allowable working stress of the material.

APPLICATIONS

Liquid storage tanks and containers, water pipes, boilers, submarine hulls, and certain airplane components are common examples of thin-walled pressure vessels.

Solved Problems

3.1. Consider a thin-walled cylinder closed at both ends by cover plates and subject to a uniform internal pressure p. The wall thickness is h and the inner radius r. Neglecting the restraining effects of the end plates, calculate the longitudinal (meridional) and circumferential (hoop) normal stresses existing in the walls due to this loading.

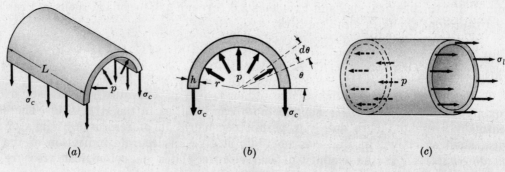

(a)   (b)   (c)

Fig. 3-3

To determine the circumferential stress σ_c let us consider a section of the cylinder of length L to be removed from the vessel. The free-body diagram of half of this section appears as in Fig. 3-3(a). Note that the body has been cut in such a way that the originally *internal* effect (σ_c) now appears as an *external* force to this free body. Figure 3-3(b) shows the forces acting on a cross-section.

The horizontal components of the radial pressures cancel one another by virtue of symmetry about the vertical centerline. In the vertical direction we have the equilibrium equation

$$\Sigma F_v = -2\sigma_c hL + \int_0^\pi pr(d\theta)(\sin\theta)L = 0$$

Integrating,

$$2\sigma_c hL = -prL\left[\cos\theta\right]_0^\pi \quad \text{or} \quad \sigma_c = \frac{pr}{h}$$

Note that the resultant vertical force due to the pressure p could have been obtained by multiplying the pressure by the horizontal *projected area* upon which the pressure acts.

To determine the longitudinal stress σ_l consider a section to be passed through the cylinder normal to its geometric axis. The free-body diagram of the remaining portion of the cylinder is shown in Fig. 3-3(c). For equilibrium,

$$\Sigma F_h = -p\pi r^2 + 2\pi rh\sigma_l = 0 \quad \text{or} \quad \sigma_l = pr/2h$$

Consequently, the circumferential stress is twice the longitudinal stress. Thus if the water in a closed pipe freezes, the pipe will rupture along a line running longitudinally along the cylinder. These rather simple expressions for stresses are not accurate in the immediate vicinity of the end closure plates.

3.2. The tank of an air compressor consists of a cylinder closed by hemispherical ends. The cylinder is 24 in. in inside diameter and is subjected to an internal pressure of 500 lb/in². If the material is a steel whose yield point is 36,000 lb/in² and a safety factor of 3.5 is used, calculate the required wall thickness of the cylinder. Neglect localized effects at the juncture of cylinder and hemisphere.

The ends of the tank are closed. Hence according to Problem 3.1 there exist a circumferential stress in the cylinder wall given by $\sigma_c = pr/h$ and a longitudinal stress given by $\sigma_l = pr/2h$.

Since the circumferential stress is twice the longitudinal stress, it is the critical one for design purposes and it must not exceed the allowable working stress of 36,000/3.5 lb/in². For the circumferential stress we thus have $36{,}000/3.5 = 500(12)/h$ or $h = 0.585$ in.

A more complete analysis would include a study of the stresses in the hemispherical ends.

3.3. A vertical steel stand-pipe, i.e. a cylindrical tank open at the top and having a vertical axis, is 8 ft in inside diameter and 80 ft high. The tank is filled with water having a specific weight of 62.4 lb/ft³. The material is structural steel having a yield point of 35,000 lb/in² and a safety factor of 2 is to be used. What is the required thickness of steel plate necessary at the bottom of the tank if the welded longitudinal seam is presumed to be as strong as the solid metal? What thickness is required if the seam is only 85 percent as strong as the solid metal?

The pressure p (in any direction) at the base of the stand-pipe is given by the formula $p = \gamma h$, where γ represents the weight of the liquid per unit volume and h denotes the height of the column of water above the base. This formula is immediately evident if we consider that the pressure on a square foot of the base numerically equals the weight of a column of water one square foot in cross-section and h feet high. Hence the pressure at the base is

$$p = 62.4(80) = 5000 \text{ lb/ft}^2 \quad \text{or} \quad p = 34.6 \text{ lb/in}^2$$

Since this pressure is hydrostatic it is acting in all directions with the same magnitude, and in particular it is acting radially against the inside wall of the stand-pipe as shown in Fig. 3-4. As may be seen from the equation $p = \gamma h$, the radial pressure decreases toward the top of the tank as shown in the sketch, but the maximum value occurs at the base and consequently that is the region that must be considered for design purposes.

Since the top of the tank is open, there is no longitudinal stress and from Problem 3.1 it is known that the circumferential stress at any point in the stand-pipe is given by $\sigma_c = pr/h$. Considering the region at the base of the tank, this equation becomes

$$\frac{35{,}000}{2} = \frac{34.6(48)}{h} \quad \text{or} \quad h = 0.095 \text{ in.}$$

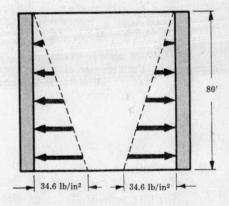

Fig. 3-4

This assumes that the longitudinal seams are as strong as the solid metal. Actually this thickness should be increased slightly to allow for the effects of corrosion.

If the longitudinal seams are only 85% as strong as the solid metal, the required thickness is $h = 0.095/0.85 = 0.112$ in.

3.4. The Space Simulator at the Jet Propulsion Laboratory in Pasadena, California, consists of a 27 ft diameter cylindrical vessel which is 85 ft high. It is made of cold-rolled stainless steel having a proportional limit of 165,000 lb/in². The minimum operating pressure of the chamber is 10^{-6} torr, where 1 torr = 1/760 of a standard atmosphere, which in turn is approximately 14.7 lb/in². Determine the required wall thickness so that a working stress based upon the proportional limit together with a safety factor of 2.5 will not be exceeded. This solution will neglect the possibility of buckling due to the external pressure, and also the effects of certain hard-load points in the Simulator to which the test specimens are attached.

From Problem 3.1 the significant stress is the circumferential stress, given by $\sigma_c = pr/h$. The pressure to be used for design is essentially the atmospheric pressure acting on the outside of the shell, which is satisfactorily represented as 14.7 lb/in² since the internal pressure of 10^{-6} torr is negligible compared to 14.7 lb/in². We thus have

$$\frac{165{,}000}{2.5} = \frac{14.7(13.5)(12)}{h} \quad \text{or} \quad h = 0.036 \text{ in.}$$

3.5. Calculate the increase in the radius of the cylinder considered in Problem 3.1 due to the internal pressure p.

Let us consider the longitudinal and circumferential loadings separately. Due to radial pressure p *only* the circumferential stress is given by $\sigma_c = pr/h$, and because $\sigma = E\epsilon$ the circumferential strain is given by $\epsilon_c = pr/Eh$.

It is to be noted that ϵ_c is a unit strain. The length over which it acts is the circumference of the cylinder which is $2\pi r$. Hence the total elongation of the circumference is

$$\Delta = \epsilon_c(2\pi r) = 2\pi pr^2/Eh$$

The final length of the circumference is thus $2\pi r + 2\pi pr^2/Eh$. Dividing this circumference by 2π we find the radius of the deformed cylinder to be $r + pr^2/Eh$, so that the increase in radius is pr^2/Eh.

Due to the axial pressure p *only*, longitudinal stresses $\sigma_l = pr/2h$ are set up. These longitudinal stresses give rise to longitudinal strains $\epsilon_l = pr/2Eh$. As in Chapter 1 an extension in the direction of loading, which is the longitudinal direction here, is accompanied by a decrease in the dimension perpendicular to the load. Thus here the circumferential dimension decreases. The ratio of the strain in the lateral direction to that in the direction of loading was defined in Chapter 1 to be Poisson's ratio, denoted by μ. Consequently the above strain ϵ_l induces a circumferential strain equal to $-\mu\epsilon_l$ and if this strain is denoted ϵ_c we have $\epsilon_c = -\mu pr/2Eh$ which tends to decrease the radius of the cylinder as shown by the negative sign.

In a manner exactly analogous to the treatment of the increase of radius due to radial loading only, the decrease of radius corresponding to the strain ϵ_c is given by $\mu pr^2/2Eh$. The resultant increase of radius due to the internal pressure p is thus

$$\Delta r = \frac{pr^2}{Eh} - \frac{\mu pr^2}{2Eh} = \frac{pr^2}{Eh}\left(1 - \frac{\mu}{2}\right)$$

3.6. Consider a closed thin-walled spherical shell subject to a uniform internal pressure p. The inside radius of the shell is r and its wall thickness is h. Derive an expression for the tensile stress existing in the wall.

For a free-body diagram, let us consider exactly half of the entire sphere. This body is acted upon by the applied internal pressure p as well as the forces that the other half of the sphere, which has been removed, exerts upon the half under consideration. Because of the symmetry of loading and deformation, these forces may be represented by circumferential tensile stresses σ_c as shown in Fig. 3-5.

This free-body diagram represents the forces acting on the hemisphere, the diagram showing only a projection of the hemisphere on a vertical plane. Actually the pressure p acts over the entire inside surface of the hemisphere and in a direction perpendicular to the surface at every point. However, as mentioned in Problem 3.1, it is permissible to consider the force exerted by this same pressure p upon the *projection* of this area which in this case is the vertical circular area denoted by a-a. This is possible because the hemisphere is symmetric about the horizontal axis and the vertical components of the pressure annul one another. Only the horizontal components produce the tensile stress σ_c. For equilibrium we have

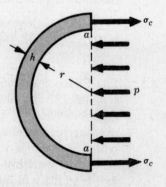

Fig. 3-5

$$\Sigma F_h \ = \ \sigma_c 2\pi r h - p\pi r^2 \ = \ 0 \quad \text{or} \quad \sigma_c \ = \ pr/2h$$

From symmetry this circumferential stress is the same in all directions at any point in the wall of the sphere.

3.7. A 60 ft diameter spherical tank is to be used to store gas. The shell plating is 0.5 in. thick and the working stress of the material is 18,000 lb/in². What is the maximum permissible gas pressure p?

From Problem 3.6 the tensile stress is uniform in all directions and is given by $\sigma_c = pr/2h$. Substituting, $18,000 = p(30 \times 12)/2(0.5)$ and $p = 50$ lb/in².

3.8. The undersea research vehicle *Alvin* has a spherical pressure hull 39.64 in. in radius and shell thickness of 1.33 in. The pressure hull is HY-100 steel having a yield point of 100,000 lb/in². Determine the depth of submergence that would set up the yield-point stress in the spherical shell. Consider sea water to weigh 64.0 lb/ft.³

From Problem 3.6 the compressive stress due to the external hydrostatic pressure is given by $\sigma_c = pr/2h$. The hydrostatic pressure corresponding to yield is thus

$$100,000 \ = \ \frac{p(39.64)}{2(1.33)} \quad \text{or} \quad p \ = \ 6720 \text{ lb/in}^2$$

Since $p = \gamma h$ where γ is the weight of the water per unit volume we have

$$6720 \ \frac{\text{lb}}{\text{in}^2} \ = \ \left(64.0 \ \frac{\text{lb}}{\text{ft}^3} \right)\left(\frac{\text{ft}^3}{1728 \text{ in}^3} \right) h \quad \text{or} \quad h \ = \ 15,100 \text{ ft}$$

It should be noted that this investigation neglects the possibility of buckling of the sphere due to hydrostatic pressure, as well as effects of entrance ports on its strength. These factors, beyond the scope of this treatment, result in a true operating depth of 6000 ft.

3.9. Consider a laminated pressure vessel composed of two thin coaxial cylinders as shown in Fig. 3-6. In the state prior to assembly there is a slight "interference" between these shells, i.e. the inner one is too large to slide into the outer one. The outer cylinder is heated, placed on the inner and allowed to cool, thus providing a "shrink fit". If both cylinders are steel and the mean diameter of the assembly is 4 in., find the tangential stresses in each shell arising from the shrinking if the initial interference (of diameters) is 0.010 in. The thickness of the inner shell is 0.10 in., that of the outer shell 0.08 in.

Laminated Pressure Vessel

Outer Cylinder

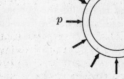
Inner Cylinder

Fig. 3-6

There is evidently an interfacial pressure p acting between the adjacent faces of the two shells. It is to be noted that there are no external applied loads. The pressure p may be considered to increase the diameter of the outer shell and decrease the diameter of the inner so that the inner shell may fit inside the outer. The radial expansion of a cylinder due to a radial pressure p was found in Problem 3.5 to be pr^2/Eh. No longitudinal forces are acting in this problem.

The increase in radius of the outer shell due to p, plus the decrease in radius of the inner one due to p, must equal the initial interference between radii, or 0.010/2 in. Thus we have

$$\frac{p(2)^2}{(30 \times 10^6)(0.10)} + \frac{p(2)^2}{(30 \times 10^6)(0.08)} = 0.005 \quad \text{or} \quad p = 1670 \text{ lb/in}^2$$

This pressure, illustrated in the above figures, acts between the cylinders after the outer one has been shrunk onto the inner one. In the inner cylinder this pressure p gives rise to a stress

$$\sigma_c = \frac{pr}{h} = -\frac{1670(2)}{0.10} = -33,440 \text{ lb/in}^2$$

In the outer cylinder the circumferential stress due to the pressure p is

$$\sigma_c' = \frac{pr}{h} = \frac{1670(2)}{0.08} = 41,700 \text{ lb/in}^2$$

If, for example, the laminated shell is subject to a uniform internal pressure, these "shrink fit" stresses would merely be added algebraically to the stresses found by the use of the simple formulas given in Problem 3.1.

3.10. The thin steel cylinder just fits over the inner copper cylinder as shown in Fig. 3-7. Find the tangential stresses in each shell due to a temperature rise of 60 F°. Do not consider the effects introduced by the accompanying longitudinal expansion. This arrangement is sometimes used for storing corrosive fluids. Take

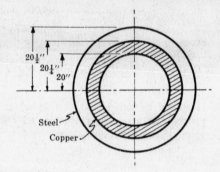

Fig. 3-7

$$E_{st} = 30 \times 10^6 \text{ lb/in}^2, \quad \alpha_{st} = 6.5 \times 10^{-6}/\text{F}°$$

$$E_{cu} = 13 \times 10^6 \text{ lb/in}^2, \quad \alpha_{cu} = 9.3 \times 10^{-6}/\text{F}°$$

The simplest approach is to first consider the two shells to be separated from one another so that they are no longer in contact.

Due to the temperature rise of 60 F° the circumference of the steel shell increases by an amount $2\pi(20.375)(60)(6.5 \times 10^{-6}) = 0.0498$ in. Also, the circumference of the copper shell increases an amount $2\pi(20.125)(60)(9.3 \times 10^{-6}) = 0.0705$ in. Thus the interference between the radii, i.e. the difference in radii, of the two shells (due to the heating) is $(0.0705 - 0.0498)/2\pi = 0.00345$ in. Again, there are no external loads acting on either cylinder.

However, from the statement of the problem the adjacent surfaces of the two shells are obviously in contact after the temperature rise. Hence there must be an interfacial pressure p between the two surfaces, i.e. a pressure tending to increase the radius of the steel shell and decrease the radius of the copper shell so that the copper shell may fit inside the steel one. Such a pressure is shown in the free-body diagrams of Fig. 3-8.

Steel Cylinder

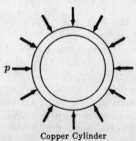

Copper Cylinder

Fig. 3-8

In Problem 3.5 the change of radius of a cylinder due to a uniform radial pressure p (with no longitudinal forces acting) was found to be pr^2/Eh. Consequently the increase of radius of the steel shell due to p, added to the decrease of radius of the copper one due to p, must equal the interference; thus

$$\frac{p(20.375)^2}{(30 \times 10^6)(0.25)} + \frac{p(20.125)^2}{(13 \times 10^6)(0.25)} = 0.00345 \quad \text{or} \quad p = 19.2 \text{ lb/in}^2$$

This interfacial pressure creates the required continuity at the common surface of the two shells when they are in contact. Using the formula for the tangential stress, $\sigma_c = pr/h$, we find the tangential stresses in the steel and copper shells to be respectively

$$\sigma_{st} = \frac{19.2(20.375)}{0.25} = 1560 \text{ lb/in}^2 \quad \text{and} \quad \sigma_{cu} = -\frac{19.2(20.125)}{0.25} = -1550 \text{ lb/in}^2$$

3.11. Consider a thin-walled conical shell containing a liquid whose weight per unit volume is γ [see Fig. 3-9(a)]. The shell is supported around its upper rim and filled with liquid to a depth H. Determine the stresses in the shell walls due to this loading. The geometric axis of the shell is vertical.

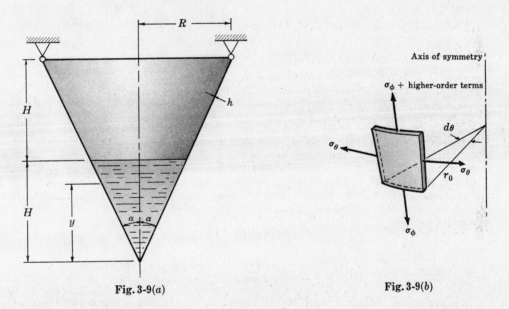

Fig. 3-9(a)   Fig. 3-9(b)

The state of stress in this shell is obviously axisymmetric. It is assumed that the shell thickness h is small compared to H and R. The stresses may be determined by consideration of the equilibrium of a shell element bounded by two closely adjacent parallel circles whose planes are normal to the vertical axis of symmetry of the cone and by two closely adjacent generators of the cone. Such an element, together with the vectors representing the stresses σ_θ in the horizontal direction and σ_ϕ in the direction of a generator, is indicated in Fig. 3-9(b). The quantity σ_θ is called *hoop stress* and σ_ϕ is termed the *meridional stress*.

In the diagram θ represents the angular coordinate measured in a horizontal plane which is normal to the vertical axis of symmetry of the shell. The radius of the cone there is r_0, which is of course a function of the location of the element with respect to its position along the axis of symmetry. Another coordinate useful for defining the geometry of the cone is r_2, which corresponds to the radius of curvature of the shell surface in a direction perpendicular to the generator. This is best illustrated by examining a section of the cone formed by passing a vertical plane through the shell axis as indicated in Fig. 3-9(c) below. It is evident that $r_0 = r_2 \cos \alpha$.

From geometry we have

$$r_0 = y \tan \alpha \quad \text{and so} \quad r_2 = \frac{y \tan \alpha}{\cos \alpha}$$

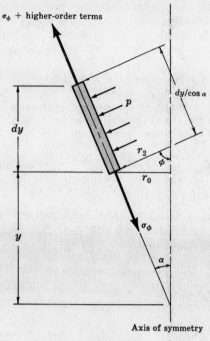

Fig. 3-9(c)   Fig. 3-9(d)

The hoop *stresses* in Fig. 3-9(b) may be visualized more clearly by looking along the axis of symmetry, as shown in Fig. 3-9(d). It is evident that each of the hoop *force* vectors $\sigma_\theta(dy/\cos\alpha)h$ makes an angle $d\theta/2$ with the tangent to the element. The resultant of these hoop forces is $2\sigma_\theta h(dy/\cos\alpha)\sin(d\theta/2)$ or, since $d\theta/2$ is small, $\sigma_\theta h(dy/\cos\alpha)\,d\theta$ acting in a horizontal plane and directed toward the geometric axis of the shell. From Fig. 3-9(c) we see that this resultant must be multiplied by $\cos\alpha$ to determine the component of this force acting in a direction normal to the shell surface. Also, it is evident that the meridional forces corresponding to Fig. 3-9(c) cancel one another. The liquid exerts a normal pressure p as indicated in the figure and it acts over an area $(r_0 d\theta)(dy/\cos\alpha)$. Thus, for equilibrium of the element in a direction normal to the surface we have

$$\sigma_\theta h\left(\frac{dy}{\cos\alpha}\right)(d\theta)\cos\alpha \;-\; pr_0(d\theta)\frac{dy}{\cos\alpha} \;=\; 0 \tag{1}$$

or
$$\sigma_\theta \;=\; \frac{pr_0}{h\cos\alpha} \;=\; \frac{py\tan\alpha}{h\cos\alpha} \;=\; \frac{pr_2}{h} \tag{2}$$

This expression holds anywhere in the conical shell. In the lower half, $0 < y < H$, we have $p = \gamma(H-y)$, so

$$\sigma_\theta \;=\; \frac{\gamma(H-y)y\tan\alpha}{h\cos\alpha} \qquad \text{for} \quad 0 < y < H \tag{3}$$

In the upper half, $H < y < 2H$, $p = 0$, so $\sigma_\theta = 0$ in that region.

The other stress σ_ϕ may be found by considering the vertical equilibrium of the conical shell. For $0 < y < H$ the weight of the liquid in the conical region abo plus that in the cylindrical region $abcd$ is held in equilibrium by the forces corresponding to σ_ϕ and we have from Fig. 3-9(e)

$$\sigma_\phi h 2\pi y\tan\alpha\cos\alpha \;-\; \gamma[\tfrac{1}{3}\pi(y\tan\alpha)^2 y + (H-y)\pi(y\tan\alpha)^2] \;=\; 0 \tag{4}$$

or
$$\sigma_\phi \;=\; \frac{\gamma\tan\alpha}{h\cos\alpha}\left(\frac{Hy}{2} - \frac{y^2}{3}\right) \qquad \text{for} \quad 0 < y < H \tag{5}$$

Similarly, for $H < y < 2H$, the weight of all the liquid is held in equilibrium by the forces corresponding to σ_ϕ so that from Fig. 3-9(f)

$$\sigma_\phi h(2\pi y)(\tan\alpha)\cos\alpha \;-\; \gamma\tfrac{1}{3}\pi r_0^2 H \;=\; 0 \tag{6}$$

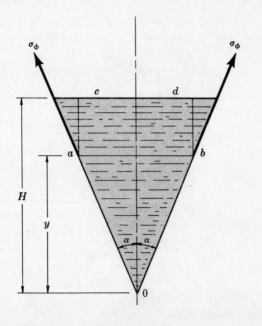

Fig. 3-9(e)

Fig. 3-9(f)

Since $r_0 = H \tan \alpha$ we get

$$\sigma_\phi = \frac{\gamma H^3 \tan \alpha}{6hy \cos \alpha} \qquad \text{for} \quad H < y < 2H \tag{7}$$

It is to be observed that the stresses associated with these axisymmetric deformations are statically determinate, i.e. it was not necessary to use any deformation relations to determine the stresses. Thus the relations are valid into the plastic range of action.

3.12. Determine the hoop stresses and meridional stresses in a thin shell of revolution subject to an internal pressure p.

This problem is readily solved as a generalization of Problem 3.11. The stresses may be determined by consideration of the equilibrium of a shell element bounded by two closely adjacent parallel circles whose planes are normal to the vertical axis of a symmetry of the shell and by two closely adjacent generators, or meridians, of the shell (see Fig. 3-1, page 40). This element is analogous to that shown in Fig. 3-9(b) of Problem 3.11 except that the vertical sides are curved rather than straight.

The hoop stresses σ_θ and the meridional stresses σ_ϕ thus appear as shown in Fig. 3-10(a). We now require two radii of curvature to describe this element. We use r_1 to denote the radius of curvature of the meridian and r_2 to denote the radius of curvature of the shell surface in a direction perpendicular to the meridian. The center of curvature corresponding to r_2 must lie on the axis of symmetry although the center for r_1 does not (in general). Figure 3-10(b) shows the hoop forces as seen by looking along

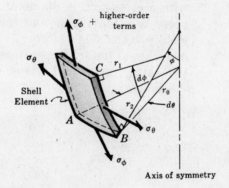

Fig. 3-10(a)

the axis of symmetry and, analogous to Problem 3.11, they have a horizontal component $2\sigma_\theta h r_1\, d\phi\, (d\theta/2)$ directed toward the shell axis. This is multiplied by $\sin\phi$ to obtain the component

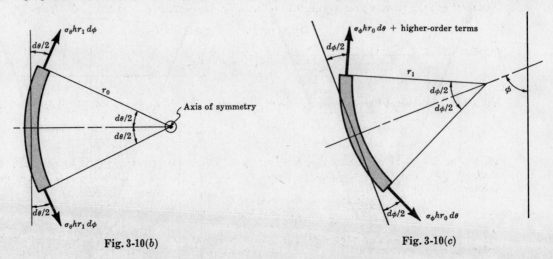

Fig. 3-10(b)   Fig. 3-10(c)

normal to the shell element. The meridional forces appear as in Fig. 3-10(c) and they have a component normal to the shell given by $\sigma_\phi h r_0\, d\theta\, d\phi$. The pressure p acts over an area $(r_0\, d\theta)(r_1\, d\phi)$ so that the equation of equilibrium in the normal direction becomes

$$\sigma_\theta h r_1\, d\theta\, d\phi\, \sin\phi + \sigma_\phi h r_0\, d\theta\, d\phi - p r_0\, d\theta\, r_1\, d\phi = 0$$

or, since $r_0 = r_2 \sin\phi$ we get

$$\frac{\sigma_\phi}{r_1} + \frac{\sigma_\theta}{r_2} = \frac{p}{h} \tag{1}$$

This fundamental equation applies to axisymmetric deformations of all thin shells of revolution. A second equation is obtained as in Problem 3.11 by consideration of the vertical equilibrium of the entire shell above some convenient parallel circle. Again, these equations are valid into the plastic range of action.

3.13. Thin toroidal shells are sometimes employed as gas storage tanks in boosters for space vehicles. One design considered by the National Aeronautics and Space Administration for possible future use employs a torus of mean diameter $2b = 70$ ft with a cross-section diameter of $2R = 5$ ft as indicated in Fig. 3-11. The internal pressure p is 20 lb/in^2 and the shell material is 2219 T87 aluminum alloy, having a yield point of 50,000 lb/in^2 at room temperature. For this material the yield point increases at lower temperatures, reaching 120 percent of the above value at $-300°$F. If a safety factor of 1.5 is employed, determine the required wall thickness.

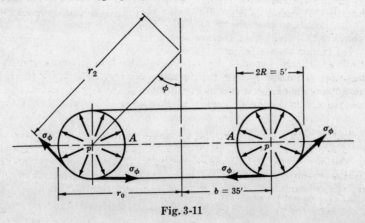

Fig. 3-11

First, we consider the vertical equilibrium of a ring-shaped portion of the toroidal shell above an arbitrary plane, as indicated by the angle ϕ. The meridional stress σ_ϕ is readily found by considering the pressure p to act on the horizontal projection of the curved area. Thus

$$2\pi r_0 \sigma_\phi h \sin \phi = \pi p(r_0^2 - b^2)$$

or since $\sin \phi = (r_0 - b)/R$

$$\sigma_\phi = \frac{pR(r_0 + b)}{2\pi r_0 h} \qquad (1)$$

From (1) it is evident that the peak value of σ_ϕ occurs at the innermost points A where

$$(\sigma_\phi)_{max} = \frac{pR}{2h}\left(\frac{2b - R}{b - R}\right) \qquad (2)$$

If $b = 0$, the torus reduces to a sphere and (2) coincides with the stresses in a sphere as found in Problem 3.6. For the given dimensions we have: $R = 30$ in., $b = 420$ in., $p = 20$ lb/in² and (2) becomes

$$\frac{50{,}000}{1.5} = \frac{20(30)(840 - 30)}{2h(420 - 30)} \qquad \text{or} \qquad h = 0.0187 \text{ in.} \qquad (3)$$

If σ_ϕ as given by (1) is substituted into (1) of Problem 3.12 (which holds for axisymmetric deformation of any thin shell of revolution) we obtain, for $r_1 = R$ and $r_2 = (b + R \sin \phi)/\sin \phi$,

$$\sigma_\theta = \frac{pR}{2h} \qquad (4)$$

at any point in the toroidal shell. Evidently the peak value of σ_ϕ as given by (2) exceeds the value of σ_θ and hence the maximum value of σ_ϕ controls the design. The required thickness is thus given by (3).

Supplementary Problems

3.14. A compressed air cylinder for laboratory use ordinarily carries approximately 2300 lb/in² pressure at the time of delivery. The outside diameter of such a cylinder is 10 in. If the steel has a yield point of 33,000 lb/in² and a safety factor of 2.5 is adequate, calculate the required wall thickness.

Ans. $h = 0.875$ in.

3.15. The research deep submersible *Aluminaut* has a cylindrical pressure hull of outside diameter 8 ft and a wall thickness of 5.5 in. It is constructed of 7079-T6 aluminum alloy, having a yield point of 60,000 lb/in². Determine the circumferential stress in the cylindrical portion of the pressure hull when the vehicle is at its operating depth of 15,000 ft below the surface of the sea. Use the mean diameter of the shell in calculations, and consider sea water to weigh 64.0 lb/ft³.

Ans. 54,800 lb/in²

3.16. A vertical cylindrical gasoline storage tank is 85 ft in diameter and is filled to a depth of 40 ft with gasoline whose specific gravity is 0.74. If the yield point of the shell plating is 35,000 lb/in² and a safety factor of 2.5 is adequate, calculate the required wall thickness at the bottom of the tank neglecting any localized bending effects there.   Ans. $h = 0.466$ in.

3.17. A spherical tank for storing gas under pressure is 80 ft in diameter and is made of structural steel 5/8 in. thick. The yield point of the material is 35,000 lb/in² and a safety factor of 2.5 is adequate. Determine the maximum permissible internal pressure, assuming the welded seams between the various plates are as strong as the solid metal. Also, determine the permissible pressure if the seams are 75 percent as strong as the solid metal.

Ans. $p = 36.5$ lb/in², $p = 27.4$ lb/in²

3.18. The Deep Submergence Rescue Vehicle for assisting sub-
marines in distress has a pressure hull consisting of three
interconnected spherical shells as indicated in Fig. 3-12.
The shells are made of HY-130 steel having an outside
diameter of 7.5 ft. and a wall thickness of 0.75 in. The yield
point of the material is 130,000 lb/in^2. Determine the circumferential stress in the spherical shells at the operating
depth of 3500 feet below the surface of the sea. Neglect
the effects of stress concentrations where the spheres are
joined and neglect the possibility of buckling due to hydrostatic pressure. Consider sea water to weigh 64.0 lb/ft^3.

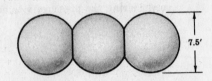

Fig. 3-12

Ans. 46,700 lb/in^2

3.19. Calculate the increase in the radius of the spherical shell mentioned in Problem 3.6 due to the
internal pressure.   Ans. $\Delta r = pr^2(1-\mu)/2Eh$

3.20. Derive an expression for the increase of volume per unit volume of a thin-walled circular cylinder
subjected to a uniform internal pressure p. The ends of the cylinder are closed by circular plates.
Assume that the radial expansion is constant along the length.

Ans. $\dfrac{\Delta V}{V} = \dfrac{pr}{Eh}\left(\dfrac{5}{2} - 2\mu\right)$

3.21. Calculate the increase of volume per unit volume of a thin-walled steel circular cylinder closed at
both ends and subjected to a uniform internal pressure of 80 lb/in^2. The wall thickness is 1/16 in.,
the radius 13.5 in. and $\mu = 1/3$. Consider $E = 30 \times 10^6$ lb/in^2.   Ans. $\Delta V/V = 1.06 \times 10^{-3}$

3.22. Consider a laminated cylinder consisting of a thin steel shell "shrunk" on an aluminum one. The
thickness of each is 0.10 in. and the mean diameter of the assembly is 4 in. The initial "interference" of the shells prior to assembly is 0.004 in. measured on a diameter. Find the tangential
stresses in each shell caused by this "shrink fit". For aluminum $E = 10 \times 10^6$ lb/in^2 and for steel
$E = 30 \times 10^6$ lb/in^2.   Ans. $\sigma_{st} = 7500$ lb/in^2, $\sigma_{al} = -7500$ lb/in^2

3.23. Consider the hemispherical vessel of radius R and thickness h filled with a liquid whose weight per
unit volume is γ. The shell is supported around its upper rim as in Fig. 3-13. Determine the
stresses in the shell due to this loading.

Ans.

$$\sigma_\phi = \frac{\gamma R^2}{3h}\left(\frac{1 - \cos^3\phi}{\sin^2\phi}\right)$$

$$\sigma_\theta = \frac{\gamma R^2}{3h}\left[3\cos\phi - \frac{1 - \cos^3\phi}{\sin^2\phi}\right]$$

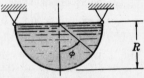

Fig. 3-13

3.24. Consider the hemispherical vessel of radius R and thickness h partially filled with a liquid whose
weight per unit volume is γ. The shell is supported around its upper rim as in Fig. 3-14. Determine the stresses in the shell.

Ans. For $\alpha < 45°$

$$\sigma_\phi = -\sigma_\theta = \frac{\gamma R^2}{h}\frac{0.0382}{\cos^2\alpha}$$

For $45° < \alpha < 90°$

$$\sigma_\phi = \frac{\gamma R^2}{h}\left[\frac{1 + \sin\alpha + \sin^2\alpha}{3(1 + \sin\alpha)} - 0.355\right]$$

$$\sigma_\theta = \frac{\gamma R^2}{h}(\sin\alpha - 0.707) - \sigma_\phi$$

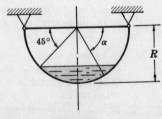

Fig. 3-14

3.25. Re-examine Problem 3.13 with all parameters as indicated there except that the shell material is now Ti-6Al-4V titanium alloy having a yield point of 126,000 lb/in² at room temperature. If a safety factor of 1.5 is used, determine the required wall thickness. *Ans.* 0.0074 in.

3.26. The Deep Submergence Rescue Vehicle mentioned in Problem 3.18 has a compressed air storage tank in the form of a toroidal shell which surrounds two of the interconnected spherical shells as indicated below. The toroidal tank is made of steel having a 150,000 lb/in² ultimate tensile strength. A safety factor of 3 based upon this strength is employed. Determine the required wall thickness of the torus to withstand an internal air pressure of 4500 lb/in². *Ans.* 0.60 in.

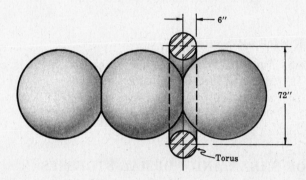

Fig. 3-15

Chapter 4

Direct Shear Stresses

DEFINITION OF SHEAR FORCE

If a plane is passed through a body, a force acting along this plane is called a *shear force* or *shearing force*. It will be denoted by F_s.

DEFINITION OF SHEAR STRESS

The shear force, divided by the area over which it acts, is called the *shear stress* or *shearing stress*. It is denoted in this book by τ. Thus

$$\tau = \frac{F_s}{A}$$

COMPARISON OF SHEAR AND NORMAL STRESSES

Let us consider a bar cut by a plane *a-a* perpendicular to its axis, as shown in Fig. 4-1. A normal stress σ is perpendicular to this plane. This is the type of stress considered in Chapters 1, 2, and 3.

A shear stress is one acting *along* the plane, as shown by the stress τ. Hence the distinction between normal stresses and shear stresses is one of *direction*.

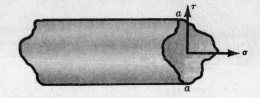

Fig. 4-1

ASSUMPTION

It is necessary to make some assumption regarding the manner of distribution of shear stresses, and for lack of any more precise knowledge it will be taken to be uniform in all problems discussed in this chapter. Thus the expression $\tau = F_s/A$ indicates an average shear stress over the area.

APPLICATIONS

Punching operations (Problem 4.3), wood test specimens (Problem 4.4), riveted joints (Problem 4.6), welded joints (Problem 4.7), and keys used to lock pulleys to shafts (Problem 4.8) are common examples of systems involving shear stresses.

DEFORMATIONS DUE TO SHEAR STRESSES

Let us consider the deformation of a plane rectangular element cut from a solid where the forces acting on the element are known to be shearing stresses τ in the directions shown in Fig. 4-2(a).

54

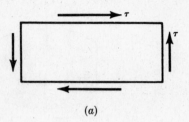

(a)　　　　　　　　　　　　　　　　　　　　(b)

Fig. 4-2

The faces of the element parallel to the plane of the paper are assumed to be load-free. Since there are no normal stresses acting on the element, the lengths of the sides of the originally rectangular element will not change when the shearing stresses assume the value τ. However, there will be a distortion of the originally right *angles* of the element, and after this distortion due to the shearing stresses the element assumes the configuration shown by the dashed lines in Fig. 4-2(b).

SHEAR STRAIN

The change of angle at the corner of an originally rectangular element is defined as the *shear strain*. It must be expressed in radian measure and is usually denoted by γ.

MODULUS OF ELASTICITY IN SHEAR

The ratio of the shear stress τ to the shear strain γ is called the *modulus of elasticity in shear* and is usually denoted by G. Thus

$$ G = \frac{\tau}{\gamma} $$

G is also known as the *modulus of rigidity*.

The units of G are the same as those of the shear stress, e.g. lb/in², since the shear strain is dimensionless The experimental determination of G and the region of linear action of τ and γ will be discussed in Chapter 5. Stress-strain diagrams for various materials may be drawn for shearing loads, just as they were drawn for normal loads in Chapter 1. They have the same general appearance as those sketched in Chapter 1 but the numerical values associated with the plots are of course different.

Solved Problems

4.1. Consider the bolted joint shown in Fig. 4-3. The force P is 7500 lb and the diameter of the bolt is 0.5 in. Determine the average value of the shearing stress existing across either of the planes *a-a* or *b-b*.

　　Lacking any more precise information we can only assume that force P is equally divided between the sections *a-a* and *b-b*. Consequently a force of 7500/2 = 3750 lb acts across either of these planes over a cross-sectional area $\frac{1}{4}\pi(0.5)^2 = 0.197$ in². Thus the average shearing stress across either plane is $\tau = \frac{1}{2}P/A = 3750/0.197 = 19,000$ lb/in².

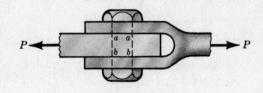

Fig. 4-3

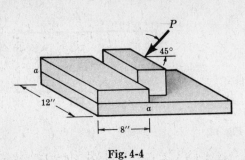

Fig. 4-4   Fig. 4-5

4.2.   Referring to Fig. 4-4, the force P tends to shear off the stop along the plane a-a. If $P = 8000$ lb, determine the average shearing stress on the plane a-a.

Only the horizontal component of P is effective in producing this shearing action. It is given by $8000 \cos 45° = 5650$ lb. The average stress across plane a-a is therefore

$$\tau = \frac{P \cos 45°}{A} = \frac{5650}{12(8)} = 59 \text{ lb/in}^2$$

4.3.   Low-carbon structural steel has a shearing ultimate strength of approximately 45,000 lb/in². Determine the force P necessary to punch a 1 in. diameter hole through a plate of this steel 3/8 in. thick. If the modulus of elasticity in shear for this material is 12×10^6 lb/in², find the shear strain at the edge of this hole when the shear stress is 21,000 lb/in².

Let us assume uniform shearing on a cylindrical surface 1 in. in diameter and 3/8 in. thick as shown in Fig. 4-5. For equilibrium the force P is $P = \tau A = \pi(1)(3/8)(45,000) = 53,100$ lb.

To determine the shear strain γ when the shear stress τ is 21,000 lb/in², we employ the definition $G = \tau/\gamma$ to obtain $\gamma = \tau/G = 21,000/12,000,000 = 0.00175$ radian.

4.4.   In the wood industries, inclined blocks of wood are sometimes used to determine the *compression-shear* strength of glued joints. Consider the pair of glued blocks A and B which are 1.5 in. deep in a direction perpendicular to the plane of the paper. Determine the shearing ultimate strength of the glue if a vertical force of 9000 lb is required to cause rupture of the joint. It is to be noted that a good glue causes a large proportion of the failure to occur in the wood.

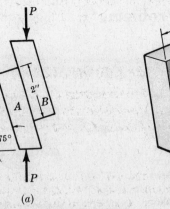

(a)

(b)

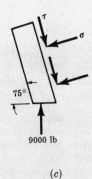

(c)

Fig. 4-6

Let us consider the equilibrium of the lower block, A. The reactions of the upper block B upon the lower one consist of both normal and shearing forces appearing as in the perspective and orthogonal views of Fig. 4-6(b) and (c).

Referring to Fig. 4-6(c) we see that for equilibrium in the horizontal direction

$$\Sigma F_h = \tau(2)(1.5) \cos 75° - \sigma(2)(1.5) \cos 15° = 0 \quad \text{or} \quad \sigma = 0.269\tau$$

For equilibrium in the vertical direction we have

$$\Sigma F_v = 9000 - \tau(2)(1.5) \sin 75° - \sigma(2)(1.5) \sin 15° = 0$$

Substituting $\sigma = 0.269\tau$ and solving, we find $\tau = 2900$ lb/in².

4.5. The shearing stress in a piece of structural steel is 15,000 lb/in². If the modulus of rigidity G is 12,000,000 lb/in², find the shearing strain γ.

By definition, $G = \dfrac{\tau}{\gamma}$. Then the shearing strain $\gamma = \dfrac{\tau}{G} = \dfrac{15,000}{12,000,000} = 0.00125$ radian.

4.6. A single rivet is used to join two plates as shown in Fig. 4-7. If the diameter of the rivet is 3/4 in. and the load P is 7000 lb, what is the average shearing stress developed in the rivet?

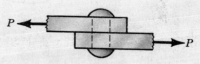

Here the average shear stress in the rivet is P/A where A is the cross-sectional area of the rivet. However, rivet holes are usually 1/16 in. larger in diameter than the rivet and it is customary to assume that the rivet fills the hole completely. Hence the shearing stress is given by

Fig. 4-7

$$\tau = \frac{7000}{\dfrac{\pi}{4}\left(\dfrac{3}{4} + \dfrac{1}{16}\right)^2} = 13,500 \text{ lb/in}^2$$

4.7. One common type of weld for joining two plates is the *fillet weld*. This weld undergoes shear as well as tension or compression and frequently bending in addition. For the two plates shown in Fig. 4-8, determine the allowable tensile force P that may be applied using an allowable working stress of 11,300 lb/in² for shear loading as indicated by the Code for Fusion Welding of the American Welding Society. Consider only shearing stresses in the weld. The load is applied midway between the two welds.

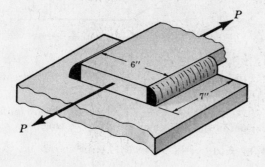

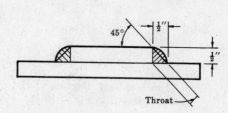

Fig. 4-8

The minimum dimension of the weld cross-section is termed the *throat*, which in this case is $\frac{1}{2}\sin 45° = 0.353$ in. The effective weld area that resists shearing is given by the length of the weld times the throat dimension, or weld area $= 7(0.353) = 2.47$ in^2 for each of the two welds. Thus the allowable tensile load P is given by the product of the working stress in shear times the area resisting shear, or $P = 11,300(2)(2.47) = 56,000$ lb.

4.8. Shafts and pulleys are usually fastened together by means of a key, as shown in Fig. 4-9(a). Consider a pulley subject to a turning moment T of 10,000 lb-in keyed by a $\frac{1}{2} \times \frac{1}{2} \times 3$ in. key to the shaft. The shaft is 2 in. in diameter. Determine the shear stress on a horizontal plane through the key.

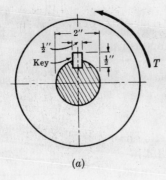 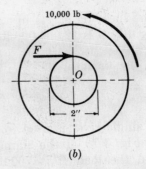

(a)   (b)

Fig. 4-9

Drawing a free-body diagram of the pulley alone, as shown in Fig. 4-9(b), we see that the applied turning moment of 10,000 lb-in must be resisted by a horizontal tangential force F exerted on the pulley by the key. For equilibrium of moments about the center of the pulley we have

$$\Sigma M_0 = 10,000 - F(1) = 0 \quad \text{or} \quad F = 10,000 \text{ lb}$$

It is to be noted that the shaft exerts additional forces, not shown, on the pulley. These act through the center O and do not enter the above moment equation. The resultant forces acting on the key appear as in Fig. 4-10(a). Actually the force F acting to the right is the resultant of distributed forces acting over the lower half of the left face. The other forces F shown likewise represent resultants of distributed force systems. The exact nature of the force distribution is not known.

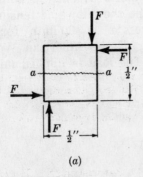 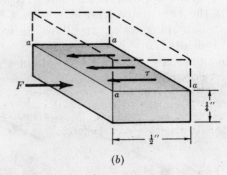

(a)   (b)

Fig. 4-10

The free-body diagram of the portion of the key below a horizontal plane a-a through its mid-section is shown in Fig. 4-10(b). For equilibrium in the horizontal direction we have

$$\Sigma F_h = 10,000 - \tau(\tfrac{1}{2})(3) = 0 \quad \text{or} \quad \tau = 6670 \text{ lb/in}^2$$

This is the horizontal shear stress in the key.

Supplementary Problems

4.9. In Problem 4.1, if the maximum allowable working stress in shear is 14,000 lb/in^2, determine the required diameter of the bolt in order that this value is not exceeded.   *Ans. d = 0.585 in.*

4.10. Consider a steel bolt 3/8 in. in diameter and subject to an axial tensile load of 2000 lb as shown in Fig. 4-11. Determine the average shearing stress in the bolt head, assuming shearing on a cylindrical surface of the same diameter as the bolt, as indicated by dashed lines.   *Ans. τ = 5420 lb/in^2*

4.11. A circular punch $\frac{3}{4}$ in. in diameter is used to punch a hole through a steel plate $\frac{1}{2}$ in. thick. If the force necessary to drive the punch through the metal is 61,000 lb, determine the maximum shearing stress developed in the material.   *Ans. τ = 51,800 lb/in^2*

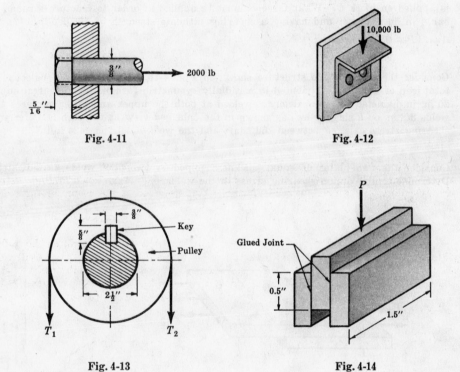

Fig. 4-11   Fig. 4-12

Fig. 4-13   Fig. 4-14

4.12. In structural practice, steel clip angles are commonly used to transfer loads from horizontal girders to vertical columns. If the reaction of the girder upon the angle is a downward force of 10,000 lb as shown in Fig. 4-12 and if two 7/8 in. diameter rivets resist this force, find the average shearing stress in each of the rivets. As in Problem 4.6, assume that the rivet fills the hole, which is 1/16 in. larger in diameter than the rivet.   *Ans. 7200 lb/in^2*

4.13. A pulley is keyed (to prevent relative motion) to a $2\frac{1}{2}$ in. diameter shaft.   The unequal belt pulls, T_1 and T_2, on the two sides of the pulley give rise to a net turning moment of 1200 lb-in. The key is 3/8 in. by 5/8 in. in cross-section and 3 in. long, as shown in Fig. 4-13. Determine the average shearing stress acting on a horizontal plane through the key.   *Ans. τ = 855 lb/in^2*

4.14. The arrangement shown in Fig. 4-14 is often used to determine the shearing strength of a glued joint. If the load P at rupture is 2500 lb, what is the average shearing stress in the joint at this instant?   *Ans. τ = 1670 lb/in^2*

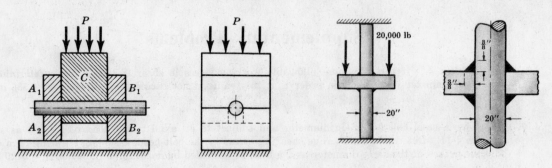

Fig. 4-15   Fig. 4-16

4.15.   Figure 4-15 shows another type of apparatus for determining shearing strengths of cylindrical
specimens.   The specimen is clamped between blocks A_1, A_2 and B_1, B_2 and a downward force P
is applied on block C.   What force P must be applied in order to fracture a round hot-rolled steel
bar $\frac{3}{4}$ in. in diameter and having a shearing ultimate strength of 105,000 lb/in²?

Ans. $P = 93{,}000$ lb

4.16.   Consider the balcony-type structure shown in Fig. 4-16.   The horizontal balcony is loaded by a
total load of 20,000 lb distributed in a radially symmetric fashion.   The central support is a shaft
20 in. in diameter and the balcony is welded at both the upper and lower surfaces to this shaft by
welds 3/8 in. on a side (or leg) as shown in the enlarged view at the right.   Determine the average
shearing stress existing between the shaft and the weld.   *Ans.* 424 lb/in²

4.17.   Consider the two plates of equal thickness joined by two fillet welds as indicated in Fig. 4-17.
Determine the maximum shearing stress in the welds.   *Ans.* $\tau = 0.707P/ab$

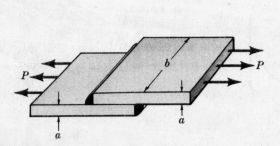

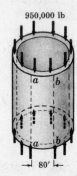

Fig. 4-17   Fig. 4-18

4.18.   A thin vertical cylindrical shell 125 ft in diameter is loaded by a uniformly distributed load along
its upper edge but only partially supported along its lower extremity as shown in Fig. 4-18.   If the
total load over the top is 950,000 lb and if 80 ft of the lower edge is unsupported, find the average
shear stress over each of the sections a-a and b-b if the shell is concrete, 8 in. thick and 22 ft high.

Ans. $\tau = 46$ lb/in²

4.19.   A copper tube 2 in. in outside diameter and of wall thickness $\frac{1}{4}$ in. fits loosely over a solid steel
circular bar $1\frac{7}{16}$ in. in diameter.   The two members are fastened together by two metal pins each
$\frac{5}{16}$ in. in diameter and passing transversely through both members, one pin being near each end of
the assembly.   At room temperature the assembly is stress-free when the pins are in position.   The temperature of the entire assembly is then raised 75 F°.   Calculate the average shear stress in the pins.
For copper $E = 13 \times 10^6$ lb/in², $\alpha = 9.3 \times 10^{-6}/\text{F}°$; for steel $E = 30 \times 10^6$ lb/in², $\alpha = 6.5 \times 10^{-6}/\text{F}°$.

Ans. $\tau = 17{,}900$ lb/in²

Chapter 5

Torsion

DEFINITION OF TORSION

Consider a bar rigidly clamped at one end and twisted at the other end by a torque $T (= Fd)$ applied in a plane perpendicular to the axis of the bar as shown in Fig. 5-1. Such a bar is said to be in *torsion*.

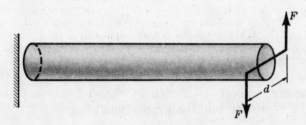

Fig. 5-1

EFFECTS OF TORSION

The effects of a torsional load applied to a bar are (1) to impart an angular displacement of one end cross-section with respect to the other end, and (2) to set up shearing stresses on any cross-section of the bar perpendicular to its axis.

TWISTING MOMENT

Occasionally a number of couples act along the length of a shaft. In that case it is convenient to introduce a new quantity, the *twisting moment*, which for any selection along the bar is defined to be the algebraic sum of the moments of the applied couples that lie to one side of the section in question. The choice of side in any case is of course arbitrary.

POLAR MOMENT OF INERTIA

For a hollow circular shaft of outer diameter D_o with a concentric circular hole of diameter D_i the *polar moment of inertia* of the cross-sectional area, usually denoted by J, is given by

$$J = \frac{\pi}{32}(D_o^4 - D_i^4)$$

The polar moment of inertia for a solid shaft is obtained by setting $D_i = 0$. See Problem 5.1. This quantity J is a mathematical property of the geometry of the cross-section which occurs in the study of the stresses set up in a circular shaft subject to torsion.

Occasionally it is convenient to rewrite the above equation in the form

$$J = \frac{\pi}{32}(D_o^2 + D_i^2)(D_o^2 - D_i^2)$$

$$= \frac{\pi}{32}(D_o^2 + D_i^2)(D_o + D_i)(D_o - D_i)$$

This last form is useful in numerical evaluation of J in those cases where the difference $(D_o - D_i)$ is small. See Problem 5.8.

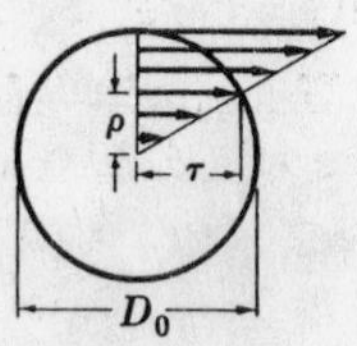

Fig. 5-2

TORSIONAL SHEARING STRESS

For either a solid or a hollow circular shaft subject to a twisting moment T the *torsional shearing stress* τ at a distance ρ from the center of the shaft is given by

$$\tau = \frac{T\rho}{J}$$

This expression is derived in Problem 5.2. For applications see Problems 5.5, 5.6, 5.7, 5.10, and 5.13. This stress distribution varies from zero at the center of the shaft (if it is solid) to a maximum at the outer fibers, as shown in Fig. 5-2. It is to be emphasized that no points of the bar are stressed beyond the proportional limit.

ASSUMPTIONS

In the derivation of the formula $\tau = T\rho/J$ it is assumed that a plane section of the shaft normal to its axis prior to loading remains plane after the torques have been applied and further that a diameter in the section before deformation remains a diameter, or straight line, after deformation. Because of the polar symmetry of a circular shaft these assumptions seem to be reasonable. However, for a shaft of noncircular cross-section these assumptions are no longer true; it is known from experimental results that the cross-sections of noncircular shafts warp out of their original planes during application of external loads.

SHEARING STRAIN

If a generator a-b is marked on the surface of the unloaded bar, then after the twisting moment T has been applied this line moves to a-b', as shown in Fig. 5-3. The angle γ, measured in radians, between the final and original positions of the generator is defined as the *shearing strain* at the surface of the bar. The same definition would hold at any interior point of the bar.

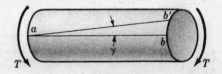

Fig. 5-3

MODULUS OF ELASTICITY IN SHEAR

The ratio of the shear stress τ to the shear strain γ is called the *modulus of elasticity in shear* and, as in Chapter 4, is given by

$$G = \frac{\tau}{\gamma}$$

Again the units of G are the same as those of shear stress, since the shear strain is dimensionless.

ANGLE OF TWIST

If a shaft of length L is subject to a constant twisting moment T along its length, then the angle θ through which one end of the bar will twist relative to the other is

$$\theta = \frac{TL}{GJ}$$

where J denotes the polar moment of inertia of the cross-section. This equation is derived in Problem 5.3. For applications see Problems 5.7, 5.9, 5.11, 5.12 and 5.13 This expression holds only for purely elastic action of the bar.

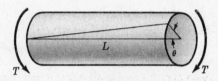

Fig. 5-4

MODULUS OF RUPTURE

This is defined as the fictitious shearing stress obtained by substituting in the equation $\tau = T\rho/J$ the maximum torque T carried by a shaft when tested to rupture. The variable ρ in this case is taken to be the outer radius of the bar. The use of this formula at the point of rupture is not of course justified because, as may be seen from Problem 5.2, it is derived for use only within the linear range of action of the material. The stress obtained by use of the formula is then not a true stress but is occasionally useful for comparative purposes.

STATICALLY INDETERMINATE PROBLEMS

Such problems frequently arise in the case of torsional loadings. One example is a shaft composed of two materials, a tube of one material surrounding a tube or solid bar of another material, with the entire assembly being subjected to a twisting moment. As usual, the applicable statics equations must be supplemented by additional equations based upon the deformations of the structure in order to furnish a number of equations equal to the number of unknowns. The unknowns in this case would be the twisting moments carried by each material. The equation based upon deformations for this case would state that the angles of twist of the various materials were equal. (See Problems 5.12 and 5.13.)

PLASTIC TORSION OF CIRCULAR BARS

As the twisting moment acting on either a solid or hollow circular bar is increased a value of the twisting moment is finally reached for which the extreme fibers of the bar have reached the yield point in shear of the material. This is the maximum possible elastic twisting moment that the bar can withstand and is denoted by T_e. A further increase in the value of the twisting moment puts the interior fibers at the yield point, with yielding progressing from the outer fibers inward. The limiting case occurs when all fibers are stressed to the yield point in shear and this represents the *fully plastic twisting moment*. It is denoted by T_p. Provided we do not consider stresses greater than the yield point in shear, this is the maximum possible twisting moment the bar can carry. For a solid circular bar subject to torsion it is shown in Problem 5.14 that $T_p = 4T_e/3$.

SHEAR FLOW

For a thin-walled tube of arbitrary cross-section the shearing force per unit length of the periphery of the tube is termed the *shear flow*. It is denoted by the symbol q and has units of force per unit length, perhaps lb/in. This concept is discussed in Problem 5.15.

ELASTIC TORSION OF THIN-WALLED CLOSED TUBES

For a thin-walled tube it is reasonable to assume that the shearing stress is constant across the thickness and also that it is tangent to the centerline between the inner and outer boundaries. The shearing stresses on such a tube appear as in Fig. 5-5.

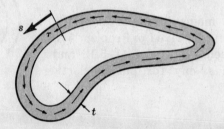

Fig. 5-5

For a thin-walled tube of either constant or variable thickness the twisting moment corresponding to a given shear flow q is

$$T = 2Aq$$

where A represents the area enclosed by the centerline of the tube. This expression is derived in Problem 5.16. The angle of twist per unit length of the tube is shown in Problem 5.17 to be

$$\theta = \frac{T}{4A^2 G} \oint \frac{ds}{t}$$

where s is the coordinate traversing the periphery of the tube and t represents the thickness. For applications see Problems 5.18 and 5.19.

Solved Problems

5.1. Derive an expression for the polar moment of inertia of the cross-sectional area of a hollow circular shaft. What does this expression become for the special case of a solid circular shaft?

Let D_o denote the outside diameter of the shaft and D_i the inside diameter. Because of the circular symmetry involved, it is most convenient to adopt the polar coordinate system shown in Fig. 5-6.

By definition, the polar moment of inertia is given by the integral

$$J = \int_A \rho^2 \, da$$

where A indicates that the integral is to be evaluated over the entire cross-sectional area.

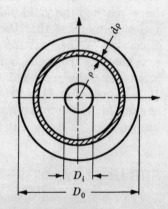

Fig. 5-6

To evaluate this integral it is desirable to select some element of area, da, so that ρ is constant for all points in the element. The thin ring-shaped element of radius ρ and radial thickness $d\rho$ shown constitutes a convenient element. It is assumed that the thickness $d\rho$ of the ring is small compared to ρ. The area of the ring-shaped element is given by $da = 2\pi\rho(d\rho)$. Hence the polar moment of inertia is given by

$$ J = \int_{\frac{1}{2}D_i}^{\frac{1}{2}D_o} \rho^2 (2\pi\rho)\, d\rho = 2\pi \left[\frac{\rho^4}{4} \right]_{\frac{1}{2}D_i}^{\frac{1}{2}D_o} = \frac{\pi}{32}(D_o^4 - D_i^4) $$

The units of this quantity are evidently (length)^4, e.g. in^4. It is not necessary to attempt to attach any physical interpretation to the quantity J. It will be found useful in problems involving the twisting of shafts.

For the special case of a solid circular shaft, the above expression becomes $J = \pi D^4/32$ where D denotes the diameter of the shaft.

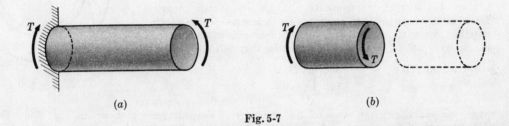

(a)   (b)

Fig. 5-7

5.2. Derive an expression relating the applied twisting moment acting on a shaft of circular cross-section and the shearing stress at any point in the shaft.

In Fig. 5-7(a) the shaft is shown loaded by the two torques T and consequently is in static equilibrium. To determine the distribution of shearing stress in the shaft, let us cut the shaft by a plane passing through it in a direction perpendicular to the geometric axis of the bar. Further, let us specify that this cutting plane is not "too close" to either end of the body, where the loads T are applied. The use of such a cutting plane is in accordance with the usual procedure followed in Strength of Materials, i.e. to cut the body in such a manner that the forces under investigation become external to the new body formed. These forces (or stresses) were of course internal effects with regard to the original, uncut body.

The free-body diagram of the portion of the shaft to the left of this plane appears as in Fig. 5-7(b). Obviously a torque T must act over the cross-section cut by the plane. This is true since the entire shaft is in equilibrium, hence any portion of it also is. The torque T acting on the cut section represents the effect of the right portion of the shaft on the left portion. Since the right portion has been removed, it must be replaced by its effect on the left portion. This effect is represented by the torque T. This torque is of course a resultant of shearing stresses distributed over the cross-section. It is now necessary to make certain assumptions in order to determine the nature of the variation of shear stress intensity over the cross-section.

One fundamental assumption is that a plane section of the shaft normal to its axis before loads are applied remains plane and normal to the axis after loading. This may be verified experimentally for circular shafts, but this assumption is not valid for shafts of non-circular cross-section.

A generator on the surface of the shaft, denoted by O_1A in Fig. 5-8, deforms into the configuration O_1B after torsion has occurred. The angle between these configurations is denoted by α. By definition, the shearing unit strain γ on the surface of the shaft is

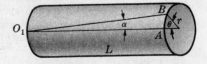

Fig. 5-8

$$ \gamma = \tan\alpha \approx \alpha $$

where the angle α is measured in radians. From the geometry of the figure,

$$\alpha = \frac{AB}{L} = \frac{r\theta}{L}; \quad \text{hence} \quad \gamma = \frac{r\theta}{L}$$

But since a diameter of the shaft prior to loading is assumed to remain a diameter after torsion has occurred, the shearing unit strain at a general distance ρ from the center of the shaft may likewise be written $\gamma_\rho = \rho\theta/L$. Consequently the shearing strains of the longitudinal fibers vary linearly as the distances from the center of the shaft.

If we assume that we are concerned only with the linear range of action of the material where the shearing stress is proportional to shearing strain, then it is evident that the shearing stresses of the longitudinal fibers vary linearly as the distances from the center of the shaft. Obviously the distribution of shearing stresses is symmetric around the geometric axis of the shaft. They have the appearance shown in Fig. 5-9. For equilibrium, the sum of the moments of these distributed shearing forces over the entire circular cross-section is equal to the applied twisting moment. Also, the sum of the moments of these forces is exactly equal to the torque T shown in Fig. 5-7(b) above.

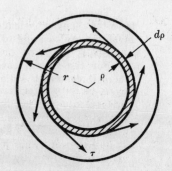

Thus we have

$$T = \int_0^r \tau\rho \, da$$

Fig. 5-9

where da represents the area of the shaded ring-shaped element shown in Fig. 5-9. However, the shearing stresses vary as the distances from the geometric axis; hence

$$\tau_\rho/\rho = \tau_r/r = \text{constant}$$

where the subscripts on the shearing stress denote the distances of the element from the axis of the shaft. Consequently we may write

$$T = \int_0^r \frac{\tau_\rho}{\rho}(\rho^2) \, da = \frac{\tau_\rho}{\rho} \int_0^r \rho^2 \, da$$

since the ratio $\dfrac{\tau_\rho}{\rho}$ is a constant. However, the expression $\displaystyle\int_0^r \rho^2 \, da$ is by definition (see Problem 5.1) the polar moment of inertia of the cross-sectional area. Values of this for solid and hollow circular shafts are derived in Problem 5.1. Hence the desired relationship is

$$T = \frac{\tau_\rho J}{\rho} \quad \text{or} \quad \tau_\rho = \frac{T\rho}{J}$$

It is to be emphasized that this expression holds *only* if no points of the bar are stressed beyond the proportional limit of the material.

5.3. Derive an expression for the angle of twist of a circular shaft as a function of the applied twisting moment. Assume that the entire shaft is acting within the elastic range of action of the material.

Let L denote the length of the shaft, J the polar moment of inertia of the cross section, T the applied twisting moment (assumed constant along the length of the bar), and G the modulus of elasticity in shear. The angle of twist in a length L is represented by θ in Fig. 5-10.

From Problem 5.2 we have at the outer fibers where $\rho = r$:

$$\gamma_r = \frac{r\theta}{L} \quad \text{and} \quad \tau_r = \frac{Tr}{J}$$

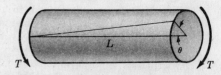

Fig. 5-10

By definition, the shearing modulus is given by $G = \dfrac{\tau}{\gamma} = \dfrac{Tr/J}{r\theta/L} = \dfrac{TL}{J\theta}$ from which $\theta = \dfrac{TL}{GJ}$.

Note that θ is expressed in radians, i.e. it is dimensionless. One consistent set of units would be to take T in lb-in, L in in., G in lb/in², and J in in⁴.

Occasionally the angle of twist in a unit length is useful. It is often denoted by ϕ and is given by $\phi = \theta/L = T/GJ$.

5.4. Derive a relationship between the twisting moment acting on a rotating shaft, the horsepower transmitted by it, and its angular velocity, which is assumed to be constant. Again, assume entirely elastic action.

Let us denote the twisting moment acting on the shaft by T, the angular velocity in rev/min by n, and the horsepower by hp. Also, let us consider a one-minute time interval. During this interval the twisting moment does an amount of work given by the product of the moment times the angular displacement in radians, or $T \times 2\pi n$. If T is measured in lb-in, then the work has these same units. By definition if work is being done at the rate of 33,000 ft-lb/min = 12(33,000) in-lb/min, it is equivalent to one horsepower. The horsepower transmitted by the shaft is consequently

$$\text{hp} = \frac{T \times 2\pi n}{12(33{,}000)} \quad \text{from which} \quad T = \frac{63{,}000 \times \text{hp}}{n}$$

where n is in rev/min and T is in lb-in.

5.5. If a twisting moment of 10,000 lb-in is impressed upon a $1\frac{3}{4}$ in. diameter shaft, what is the maximum shearing stress developed? Also, what is the angle of twist in a 4 ft length of the shaft? The material is steel for which $G = 12 \times 10^6$ lb/in². Assume entirely elastic action.

From Problem 5.1 the polar moment of inertia of the cross-sectional area is

$$J = \frac{\pi}{32}(D_o)^4 = \frac{\pi}{32}\left(\frac{7}{4}\right)^4 = 0.92 \text{ in}^4$$

The torsional shearing stress τ at any distance ρ from the center of the shaft was shown in Problem 5.2 to be $\tau_\rho = T\rho/J$. The maximum shear stress is developed at the outer fibers and there at $\rho = 7/8$ in.

$$\tau_{max} = \frac{10{,}000(7/8)}{0.92} = 9500 \text{ lb/in}^2$$

Hence the shear stress varies linearly from zero at the center of the shaft to 9500 lb/in² at the outer fibers as shown in Fig. 5-11.

The angle of twist θ in a 4 ft length of the shaft is

$$\theta = \frac{TL}{GJ} = \frac{10{,}000(48 \text{ in.})}{12 \times 10^6(0.92)} = 0.0435 \text{ radian}$$

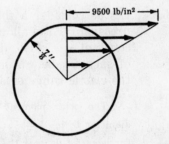

Fig. 5-11

5.6. Consider a solid circular shaft and also a hollow circular shaft whose inside diameter is 3/4 of the outside. Compare the weights of equal lengths of these two shafts required to transmit a given torsional load if the maximum shear stresses developed in the two shafts are equal. Assume entirely elastic action.

For the solid shaft of diameter d the shearing stress is given by $\tau_\rho = T\rho/J$ and the maximum shearing stress occurs at the outer fibers where $\rho = d/2$. Hence

$$\tau_{max} = \frac{T(d/2)}{(\pi/32)d^4} = \frac{16T}{\pi d^3} \quad \text{or} \quad \frac{T}{\tau_{max}} = \frac{\pi d^3}{16}$$

For the hollow shaft of outer diameter D the maximum shearing stress still occurs at the outer fibers where $\rho = D/2$. Then

$$\tau_{max} = \frac{T(D/2)}{\frac{\pi}{32}[D^4 - (\frac{3}{4}D)^4]} = \frac{16T}{\pi(0.684)D^3} \quad \text{or} \quad \frac{T}{\tau_{max}} = \frac{\pi(0.684)D^3}{16}$$

But the ratio T/τ_{max} is the same for both shafts; hence $0.684D^3 = d^3$ or $D = 1.135d$. Thus

$$\text{ratio of weights} = \frac{D^2 - (3D/4)^2}{d^2} = \frac{0.4375D^2}{d^2} = \frac{0.4375(1.135d)^2}{d^2} = 0.563$$

The hollow shaft weighs only 56.3% as much as the solid one, which illustrates the efficiency of hollow shafting as compared to solid.

5.7. A hollow steel shaft 10 ft long must transmit a torque of 250,000 lb-in. The total angle of twist in this length is not to exceed 2.5° and the allowable shearing stress is 12,000 lb/in². Determine the inside and outside diameters of the shaft if $G = 12 \times 10^6$ lb/in². The action is elastic.

Let d_o and d_i designate the outside and inside diameters of the shaft respectively. From Problem 5.3 the angle of twist θ is given by $\theta = TL/GJ$, where θ is expressed in radians. Consequently in the 10 ft length we have

$$2.5 \text{ deg} \times \frac{1 \text{ rad}}{57.3 \text{ deg}} = \frac{250,000(120 \text{ in.})}{12 \times 10^6 \times \frac{\pi}{32}(d_o^4 - d_i^4)} \quad \text{or} \quad d_o^4 - d_i^4 = 583.65$$

The maximum shearing stress occurs at the outer fibers where $\rho = d_o/2$. Then

$$\tau_{max} = \frac{T(d_o/2)}{\frac{\pi}{32}(d_o^4 - d_i^4)}, \quad 12,000 = \frac{250,000(d_o/2)}{\frac{\pi}{32}(d_o^4 - d_i^4)} \quad \text{and} \quad d_o^4 - d_i^4 = 106.10d_o$$

Thus $106.10d_o = 583.65$ or $d_o = 5.50$ in. Substituting, $d_i = 4.27$ in.

5.8. Let us consider a thin-walled tube subject to torsion. Derive an approximate expression for the allowable twisting moment if the working stress in shear is a given constant τ_w. Also, derive an approximate expression for the strength-weight ratio of such a tube. It is assumed the tube does not buckle, and the material is within the elastic range of action.

The polar moment of inertia of a hollow circular shaft of outer diameter D_o and inner diameter D_i is $J = \frac{\pi}{32}(D_o^4 - D_i^4)$. If R denotes the outer radius of the tube, then $D_o = 2R$ and further, if t denotes the wall thickness of the tube, then $D_i = 2R - 2t$.

The polar moment of inertia J may be written in the alternate form

$$J = \frac{\pi}{32}[(2R)^4 - (2R - 2t)^4] = \frac{\pi}{2}[R^4 - (R - t)^4] = \frac{\pi}{2}(4R^3t - 6R^2t^2 + 4Rt^3 - t^4)$$

$$= \frac{\pi}{2}R^4[4(t/R) - 6(t/R)^2 + 4(t/R)^3 - (t/R)^4]$$

Neglecting squares and higher powers of the ratio t/R, since we are considering a thin-walled tube, this becomes, approximately, $J = 2\pi R^3 t$.

The ordinary torsion formula is $T = \tau_w J/R$. For a thin-walled tube this becomes, for the allowable twisting moment, $T = 2\pi R^2 t \tau_w$.

The weight W of the tube is $W = \gamma LA$ where γ is the specific weight of the material, L the length of the tube and A the cross-sectional area of the tube. The area is given by

$$A = \pi[R^2 - (R-t)^2] = \pi(2Rt - t^2) = \pi R^2[2t/R - (t/R)^2]$$

Again neglecting the square of the ratio t/R for a thin tube, this becomes $A = 2\pi Rt$.

The strength-weight ratio is defined to be T/W.　This is given by

$$\frac{T}{W} = \frac{2\pi R^2 t \tau_w}{2\pi RtL\gamma} = \frac{R\tau_w}{L\gamma}$$

The ratio is of considerable importance in aircraft design.

5.9.　A solid circular shaft has a slight taper extending uniformly from one end to the other.　Denote the radius at the small end by a, that at the large end by b.　Determine the error committed if the angle of twist for a given length is calculated using the mean radius of the shaft.　The radius at the larger end is 1.2 times that at the smaller end.

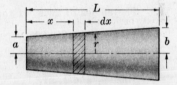

Fig. 5-12

Let us set up a coordinate system with the variable x denoting the distance from the small end of the shaft. The radius at a section at the distance x from the small end is

$$r = a + \frac{(b-a)x}{L}$$

where L is the length of the bar.

Provided the angle of taper is small, it is sufficient to consider the angle $d\theta$ through which the shaded element of length dx is twisted.　This is obtained by applying the expression $\theta = TL/GJ$ to the element of length dx and radius $r = a + \dfrac{(b-a)x}{L}$. For such an element the polar moment of inertia is

$$J = \frac{\pi}{32}D^4 = \frac{\pi}{2}r^4 = \frac{\pi}{2}\left[a + \frac{(b-a)x}{L}\right]^4$$

Thus

$$d\theta = \frac{T\,dx}{G\dfrac{\pi}{2}\left[a + \dfrac{(b-a)x}{L}\right]^4}$$

The angle of twist in the length L is found by integrating the last equation.　Thus

$$\theta = \frac{2T}{G\pi}\int_0^L \frac{dx}{\left[a + \dfrac{(b-a)x}{L}\right]^4} = \frac{2T}{G\pi}\left(-\frac{1}{3}\right)\left(\frac{L}{b-a}\right)\left[\frac{1}{\left[a + \dfrac{(b-a)x}{L}\right]^3}\right]_0^L = \frac{2TL}{3G\pi(b-a)}\left(-\frac{1}{b^3} + \frac{1}{a^3}\right)$$

If $b = 1.2a$, this becomes $\theta = 1.40433TL/G\pi a^4$.　For a solid shaft of radius $1.1a$

$$\theta_1 = \frac{TL}{G\dfrac{\pi}{2}(1.1a)^4} = \frac{1.36602TL}{G\pi a^4}$$

Using these values of θ and θ_1 we find

$$\text{percent error} = \frac{0.03831}{1.40433} \times 100 = 2.73\%$$

5.10.　A solid circular shaft has a uniform diameter of 2 in. and is 10 ft long.　At its mid-point 65 hp is delivered to the shaft by means of a belt passing over a pulley.　This power is used to drive two machines, one at the left end of the shaft consuming 25 hp and one at the right end consuming the remaining 40 hp. Determine the maximum shearing stress in the shaft and also the relative angle of twist between the two extreme ends of the shaft. The shaft turns at 200 rpm and the material is steel for which $G = 12 \times 10^6$ lb/in². Assume elastic action.

In the left half of the shaft we have 25 hp which corresponds to a torque T_1 given by

$$T_1 = \frac{63,000 \times \text{hp}}{n} = \frac{63,000(25)}{200} = 7880 \text{ lb-in}$$

Similarly, in the right half we have 40 hp corresponding to a torque T_2 given by

$$T_2 = \frac{63,000(40)}{200} = 12,600 \text{ lb-in}$$

The maximum shearing stress consequently occurs in the outer fibers in the right half and is given by the ordinary torsion formula:

$$\tau_\rho = \frac{T\rho}{J} \quad \text{or} \quad \tau = \frac{12,600(1)}{\frac{\pi}{32}(2)^4} = 8000 \text{ lb/in}^2$$

The angles of twist of the left and right ends relative to the center are respectively

$$\theta_1 = \frac{7880(60)}{12 \times 10^6 \frac{\pi}{32}(2)^4} = 0.0250 \text{ rad} \quad \text{and} \quad \theta_2 = \frac{12,600(60)}{12 \times 10^6 \frac{\pi}{32}(2)^4} = 0.0401 \text{ rad}$$

Since θ_1 and θ_2 are in the same direction, the relative angle of twist between the two ends of the shaft is $\theta = \theta_2 - \theta_1 = 0.015$ rad.

5.11. Consider two solid circular shafts connected by 2 in. and 10 in. pitch diameter gears as in Fig. 5-13(a). The shafts are assumed to be supported by the bearings in such a manner that they undergo no bending. Find the angular rotation of D, the right end of one shaft, with respect to A, the left end of the other, caused by the torque of 2500 lb-in applied at D. The left shaft is steel for which $G = 12 \times 10^6$ lb/in² and the right is brass for which $G = 5 \times 10^6$ lb/in². Assume elastic action.

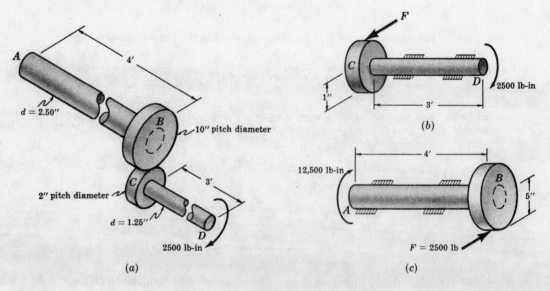

Fig. 5-13

A free-body diagram of the right shaft [Fig. 5-13(b)] reveals that a tangential force F must act on the smaller gear. For equilibrium, $F = 2500$ lb.

The angle of twist of the right shaft is

$$\theta_1 = \frac{TL}{GJ} = \frac{2500(36)}{5 \times 10^6 \frac{\pi}{32}(1.25)^4} = 0.0750 \text{ rad}$$

A free-body diagram of the left shaft is shown in Fig. 5-13(c). The force F is equal and opposite to that acting on the small gear C. This force F acts 5 in. from the center line of the left shaft, hence it imparts a torque of $5(2500) = 12,500$ lb-in to the shaft AB. Because of this torque there is a rotation of end B with respect to end A given by the angle θ_2, where

$$\theta_2 = \frac{12,500(48)}{12 \times 10^6 \frac{\pi}{32}(2.5)^4} = 0.0130 \text{ rad}$$

It is to be carefully noted that this angle of rotation θ_2 induces a *rigid body* rotation of the entire shaft CD because of the gears. In fact, the rotation of CD will be in the same ratio to that of AB as the ratio of the pitch diameters, or 5:1. Thus a rigid body rotation of 5(0.0130) rad is imparted to shaft CD. Superposed on this rigid body movement of CD is the angular displacement of D with respect to C previously denoted by θ_1.

Hence the resultant angle of twist of D with respect to A is $\theta = 5(0.0130) + 0.075 = 0.140$ rad.

5.12. Determine the reactive torques at the fixed ends of the circular shaft loaded by the couples shown in Fig. 5-14(a). The cross-section of the bar is constant along the length. Assume elastic action.

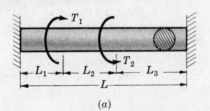

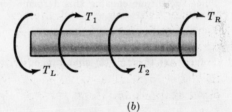

(a)   (b)

Fig. 5-14

Let us assume that the reactive torques T_L and T_R are positive in the directions shown in Fig. 5-14(b). From statics we have

$$T_L - T_1 + T_2 - T_R = 0 \qquad (1)$$

This is the only equation of static equilibrium and it contains two unknowns. Hence this problem is statically indeterminate and it is necessary to augment this equation with another equation based on the deformations of the system.

The variation of torque with length along the bar may be represented by the following plot:

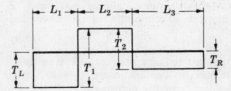

Fig. 5-15

The free-body diagram of the left region of length L_1 appears as in Fig. 5-16(a).

Working from left to right along the shaft, the twisting moment in the central region of length L_2 is given by the algebraic sum of the torques to the left of this section, i.e. $T_1 - T_L$. The free-body diagram of this region appears as in Fig. 5-16(b).

Finally, the free-body diagram of the right region of length L_3 appears as in Fig. 5-16(c).

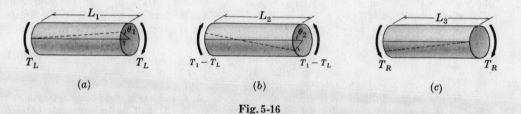

$$(a) \qquad\qquad\qquad (b) \qquad\qquad\qquad (c)$$

Fig. 5-16

Let θ_1 denote the angle of twist at the point of application of T_1, and θ_2 the angle at T_2. Then from a consideration of the regions of lengths L_1 and L_3 we immediately have

$$\theta_1 = \frac{T_L L_1}{GJ} \tag{2}$$

$$\theta_2 = \frac{T_R L_3}{GJ} \tag{3}$$

The original position of a generator on the surface of the shaft is shown by a solid line in Fig. 5-16, and the deformed position by a dashed line. Consideration of the central region of length L_2 reveals that the angle of twist of its right end with respect to its left end is $\theta_1 + \theta_2$. Hence, since the torque causing this deformation is $T_1 - T_L$, we have

$$\theta_1 + \theta_2 = \frac{(T_1 - T_L)L_2}{GJ} \tag{4}$$

Solving (1) through (4) simultaneously, we find

$$T_L = T_1 \frac{L_2 + L_3}{L} - T_2 \frac{L_3}{L} \qquad \text{and} \qquad T_R = -T_1 \frac{L_1}{L} + T_2 \frac{L_1 + L_2}{L}$$

It is of interest to examine the behavior of a generator on the surface of the shaft. Originally it was, of course, straight over the entire length L, but after application of T_1 and T_2 it has the appearance shown by the broken line in Fig. 5-17.

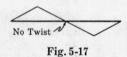

Fig. 5-17

5.13. Consider a composite shaft fabricated from a 2 in. diameter solid aluminum alloy, $G = 4 \times 10^6$ lb/in^2, surrounded by a hollow steel circular shaft of outside diameter 2.5 in. and inside diameter 2 in., $G = 12 \times 10^6$ lb/in^2. The two metals are rigidly connected at their juncture. If the composite shaft is loaded by a twisting moment of 14,000 lb-in, calculate the shearing stress at the outer fibers of the steel and also at the extreme fibers of the aluminum. The action is elastic.

Let T_1 = torque carried by the aluminum shaft and T_2 = torque carried by the steel. For static equilibrium of moments about the geometric axis we have

$$T_1 + T_2 = T = 14,000$$

where T = external applied twisting moment. This is the only equation from statics available in this problem. Since it contains two unknowns, T_1 and T_2, it is necessary to supplement it with an additional equation coming from the deformations of the shaft. The structure is thus statically indeterminate.

Such an equation is easily found, since the two materials are rigidly joined; hence their angles of twist must be equal. In a length L of the shaft we have, using the formula $\theta = TL/GJ$,

$$\frac{T_1 L}{4 \times 10^6 \frac{\pi}{32}(2)^4} = \frac{T_2 L}{12 \times 10^6 \frac{\pi}{32}[(2.5)^4 - (2)^4]} \quad \text{or} \quad T_1 = 0.231 T_2$$

This equation, together with the statics equation, may be solved simultaneously to yield

$$T_1 = 2600 \text{ lb-in (carried by aluminum)} \quad \text{and} \quad T_2 = 11,400 \text{ lb-in (carried by steel)}$$

The shearing stresses at the extreme fibers of the steel and of the aluminum are respectively

$$\tau_2 = \frac{11,400(1.25)}{\frac{\pi}{32}[(2.5)^4 - (2)^4]} = 6300 \text{ psi} \quad \text{and} \quad \tau_1 = \frac{2600(1)}{\frac{\pi}{32}(2)^4} = 1650 \text{ psi}$$

5.14. Consider a bar of solid circular cross-section subject to torsion. The material is considered to be elastic-perfectly plastic, i.e. the shear stress-strain diagram has the appearance indicated in Fig. 5-18(a). Determine the distance from the center at which plastic flow begins in terms of the twisting moment. Also determine the twisting moment for fully plastic action of the cross-section.

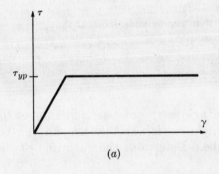
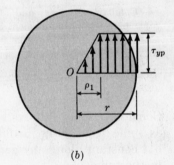

$$(a) \qquad\qquad\qquad\qquad\qquad (b)$$

Fig. 5-18

Even though torsion of the bar has caused the outer portion to have yielded it is still realistic to assume that plane sections of the bar normal to its axis prior to loading remain plane after the torques have been applied, and further that a diameter in the section before deformation remains a diameter, or straight line, after deformation. Consequently the shearing strains of the longitudinal fibers vary linearly as the distances from the center of the bar.

Let us assume that plastic action begins at a distance ρ_1 from the center of the bar, so that the stress distribution appears as in Fig. 5-18(b). Thus, the shearing stresses vary linearly as the distance of the fiber from the center up to the point ρ_1 after which they are constant and equal to the yield point in shear.

From Fig. 5-18(b) we have for $\rho < \rho_1$:

$$\frac{\tau}{\rho} = \frac{\tau_{yp}}{\rho_1} \quad \text{or} \quad \tau = \left(\frac{\rho}{\rho_1}\right)\tau_{yp}$$

and for $\rho > \rho_1$: $\tau = \tau_{yp} = $ constant. Thus the twisting moment is

$$T = \int_0^r \tau\rho\, da$$

where da refers to the ring-shaped element shown in Fig. 5-9 of Problem 5.2. Using the above values of shearing stress in the inner elastic region and outer plastic region we have

$$T = \int_0^{\rho_1} \left(\frac{\rho}{\rho_1}\right) \tau_{yp} \rho \, da + \int_{\rho_1}^{r} \tau_{yp} \rho \, da = \frac{\tau_{yp}}{\rho_1} \int_0^{\rho_1} \rho^2 \, da + \tau_{yp} \int_{\rho_1}^{r} \rho \, da$$

$$= \frac{\tau_{yp}}{\rho_1} \int_0^{\rho_1} \rho^2 2\pi \rho \, d\rho + \tau_{yp} \int_{\rho_1}^{r} \rho 2\pi \rho \, d\rho = \tau_{yp}\left(\frac{\pi}{2} - \frac{2\pi}{3}\right) \rho_1^3 + \frac{2\pi}{3} \tau_{yp} r^3$$

Solving for ρ_1:

$$\rho_1 = \left[4r^3 - \frac{6T}{\pi\tau_{yp}}\right]^{1/3} \tag{1}$$

as the distance from the center at which plastic flow begins. For fully plastic action, i.e. $\tau = \tau_{yp}$ at all points of the cross-section, we set $\rho_1 = 0$ to obtain the fully plastic twisting moment T_p:

$$T_p = \tfrac{2}{3}\pi r^3 \tau_{yp} \tag{2}$$

But from Problem 5.2 if only the outer fibers of the bar are stressed to the yield point of the material and all interior fibers are in the elastic range of action we have the maximum possible elastic twisting moment T_e:

$$T_e = \frac{\tau_{yp}}{2} \pi r^3 \tag{3}$$

Comparison of (2) and (3) indicates that $T_p = 4T_e/3$, i.e. fully plastic action permits applicaption of a twisting moment $33\tfrac{1}{3}$ percent greater than the twisting moment that just causes plastic action to begin in the outer fibers.

5.15.   Consider a long, thin-walled, closed tube of arbitrary cross-section subject to torsion as indicated in Fig. 5-19(a). The wall thickness is variable around the periphery but constant along any generator of the tube. Determine the variation of shear flow in the tube.

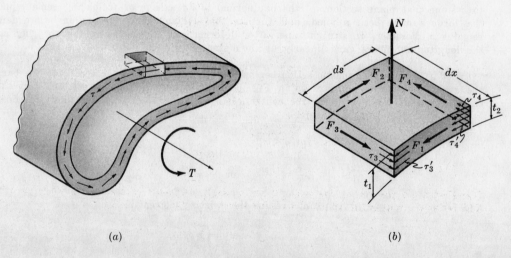

(a)   (b)

Fig. 5-19

Unlike the circular cross-sections considered up to now, the cross-sections of this tube may *warp* during deformation, i.e. distort out of the planes originally normal to the longitudinal axis of the tube. We shall assume that there are no restraints against this warping so that no longitudinal normal stresses arise. Since the tube is thin-walled it is logical to assume that the shearing stress is tangent to the centerline between the inner and outer boundaries and further that it is constant across the thickness. The shear stress thus has the appearance indicated in Fig. 5-19(a).

Let us isolate an infinitesimal element from the tube as shown in Fig. 5-19(b). There will be a torsional shearing force F_1 acting on the forward face as indicated and a counterpart force F_2 on the back face. Also, there are shearing forces F_3 and F_4 acting along the longitudinal faces as indicated. Since warping is unrestrained the element is in equilibrium under the action of only these four forces. For tangential equilibrium we have $F_1 = F_2$ and for longitudinal equilibrium we have $F_3 = F_4$. Assuming all shearing stresses to be constant over the small wall thickness and letting τ_3 represent the shearing stress corresponding to F_3 and τ_4 that corresponding to F_4, we have $\tau_3 t_1\,dx = \tau_4 t_2\,dx$ from which

$$\tau_3 t_1 \;=\; \tau_4 t_2 \tag{1}$$

The shearing stresses on the forward face at the corners are designated by τ_3' and τ_4' as indicated in Fig. 5-19(b). Counterpart stresses exist on the back face but are not shown. An equation for equilibrium of moments about the N-axis normal to the surface of the element indicates that $\tau_3 = \tau_3'$ and $\tau_4 = \tau_4'$, i.e. the shearing stresses on two perpendicular planes through a point are of equal magnitude. This point is discussed in greater detail in Problem 8.17. Thus, (1) becomes

$$\tau_3' t_1 \;=\; \tau_4' t_2 \tag{2}$$

Thus the product of shearing stress and thickness is constant for the tube. This product is termed *shear flow* and is usually denoted by the symbol q, i.e. $q = \tau t$. Thus, shear flow is *constant* for a thin-walled tube subject to torsion even though the tube is of variable wall thickness.

Shear flow is actually shear force per unit length of arc around the periphery of the tube and has the units of force per unit length, perhaps lb per in.

5.16. Determine the twisting moment acting on a thin-walled closed tube of arbitrary cross-section in terms of the shear flow and the geometric characteristics of the cross-section.

Let us consider again the transverse cross-section indicated in Problem 5.15 and determine the moment about an arbitrary point O of the shear force F_1 acting on the infinitesimal element shown in Fig. 5-19(b). The force F_1 can be represented as $F_1 = q\,ds$ where q is the shear flow, i.e. the shear force per unit length of the periphery. A view along the geometric axis appears in Fig. 5-20.

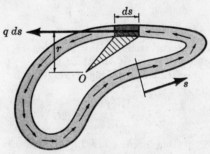

Fig. 5-20

The shearing force $q\,ds$ on the element has a twisting moment about point O given by $dT = rq\,ds$. The resultant moment of all such shearing forces over the cross-section is

$$T \;=\; q \oint r\,ds \tag{1}$$

where we have been able to take the shear flow q outside the integral since in Problem 5.15 it was demonstrated that it is constant for a thin-walled tube subject to torsion.

However, the product $r\,ds$ is twice the area of the shaded triangle having the base ds and altitude r. Thus, the integral in (1) is merely twice the area enclosed by the centerline of the tube. This expression, called *Bredt's law*, is usually written

$$T = 2Aq \qquad (2)$$

where A represents the area enclosed by the centerline of the tube. It is to be emphasized that this is *not* the area around the periphery on which forces such as $q\,ds$ act. Equation (2) applies only if the tube is continuous around the periphery, i.e. it must have no longitudinal slits.

5.17. Determine the angle of twist per unit length of a thin-walled closed tube of arbitrary cross-section in terms of the twisting moment, the shear modulus, and the geometric characteristics of the cross-section.

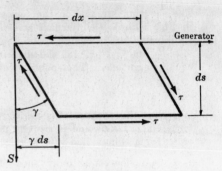

Fig. 5-21

Unlike the symmetric case considered in Problem 5.1, the shearing strains for such a tube do not vary linearly as the distance of the element from the geometric axis. It is best to approach this problem on the basis of conservation of energy. Let us examine the distortion of the element indicated in Fig. 5-19(b) of Problem 5.15. A view of the distorted element looking along the N-axis appears as shown in Fig. 5-21. Here, as before, γ denotes the shear strain. For the purpose of this problem it will be satisfactory to consider the element to be of constant thickness t. The only force doing work to produce the distorted configuration shown is the shear force on the bottom face of the element. The force has an initial value of zero and a final value of $\tau t\,dx$, thus having a mean value of $(\tau/2)t\,dx$. The work that this mean force does is $(\tau/2)t\,dx\,\gamma\,ds$. But since $G = \tau/\gamma$, the work done by all such forces around the periphery of a tube of unit length is

$$\oint \frac{\tau^2}{2G} t\,ds \qquad \text{or} \qquad \oint \frac{(\tau t)^2}{2G} \frac{ds}{t}$$

But from Problem 5.15, $q = \tau t$ is constant, hence may be taken through the integral to yield

$$\frac{q^2}{2G} \oint \frac{ds}{t} \qquad (1)$$

The twisting moment starts from an initial value of zero and has a final value of T. If θ represents the angle of twist per unit length of the tube, the work done by the external twisting moment is simply $T\theta/2$. Equating this to the work done by the internal forces as given by (1) we have

$$\frac{T\theta}{2} = \frac{q^2}{2G} \oint \frac{ds}{t} \qquad (2)$$

Substituting the value of q from Bredt's law in Problem 5.16 and solving for θ we obtain

$$\theta = \frac{T}{4A^2G} \oint \frac{ds}{t} \qquad (3)$$

where the integration is extended around the periphery of the tube.

5.18. Determine the angle of twist per unit length of the thin-walled tube of rectangular cross-section shown in Fig. 5-22.

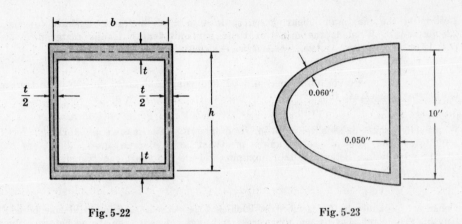

Fig. 5-22   Fig. 5-23

The desired angle of twist is given by (3) of Problem 5.17. The area A is merely bh, i.e. the area enclosed by the centerline of the tube. In each straight region the wall thickness is constant and thus the integral in (3) may be replaced by a finite summation. Thus we have

$$\theta = \frac{T}{4A^2G} \oint \frac{ds}{t} = \frac{T}{4b^2h^2G}\left[\frac{2b}{t} + \frac{2h}{t/2}\right] = \frac{T(2b+4h)}{4b^2h^2Gt}$$

5.19. Consider the thin-walled section shown in Fig. 5-23 which represents the leading edge of an airplane wing. The length of the curved portion is 32 in. and the enclosed area is 100 in². The material is 2024 T-4 aluminum alloy for which the allowable shearing stress is taken to be 10,000 lb/in² and $G = 4 \times 10^6$ lb/in². Determine the allowable twisting moment and also the angle of twist per unit length corresponding to this twisting moment.

The shear flow q was found in Problem 5.15 to be $q = \tau t$. The allowable shearing stress is found in the vertical region and consideration of this region indicates the shear flow to be

$$q = \tau t = 10,000(0.05) = 500 \text{ lb/in.}$$

The shearing stress in the curved region is also found from the same equation and is given by

$$\tau_{\text{curve}} = 500/0.06 = 8300 \text{ lb/in}^2$$

The allowable twisting moment is given by (2) of Problem 5.16 to be

$$T = 2Aq = 2(100)(500) = 100,000 \text{ lb-in}$$

The angle of twist per unit length is now found from (3) of Problem 5.17 to be

$$\theta = \frac{T}{4A^2G} \oint \frac{ds}{t} = \frac{100,000}{4(100)^2(4\times10^6)}\left[\frac{32}{0.060} + \frac{10}{0.050}\right] = 0.000436 \text{ rad/in}$$

Supplementary Problems

5.20. If a solid circular shaft of 1.25 in. diameter is subject to a torque T of 2,500 lb-in causing an angle of twist of 3.12 degrees in a 5 ft length, determine the shear modulus of the material.
Ans. $G = 11.5 \times 10^6$ lb/in²

5.21. Let us consider a hollow circular shaft of outside diameter 5 in. and inside diameter 3 in. By experiment it is determined that the shearing stress at the inside fibers is 7000 lb/in². What is the shearing stress at the outside fibers?   *Ans.*   11,700 lb/in²

5.22. Determine the maximum shearing stress in a 4 in. diameter solid shaft carrying a torque of 228,000 lb-in. What is the angle of twist per unit length if the material is steel for which $G = 12 \times 10^6$ lb/in²? *Ans.* 18,100 lb/in², 0.000755 rad/in

5.23. Determine the maximum horsepower a solid steel shaft 2.25 in. in diameter can transmit at 250 rpm if the working stress in shear is 11,000 lb/in². *Ans.* 98 hp

5.24. A propeller shaft in a ship is 14 in. in diameter. The allowable working stress in shear is 7500 lb/in² and the allowable angle of twist is 1 degree in 15 diameters of length. If $G = 12 \times 10^6$ lb/in², determine the maximum torque the shaft can transmit.
Ans. 3,780,000 lb-in

5.25. Consider the same shaft described in Problem 5.24 but with a 7 in. axial hole bored throughout its length. The conditions on working stress and angle of twist remain as before. By what percentage is the torsional load carrying capacity reduced? By what percentage is the weight of the shaft reduced? *Ans.* 6.25%, 25%

5.26. Determine the diameter of a solid steel shaft that will transmit 200 hp at a speed of 250 rpm if the allowable shearing stress is 12,000 lb/in². Also, determine the dimensions of a hollow steel shaft whose inside diameter is three-fourths of its outside diameter for these same conditions. What is the ratio of the angles of twist per unit length for these two shafts?
Ans. diameter = 2.77 in., outer diameter = 3.15 in., ratio = 0.88

5.27. Consider a solid circular shaft transmitting 1800 hp at 350 rpm. Determine the necessary diameter of the shaft so that (*a*) it does not twist through an angle greater than 1 degree in a length of 20 diameters, and also (*b*) the shear stress does not exceed 9000 lb/in². The shaft is steel for which $G = 12 \times 10^6$ lb/in². *Ans.* 6.80 in.

5.28. A compound shaft is composed of a 24 in. length of solid copper 4 in. in diameter, joined to a 32 in. length of solid steel 4.5 in. in diameter. A torque of 120,000 lb-in is applied to each end of the shaft. Find the maximum shear stress in each material and the total angle of twist of the entire shaft. For copper $G = 6 \times 10^6$ lb/in², for steel $G = 12 \times 10^6$ lb/in².
Ans. in the copper, 9520 lb/in²; in the steel, 6700 lb/in²; $\theta = 0.027$ rad

5.29. In Fig. 5-24 the vertical shaft and pulleys keyed to it may be considered to be weightless. The shaft rotates with a uniform angular velocity. The known belt pulls are indicated and the three pulleys are rigidly keyed to the shaft. If the working stress in shear is 7500 lb/in² determine the necessary diameter of a solid circular shaft. Neglect bending of the shaft because of the proximity of the bearings to the pulleys. *Ans.* 1.21 in.

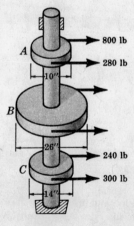

Fig. 5-24

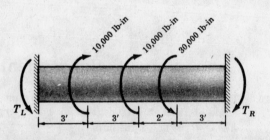

Fig. 5-25

5.30. Determine the reactive torques at the fixed ends of the circular shaft loaded by the three couples shown in Fig. 5-25. The cross-section of the bar is constant along the length.

Ans. T_L = 3600 lb-in, T_R = 13,600 lb-in

5.31. A hollow steel shaft has an outside diameter of 4 in. and an inside diameter of 3 in. Determine the maximum torque the shaft can transmit in fully plastic action if the yield point of the material in shear is 22,000 lb/in².   *Ans.*   214,000 lb-in

5.32. The thin-walled section shown in Fig. 5-26 is a portion of an airplane wing. The allowable shearing stress is 9000 lb/in².   The thicknesses of the four portions of this box section are uniform within each member and have the values indicated.   Also, the mean height of the section is 11.4 in.   Determine the maximum twisting moment that can be transmitted and also the angle of twist per inch of length of the section.   Take $G = 4 \times 10^6$ lb/in².

Ans.   460,000 lb-in; 0.000208 rad/in

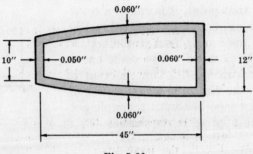

Fig. 5-26

Chapter 6

Shearing Force and Bending Moment

DEFINITION OF A BEAM

A bar subject to forces or couples that lie in a plane containing the longitudinal axis of the bar is called a *beam*. The forces are understood to act perpendicular to the longitudinal axis.

CANTILEVER BEAMS

If a beam is supported at only one end and in such a manner that the axis of the beam cannot rotate at that point, it is called a *canti-lever beam*. This type of beam is illustrated in Fig. 6-1. The left end of the bar is free to deflect but the right end is rigidly clamped. The right end is usually said to be "restrained". The reaction of the supporting wall at the right upon the beam consists of a vertical force together with a couple acting in the plane of the applied loads shown.

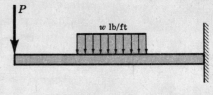

Fig. 6-1

SIMPLE BEAMS

A beam that is freely supported at both ends is called a *simple beam*. The term "freely supported" implies that the end supports are capable of exerting only forces upon the bar and are not capable of exerting any moments. Thus there is no restraint offered to the angular rotation of the ends of the bar at the supports as the bar deflects under the loads. Two simple beams are sketched in Fig. 6-2.

It is to be observed that at least one of the supports must be capable of undergoing horizontal movement so that no force will exist in the direction of the axis of the beam. If neither end were free to move horizontally, then some axial force would arise in the beam as it deforms under load. Problems of this nature are not considered in this book.

The beam of Fig. 6-2(a) is said to be subject to a concentrated force; that of Fig. 6-2(b) is loaded by a uniformly distributed load as well as a couple.

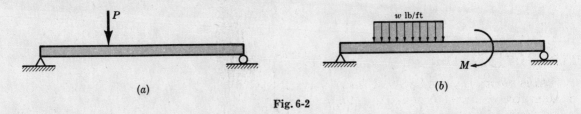

(a)   (b)

Fig. 6-2

OVERHANGING BEAMS

A beam freely supported at two points and having one or both ends extending beyond these supports is termed an *overhanging beam*. Two examples are given in Fig. 6-3.

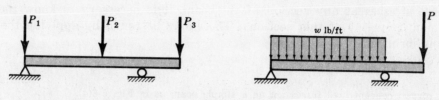

Fig. 6-3

STATICALLY DETERMINATE BEAMS

All the beams considered above, the cantilevers, simple beams, and overhanging beams, are ones in which the reactions of the supports may be determined by use of the equations of static equilibrium. The values of these reactions are independent of the deformations of the beam. Such beams are said to be statically determinate.

STATICALLY INDETERMINATE BEAMS

If the number of reactions exerted upon the beam exceeds the number of equations of static equilibrium, then the statics equations must be supplemented by equations based upon the deformations of the beam. In this case the beam is said to be statically indeterminate.

A cantilever-type beam that is supported at the extreme end [Fig. 6-4(*a*)], a beam rigidly clamped at both ends [Fig. 6-4(*b*)], and a beam extending over three or more supports [Fig. 6-4(*c*)] are all examples of indeterminate beams. This type of beam will be discussed in Chapter 12.

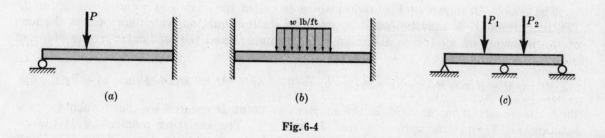

(*a*)   (*b*)   (*c*)

Fig. 6-4

TYPES OF LOADING

Loads commonly applied to a beam may consist of concentrated forces (applied at a point), uniformly distributed loads, in which case the magnitude is expressed as a certain number of pounds per foot of length of the beam, or uniformly varying loads. This last type of load is exemplified in Fig. 6-5.

A beam may also be loaded by an applied couple. The magnitude of the couple is usually expressed in lb-ft or lb-in.

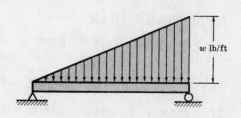

Fig. 6-5

INTERNAL FORCES AND MOMENTS IN BEAMS

When a beam is loaded by forces and couples, internal stresses arise in the bar. In general, both normal and shearing stresses will occur. In order to determine the magnitude of these stresses at any section of the beam, it is necessary to know the resultant force and moment acting at that section. These may be found by applying the equations of static equilibrium.

Example 1.

Suppose several concentrated forces act on a simple beam as in Fig. 6-6(a).

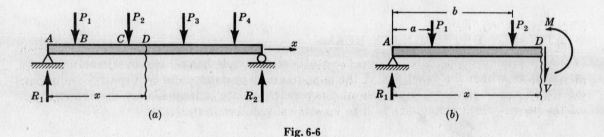

Fig. 6-6

It is desired to study the internal stresses across the section at D, located a distance x from the left end of the beam. To do this let us consider the beam to be cut at D and the portion of the beam to the right of D removed. The portion removed must then be replaced by the effect it exerted upon the portion to the left of D and this effect will consist of a vertical shearing force together with a couple, as represented by the vectors V and M respectively in the free-body diagram of the left portion of the beam shown in Fig. 6-6(b).

The force V and the couple M hold the left portion of the bar in equilibrium under the action of the forces R_1, P_1, P_2. The quantities V and M are taken to be positive if they have the senses indicated above.

RESISTING MOMENT

The couple M shown in Fig. 6-6(b) above is called the *resisting moment* at section D. The magnitude of M may be found by use of a statics equation which states that the sum of the moments of all forces about an axis through D and perpendicular to the plane of the page is zero. Thus

$$\Sigma M_0 = M - R_1 x + P_1(x-a) + P_2(x-b) = 0 \quad \text{or} \quad M = R_1 x - P_1(x-a) - P_2(x-b)$$

Thus the resisting moment M is the moment at point D created by the moments of the reaction at A and the applied forces P_1 and P_2. The resisting moment M is the resultant couple due to stresses that are distributed over the vertical section at D. These stresses act in a horizontal direction and are tensile in certain portions of the cross-section and compressive in others. Their nature will be discussed in detail in Chapter 8.

RESISTING SHEAR

The vertical force V shown in Fig. 6-6(b) is called the *resisting shear* at section D. For equilibrium of forces in the vertical direction,

$$\Sigma F_v = R_1 - P_1 - P_2 - V = 0 \quad \text{or} \quad V = R_1 - P_1 - P_2$$

This force V is actually the resultant of shearing stresses distributed over the vertical section at D. The nature of these stresses will be studied in Chapter 8.

BENDING MOMENT

The algebraic sum of the moments of the external forces to one side of the section D about an axis through D is called the *bending moment* at D. This is represented by

$$R_1 x - P_1(x-a) - P_2(x-b)$$

for the loading considered above. This quantity is considered in Problems 6.1–6.12. Thus the bending moment is opposite in direction to the resisting moment but is of the same magnitude. It is usually denoted by M also. Ordinarily the bending moment rather than the resisting moment is used in calculations because it can be represented directly in terms of the external loads.

SHEARING FORCE

The algebraic sum of all the vertical forces to one side, say the left side, of section D is called the *shearing force* at that section. This is represented by $R_1 - P_1 - P_2$ for the above loading. The shearing force is opposite in direction to the resisting shear but of the same magnitude. Usually it is denoted by V. It is ordinarily used in calculations, rather than the resisting shear. This quantity is considered in Problems 6.1–6.12.

SIGN CONVENTIONS

The customary sign conventions for shearing force and bending moment are represented in Fig. 6-7.

Fig. 6-7

Thus a force that tends to bend the beam so that it is concave upward is said to produce a positive bending moment. A force that tends to shear the left portion of the beam upward with respect to the right portion is said to produce a positive shearing force.

An easier method for determining the algebraic sign of the bending moment at any section is to say that upward external forces produce positive bending moments, downward forces yield negative bending moments.

SHEAR AND MOMENT EQUATIONS

Usually it is convenient to introduce a coordinate system along the beam, with the origin at one end of the beam. It will be desirable to know the shearing force and bending moment at all sections along the beam and for this purpose two equations are written, one specifying the shearing force V as a function of the distance, say x, from one end of the beam, the other giving the bending moment M as a function of x.

SHEARING FORCE AND BENDING MOMENT DIAGRAMS

The plots of these equations for V and M are known as shearing force and bending moment diagrams respectively. In these plots the abscissas (horizontals) indicate the position of the section along the beam and the ordinates (verticals) represent the values of the shearing force and bending moment respectively. Thus these diagrams represent graphically the variation of shearing force and bending moment at any section along the length of the bar. From these plots it is quite easy to determine the maximum value of each of these quantities.

RELATIONS BETWEEN LOAD INTENSITY, SHEARING FORCE AND BENDING MOMENT

A simple beam with a varying load indicated by $w(x)$ is sketched in Fig. 6-8. The coordinate system with origin at the left end A is established and distances to various sections in the beam are denoted by the variable x.

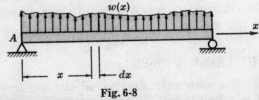

Fig. 6-8

For any value of x the relationship between the load $w(x)$ and the shearing force V is

$$w = \frac{dV}{dx}$$

and the relationship between shearing force and bending moment M is

$$V = \frac{dM}{dx}$$

These relations are derived in Problem 6.1. For applications see Problems 6.4 and 6.8–6.12.

SINGULARITY FUNCTIONS

For ease in treating problems involving concentrated forces and concentrated moments we introduce the function

$$f_n(x) = \langle x - a \rangle^n$$

where for $n > 0$ the quantity in pointed brackets is zero if $x < a$ and is the usual $(x - a)$ if $x > a$. This is the *singularity* or *half-range* function. Thus, if the argument is positive the pointed brackets behave just as ordinary parentheses. Problem 6.7 illustrates this function for certain values of n which serve to represent most common beam loadings. For applications see Problems 6.8–6.12.

Solved Problems

6.1. Derive relationships between load intensity, shearing force and bending moment at any point in a beam.

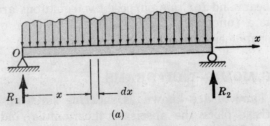

(a)

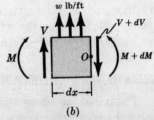

(b)

Fig. 6-9

Let us consider a beam subject to any type of transverse load of the general form shown in Fig. 6-9(a). Simple supports are illustrated but the following consideration holds for all types of beams. We will isolate from the beam the element of length dx shown and draw a free-body diagram of it. The shearing force V acts on the left side of the element, and in passing through the distance dx the shearing force will in general change slightly to an amount $V + dV$. The bending moment M acts on the left side of the element and it changes to $M + dM$ on the right side. Since dx is extremely small, the applied load may be taken as uniform over the top of the beam and equal to w lb/ft. The free-body diagram of this element thus appears as in Fig. 6-9(b). For equilibrium of moments about O, we have

$$\Sigma M_0 \;=\; M - (M + dM) + V\,dx + w\,dx\,(dx/2) \;=\; 0 \qquad \text{or} \qquad dM \;=\; V\,dx + \tfrac{1}{2}w(dx)^2$$

Since the last term consists of the product of two differentials, it is negligible compared with the other forms involving only one differential. Hence

$$dM \;=\; V\,dx \qquad \text{or} \qquad V \;=\; dM/dx$$

Thus the shearing force is equal to the rate of change of the bending moment with respect to x.

This equation will prove to be of considerable value in drawing shearing force and bending moment diagrams for the more complicated types of loading. For example, from this equation it is evident that if the shearing force is positive at a certain section of the beam then the slope of the bending moment diagram is also positive at that point. Also, it demonstrates that an abrupt change in shear, corresponding to a concentrated force, is accompanied by an abrupt change in the slope of the bending moment diagram.

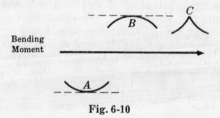

Fig. 6-10

Further, at those points where the shear is zero, the slope of the bending moment diagram is zero. At these points where the tangent to the moment diagram is horizontal, the moment may have a maximum or minimum value. This follows from the usual calculus technique of obtaining maximum or minimum values of a function by equating the first derivative of the function to zero. Thus in Fig. 6-10 if the curves shown represent portions of a bending moment diagram then critical values may occur at points A and B.

To establish the direction of concavity at a point such as A or B, we may form the second derivative of M with respect to x, i.e. d^2M/dx^2. If the value of this second derivative is positive, then the moment diagram is concave upward, as at A, and the moment assumes a minimum value. If the second derivative is negative the moment diagram is concave downward, as at B, and the moment assumes a maximum value.

However, it is to be carefully noted that the calculus method of obtaining critical values by use of the first derivative does not indicate possible maximum values at a cusp-like point in the moment diagram, if one occurs, such as that shown at C. If such a point is present, the moment there must be determined numerically and then compared to other values that are possibly critical.

Lastly, for vertical equilibrium of the element we have

$$w\,dx + V - (V + dV) \;=\; 0 \qquad \text{or} \qquad w \;=\; dV/dx$$

This relation will be of value in establishing shearing force diagrams.

6.2. For the cantilever beam subject to the uniformly distributed load of w lb per ft of length, as shown below in Fig. 6-11(a), write equations for the shearing force and bending moment at any point along the length of the bar. Also sketch the shearing force and bending moment diagrams.

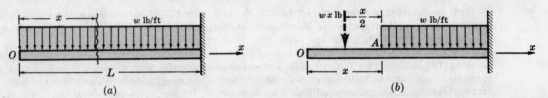

Fig. 6-11

It is not necessary to determine the reactions at the supporting wall. We shall choose the axis of the beam as the x-axis of a coordinate system with origin O at the left end of the bar. To determine the shearing force and bending moment at any section of the beam a distance x from the free end, we may replace the portion of the distributed load to the left of this section by its resultant. As shown by the dashed vector in Fig. 6-11(b), the resultant is a downward force of wx lb acting midway between O and the section x. Note that none of the load to the right of the section is included in calculating this resultant. Such a resultant force tends to shear the portion of the bar to the left of the section downward with respect to the portion to the right. By our sign convention this constitutes negative shear.

The shearing force at this section x is defined to be the sum of the forces to the left of the section. In this case, the sum is wx lb acting downward; hence

$$V = -wx \text{ lb}$$

This equation indicates that the shear is zero at $x = 0$ and when $x = L$ it is $-wL$. Since V is a first-degree function of x, the shearing force plots as a straight line connecting these values at the ends of the beam. It has the appearance shown in Fig. 6-12(a). The ordinate to this inclined line at any point represents the shearing force at that same point.

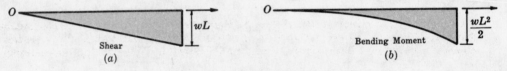

Fig. 6-12

The bending moment at this same section x is defined to be the sum of the moments of the forces to the left of this section about an axis through point A and perpendicular to the plane of the page. This sum of the moments is given by the moment of the resultant, wx lb about an axis through A; it is

$$M = -wx(x/2) \text{ lb-ft}$$

The minus sign is necessary because downward loads indicate negative bending moments. By this equation the bending moment is zero at the left end of the bar and $-wL^2/2$ at the clamped end when $x = L$. The variation of bending moment is parabolic along the bar and may be plotted as in Fig. 6-12(b). The ordinate to this parabola at any point represents the bending moment at that same point.

It is to be noted that a downward uniform load as considered here leads to a bending moment diagram that is concave downward. This could be established by taking the second derivative of M with respect to x, the derivative in this particular case being $-w$. Since the second derivative is negative, the rules of calculus tell us that the curve must be concave downward.

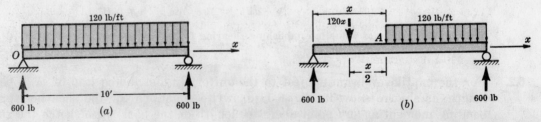

Fig. 6-13

6.3. Consider a simply supported beam 10 ft long and subject to a uniformly distributed vertical load of 120 lb per ft of length, as shown in Fig. 6-13(a). Draw shearing force and bending moment diagrams.

The total load on the beam is 1200 lb, and from symmetry each of the end reactions is 600 lb. We shall now consider any cross-section of the beam at a distance x from the left end. The shearing force at this section is given by the algebraic sum of the forces to the left of this section and these forces consist of the 600 lb reaction and the distributed load of 120 lb/ft extending over a length x ft. We may replace the portion of the distributed load to the left of the section at x by its resultant, which is $120x$ lb acting downward as shown by the dashed vector in Fig. 6-13(b). None of the load to the right of x is included in this resultant. The shearing force at x is then given by

$$V = 600 - 120x \text{ lb}$$

Since there are no concentrated loads acting on the beam, this equation is valid at all points along its length. Evidently the shearing force varies linearly from $V = 600$ lb at $x = 0$ to $V = 600 - 1200 = -600$ lb at $x = 10$ ft. The variation of shearing force along the length of the bar may then be represented by a straight line connecting these two end-point values. The shear diagram is shown in Fig. 6-14(a). The shear is zero at the center of the beam.

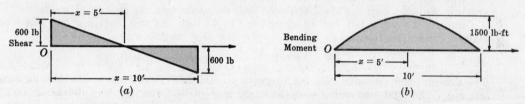

Fig. 6-14

The bending moment at the section x is given by the algebraic sum of the moments of the 600 lb reaction and the distributed load of $120x$ lb about an axis through A perpendicular to the plane of the paper. Remembering that upward forces give positive bending moments, we have

$$M = 600x - 120x(x/2) \text{ lb-ft}$$

Again, this equation holds along the entire length of the beam. It is to be noted that since the load is uniformly distributed the resultant indicated by the dashed vector acts at a distance $x/2$ from A, i.e. at the midpoint of the uniform load to the left of the section x where the bending moment is being calculated. From the above equation it is evident that the bending moment is represented by a parabola along the length of the beam. Since the bar is simply supported the moment is zero at either end and, because of the symmetry of loading, the bending moment must be a maximum at the center of the beam where $x = 5$ ft. The bending moment at that point is

$$M_{x=5} = 600(5) - 60(5)^2 = 1500 \text{ lb-ft}$$

The parabolic variation of bending moment along the length of the bar may thus be represented by the ordinates to the bending moment diagram shown in Fig. 6-14(b).

6.4. The simply supported beam shown in Fig. 6-15(a) carries a vertical load that increases uniformly from zero at the left end to a maximum value of 600 lb/ft of length at the right end. Draw the shearing force and bending moment diagrams.

For the purpose of determining the reactions R_1 and R_2 the entire distributed load may be replaced by its resultant which will act through the centroid of the triangular loading diagram. Since

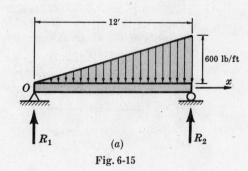

Fig. 6-15

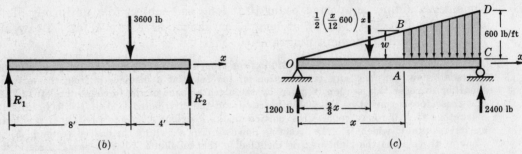

Fig. 6-15

the load varies from 0 at the left end to 600 lb/ft at the right end, the average intensity is 300 lb/ft acting over a length of 12 ft. Hence the total load is 3600 lb applied 8 ft to the right of the left support. The free-body diagram to be used in determining the reactions is shown in Fig. 6-15(b). Applying the equations of static equilibrium to this bar, we find $R_1 = 1200$ lb and $R_2 = 2400$ lb.

However, this resultant cannot be used for the purpose of drawing shear and moment diagrams. We must consider the distributed load and determine the shear and moment at a section a distance x from the left end as shown in Fig. 6-15(c). At this section x the load intensity w may be found from the similar triangles OAB and OCD as follows:

$$w/x = 600/12 \quad \text{or} \quad w = (x/12)600 \text{ lb/ft}$$

The average load intensity over the length x is $\frac{1}{2}(x/12)600$ lb/ft because the load is zero at the left end. The total load acting over the length x is the average intensity of loading multiplied by

the length, or $\frac{1}{2}\left(\frac{x}{12}600\right)x$ lb. This acts through the centroid of the triangular region OAB shown,

i.e. through a point located a distance $\frac{2}{3}x$ from O. The resultant of this portion of the distributed load is indicated by the dashed vector in Fig. 6-15(c). No portion of the load to the right of the section x is included in this resultant force.

The shearing force and bending moment at A are now readily found to be

$$V = 1200 - \frac{1}{2}\left(\frac{x}{12}600\right)x = 1200 - 25x^2$$

$$M = 1200x - \frac{1}{2}\left(\frac{x}{12}600\right)x\left(\frac{x}{3}\right) = 1200x - \frac{25}{3}x^3$$

These equations are true along the entire length of the beam. The shearing force thus plots as a parabola, having a value 1200 lb when $x = 0$ and -2400 lb when $x = 12$ ft. The bending moment is a third-degree polynomial. It vanishes at the ends and assumes a maximum value where the shear is zero. This is true because $V = dM/dx$, hence the point of zero shear must be the point where the tangent to the moment diagram is horizontal. This point of zero shear may be found by setting $V = 0$:

$$0 = 1200 - 25x^2 \quad \text{or} \quad x = 6.94 \text{ ft}$$

The bending moment at this point is found by substitution in the general expression given above:

$$M_{x=6.94} = 1200(6.94) - \frac{25}{3}(6.94)^3 = 5520 \text{ lb-ft}$$

The plots of the shear and moment equations appear in Fig. 6-16.

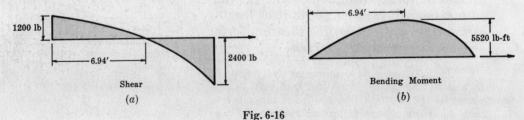

Fig. 6-16

6.5.   The horizontal beam AD is loaded by a uniformly distributed load of 400 lb per foot of length and is also subject to the concentrated force of 3000 lb applied as shown. Draw the shearing force diagram and also draw the bending moment diagram in parts.

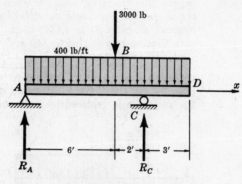

From statics the following equilibrium equations may be written:

$$\Sigma M_A = 8R_C - 3000(6) - 400(11)(5.5) = 0$$

$$\Sigma F_v = R_A + R_C - 3000 - 400(11) = 0$$

from which $R_C = 5275$ lb and $R_A = 2125$ lb.

Fig. 6-17

An x-axis is introduced with the origin at point A. Three regions must be considered in writing equations for the shearing force, namely AB, BC, CD. The equations for shearing force are

$$V = 2125 - 400x \text{ lb} \qquad \text{for} \quad 0 < x < 6 \text{ ft} \qquad (1)$$

$$V = 2125 - 400x - 3000 \text{ lb} \qquad \text{for} \quad 6 < x < 8 \text{ ft} \qquad (2)$$

$$V = 2125 - 400x - 3000 + 5275 \text{ lb} \quad \text{for} \quad 8 < x < 11 \text{ ft} \qquad (3)$$

From (1), the shear at $x = 0$ is 2125 lb. Also, the shear just to the left of the 3000 lb load is found by substituting $x = 6$ ft in (1); the result is -275 lb. The shear immediately to the right of the 3000 lb load is found by substituting $x = 6$ ft in (2); this yields

$$V_{x=6} = 2125 - 400(6) - 3000 = -3275 \text{ lb}$$

The shear just to the left of point C is found by substituting $x = 8$ ft in (2); this gives

$$V_{x=8} = 2125 - 400(8) - 3000 = -4075 \text{ lb}$$

The shear immediately to the right of point C is found by substituting $x = 8$ ft in (3); the result is

$$V_{x=8} = 2125 - 400(8) - 3000 + 5275 = 1200 \text{ lb}$$

From (1), (2) and (3) it is evident that the shearing force diagram plots as a straight line in each of the three regions. The values of the shearing force at the endpoints of these intervals have just been established, so these values may be plotted and then connected by straight lines to give the shearing force diagram shown in Fig. 6-18.

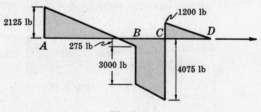

Fig. 6-18

The bending moment diagram will be plotted in a different manner than previously. The technique will be to consider each loading on the bar separately and plot the bending moment due to it alone as if no other forces were acting on the structure. The moment diagram is then said to be plotted in *parts*. As will be seen in a later chapter on the deflections of beams this method is often very convenient, although the choice as to whether to use it or the conventional type of plot presented in earlier problems depends upon the purpose for which the diagram is being drawn. More will be said about this later.

Let us work from left to right along the beam. The moment diagram may be considered to consist of four parts, one due to the reaction R_A, another due to the uniformly distributed load, a third due to the 3000 lb force, and the last due to the reaction R_C. At any section a distance x from the point A the bending moment due to R_A only is simply $2125x$ lb-ft. This value is positive because R_A acts upward. This same expression holds for all values of x along the bar. This is a first-

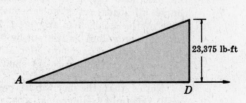

Fig. 6-19

degree function of x, hence the bending moment due to R_A only plots as a straight line. At $x = 0$, the moment is zero, and substituting $x = 11$ ft in the above expression it is seen that the bending moment at point D is 23,375 lb-ft. The bending moment at any section x due to this force only may thus be represented by the ordinates to the triangle shown in Fig. 6-19.

The uniformly distributed load will be treated next. All other loads are temporarily disregarded and the bending moment at any section x due to the uniform load is calculated. This proceeds in the same manner as discussed before, i.e. the portion of the load to the left of the section x is replaced by its resultant, indicated by the dashed vector in Fig. 6-20(a).

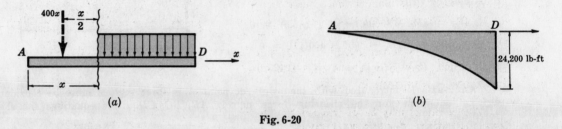

(a)   (b)

Fig. 6-20

Due to the distributed load only, the bending moment at any section x anywhere along the beam is given by

$$-400x(x/2) \text{ lb-ft}$$

When $x = 0$ this expression vanishes, and when $x = 11$ ft it is equal to $-24,200$ lb-ft. It plots as a parabola since the expression is of the second degree, as shown in Fig. 6-20(b).

As we work from left to right along the beam, the influence of the 3000 lb load is not apparent until we pass to the right of point B. After that, at any section x the bending moment due to this force alone, temporarily disregarding all other forces, is given by

$$-3000(x - 6) \text{ lb-ft} \qquad \text{for} \qquad 6 < x < 11 \text{ ft}$$

It is to be noted that x is always measured from the point A. When $x = 6$ ft the bending moment due to this force only vanishes, and when $x = 11$ ft this expression has the value $-15,000$ lb-ft. It is a first-degree expression in x, hence the bending moment due to this force alone plots as a straight line in the region BD, as shown in Fig. 6-21.

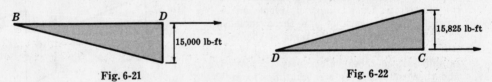

Fig. 6-21   Fig. 6-22

The bending moment diagram due to R_C only may be formed in an analogous manner. As soon as we consider sections anywhere in the region CD, the force R_C will give rise to a bending moment. Due to this force only, there is a bending moment of $5275(x - 8)$ lb-ft for $8 < x < 11$ ft. When $x = 8$ ft this value is zero, and when $x = 11$ ft the bending moment due to R_C only is $5275(11 - 8) = 15,825$ lb-ft. This is a first-degree expression in x and hence the moment diagram due to R_C only also appears as a triangle as shown in Fig. 6-22.

The moment diagrams due to the various loadings have each been obtained as if there were only that one load acting on the beam. Actually of course all loads act simultaneously, so the true value of the moment at any point is the algebraic sum of the values indicated by the above four plots. It is customary to plot all of these individual diagrams together, as shown in Fig. 6-23.

Notice that the horizontal bases of the two small triangular diagrams are shifted so that there is no overlapping of the various figures.

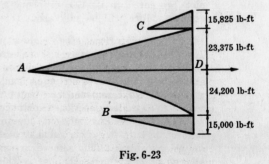

Fig. 6-23

This is not mandatory but makes for ease of interpretation. The algebraic sum of the four ordinates at D equals zero, which is necessary because that is a free end. The composite type of diagram discussed in previous problems can now be obtained by summing ordinates at every point of the above diagram. Between A and B, only two quantities would be included in the sum; between B and C, three quantities; and between C and D, four.

6.6. The beam AE is simply supported at B and D and overhangs both ends. It is subject to a uniformly distributed load of 600 lb per ft of length as well as a couple of magnitude 10,000 lb-ft applied at C. Draw both the shearing force diagram and the bending moment diagram in parts.

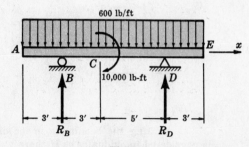

Fig. 6-24

The reactions may be determined by the following equations of static equilibrium:

$$\Sigma M_B = 8R_D - 10,000 - 600(14)(4) = 0 \qquad R_D = 5450 \text{ lb}$$

$$\Sigma F_v = R_B + 5450 - 600(14) = 0 \qquad R_B = 2950 \text{ lb}$$

The x-axis is introduced with its origin at point A. In the region AB, the shearing force at any section a distance x from point A is given by the resultant of the distributed load to the left of this section. This resultant is evidently a force of $600x$ lb acting downward. Thus we have

$$V = -600x \qquad \text{for} \qquad 0 < x < 3 \text{ ft} \tag{1}$$

Substituting $x = 3$ ft, this equation yields a shearing force at that point of -1800 lb. The shear at $x = 0$ is, of course, zero.

As soon as we pass to the right of B the reaction R_B appears in the shearing force equation. For any section a distance x from A the shearing force in the region BD is obtained by summing the applied forces to the left of this section. This sum is given by

$$V = -600x + 2950 \text{ lb} \qquad \text{for} \qquad 3 < x < 11 \text{ ft} \tag{2}$$

Note that the applied couple at C does not enter the equations for shearing force because the couple does not have any force effect in any direction. It does, however, enter the equation indirectly since it influences the values of the reactions R_B and R_D. Substituting $x = 3$ ft and $x = 11$ ft in (2),

$$V_{x=3} = 1150 \text{ lb} \qquad \text{and} \qquad V_{x=11} = -3650 \text{ lb}$$

In considering values of x greater than 11 ft, the reaction R_D must be included in the equation for shearing force. Summing forces to the left of a section x in the region DE, we find

$$V = -600x + 2950 + 5450 \text{ lb} \qquad \text{for} \qquad 11 < x < 14 \text{ ft} \tag{3}$$

Substituting $x = 11$ ft and $x = 14$ ft in (3), we find

$$V_{x=11} = 1800 \text{ lb} \qquad \text{and} \qquad V_{x=14} = 0 \text{ lb}$$

The shearing force at any point along the bar is defined by one of the three equations (1), (2), or (3), depending upon the region in which the point x lies. Since V is a first-degree function of x in each of these regions, the shearing force diagram plots as a straight line in each of these three regions. The values of the ordinates at the end points of each of these regions have already been obtained by substitution. In AB these end-point values were 0 and -1800 lb. In BD they were

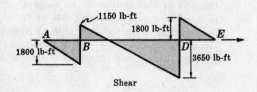

Fig. 6-25

1150 lb and −3650 lb.   Finally, in *DE* they were found to be 1800 lb and 0.   These values may be plotted at the corresponding points along the beam and the ordinates representing these values connected by a straight line in each region.   In this manner Fig. 6-25 is obtained.

The magnitude of the vertical *jump* at each of the points *B* and *D* is of course equal to the value of the concentrated reactions R_B and R_D at each of these points.

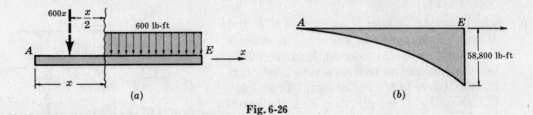

Fig. 6-26

In plotting the bending moment diagram in parts, each of the loads, including the reactions, is considered individually as if there were no other loads acting on the beam.   Beginning with the uniform load of 600 lb/ft, a section at a distance *x* from the left end *A* is considered and the bending moment at this section due to the distributed load only is calculated.   The resultant of the distributed forces lying to the left of this section is indicated by the dashed vector in Fig. 6-26(*a*).   The moment of this resultant about an axis through the section at *x* and perpendicular to the plane of the page is

$$M = -600x\,(x/2) = -300x^2 \text{ lb-ft} \quad \text{for} \quad 0 < x < 14 \text{ ft}$$

Thus the bending moment diagram representing only the distributed load is parabolic.   At $x = 0$ the moment is 0 and at the right end, $x = 14$ ft, the above equation yields

$$M_{x=14} = -300(14)^2 = -58,800 \text{ lb-ft}$$

This part of the bending moment diagram thus has the appearance indicated in Fig. 6-26(*b*).

Since we are working from left to right along the beam, the moment due to the reaction R_B does not come into consideration until we consider values of *x* greater than 3 ft.   Then, due to this load only, the moment of this 2950 lb force about an axis through the section *x* is given by

$$M = 2950(x - 3) \text{ lb-ft} \quad \text{for} \quad 3 < x < 14 \text{ ft}$$

Since this is a first-degree function of *x* the bending moment due to R_B only plots as a straight line.   According to this equation, the bending moment is zero at $x = 3$ ft and is $2950(14 - 3) = 32,450$ lb-ft at the point *E*.   These two end-point values may now be connected by a straight line to yield the moment diagram, Fig. 6-27, due to R_B only.

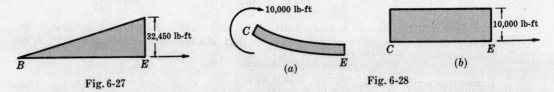

Fig. 6-27   (*a*)   **Fig. 6-28**

Progressing to the right along the beam, we next consider the applied couple of 10,000 lb-ft at the point *C*.   For sections located at a distance *x* from point *A*, where *x* lies to the right of point *C*, this applied couple appears in the bending moment diagram.   While it is true that the moment of this couple is the same about all points in the plane, it does not appear in the bending moment diagram until we consider values of *x* greater than 6 ft because the bending moment takes into account only the moments of those forces and couples to the left of the section *x* being considered.   This applied couple produces curvature as shown in Fig. 6-28(*a*).   According to our sign convention this constitutes positive bending.   Hence we have for the applied couple only

$$M = 10,000 \text{ lb-ft} \quad \text{for} \quad 6 < x < 14 \text{ ft}$$

This constant value plots as a horizontal straight line, as shown in Fig. 6-28(*b*).

Finally, the reaction R_D appears in the calculation of the bending moment at sections lying to the right of point D. For such a section at the distance x to the right of point A, the bending moment due to R_D only is

$$M = 5450(x-11) \text{ lb-ft} \qquad \text{for} \qquad 11 < x < 14 \text{ ft}$$

This too plots as a straight line. At point D this moment is zero, and substituting $x = 14$ ft we find $M = 16{,}350$ lb-ft at point E. Connecting these two end-point values by a straight line yields the moment diagram due to R_D only, Fig. 6-29.

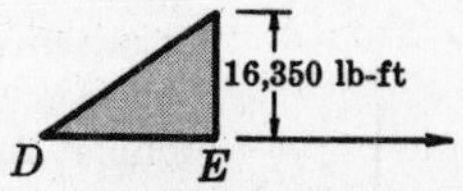

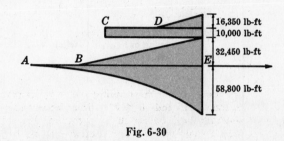

Fig. 6-29

The four parts of the bending moment diagram are finally plotted together, as shown in Fig. 6-30. The horizontal bases of the parts have been displaced vertically so as to avoid overlapping of the various diagrams.

The analysis could just as well have been carried out by working from right to left. The resulting diagram by parts would have had an entirely different appearance.

Fig. 6-30

6.7. Discuss singularity functions as a method for representing shearing forces and bending moments.

The techniques discussed in the preceding problems are adequate if the loadings are continuously varying over the length of the beam. However, if concentrated forces or moments are present, a distinct pair of shearing force and bending moment equations must be written for each region between such concentrated forces or moments. Although this presents no fundamental difficulties, it usually leads to very cumbersome results. As we shall see in a later chapter, these results are particularly unwieldly to work with in dealing with deflections of beams.

At least some compactness of representation may be achieved by introduction of so-called *singularity* or *half-range* functions. Such functions were applied to beam analysis by Macauley in 1919 and this technique of analysis sometimes bears the name of Macauley's method, although the functions were actually used in the 19th century by A. Clebsch. Let us introduce, by definition, the pointed brackets $\langle x - a \rangle$ and define this quantity to be zero if $(x - a) < 0$, i.e. $x < a$, and to be simply $(x - a)$ if $(x - a) > 0$, i.e. $x > a$. That is, a half-range function is defined to have a value only when the argument is positive. When the argument is positive, the pointed brackets behave just as ordinary parentheses. The singularity function

$$f_n(x) = \langle x - a \rangle^n$$

obeys the integration law

$$\int_{-\infty}^{x} \langle y - a \rangle^n \, dy = \frac{\langle x - a \rangle^{n+1}}{n+1} \qquad \text{for} \quad n \geqq 0 \tag{1}$$

It is also necessary to define the integration law for the cases of $n = -1$ and $n = -2$ and this we do as follows:

$$\int_{-\infty}^{x} \langle y - a \rangle^{-1} \, dy = \langle x - a \rangle^0 = \begin{cases} 0 & \text{when } x < a \\ 1 & \text{when } x > a \end{cases} \tag{2}$$

$$\int_{-\infty}^{x} \langle y - a \rangle^{-2} \, dy = \langle x - a \rangle^{-1} \tag{3}$$

The functions $\langle x - a \rangle^{-1}$ and $\langle x - a \rangle^{-2}$ are zero everywhere except at $x = a$ where they are infinite in such a way that (2) and (3) hold.

The five diagrams in the following Fig. 6-31 illustrate the singularity functions of different degrees required to represent common beam loadings.

Type of loading	Singularity function	Pictorial representation
Concentrated moment (doublet function)	$w(x) = M_0\langle x - a\rangle^{-2}$	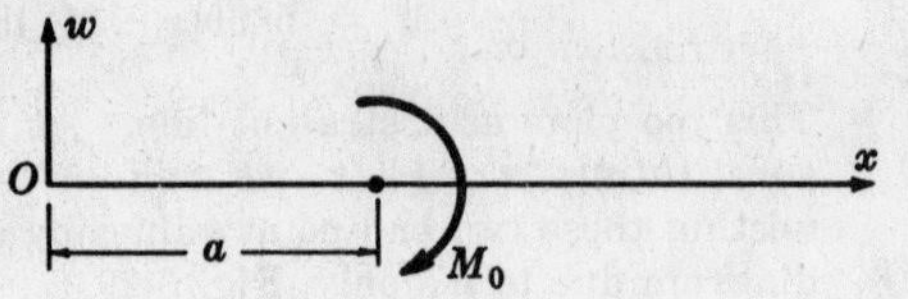
Concentrated force (impulse or Dirac delta function)	$w(x) = F_0\langle x - a\rangle^{-1}$	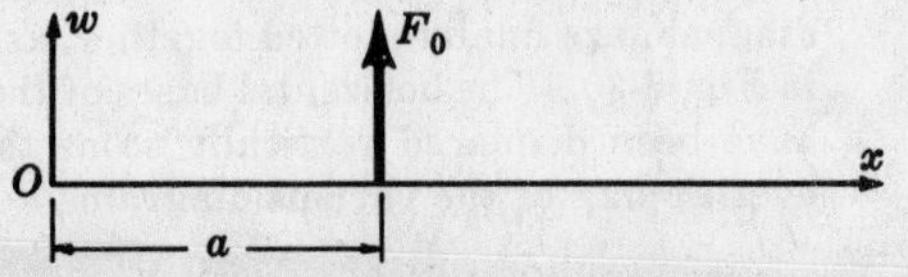
Uniformly distributed load (step function)	$w(x) = w_0\langle x - a\rangle^0$	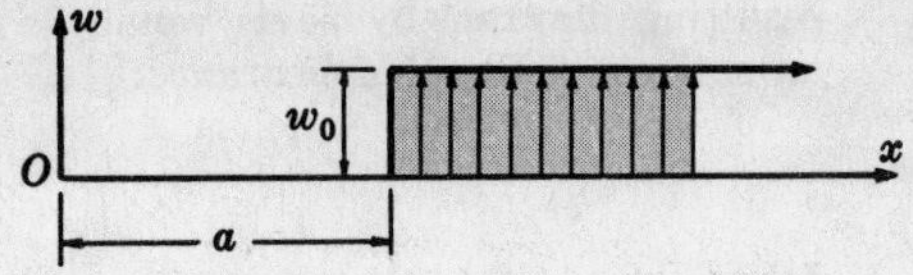
Linearly varying load	$w(x) = \dfrac{dw}{dx}\langle x - a\rangle^1$	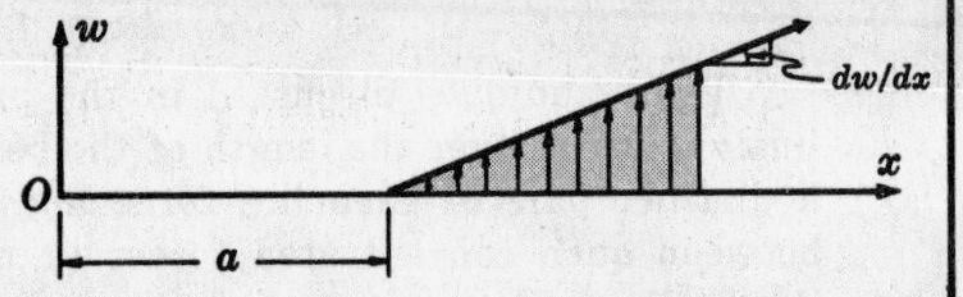
Quadratically varying load	$w(x) = \dfrac{C\langle x - a\rangle^2}{2}$	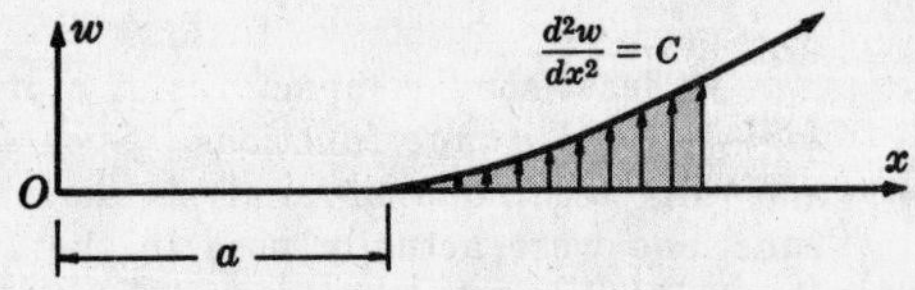

Fig. 6-31

Once the loading on a beam has been represented by the above singularity functions it is then possible to obtain the shearing forces and bending moments at all points along the length of the beam through use of the above integration rules.

6.8. Consider a simply supported beam subject to a single concentrated load of 4000 lb as shown in Fig. 6-32(a). Using singularity functions, write equations for the shearing force and bending moment at any position in the beam and plot the shear and moment diagrams.

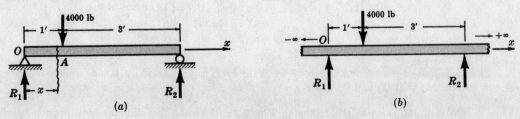

(a)   (b)

Fig. 6-32

The reactions R_1 and R_2 are shown as concentrated forces. It is possible to imagine the beam extending to infinity in both directions. The beam under study will then correspond to a portion of this hypothetical beam, as indicated in Fig. 6-32(b).

To the left of R_1 and to the right of R_2 there are no forces on the infinitely long beam and, thus, the shearing force and bending moment must be zero everywhere in these regions. Using Fig. 6-31 of Problem 6.7, we may write the load-intensity function

$$w(x) = R_1\langle x\rangle^{-1} - 4000\langle x-1\rangle^{-1} + R_2\langle x-4\rangle^{-1} \tag{1}$$

This functional representation holds for all values of x provided the pointed brackets are interpreted as in Problem 6.7. From Problem 6.1,

$$w = dV/dx \quad \text{from which} \quad V = \int_{-\infty}^{x} w\,dx$$

Using the integration laws of Problem 6.7, we find the shear V to be

$$V(x) = R_1\langle x\rangle^0 - 4000\langle x-1\rangle^0 + R_2\langle x-4\rangle^0 \tag{2}$$

Again from Problem 6.1,

$$V = dM/dx \quad \text{from which} \quad M = \int_{-\infty}^{x} V\,dx$$

and from (2) we obtain
$$M(x) = R_1\langle x\rangle^1 - 4000\langle x-1\rangle^1 + R_2\langle x-4\rangle^1 \tag{3}$$

We may determine R_1 and R_2 from (2) and (3) by noting that for x slightly larger than 4, i.e. for points just to the right of the reaction R_2, $V(x)$ vanishes. Thus from (2),

$$R_1 - 4000 + R_2 = 0$$

and from (3), since the bending moment also vanishes at such a point,

$$R_1(4) - 4000(4-1) + 0 = 0$$

It follows that $R_1 = 3000$ lb, $R_2 = 1000$ lb. These results could of course have been obtained from equilibrium considerations.

The shearing force and bending moment diagrams may now be drawn from (2) and (3) after substituting the numerical values of R_1 and R_2. These diagrams are indicated in Fig. 6-33.

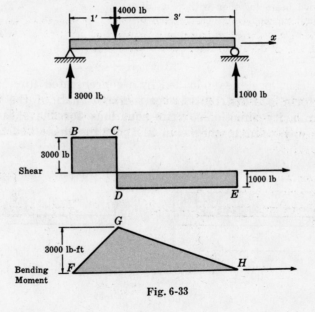

Fig. 6-33

6.9. Consider a cantilever beam loaded only by the couple of 200 lb-ft applied as shown in Fig. 6-34(a). Using singularity functions, write equations for the shearing force and bending moment at any position in the beam and plot the shear and moment diagrams.

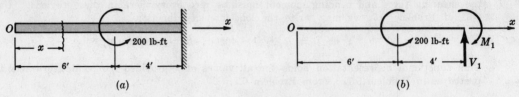

Fig. 6-34

A free-body diagram is indicated in Fig. 6-34(b), where V_1 and M_1 denote the reaction of the supporting wall. The load intensity function is

$$w(x) = -200\langle x - 6\rangle^{-2} + V_1\langle x - 10\rangle^{-1} + M_1\langle x - 10\rangle^{-2} \qquad (1)$$

Since $V = \displaystyle\int_{-\infty}^{x} w\,dx$, we have

$$V(x) = -200\langle x - 6\rangle^{-1} + V_1\langle x - 10\rangle^{0} + M_1\langle x - 10\rangle^{-1} \qquad (2)$$

Considering a point just to the right of V_1 on the hypothetical beam of infinite length, we immediately have $V_1 = 0$. Thus

$$V(x) = -200\langle x - 6\rangle^{-1} + M_1\langle x - 10\rangle^{-1}$$

which vanishes, according to our definition of the function $\langle x - a\rangle^{-1}$ in Problem 6.7.

Next, since $M = \displaystyle\int_{-\infty}^{x} V\,dx$, we have

$$M(x) = -200\langle x - 6\rangle^{0} + M_1\langle x - 10\rangle^{0}$$

Again, consideration of the point just to the right of the wall yields $M_1 = 200$ lb-ft.

The loaded beam together with plots of the shear and moment equations is shown in Fig. 6-35.

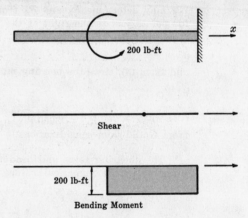

Fig. 6-35

6.10. Consider a cantilever beam loaded by a concentrated force at the free end together with a uniform load distributed over the right half of the beam [see Fig. 6-36(a)]. Using singularity functions, write equations for the shearing force and bending moment at any point in the beam and plot the shear and moment diagrams.

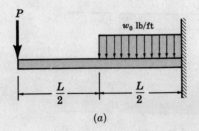

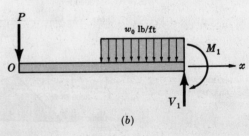

Fig. 6-36

The free-body diagram appears as indicated in Fig. 6-36(b). It is not necessary to determine the reactions V_1 and M_1. The load intensity function is

$$w(x) = -P\langle x \rangle^{-1} - w_0 \left\langle x - \frac{L}{2} \right\rangle^0 + V_1 \langle x - L \rangle^{-1} + M_1 \langle x - L \rangle^{-2} \tag{1}$$

Since $\quad V = \displaystyle\int_{-\infty}^{x} w\, dx, \quad$ we have

$$V(x) = -P\langle x \rangle^0 - w_0 \left\langle x - \frac{L}{2} \right\rangle^1 + V_1 \langle x - L \rangle^0 + M_1 \langle x - L \rangle^{-1} \tag{2}$$

Considering a point just to the right of V_1 on the hypothetical beam of infinite length, we have from (2)

$$V_1 = P + \frac{w_0 L}{2}$$

Lastly, since $\quad M = \displaystyle\int_{-\infty}^{x} V\, dx, \quad$ we have

$$M(x) = -P\langle x \rangle^1 - \frac{w_0}{2} \left\langle x - \frac{L}{2} \right\rangle^2 + \left(P + \frac{w_0 L}{2} \right) \langle x - L \rangle^1 + M_1 \langle x - L \rangle^0 \tag{3}$$

Again, consideration of the point just to the right of the support yields

$$M_1 = PL + \frac{w_0 L^2}{2}$$

The loaded beam together with plots of the shear and moment equations is shown in Fig. 6-37.

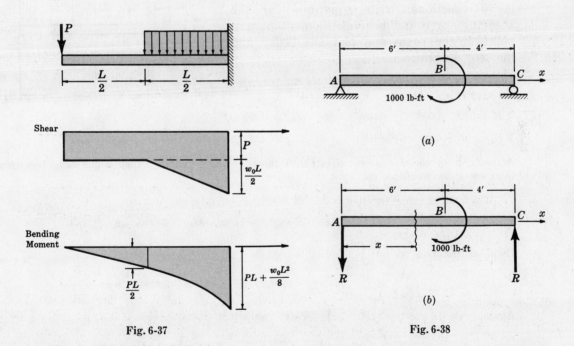

Fig. 6-37   Fig. 6-38

6.11. In Fig. 6-38(a) a simply supported beam is loaded by the couple of 1000 lb-ft. Using singularity functions, write equations for the shearing force and bending moment at any point in the beam and plot the shear and moment diagrams.

The beam is loaded by one couple, and the only possible manner in which equilibrium may be created is for the reactions at the supports A and C to constitute another couple. Thus these reactions appear as in Fig. 6-38(b). For equilibrium,

$$\Sigma M_A = 10R - 1000 = 0 \quad \text{from which} \quad R = 100 \text{ lb}$$

Thus the two forces R shown constitute the reactions necessary for equilibrium.

98   SHEARING FORCE AND BENDING MOMENT   [CHAP. 6

The load intensity function is

$$w(x) = -100\langle x\rangle^{-1} + 1000\langle x-6\rangle^{-2} + 100\langle x-10\rangle^{-1} \quad (1)$$

Since $V = \int_{-\infty}^{x} w\,dx$, we obtain

$$V(x) = -100\langle x\rangle^{0} + 1000\langle x-6\rangle^{-1} + 100\langle x-10\rangle^{0} \quad (2)$$

Again, since $M = \int_{-\infty}^{x} V\,dx$, we get

$$M(x) = -100\langle x\rangle^{1} + 1000\langle x-6\rangle^{0} + 100\langle x-10\rangle^{1} \quad (3)$$

Shear and moment diagrams are plotted in Fig. 6-39. From these it is evident that when a couple acts on a bar the bending moment diagram exhibits an abrupt jump or discontinuity at the point where the couple is applied.

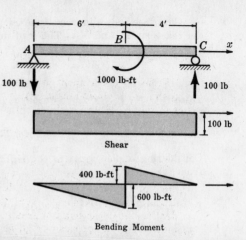

Fig. 6-39

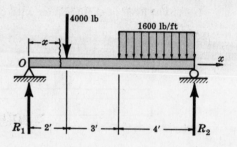

6.12. A simply supported beam is subject to a concentrated force of 4000 lb together with a distributed load of 1600 lb per ft of length applied as indicated in Fig. 6-40. Using singularity functions, write equations for the shearing force and bending moment at any point in the beam and plot the shear and moment diagrams.

Let us first use statics principles to determine the reactions R_1 and R_2. We may write

$$\Sigma M_0 = 9R_2 - 4000(2) - 6400(7) = 0, \quad R_2 = 5870\text{ lb}$$

$$\Sigma F_v = R_1 - 4000 - 6400 + 5870 = 0, \quad R_1 = 4530\text{ lb}$$

Fig. 6-40

Actually, it is not mandatory to establish these at the beginning; they could have been found later using the technique employed in Problem 6.8.

The load intensity function is

$$w(x) = 4530\langle x\rangle^{-1} - 4000\langle x-2\rangle^{-1} - 1600\langle x-5\rangle^{0} + 5870\langle x-9\rangle^{-1} \quad (1)$$

Using the relation $V = \int_{-\infty}^{x} w\,dx$ we obtain

$$V(x) = 4530\langle x\rangle^{0} - 4000\langle x-2\rangle^{0} - 1600\langle x-5\rangle^{1} + 5870\langle x-9\rangle^{0} \quad (2)$$

Again, from the relation $M = \int_{-\infty}^{x} V\,dx$ we obtain

$$M(x) = 4530\langle x\rangle^{1} - 4000\langle x-2\rangle^{1} - 800\langle x-5\rangle^{2} + 5870\langle x-9\rangle^{1} \quad (3)$$

The shearing force diagram is now readily plotted. In the left and central portions of the bar it is represented by two horizontal straight lines having ordinates of 4530 lb and 530 lb respectively. In the right region it is represented by an inclined straight line joining the ordinates of 530 at $x=5$ and -5870 at $x=9$. It is shown in Fig. 6-41.

The point where the shear is zero under the distributed load is found by setting $V=0$ in the shear equation for that region. Doing this we find from (2)

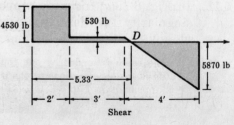

Fig. 6-41

$$4530 - 4000 - 1600(x - 5) = 0 \quad \text{from which} \quad x = 5.33 \text{ ft}$$

This is point D in the above shear diagram.

From (3) the moment is zero at $x = 0$ and is 9060 lb-ft at $x = 2$ ft. The moment at $x = 5$ is found to be 10,650 lb-ft. It is of interest to calculate the moment at $x = 5.33$ ft, where the shear is zero. From (3) the moment there is

$$M_{x=5.33} = 4530(5.33) - 4000(3.33) - 800(0.33)^2 = 10{,}810 \text{ lb-ft}$$

The moment diagram may now be plotted. It consists of two straight lines in the left and central regions and a parabola in the right portion. This parabola has a horizontal tangent at $x = 5.33$ ft and evidently that is the point of maximum moment. The moment diagram appears in Fig. 6-42

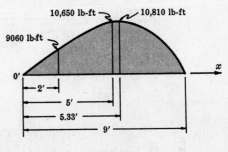

It may be noted from the shear diagram that there is no jump in shear at $x = 5$ ft. Since $V = dM/dx$ at all points along the bar, then there is no jump in the slope of the moment diagram at this point. Hence the straight line and the parabola in the moment diagram have a common tangent at $x = 5$ ft.

Fig. 6-42

Supplementary Problems

For the cantilever beams of Problems 6.13–6.15 loaded as shown, write equations for the shearing force and bending moment at any point along the length of the beam. Also, draw the shearing force and bending moment diagrams.

6.13.

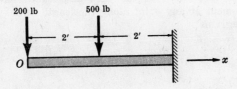

Ans.

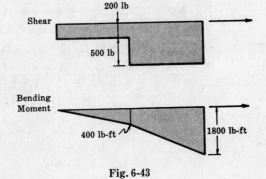

$V = -200$ lb for $0 < x < 2$ ft
$V = -700$ lb for $2 < x < 4$ ft
$M = -200x$ lb-ft for $0 < x < 2$ ft
$M = -200x - 500(x - 2)$ lb-ft for $2 < x < 4$ ft

Fig. 6-43

6.14.

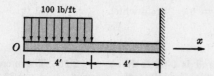

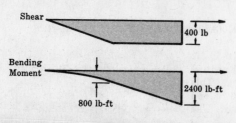

Ans.

$$V = -100x \text{ lb for } 0 < x < 4 \text{ ft}$$
$$V = -400 \text{ lb for } 4 < x < 8 \text{ ft}$$
$$M = -50x^2 \text{ lb-ft for } 0 < x < 4 \text{ ft}$$
$$M = -400(x-2) \text{ lb-ft for } 4 < x < 8 \text{ ft}$$

Fig. 6-44

6.15.

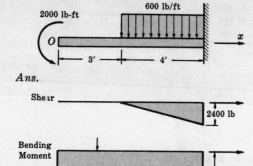

Ans.

$$V = 0 \text{ for } 0 < x < 3 \text{ ft}$$
$$V = -600(x-3) \text{ lb for } 3 < x < 7 \text{ ft}$$
$$M = -2000 \text{ lb-ft for } 0 < x < 3 \text{ ft}$$
$$M = -2000 - 300(x-3)^2 \text{ lb-ft for } 3 < x < 7 \text{ ft}$$

Fig. 6-45

For the beams of Problems 6.16–6.22 simply supported at the ends and loaded as shown write equations for the shearing force and bending moment at any point along the length of the beam. Also, draw the shearing force and bending moment diagrams.

6.16.

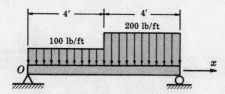

Ans.

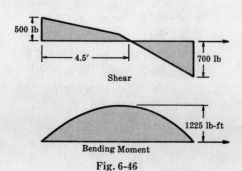

$$V = 500 - 100x \text{ lb for } 0 < x < 4 \text{ ft}$$
$$V = 100 - 200(x-4) \text{ lb for } 4 < x < 8 \text{ ft}$$
$$M = 500x - 50x^2 \text{ lb-ft for } 0 < x < 4 \text{ ft}$$
$$M = 500x - 400(x-2) - 100(x-4)^2 \text{ lb-ft}$$
$$\text{for } 4 < x < 8 \text{ ft}$$

Fig. 6-46

6.17.

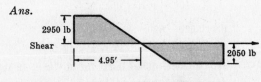

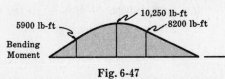

Ans.

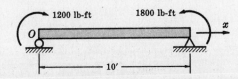

Fig. 6-47

$V = 2950$ lb for $0 < x < 2$ ft

$V = 2950 - 1000(x-2)$ lb for $2 < x < 7$ ft

$V = -2050$ lb for $7 < x < 11$ ft

$M = 2950x$ lb-ft for $0 < x < 2$ ft

$M = 2950x - 500(x-2)^2$ lb-ft for $2 < x < 7$ ft

$M = 2050z$ lb-ft for $0 < z < 4$ ft

6.18.

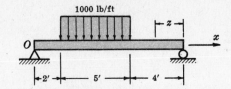

Ans.

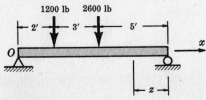

Fig. 6-48

$V = 60$ lb for $0 < x < 10$ ft

$M = 1200 + 60x$ lb-ft for $0 < x < 10$ ft

6.19.

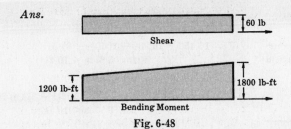

Ans.

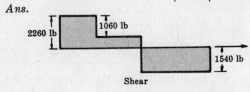

Fig. 6-49

$V = 2260$ lb for $0 < x < 2$ ft

$V = 1060$ lb for $2 < x < 5$ ft

$V = -1540$ lb for $5 < x < 10$ ft

$M = 2260x$ lb-ft for $0 < x < 2$ ft

$M = 2260x - 1200(x-2)$ lb-ft for $2 < x < 5$ ft

$M = 1540z$ lb-ft for $0 < z < 5$ ft

6.20.

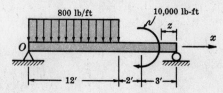

Ans.

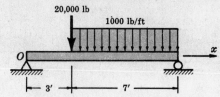

Fig. 6-50

$V = 5620 - 800x$ lb for $0 < x < 12$ ft

$V = -3980$ lb for $12 < x < 17$ ft

$M = 5620x - 400x^2$ lb-ft for $0 < x < 12$ ft

$M = 5620x - 9600(x - 6)$ lb-ft for $12 < x < 14$ ft

$M = 3980z$ for $0 < z < 3$ ft

6.21.

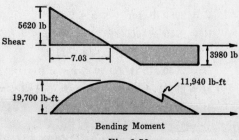

Ans.

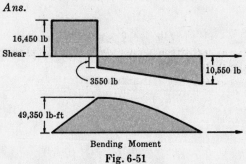

Fig. 6-51

$V = 16{,}450$ lb for $0 < x < 3$ ft

$V = 16{,}450 - 20{,}000 - 1000(x - 3)$ lb
$\qquad$ for $3 < x < 10$ ft

$M = 16{,}450x$ lb-ft for $0 < x < 3$ ft

$M = 16{,}450x - 20{,}000(x - 3) - 500(x - 3)^2$ lb-ft
$\qquad$ for $3 < x < 10$ ft

6.22.

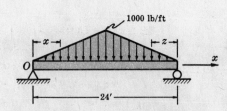

Ans.

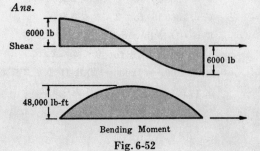

Fig. 6-52

$V = 6000 - \dfrac{x^2}{24}(1000)$ lb for $0 < x < 12$ ft

$V = -6000 + \dfrac{z^2}{24}(1000)$ lb for $0 < z < 12$ ft

$M = 6000x - \dfrac{x^3}{72}(1000)$ lb-ft for $0 < x < 12$ ft

$M = 6000z - \dfrac{z^3}{72}(1000)$ lb-ft for $0 < z < 12$ ft

For the simply supported beams of Problems 6.23–6.24 with overhanging ends and loaded as shown, draw the shearing force diagram and also draw the bending moment diagram in parts.

6.23.   **6.24.**

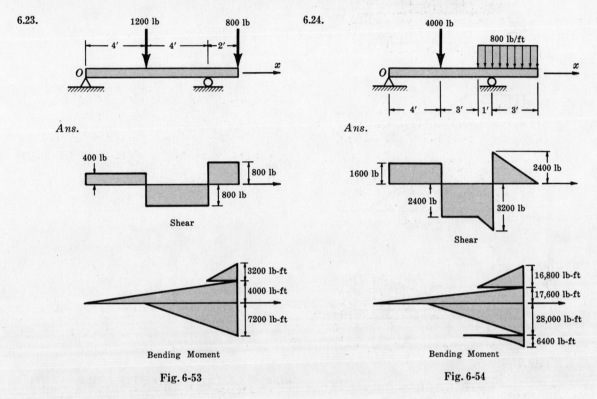

Ans.   *Ans.*

Fig. 6-53   Fig. 6-54

For Problems 6.25–6.28 use singularity functions to write the equations for shearing force and bending moment at any point in the beam. Plot the corresponding diagrams.

6.25.

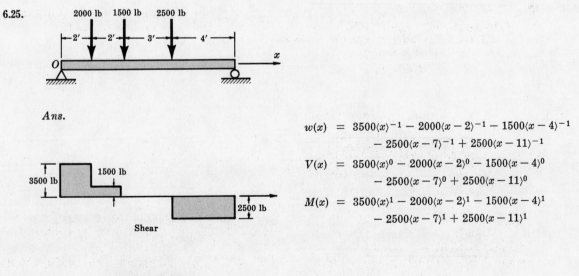

Ans.

$$w(x) = 3500\langle x\rangle^{-1} - 2000\langle x-2\rangle^{-1} - 1500\langle x-4\rangle^{-1}$$
$$- 2500\langle x-7\rangle^{-1} + 2500\langle x-11\rangle^{-1}$$

$$V(x) = 3500\langle x\rangle^{0} - 2000\langle x-2\rangle^{0} - 1500\langle x-4\rangle^{0}$$
$$- 2500\langle x-7\rangle^{0} + 2500\langle x-11\rangle^{0}$$

$$M(x) = 3500\langle x\rangle^{1} - 2000\langle x-2\rangle^{1} - 1500\langle x-4\rangle^{1}$$
$$- 2500\langle x-7\rangle^{1} + 2500\langle x-11\rangle^{1}$$

Fig. 6-55

6.26.

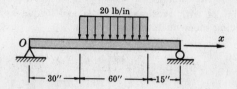

Ans.

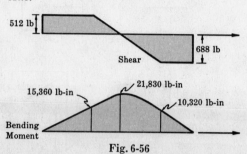

Fig. 6-56

$$w(x) = 512\langle x \rangle^{-1} - 20\langle x-30 \rangle^0 + 20\langle x-90 \rangle^0$$
$$+ 688\langle x-105 \rangle^{-1}$$

$$V(x) = 512\langle x \rangle^0 - 20\langle x-30 \rangle^1 + 20\langle x-90 \rangle^1$$
$$+ 688\langle x-105 \rangle^0$$

$$M(x) = 512\langle x \rangle^1 - 10\langle x-30 \rangle^2 + 10\langle x-90 \rangle^2$$
$$+ 688\langle x-105 \rangle^1$$

6.27.

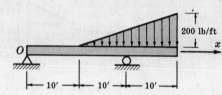

Ans.

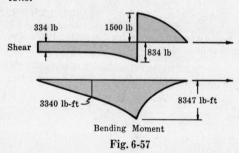

Fig. 6-57

$$w(x) = -334\langle x \rangle^{-1} - 10\langle x-10 \rangle^1 + 2334\langle x-30 \rangle^{-1}$$
$$V(x) = -334\langle x \rangle^0 - 5\langle x-10 \rangle^2 + 2334\langle x-20 \rangle^0$$
$$M(x) = -334\langle x \rangle^1 - \tfrac{5}{3}\langle x-10 \rangle^3 + 2334\langle x-20 \rangle^1$$

6.28.

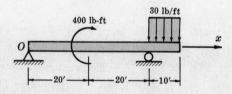

Ans.

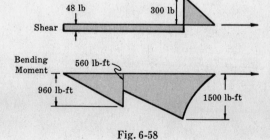

Fig. 6-58

$$w(x) = -48\langle x \rangle^{-1} + 400\langle x-20 \rangle^{-2} + 348\langle x-40 \rangle^{-1}$$
$$- 30\langle x-40 \rangle^0$$

$$V(x) = -48\langle x \rangle^0 + 400\langle x-20 \rangle^{-1} + 348\langle x-40 \rangle^0$$
$$- 30\langle x-40 \rangle^1$$

$$M(x) = -48\langle x \rangle^1 + 400\langle x-20 \rangle^0 + 348\langle x-40 \rangle^1$$
$$- 15\langle x-40 \rangle^2$$

Chapter 7

Centroids, Moments of Inertia, and Products of Inertia of Plane Areas

FIRST MOMENT OF AN ELEMENT OF AREA

The first moment of an element of area about any axis in the plane of the area is given by the product of the area of the element and the perpendicular distance between the element and the axis. For example, in Fig. 7-1 the first moment dQ_x of the element da about the x-axis is given by

$$dQ_x = y\, da$$

About the y-axis the first moment is

$$dQ_y = x\, da$$

For applications, see Problem 7.1.

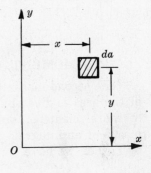

Fig. 7-1

FIRST MOMENT OF A FINITE AREA

The first moment of a finite area about any axis in the plane of the area is given by the summation of the first moments about that same axis of all the elements of area contained in the finite area. This is frequently evaluated by means of an integral. If the first moment of the finite area is denoted by Q_x, then

$$Q_x = \int dQ_x$$

For applications, see Problems 7.1, 7.2, and 7.3.

CENTROID OF AN AREA

The centroid of an area is defined by the equations

$$\bar{x} = \frac{\int x\, da}{A} = \frac{Q_y}{A}, \qquad \bar{y} = \frac{\int y\, da}{A} = \frac{Q_x}{A}$$

where A denotes the area. For applications, see Problems 7.1, 7.2, 7.3, 7.10, 7.12, 7.13, 7.14, and 7.18.

The centroid of an area is the point at which the area might be considered to be concentrated and still leave unchanged the first moment of the area about any axis. For example, a thin metal plate will balance in a horizontal plane if it is supported at a point directly under its center of gravity.

The centroids of a few areas are obvious. In a symmetrical figure such as a circle or square, the centroid coincides with the geometric center of the figure.

It is common practice to denote a centroid distance by a bar over the coordinate distance. Thus $\bar{x}$ indicates the x-coordinate of the centroid.

105

SECOND MOMENT, OR MOMENT OF INERTIA, OF AN ELEMENT OF AREA

The second moment, or moment of inertia, of an element of area about any axis in the plane of the area is given by the product of the area of the element and the square of the perpendicular distance between the element and the axis. In the above figure, the moment of inertia dI_x of the element about the x-axis is

$$dI_x = y^2\, da$$

About the y-axis the moment of inertia is

$$dI_y = x^2\, da$$

SECOND MOMENT, OR MOMENT OF INERTIA, OF A FINITE AREA

The second moment, or moment of inertia, of a finite area about any axis in the plane of the area is given by the summation of the moments of inertia about that same axis of all of the elements of area contained in the finite area. This, too, is frequently found by means of an integral. If the moment of inertia of the finite area about the x-axis is denoted by I_x then we have

$$I_x = \int dI_x = \int y^2\, da \qquad I_y = \int dI_y = \int x^2\, da$$

For applications, see Problems 7.4, 7.6, 7.7, and 7.9.

UNITS

The units of moment of inertia are the fourth power of a length, perhaps in^4 or ft^4.

PARALLEL-AXIS THEOREM FOR MOMENT OF INERTIA OF A FINITE AREA

The parallel-axis theorem for moment of inertia of a finite area states that the moment of inertia of an area about any axis is equal to the moment of inertia about a parallel axis through the centroid of the area plus the product of the area and the square of the perpendicular distance between the two axes. For the area shown in Fig. 7-2, the axes x_G and y_G pass through the centroid of the plane area. The x- and y-axes are parallel axes located at distances x_1 and y_1 from the centroidal axes. Let A denote the area of the figure, I_{x_G} and I_{y_G} the moments of inertia about the axes through the centroid, and I_x and I_y the moments of inertia about the x- and y-axes. Then we have

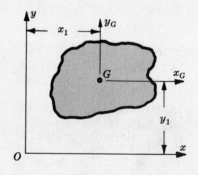

$$I_x = I_{x_G} + A(y_1)^2$$

$$I_y = I_{y_G} + A(x_1)^2$$

Fig. 7-2

This relation is derived in Problem 7.5. For applications, see Problems 7.6, 7.8, 7.10, 7.11, 7.12, 7.13, 7.14, and 7.20.

COMPOSITE AREAS

The moment of inertia of a composite area is the summation of the moments of inertia

of the component areas making up the whole. This frequently eliminates the necessity for integration if the area can be broken down into rectangles, triangles, circles, etc., for each of which the moment of inertia is known. See Problems 7.10 through 7.14.

RADIUS OF GYRATION

If the moment of inertia of an area A about the x-axis is denoted by I_x, then the radius of gyration r_x is defined by

$$r_x = \sqrt{I_x/A}$$

Similarly, the radius of gyration with respect to the y-axis is given by

$$r_y = \sqrt{I_y/A}$$

Since I is in units of length to the fourth power, and A is in units of length to the second power, then the radius of gyration has the units of length, say in. or ft. It is frequently useful for comparative purposes but has no physical significance. See Problems 7.12 and 7.13.

PRODUCT OF INERTIA OF AN ELEMENT OF AREA

The product of inertia of an element of area with respect to the x-y axes in the plane of the area is given by

$$dI_{xy} = xy \, da$$

where x and y are coordinates of the elemental area as shown in Fig. 7-1, page 105.

PRODUCT OF INERTIA OF A FINITE AREA

The product of inertia of a finite area with respect to the x-y axes in the plane of the area is given by the summation of the products of inertia about those same axes of all elements of area contained within the finite area. Thus

$$I_{xy} = \int xy \, da$$

From this, it is evident that I_{xy} may be positive, negative, or zero. For applications, see Problems 7.15 and 7.17.

UNITS

The units of product of inertia are the fourth power of a length, perhaps in^4 or ft^4.

PARALLEL-AXIS THEOREM FOR PRODUCT OF INERTIA OF A FINITE AREA

The parallel-axis theorem for product of inertia of a finite area states that the product of inertia of an area with respect to the x-y axes is equal to the product of inertia about a set of parallel axes passing through the centroid of the area plus the product of the area and the two perpendicular distances from the centroid to the x-y axes. For the area shown in Fig. 7-2, the axes x_G and y_G pass through the centroid of the plane area. The x- and y-axes are parallel axes located at distances x_1 and y_1 from the centroidal axes. Let A represent the area of the figure and $I_{x_G y_G}$ be the product of inertia about the axes through the centroid. Then we have

$$I_{xy} = I_{x_G y_G} + A x_1 y_1$$

This relation is derived in Problem 7.16.   For applications see Problems 7.17, 7.18, and 7.20.

COMPOSITE AREAS

The product of inertia of a composite area is the summation of the products of inertia of the component areas making up the whole.   This may eliminate the need for integration if the area can be broken down into rectangles, triangles, etc.   See Problems 7.17, 7.18, and 7.20.

PRINCIPAL MOMENTS OF INERTIA

At any point in the plane of an area there exist two perpendicular axes about which the moments of inertia of the area are maximum and minimum for that point.   These maximum and minimum values of moment of inertia are termed *principal moments of inertia* and are given by

$$(I_{x_1})_{\max} \;=\; \left(\frac{I_x + I_y}{2}\right) + \sqrt{\left(\frac{I_x - I_y}{2}\right)^2 + (I_{xy})^2}$$

$$(I_{x_1})_{\min} \;=\; \left(\frac{I_x + I_y}{2}\right) - \sqrt{\left(\frac{I_x - I_y}{2}\right)^2 + (I_{xy})^2}$$

These expressions are derived in Problem 7.19.   For application, see Problem 7.20.

PRINCIPAL AXES

The pair of perpendicular axes through a selected point about which the moments of inertia of a plane area are maximum and minimum are termed *principal axes*.   For application, see Problem 7.20.

The product of inertia vanishes if the axes are principal axes.   Also, from the integral defining product of inertia of a finite area, it is evident that if either the x-axis, or the y-axis, or both, are axes of symmetry, the product of inertia vanishes.   Thus, axes of symmetry are principal axes.

Solved Problems

7.1. Locate the centroid of a triangle.

Let us introduce the coordinate system shown in Fig. 7-3.

The y-coordinate of the centroid is defined by the equation

$$\bar{y} = \frac{\int y \, da}{A}$$

It is simplest to choose an element such that y is constant for all points in the element. The horizontal shaded area satisfies this condition and the area da of the element is $s \, dy$. Thus

$$\bar{y} = \frac{\int y \, s \, dy}{A}$$

The product $y \, s \, dy$ represents the first moment of the shaded element about the x-axis.

From similar triangles, $\dfrac{s}{b} = \dfrac{h - y}{h}$. Substituting this value of s in the above integral,

$$\bar{y} = \frac{\int_0^h y \frac{b}{h}(h - y)\, dy}{\frac{1}{2}bh} = \frac{2}{h^2}\int_0^h (hy - y^2)\, dy$$

$$= \frac{2}{h^2}\{h[y^2/2]_0^h - [y^3/3]_0^h\} = \frac{2}{h^2}\left(\frac{h^3}{2} - \frac{h^3}{3}\right) = \frac{1}{3}h$$

Knowing the distance of the centroid above each side of the triangle, we can locate it uniquely.

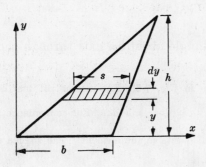

Fig. 7-3

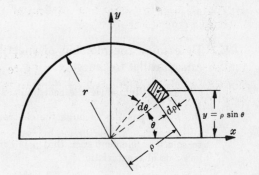

Fig. 7-4

7.2. Locate the centroid of a semicircle.

The polar coordinate system shown in Fig. 7-4 will be a logical choice for such a contour. The shaded element of area is approximately a rectangle and its area is given by $\rho \, d\theta \, d\rho$. The y-coordinate of the centroid is given by

$$\bar{y} = \frac{\int y \, da}{\int da} = \frac{\int_0^\pi \int_0^r (\rho \sin\theta)(\rho \, d\theta \, d\rho)}{\int_0^\pi \int_0^r \rho \, d\theta \, d\rho} = \frac{\int_0^\pi [\rho^3/3]_0^r \sin\theta \, d\theta}{\int_0^\pi [\rho^2/2]_0^r \, d\theta} =$$

$$= \quad \frac{\dfrac{r^3}{3} \displaystyle\int_0^{\pi} \sin\theta\, d\theta}{\dfrac{r^2}{2} \displaystyle\int_0^{\pi} d\theta} \quad = \quad \frac{2r}{3\pi}[-\cos\theta]_0^{\pi} \quad = \quad \frac{4r}{3\pi}$$

Of course, by symmetry, $\bar{x} = 0$.

7.3. Locate the centroid of the shaded area remaining after one corner and the semicircular area have been removed from the originally rectangular area in Fig. 7-5.

The shaded area in Fig. 7-5 consists of (1) a rectangle 6 in $\times$ 12 in, minus (2) a triangle 6 in $\times$ 3 in, minus (3) a semicircular area. Since the centroids of (2) and (3) were determined in Problems 7.1 and 7.2 respectively, integration is not necessary and a finite summation may be used.

The y-coordinate of the centroid is given by $\bar{y} = \dfrac{\displaystyle\int y\, da}{A}$. The numerator, representing the first moment of the shaded area about the x-axis, may be evaluated as the first moment of the rectangle, minus that of the triangle, minus that of the semicircle. Thus

$$\bar{y} \quad = \quad \frac{12(6)(3) - \frac{1}{2}(3)(6)(4) - \frac{1}{2}\pi(2)^2\left[6 - \dfrac{4(2)}{3\pi}\right]}{12(6) - \frac{1}{2}(3)(6) - \frac{1}{2}\pi(2)^2} \quad = \quad 2.60 \text{ in.}$$

Similarly, the x-coordinate of the centroid may be located by $\bar{x} = \dfrac{\displaystyle\int x\, da}{A}$. The numerator here represents the first moment of the rectangle, minus that of the triangle, minus that of the semicircle about the y-axis. Thus

$$\bar{x} \quad = \quad \frac{12(6)(6) - \frac{1}{2}(3)(6)(1) - \frac{1}{2}\pi(2)^2(8)}{12(6) - \frac{1}{2}(3)(6) - \frac{1}{2}\pi(2)^2} \quad = \quad 6.58 \text{ in.}$$

7.4. Determine the moment of inertia of a rectangle about an axis through the centroid and parallel to the base.

Let us introduce the coordinate system shown in Fig. 7-6. The moment of inertia I_{x_G} about the x-axis passing through the centroid is given by $I_{x_G} = \displaystyle\int y^2\, da$. For convenience it is logical to select an element such that y is constant for all points in the element. The shaded area shown has this characteristic.

$$I_{x_G} \quad = \quad \int_{-h/2}^{h/2} y^2 b\, dy \quad = \quad b\left[\frac{y^3}{3}\right]_{-h/2}^{h/2} \quad = \quad \tfrac{1}{12}bh^3$$

This quantity has the dimension of a length to the fourth power, perhaps in^4.

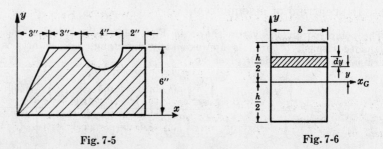

Fig. 7-5   Fig. 7-6

7.5.   Derive the parallel-axis theorem for moments of inertia of a plane area.

Let us consider the plane area A shown in Fig. 7-7. The axes x_G and y_G pass through its centroid, whose location is presumed to be known. The axes x and y are located at known distances y_1 and x_1 respectively, from the axes through the centroid.

For the element of area da the moment of inertia about the x-axis is given by

$$dI_x = (y_1 + y')^2\, da$$

For the entire area A the moment of inertia about the x-axis is

$$I_x = \int dI_x = \int (y_1 + y')^2\, da = \int (y_1)^2\, da + 2\int y_1 y'\, da + \int (y')^2\, da$$

The first integral on the right is equal to $y_1^2 \int da = y_1^2 A$ because y_1 is a constant. The second integral on the right is equal to $2y_1 \int y'\, da = 2y_1(0) = 0$ because the axis from which y' is measured passes through the centroid of the area. The third integral on the right is equal to I_{x_G}, i.e. the moment of inertia of the area about the horizontal axis through the centroid.   Thus

$$I_x = I_{x_G} + A(y_1)^2$$

A similar consideration in the other direction would show that

$$I_y = I_{y_G} + A(x_1)^2$$

This is the parallel-axis theorem for plane areas. It is to be noted that one of the axes involved in each equation must pass through the centroid of the area. In words, this may be stated as follows: The moment of inertia of an area with reference to an axis not through the centroid of the area is equal to the moment of inertia about a parallel axis through the centroid of the area plus the product of the same area and the square of the distance between the two axes.

The moment of inertia always has a positive value, with a minimum value for axes through the centroid of the area in question.

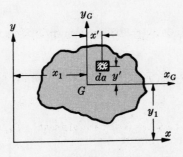

Fig. 7-7

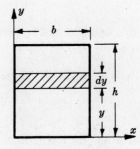

Fig. 7-8

7.6.   Find the moment of inertia of a rectangle about an axis coinciding with the base.

The coordinate system shown in Fig. 7-8 is convenient. By definition the moment of inertia about the x-axis is given by $I_x = \int y^2\, da$. For the element shown y is constant for all points in the element.   Hence

$$I_x = \int_0^h y^2 b\, dy = b[y^3/3]_0^h = \tfrac{1}{3}bh^3$$

This solution could also have been obtained by applying the parallel-axis theorem to the result obtained in Problem 7.4. This states that the moment of inertia about the base is equal to the moment of inertia about the horizontal axis through the centroid plus the product of the area and the square of the distance between these two axes.   Thus

$$I_x = I_{x_G} + A(y_1)^2 = \tfrac{1}{12}bh^3 + bh\left(\frac{h}{2}\right)^2 = \tfrac{1}{3}bh^3$$

7.7. Determine the moment of inertia of a triangle about an axis coinciding with the base.

Let us introduce the coordinate system shown in Fig. 7-9. The moment of inertia about the horizontal base is

$$I_x = \int y^2\, da$$

For the shaded element shown the quantity y is constant for all points in the element.   Thus

$$I_x = \int_0^h y^2 s\, dy$$

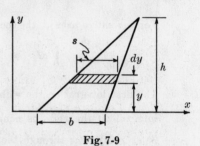

Fig. 7-9

By similar triangles, $s/b = (h - y)/h$, so that

$$I_x = \int_0^h y^2 \frac{b}{h}(h - y)\, dy = \frac{b}{h}\left[h\int_0^h y^2\, dy - \int_0^h y^3\, dy \right] = \frac{1}{12}bh^3$$

7.8. Determine the moment of inertia of a triangle about an axis through the centroid and parallel to the base.

Let the x_G-axis pass through the centroid and take the x-axis to coincide with the base as shown in Fig. 7-10.

From Problem 7.1 the x_G-axis is located a distance of $h/3$ above the base. Also, the parallel-axis theorem tells us that

$$I_x = I_{x_G} + A(y_1)^2$$

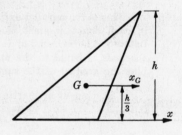

Fig. 7-10

But I_x was determined in Problem 7.7, and A and y_1 ($= h/3$) are known. Hence we may solve for the desired unknown, I_{x_G}.   Substituting,

$$\frac{1}{12}bh^3 = I_{x_G} + \frac{1}{2}bh\left(\frac{h}{3}\right)^2 \qquad \text{or} \qquad I_{x_G} = \frac{1}{36}bh^3$$

7.9. Determine the moment of inertia of a circle about a diameter.

Let us select the shaded element of area shown in Fig. 7-11 and work with the polar coordinate system. The radius of the circle is r.

To find I_x we have the definition $I_x = \int y^2\, da$.

But $y = \rho \sin \theta$ and $da = \rho\, d\theta\, d\rho$.   Hence

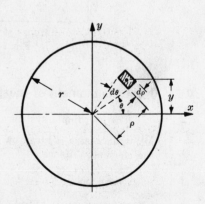

$$I_x = \int_0^{2\pi}\int_0^r \rho^2 \sin^2 \theta\; \rho\, d\theta\, dp = \int_0^{2\pi} \sin^2 \theta\, d\theta\, [\tfrac{1}{4}\rho^4]_0^r$$

$$= \frac{r^4}{4}\int^{2\pi} \sin^2 \theta\, d\theta = \frac{\pi r^4}{4}$$

Fig. 7-11

If D denotes the diameter of the circle, then $D = 2r$ and $I_x = \pi D^4/64$. This is half the value of the polar moment of inertia of a solid circular area (see Problem 5.1).

Hence the moment of inertia of a semicircular area about an axis coinciding with its base is

$$I_x = \frac{1}{2} \frac{\pi D^4}{64} = \frac{\pi D^4}{128}$$

7.10. Determine the moment of inertia of the T-section shown in Fig. 7-12, about a horizontal axis passing through the centroid.

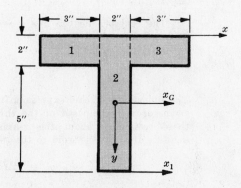

It is first necessary to locate the centroid of the area. To do this, we introduce the x-y coordinate system shown. By definition, the y-coordinate of the centroid is given by

$$\bar{y} = \frac{\int y\,da}{A}$$

The numerator of this expression represents the first moment of the entire area about the x-axis. This may be calculated by multiplying the area of each of the three component rectangles 1, 2, and 3 by the distance from the x-axis to the centroid of the particular rectangle. Thus

Fig. 7-12

$$\bar{y} = \frac{3(2)(1) + 7(2)(3.5) + 3(2)(1)}{3(2) + 7(2) + 3(2)} = +2.35 \text{ in.}$$

Hence the centroid is located 2.35 in. below the x-axis. The horizontal axis passing through this point is denoted by x_G in the above figure.

There are several techniques for determining the required moment of inertia. One is to calculate the moment of inertia of the entire area about the x-axis, then use the parallel-axis theorem to transfer this result to the x_G-axis.

The moment of inertia about the x-axis is found as the sum of the moments of inertia about this same axis of each of the three component rectangles. The expression for the moment of inertia of a rectangle about an axis coinciding with its base was derived in Problem 7.6. Note that it is easiest to subdivide the T-section into the three rectangles shown, rather than in any other manner, because the moment of inertia of each about the x-axis is known from Problem 7.6. Thus

$$I_x = (1/3)(3)(2)^3 + (1/3)(2)(7)^3 + (1/3)(3)(2)^3 = 245 \text{ in}^4$$

The parallel-axis theorem may now be used to find the moment of inertia of the entire figure about the x_G-axis. Thus

$$I_x = I_{x_G} + A(y_1)^2, \quad 245 = I_{x_G} + 26(2.35)^2 \quad \text{and} \quad I_{x_G} = 101 \text{ in}^4$$

7.11. Determine the moment of inertia of the T-section of Problem 7.10 about a horizontal axis x_1 through its lower extremity.

This axis is located $7 - 2.35 = 4.65$ in. below the horizontal axis through the centroid. The parallel-axis theorem may be used to transfer the known moment of inertia from the x_G-axis to the x_1-axis. Thus

$$I_{x_1} = I_{x_G} + A(y_1)^2 = 101 + 26(4.65)^2 = 664 \text{ in}^4$$

Note carefully that the parallel-axis theorem can only be used if one of the two axes concerned passes through the centroid of the area. For example, it is *not* permissible to transfer from the

x-axis to the x_1-axis merely by adding the product of the area and the square of the distance between these axes. The reason this is not valid is that neither of these axes passes through the centroid of the figure.

7.12. Determine the moment of inertia and also the radius of gyration of the channel section in Fig. 7-13 about a horizontal axis through the centroid.

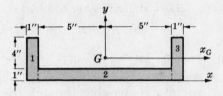

The centroid lies on the y-axis and its location is given by

Fig. 7-13

$$\bar{y} = \frac{\int y \, da}{A}$$

The numerator of this expression represents the first moment of the area about the x-axis. The entire area is composed of the three component rectangles shown. The first moment of each of these rectangles about the x-axis is given by the product of its area and the perpendicular distance from its centroid to the x-axis. Thus

$$\bar{y} = \frac{1(5)(2.5) + 10(1)(0.5) + 1(5)(2.5)}{1(5) + 10(1) + 1(5)} = 1.5 \text{ in.}$$

The horizontal axis passing through the centroid is denoted by x_G in the above figure.

It is convenient to first determine the moment of inertia with respect to the x-axis. For each of the three component rectangles the moment of inertia about an axis through its base was found in Problem 7.6 to be $I_x = bh^3/3$. For the entire figure,

$$I_x = \tfrac{1}{3}(1)(5)^3 + \tfrac{1}{3}(10)(1)^3 + \tfrac{1}{3}(1)(5)^3 = 86.6 \text{ in}^4$$

From the parallel-axis theorem,

$$I_x = I_{x_G} + A(y_1)^2, \quad 86.6 = I_{x_G} + 20(1.5)^2 \quad \text{and} \quad I_{x_G} = 41.6 \text{ in}^4$$

The radius of gyration with respect to the x_G-axis is $r_{x_G} = \sqrt{I_{x_G}/A} = \sqrt{41.6/20} = 1.45 \text{ in.}$

7.13. Determine the moment of inertia and also the radius of gyration of the I-section shown in Fig. 7-14, about a horizontal axis passing through the centroid.

To locate the centroid, which lies on the y-axis, we have

$$\bar{y} = \frac{\int y \, da}{A}$$

The entire section is divided into the five component rectangles shown and the numerator of the above fraction may then be evaluated by a numerical summation. Thus

$$\bar{y} = \frac{4(2)(1) + 11(2)(5.5) + 4(2)(1) + 3(2)(10) + 3(2)(10)}{4(2) + 11(2) + 4(2) + 3(2) + 3(2)}$$

$$= 5.14 \text{ in.}$$

The horizontal axis passing through the centroid is denoted by x_G in the figure.

We shall first determine the moment of inertia with respect to the x-axis. For the rectangles 1, 2, and 3 the moment of inertia about this axis is given by $I_x = bh^3/3$.

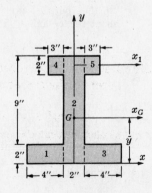

Fig. 7-14

For the rectangles 4 and 5 it is first necessary to determine the moment of inertia about a horizontal axis x_1 passing through the centroid of these rectangles, then apply the parallel-axis theorem to transfer this result to the x-axis.

For the entire figure we thus have

$$I = \tfrac{1}{3}(4)(2)^3 + \tfrac{1}{3}(2)(11)^3 + \tfrac{1}{3}(4)(2)^3 + [\tfrac{1}{12}(3)(2)^3 + 3(2)(10)^2]2 = 2113 \text{ in}^4$$

From the parallel-axis theorem,

$$I_x = I_{x_G} + A(y_1)^2, \quad 2113 = I_{x_G} + 50(5.14)^2 \quad \text{and} \quad I_{x_G} = 793 \text{ in}^4$$

The radius of gyration with respect to the x_G-axis is

$$r_{x_G} = \sqrt{I_{x_G}/A} = \sqrt{793/50} = 3.99 \text{ in.}$$

7.14. Determine the moment of inertia of the hollowed rectangular area of Fig. 7-15 about a horizontal axis through its centroid.

The centroid lies on the y-axis and its location is given by

$$\bar{y} = \frac{\displaystyle\int y \, da}{A}$$

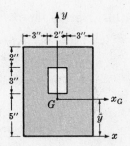

The numerator may be evaluated as the first moment of the entire 8 in $\times$ 10 in rectangle about the x-axis, minus the first moment of the 2 in $\times$ 3 in rectangle that has been removed. Thus

$$\bar{y} = \frac{8(10)(5) - 2(3)(6.5)}{8(10) - 2(3)} = 4.88 \text{ in.}$$

Fig. 7-15

The horizontal axis through the centroid is denoted by x_G.

We shall first compute the moment of inertia of the entire 8 in $\times$ 10 in rectangle about the x_G-axis. This is done by finding its moment of inertia about a horizontal axis through the centroid of this rectangle (assuming that the 2 in $\times$ 3 in hole is not present), then transferring this result to the x_G-axis. For the entire 8 in $\times$ 10 in rectangle this application of the parallel-axis theorem gives

$$I'_{x_G} = (1/12)(8)(10)^3 + 8(10)(5 - 4.88)^2 = 668 \text{ in}^4$$

Similarly for the 2 in $\times$ 3 in rectangle that has been removed, the moment of inertia of it with respect to the x_G-axis is found by computing the moment of inertia with respect to a horizontal axis through the centroid of this 2 in $\times$ 3 in area then transferring this result to the x_G-axis. This yields

$$I''_{x_G} = (1/12)(2)(3)^3 + 2(3)(6.5 - 4.88)^2 = 20.3 \text{ in}^4$$

Consequently the moment of inertia of the hollow rectangular area is

$$I_x = I'_{x_G} - I''_{x_G} = 668 - 20.3 = 647.7 \text{ in}^4$$

7.15. Determine the product of inertia of a rectangle with respect to the x-y axes indicated in Fig. 7-16.

We employ the definition $I_{xy} = \displaystyle\int xy \, da$ and consider the shaded element shown. Integrating:

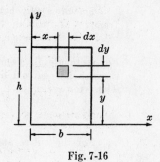

$$I_{xy} = \int_{y=0}^{y=h} \int_{x=0}^{x=b} xy \, dx \, dy = \int_{y=0}^{y=h} \left[\frac{x^2}{2}\right]_0^b y \, dy$$

$$= \frac{b^2}{2}\left[\frac{y^2}{2}\right]_0^h = \frac{b^2 h^2}{4} \qquad\qquad (1)$$

Fig. 7-16

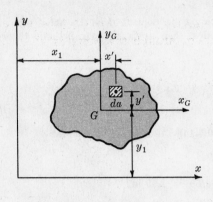

Fig. 7-17

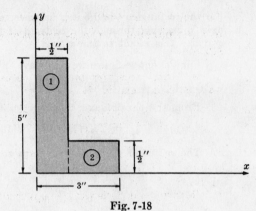

Fig. 7-18

7.16. Derive the parallel-axis theorem for product of inertia of a plane area.

In Fig. 7-17, the axes x_G and y_G pass through the centroid of the area A. The axes x and y are located the known distances y_1 and x_1, respectively, from the axes through the centroid.

For the element of area da the product of inertia with respect to the x-y axes is given by

$$dI_{xy} = (x_1 + x')(y_1 + y')\, dx\, dy$$

For the entire area the product of inertia with respect to the x-y axes becomes

$$I_{xy} = \int dI_{xy} = \iint (x_1 + x')(y_1 + y')\, dx\, dy$$

$$= \iint x_1 y_1\, dx\, dy + \iint x' y_1\, dx\, dy + \iint x_1 y'\, dx\, dy + \iint x' y'\, dx\, dy$$

The first integral on the right side equals $x_1 y_1 A$ since x_1 and y_1 are constants. The second and third integrals vanish because x' and y' are measured from the axes through the centroid of the area A. The fourth integral is equal to $I_{x_G y_G}$, i.e. the product of inertia of the area with respect to axes through its centroid and parallel to the x-y axes. Thus, we have

$$I_{xy} = x_1 y_1 A + I_{x_G y_G} \tag{1}$$

This is the parallel-axis theorem for product of inertia of a plane area. It is to be noted that the x_G-y_G axes must pass through the centroid of the area. Also, x_1 and y_1 are positive only when the x-y coordinates have the location relative to the x_G-y_G system indicated in Fig. 7-17. Thus, care must be taken with regard to the algebraic signs of x_1 and y_1.

7.17. Determine I_{xy} for the angle section indicated in Fig. 7-18.

The area may be divided into the component rectangles as shown. For rectangle 1 we have from (1) of Problem 7.15:

$$(I_{xy})_1 = \tfrac{1}{4}(\tfrac{1}{2})^2(5)^2 = 1.56 \text{ in}^4$$

For rectangle 2 we employ (1) of Problem 7.16. The product of inertia of rectangle 2 about axes through its centroid and parallel to the x-y axes vanishes because these are axes of symmetry. Thus, for rectangle 2, $I_{x_G y_G} = 0$. The parallel-axis theorem of Problem 7.16 thus becomes

$$(I_{xy})_2 = (1\tfrac{3}{4})(\tfrac{1}{4})(2\tfrac{1}{2})(\tfrac{1}{2}) + 0 = 0.55 \text{ in}^4$$

For the entire angle section we thus have

$$I_{xy} = 1.56 + 0.55 = 2.11 \text{ in}^4$$

7.18. Determine the product of inertia of the angle section of Problem 7.17 with respect to axes parallel to the x- and y-axes and passing through the centroid of the angle section.

It is first necessary to locate the centroid of the area, i.e. we must find $\bar{x}$ and $\bar{y}$. We have

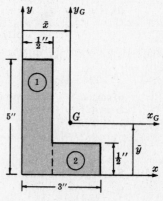

$$\bar{x} = \frac{5(\tfrac{1}{2})(\tfrac{1}{4}) + 2.5(\tfrac{1}{2})(1.75)}{5(\tfrac{1}{2}) + 2.5(\tfrac{1}{2})} = 0.752 \text{ in.}$$

$$\bar{y} = \frac{5(\tfrac{1}{2})(2.5) + 2.5(\tfrac{1}{2})(\tfrac{1}{4})}{5(\tfrac{1}{2}) + 2.5(\tfrac{1}{2})} = 1.75 \text{ in.}$$

Now we employ the parallel-axis theorem of Problem 7.16, i.e.

$$I_{xy} = x_1 y_1 A + I_{x_G y_G}$$

In Problem 7.17 we found $I_{xy} = 2.11$ in⁴. Thus

$$2.11 = 0.752(1.75)(3.75) + I_{x_G y_G} \quad \text{whence} \quad I_{x_G y_G} = -2.83 \text{ in}^4$$

Fig. 7-19

7.19. Consider a plane area A and assume that I_x, I_y and I_{xy} are known. Determine the moments of inertia I_{x_1} and I_{y_1} as well as the product of inertia $I_{x_1 y_1}$ for the set of orthogonal axes x_1-y_1 oriented as shown in Fig. 7-20. Determine also the maximum and minimum values of I_{x_1}.

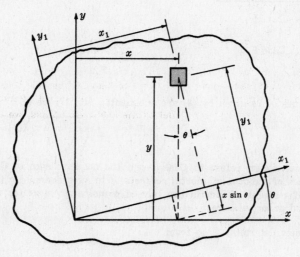

Fig. 7-20

The moment of inertia of the area with respect to the x_1-axis is

$$I_{x_1} = \int y_1^2 \, da = \int (y \cos \theta - x \sin \theta)^2 \, da$$

$$= \cos^2 \theta \int y^2 \, da + \sin^2 \theta \int x^2 \, da - 2 \sin \theta \cos \theta \int xy \, da$$

$$= I_x \cos^2 \theta + I_y \sin^2 \theta - 2 I_{xy} \sin \theta \cos \theta$$

$$= I_x \left(\frac{1 + \cos 2\theta}{2} \right) + I_y \left(\frac{1 - \cos 2\theta}{2} \right) - I_{xy} \sin 2\theta$$

Or

$$I_{x_1} = \left(\frac{I_x + I_y}{2}\right) + \left(\frac{I_x - I_y}{2}\right)\cos 2\theta - I_{xy}\sin 2\theta \qquad (1)$$

Analogously, I_{y_1} may be obtained from (1) by replacing θ by $\theta + \pi/2$ to yield

$$I_{y_1} = \left(\frac{I_x + I_y}{2}\right) - \left(\frac{I_x - I_y}{2}\right)\cos 2\theta + I_{xy}\sin 2\theta \qquad (2)$$

The value of θ that renders I_{x_1} maximum or minimum is found by setting the derivative of equation (1) with respect to θ equal to zero. Thus, since I_x, I_y, and I_{xy} are constants we have from (1):

$$\frac{dI_{x_1}}{d\theta} = -(I_x - I_y)\sin 2\theta - 2I_{xy}\cos 2\theta = 0$$

Solving:
$$\tan 2\theta = -\frac{I_{xy}}{\left(\dfrac{I_x - I_y}{2}\right)} \qquad (3)$$

Equation (3) has the following convenient graphical interpretation:

Case I   Case II

Fig. 7-21

If now the values of 2θ given by (3) are substituted into (1) we obtain

$$(I_{x_1})_{\substack{\max \\ \min}} = \left(\frac{I_x + I_y}{2}\right) \pm \sqrt{\left(\frac{I_x - I_y}{2}\right)^2 + (I_{xy})^2} \qquad (4)$$

where the positive sign refers to Case I and the negative sign to Case II. These maximum and minimum values of moment of inertia correspond to axes defined by (3). The maximum and minimum values of moment of inertia are termed *principal moments of inertia* and the corresponding axes are termed *principal axes*.

We may now determine $I_{x_1 y_1}$ from

$$I_{x_1 y_1} = \int x_1 y_1\, da$$

$$= \int (x \cos\theta + y \sin\theta)(y \cos\theta - x \sin\theta)\, da$$

$$= \cos^2\theta \int xy\, da - \sin^2\theta \int xy\, da$$

$$\qquad + \sin\theta \cos\theta \int y^2\, da - \sin\theta \cos\theta \int x^2\, da$$

$$= I_{xy}(\cos^2\theta - \sin^2\theta) + (I_x - I_y)\sin\theta \cos\theta$$

$$= \left(\frac{I_x - I_y}{2}\right)\sin 2\theta + I_{xy}\cos 2\theta \qquad (5)$$

From (5), $I_{x_1 y_1}$ vanishes if

$$\tan 2\theta = -\frac{I_{xy}}{\left(\dfrac{I_x - I_y}{2}\right)}$$

which is identical to condition (3). Since (3) defined principal axes, it follows that the product of inertia vanishes for principal axes.

7.20. A structural aluminum 6 Z 5.42 section has the nominal dimensions indicated in Fig. 7-22. Determine I_x, I_y, I_{xy} and also the maximum and minimum values of the moment of inertia with respect to axes through the point O.

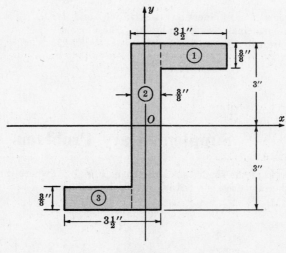

Fig. 7-22

The section may be divided into the component rectangles 1, 2, and 3 as indicated. The result obtained in Problem 7.4, together with the parallel-axis theorem given in Problem 7.5, may be used to determine I_x and I_y:

$$I_x = \tfrac{1}{12}(\tfrac{3}{8})(6)^3 + 2[\tfrac{1}{12}(3\tfrac{1}{8})(\tfrac{3}{8})^3 + (3\tfrac{1}{8})(\tfrac{3}{8})(2\tfrac{13}{16})^2] = 25.27 \text{ in}^4$$

$$I_y = \tfrac{1}{12}(6)(\tfrac{3}{8})^3 + 2[\tfrac{1}{12}(\tfrac{3}{8})(3\tfrac{1}{8})^3 + (\tfrac{3}{8})(3\tfrac{1}{8})(1\tfrac{3}{4})^2] = 9.08 \text{ in}^4$$

The product of inertia with respect to the x-y axes may be determined through use of the parallel-axis theorem for product of inertia as given in Problem 7.16. It is to be noted that the product of inertia of each of the component rectangles about axes through the centroid of each component and parallel to the x-y axes vanishes because these are axes of symmetry. Hence, from (1) of Problem 7.16 we have for the entire Z-section

$$I_{xy} = 2[(\tfrac{7}{4})(2\tfrac{13}{16})(3\tfrac{1}{8})(\tfrac{3}{8})] = 11.6 \text{ in}^4$$

The maximum and minimum values of moment of inertia with respect to axes through the point O may be found from (4) of Problem 7.19. From that equation

$$(I_{x_1})^{\text{max}}_{\text{min}} = \left(\frac{I_x + I_y}{2}\right) \pm \sqrt{\left(\frac{I_x - I_y}{2}\right)^2 + (I_{xy})^2}$$

$$= \left(\frac{25.27 + 9.08}{2}\right) \pm \sqrt{\left(\frac{25.27 - 9.08}{2}\right)^2 + (11.6)^2}$$

$$(I_{x_1})_{max} = 31.38 \text{ in}^4 \qquad\qquad (1)$$

$$(I_{x_1})_{min} = 2.98 \text{ in}^4 \qquad\qquad (2)$$

The orientation of these principal moments of inertia is found from (3) of Problem 7.19 to be

$$\tan 2\theta = -\frac{I_{xy}}{\left(\dfrac{I_x - I_y}{2}\right)}$$

$$= -\frac{11.6}{\left(\dfrac{25.27 \div 9.08}{2}\right)}$$

$$\theta = -27°20', \quad -117°20' \qquad (3)$$

The principal moments of inertia given in (1) and (2) correspond to the principal axes given by (3). These principal axes are represented by the dashed lines in Fig. 7-23.

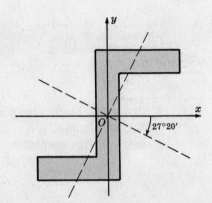

Fig. 7-23

Supplementary Problems

7.21. Locate the centroid of the shaded area shown in Fig. 7-24, where the rectangular portion has been removed from the semicircle.   *Ans.* $\bar{x} = 0$, $\bar{y} = 2.80$ in.

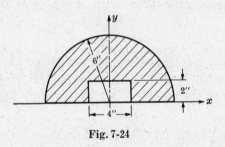

Fig. 7-24

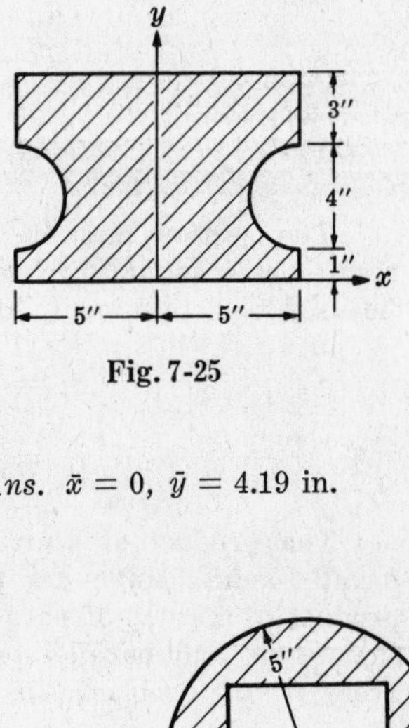

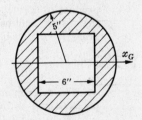

Fig. 7-25

7.22. Locate the centroid of the shaded area shown in Fig. 7-25.   *Ans.* $\bar{x} = 0$, $\bar{y} = 4.19$ in.

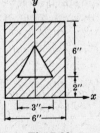

Fig. 7-26

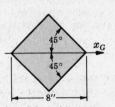

Fig. 7-27

Fig. 7-28

7.23. Locate the centroid of the shaded area remaining after the equilateral triangle has been removed from the rectangle as shown in Fig. 7-26.   *Ans.* $\bar{x} = 0$, $\bar{y} = 4.10$ in.

7.24. Determine the moment of inertia of an equilateral triangle 6 in. on a side about an axis through its centroid and parallel to the base.   *Ans.* $I_{x_G} = 23.4 \text{ in}^4$

7.25. Determine the moment of inertia of one quadrant of a circle of radius 2 in. about a diameter coinciding with one side of the quadrant.   *Ans.* $I = 3.14 \text{ in}^4$

7.26. Determine the moment of inertia of the diamond-shaped figure shown in Fig. 7-27, with respect to the horizontal axis of symmetry.   *Ans.* $I_{x_G} = 85.4 \text{ in}^4$

7.27. Refer to Fig. 7-28.   Determine the moment of inertia about the x_G-axis of the shaded area remaining after the square area has been removed from the circle.   The x_G-axis is an axis of symmetry.   *Ans.* $I_{x_G} = 383 \text{ in}^4$

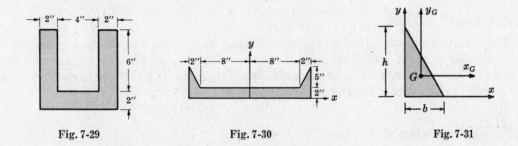

Fig. 7-29   Fig. 7-30   Fig. 7-31

7.28. Determine the moment of inertia of a channel-type section about a horizontal axis through the centroid.   Refer to Fig. 7-29.   What is the radius of gyration about this same axis?

Ans. $I_{x_G} = 231 \text{ in}^4$, $r_{x_G} = 2.40 \text{ in.}$

7.29. Locate the centroid of the channel-type section shown in Fig. 7-30 and determine the moment of inertia of the cross-sectional area about a horizontal axis through the centroid.

Ans. $\bar{y} = 1.53 \text{ in.}$, $I_{x_G} = 84 \text{ in}^4$

7.30. Determine the product of inertia of a triangle with respect to the x-y axes indicated in Fig. 7-31.
Ans. $b^2h^2/24$

7.31. Determine the product of inertia of the triangle shown in Fig. 7-31 with respect to the axes x_G and y_G passing through the centroid.   *Ans.* $-b^2h^2/72$

7.32. For the plane area in Fig. 7-32 determine the moments of of inertia and product of inertia with respect to the x_G-y_G axes passing through the centroid.   Also, determine the principal moments of inertia with respect to the centroid.

Ans. $I_{x_G} = 1021 \text{ in}^4$

  $I_{y_G} = 372 \text{ in}^4$

  $I_{x_G y_G} = -138 \text{ in}^4$

  $(I_{x_1})_{max} = 1057 \text{ in}^4$

  $(I_{x_1})_{min} = 336 \text{ in}^4$

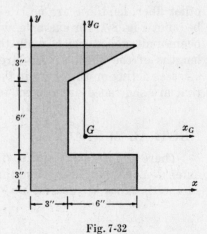

Fig. 7-32

Chapter 8

Stresses in Beams

TYPES OF LOADS ACTING ON BEAMS

Either forces or couples that lie in a plane containing the longitudinal axis of the beam may act upon the member. The forces are understood to act perpendicular to the longitudinal axis, and the plane containing the forces is assumed to be a plane of symmetry of the beam.

EFFECTS OF LOADS

The effects of these forces and couples acting on a beam are (a) to impart deflections perpendicular to the longitudinal axis of the bar and (b) to set up both normal and shearing stresses on any cross-section of the beam perpendicular to its axis. Beam deflections will be considered in Chapters 9, 10, and 11.

TYPES OF BENDING

If couples are applied to the ends of the beam and no forces act on the bar, then the bending is termed *pure bending*. For example, in Fig. 8-1 the portion of the beam between the two downward forces is subject to pure bending. Bending produced by forces that do not form couples is called *ordinary bending*. A beam subject to pure bending has only normal stresses with no shearing stresses set up in it; a beam subject to ordinary bending has both normal and shearing stresses acting within it.

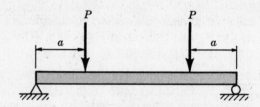

Fig. 8-1

NATURE OF BEAM ACTION

It is convenient to imagine a beam to be composed of an infinite number of thin longitudinal rods or fibers. Each longitudinal fiber is assumed to act independently of every other fiber, i.e. there are no lateral pressures or shearing stresses between the fibers. The beam of Fig. 8-1, for example, will deflect downward and the fibers in the lower part of the beam undergo extension, while those in the upper part are shortened. These changes in the lengths of the fibers set up stresses in the fibers. Those that are extended have tensile stresses acting on the fibers in the direction of the longitudinal axis of the beam, while those that are shortened are subject to compressive stresses.

NEUTRAL SURFACE

There always exists one surface in the beam containing fibers that do not undergo any extension or compression, and thus are not subject to any tensile or compressive stress. This surface is called the *neutral surface* of the beam.

NEUTRAL AXIS

The intersection of the neutral surface with any cross-section of the beam perpendicular to its longitudinal axis is called the *neutral axis*. All fibers on one side of the neutral axis are in a state of tension, while those on the opposite side are in compression.

BENDING MOMENT

The algebraic sum of the moments of the external forces to one side of any cross-section of the beam about an axis through that section is called the bending moment at that section. This concept was discussed in Chapter 6.

ELASTIC BENDING OF BEAMS

The following remarks apply *only if* all fibers in the beam are acting within the elastic range of action of the material.

NORMAL STRESSES IN BEAMS

For any beam having a longitudinal plane of symmetry and subject to a bending moment M at a certain cross-section, the normal stress acting on a longitudinal fiber at a distance y from the neutral axis of the beam (see Fig. 8-2) is given by

$$\sigma = \frac{My}{I}$$

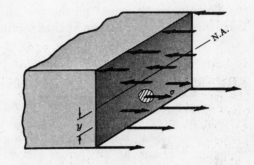

Fig. 8-2

where I denotes the moment of inertia of the cross-sectional area about the neutral axis. This quantity was discussed in Chapter 7. The der-ivation of this equation is discussed in detail in Problem 8.1. For applications see Problems 8.2–8.16. These stresses vary from zero at the neutral axis of the beam to a maximum at the outer fibers as shown. The stresses are tensile on one side of the neutral axis, compressive on the other. These stresses are also called bending, flexural, or fiber stresses.

LOCATION OF THE NEUTRAL AXIS

When the beam action is entirely elastic the neutral axis passes through the centroid of the cross-section. Hence, the moment of inertia I appearing in the above equation for normal stress is the moment of inertia of the cross-sectional area about an axis through the centroid of the cross-section of the beam.

SECTION MODULUS

At the outer fibers of the beam the value of the coordinate y is frequently denoted by the symbol c. In that case the maximum normal stresses are given by

$$\sigma = \frac{Mc}{I} \quad \text{or} \quad \sigma = \frac{M}{I/c}$$

The ratio I/c is called the *section modulus* and is usually denoted by the symbol Z. The units are in^3 or ft^3. The maximum bending stresses may then be represented as

$$\sigma = \frac{M}{Z}$$

This form is convenient because values of Z are available in handbooks for a wide range of standard structural steel shapes. See Problems 8.5, 8.9, and 8.13.

ASSUMPTIONS

In the derivation of the above expression for normal stresses it is assumed that a plane section of the beam normal to its longitudinal axis prior to loading remains plane after the forces and couples have been applied. Further, it is assumed that the beam is initially straight and of uniform cross-section and that the moduli of elasticity in tension and compression are equal. Again, it is to be emphasized that no fibers of the beam are stressed beyond the proportional limit.

SHEARING FORCE

The algebraic sum of all the vertical forces to one side of any cross-section of the beam is called the shearing force at that section. This concept was discussed in Chapter 6.

SHEARING STRESSES IN BEAMS

For any beam subject to a shearing force V (expressed in pounds) at a certain cross-section, both vertical and horizontal shearing stresses τ are set up. The magnitudes of the vertical shearing stresses at any cross-section are such that these stresses have the shearing force V as a resultant. In the cross-section of the beam shown in Fig. 8-3, the vertical plane of symmetry contains the applied forces and the neutral axis passes through the centroid of the section. The coordinate y is measured from the neutral axis. The moment of inertia of the *entire* cross-sectional area about the neutral axis is denoted by I. The shearing stress on all fibers a distance y_0 from the neutral axis is given by the formula

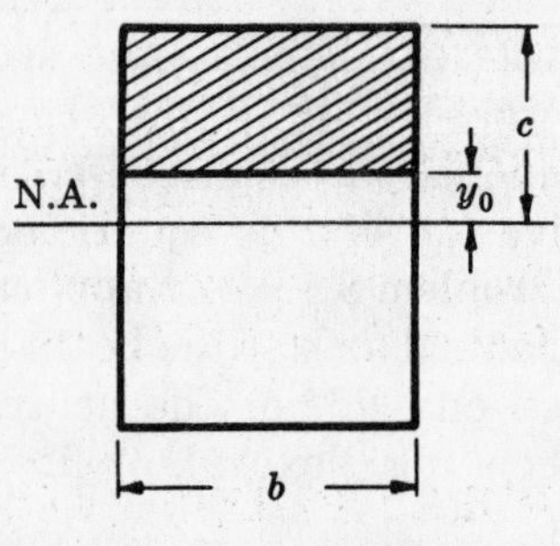

Fig. 8-3

$$\tau = \frac{V}{Ib} \int_{y_0}^{c} y \, da \qquad (1)$$

where b denotes the width of the beam at the location where the shearing stress is being calculated. This expression is derived in Problem 8.17. For applications see Problems 8.18–8.21. The integral in (1) represents the first moment of the shaded area of the cross-section about the neutral axis. This quantity was discussed in detail in Chapter 7. More generally, the integral always represents the first moment about the neutral axis of that part of the cross-sectional area of the beam between the horizontal plane on which the shearing stress τ occurs and the outer fibers of the beam, i.e. the area between y_0 and c.

From (1) it is evident that the maximum shearing stress always occurs at the neutral axis of the beam, whereas the shearing stress at the outer fibers is always zero. In contrast, the normal stress varies from zero at the neutral axis to a maximum at the outer fibers.

In a beam of rectangular cross-section the above equation for shearing stress becomes

$$\tau \;=\; \frac{V}{2I}\left(\frac{h^2}{4} - y_0^2\right)$$

where τ denotes the shearing stress on a fiber at a distance y_0 from the neutral axis and h denotes the depth of the beam. The distribution of vertical shearing stress over the rectangular cross-section is thus parabolic, varying from zero at the outer fibers to a maximum at the neutral axis. This expression is derived in Problem 8.18. For application see Problem 8.19.

Both the above equations for shearing stress give the vertical and also the horizontal shearing stresses at a point, as discussed in Problem 8.17, since the intensities of shearing stresses in these two directions are always equal.

PLASTIC BENDING OF BEAMS

The following remarks apply if some or all of the fibers of the beam are stressed to the yield point of the material.

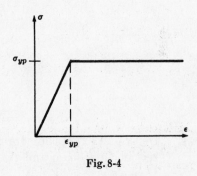

We shall consider a simplified stress-strain curve such as that of Fig. 8-4, where it is assumed that the proportional limit and the yield point coincide. The yield region, i.e. the horizontal plateau of the curve, is assumed to extend indefinitely. This conventionalized representation of ductile material behavior is termed *elastic-perfectly plastic* behavior. Here, σ_{yp} denotes the yield point of the material and ϵ_{yp} represents the strain corresponding to that stress. We shall assume that material properties are identical in tension and compression.

Fig. 8-4

ELASTO-PLASTIC ACTION

For sufficiently large bending moments in a beam the interior fibers will be stressed in the elastic range of action, whereas the outer fibers will have reached the yield point of the material. Such a stress distribution may be as indicated in Fig. 8-5.

FULLY PLASTIC ACTION

As bending moments continue to increase, a limiting case is approached in which all fibers are stressed to the yield point of the material. This stress distribution appears in Fig. 8-6.

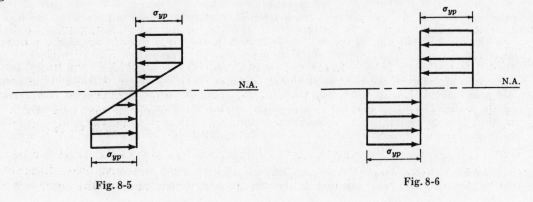

Fig. 8-5   Fig. 8-6

LOCATION OF NEUTRAL AXIS

When beam action is entirely elastic the neutral axis passes through the centroid of the cross-section. However, as plastic action spreads from the outer fibers inward, the neutral axis shifts from this location to another, which is determined by realizing that the resultant normal force over any cross-section vanishes. This is illustrated in Problem 8.29. In the limiting case of fully plastic action the neutral axis assumes a position such that the total cross-sectional area is divided into two equal parts. This is discussed in Problem 8.27.

FULLY PLASTIC MOMENT

The bending moment corresponding to fully plastic action is termed the *fully plastic moment* and will be denoted by M_p. For the stress-strain diagram assumed here no greater moment can be developed.

For a beam of rectangular cross-section the fully plastic moment is shown in Problem 8.23 to be $M_p = bh^2\sigma_{yp}/4$ where b represents the width of the beam and h its depth.

Solved Problems

ELASTIC BENDING OF BEAMS

8.1.   Derive an expression for the relationship between the bending moment acting at any section in a beam and the bending stress at any point in this same section. Assume Hooke's law holds.

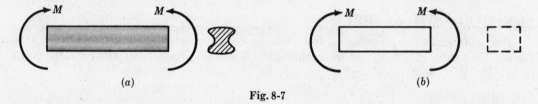

$$(a) \qquad\qquad\qquad\qquad (b)$$

Fig. 8-7

The beam shown in Fig. 8-7(a) is loaded by the two couples M and consequently is in static equilibrium. Since the bending moment has the same value at all points along the bar, the beam is said to be in a condition of *pure bending*. To determine the distribution of bending stress in the beam, let us cut the beam by a plane passing through it in a direction perpendicular to the geometric axis of the bar. In this manner the forces under investigation become external to the new body formed, even though they were internal effects with regard to the original uncut body.

The free-body diagram of the portion of the beam to the left of this cutting plane now appears as in Fig. 8-7(b). Evidently a moment M must act over the cross-section cut by the plane so that the left portion of the beam will be in static equilibrium. The moment M acting on the cut section represents the effect of the right portion of the beam on the left portion. Since the right portion has been removed, it must be replaced by its effect on the left portion and this effect is represented by the moment M. This moment is the resultant of the moments of forces acting perpendicular to the cut cross-section and in the plane of the page. It is now necessary to make certain assumptions in order to determine the nature of the variation of these forces over the cross-section.

It is convenient to consider the beam to be composed of an infinite number of thin longitudinal rods or fibers. It is assumed that every longitudinal fiber acts independently of every other fiber; that is, there are no lateral pressures or shearing stresses between adjacent fibers. Thus each fiber is subject only to axial tension or compression. Further, it is assumed that a plane section of the beam normal to its axis before loads are applied remains plane and normal to the axis after loading. Lastly, it is assumed that the material follows Hooke's law and that the moduli of elasticity in tension and compression are equal.

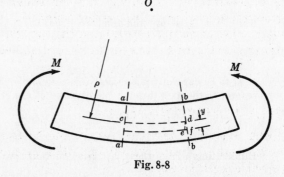

Let us next consider two adjacent cross-sections aa and bb marked on the side of the beam, as shown in Fig. 8-8. Prior to loading, these sections are parallel to each other. After the applied moments have acted on the beam, these sections are still planes but they have rotated with respect to each other to the positions shown, where O represents the center of curvature of the beam. Evidently the fibers on the upper surface of the beam are in a state of compression, while those on the lower surface have been extended slightly and are thus in

Fig. 8-8

tension. The line cd is the trace of the surface in which the fibers do not undergo any strain during bending and this surface is called the neutral surface, and its intersection with any cross-section is called the neutral axis. The elongation of the longitudinal fiber at a distance y (measured positive downward) may be found by drawing line de parallel to aa. If ρ denotes the radius of curvature of the bent beam, then from the similar triangles cOd and edf we find the strain of this fiber to be

$$\epsilon \;=\; \frac{\overline{ef}}{\overline{cd}} \;=\; \frac{\overline{de}}{\overline{cO}} \;=\; \frac{y}{\rho} \tag{1}$$

Thus, the strains of the longitudinal fibers are proportional to the distance y from the neutral axis.

Since Hooke's law holds, and therefore $E = \sigma/\epsilon$, or $\sigma = E\epsilon$, it immediately follows that the stresses existing in the longitudinal fibers are proportional to the distance y from the neutral axis, or

$$\sigma \;=\; \frac{Ey}{\rho} \tag{2}$$

Let us consider a beam of rectangular cross-section, although the derivation actually holds for any cross-section which has a longitudinal plane of symmetry. In this case, these longitudinal, or bending, stresses appear as in Fig. 8-9.

Let da represent an element of area of the cross-section at a distance y from the neutral axis. The stress acting on da is given by the above expression and consequently the force on this element is the product of the stress and the area da, i.e.

$$dF \;=\; \frac{Ey}{\rho}\,da \tag{3}$$

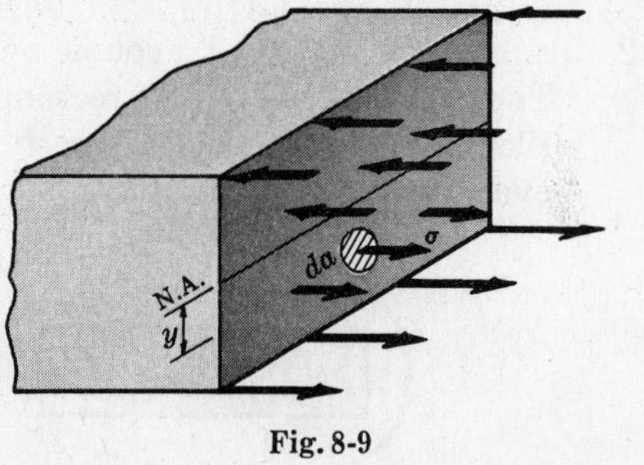

Fig. 8-9

However, the resultant longitudinal force acting over the cross-section is zero (for the case of pure bending) and this condition may be expressed by the summation of all forces dF over the cross-section. This is done by integration:

$$\int \frac{Ey}{\rho}\,da \;=\; \frac{E}{\rho}\int y\,da \;=\; 0 \tag{4}$$

Evidently $\int y\,da = 0$. However, this integral represents the first moment of the area of the cross-section with respect to the neutral axis, since y is measured from that axis. But, from Chapter 7 we may write $\int y\,da = \bar{y}A$, where $\bar{y}$ is the distance from the neutral axis to the centroid of the cross-sectional area. From this, $\bar{y}A = 0$; and since A is not zero, then $\bar{y} = 0$. Thus the neutral axis always passes through the centroid of the cross-section, provided Hooke's law holds.

The moment of the elemental force dF about the neutral axis is given by

$$dM \;=\; y\,dF \;=\; y\left(\frac{Ey}{\rho}\,da\right) \tag{5}$$

The resultant of the moments of all such elemental forces summed over the entire cross-section must be equal to the bending moment M acting at that section and thus we may write

$$M = \int \frac{Ey^2}{\rho} \, da \tag{6}$$

But $I = \int y^2 \, da$ and thus we have

$$M = \frac{EI}{\rho} \tag{7}$$

It is to be carefully noted that this moment of inertia of the cross-sectional area is computed with respect to the axis through the centroid of the cross-section. But previously we had

$$\sigma = \frac{Ey}{\rho} \tag{8}$$

Eliminating ρ from these last two equations, we obtain

$$\sigma = \frac{My}{I} \tag{9}$$

This formula gives the so-called bending or flexural stresses in the beam. In it, M is the bending moment at any section, I the moment of inertia of the cross-sectional area about an axis through the centroid of the cross-section, and y the distance from the neutral axis (which passes through the centroid) to the fiber on which the stress σ acts.

The value of y at the outer fibers of the beam is frequently denoted by c. At these fibers the bending stresses are maximum and there we may write

$$\sigma = \frac{Mc}{I} \tag{10}$$

8.2.   A beam is loaded by a couple of 12,000 lb-in at each of its ends, as shown in Fig. 8-10. The beam is steel and of rectangular cross-section 1 in. wide by 2 in. deep. Determine the maximum bending stress in the beam and indicate the variation of bending stress over the depth of the beam.

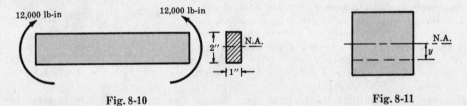

Fig. 8-10   Fig. 8-11

From Problem 8.1, bending takes place about the horizontal neutral axis denoted by N. A. This axis passes through the centroid of the cross-section. The moment of inertia of the shaded rectangular cross-section about this axis was found in Problem 7.4 to be

$$I = \tfrac{1}{12}bh^3 = \tfrac{1}{12}(1)(2)^3 = 0.667 \text{ in}^4$$

Also from Problem 8.1, the bending stress at a distance y from the neutral axis is given by $\sigma = My/I$, where y is illustrated in Fig. 8-11. Thus, all longitudinal fibers of the beam at the distance y from the neutral axis are subject to the same bending stress given by the above formula.

Since M and I are constant along the length of the bar, evidently the maximum bending stress occurs on those fibers where y takes on its maximum value. These are the fibers along the upper and lower surfaces of the beam, and from inspection it is obvious that for the direction of loading shown the upper fibers are in compression and the lower fibers in tension. For the lower fibers, $y = 1$ in. and the maximum bending stress is

$$\sigma = \frac{12,000(1)}{0.667} = 18,000 \text{ lb/in}^2$$

For the fibers along the upper surface y may be considered to be negative and we have

$$\sigma = \frac{12,000(-1)}{0.667} = -18,000 \text{ lb/in}^2$$

Thus the peak stresses are 18,000 lb/in² in tension for all fibers along the lower surface of the beam and 18,000 lb/in² in compression for all fibers along the upper surface. According to the formula $\sigma = My/I$, the bending stress varies linearly from zero at the neutral axis to a maximum at the outer fibers and hence the variation over the depth of the beam may be plotted as in Fig. 8-12.

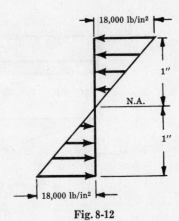

Fig. 8-12

8.3. A beam of circular cross-section is 7 in. in diameter. It is simply supported at each end and loaded by two concentrated loads of 20,000 lb each, applied 12 in. from the ends of the beam. Determine the maximum bending stress in the beam.

Here the moment is not constant along the length of the beam, as it was in Problem 8.2. The loading is illustrated in Fig. 8-13 together with the bending moment diagram obtained by the methods of Chapter 6. It is to be noted that the portion of the beam between the two downward loads of 20,000 lb is in a condition termed *pure bending* and everywhere in that region the bending moment is equal to $20,000(12) = 240,000$ lb-in.

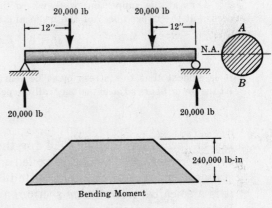

Fig. 8-13

From Problem 7.9 the moment of inertia of the shaded circular cross-section about the neutral axis, which passes through the centroid of the circle, is
$I = \pi D^4/64 = \pi(7)^4/64 = 118 \text{ in}^4.$

The bending stress at a distance y from the horizontal neutral axis shown is $\sigma = My/I$. Evidently the maximum bending stresses occur along the fibers located at the ends of a vertical diameter and designated as A and B. This maximum stress is the same at all such points between the applied loads. At point B, $y = 3.5$ in. and the stress becomes

$$\sigma = \frac{240,000(3.5)}{118} = 7120 \text{ lb/in}^2 \text{ tension}$$

At point A the stress is 7120 lb/in² compression.

8.4. A steel cantilever beam 16 ft 8 in. in length is subjected to a concentrated load of 320 lb acting at the free end of the bar. The beam is of rectangular cross-section, 2 in. wide by 3 in. deep. Determine the magnitude and location of the maximum tensile and compressive bending stresses in the beam.

The bending moment diagram for this type of loading, determined by the techniques of Chapter 6, is triangular with a maximum ordinate at the suporting wall, as shown below in Fig. 8.14(a). The maximum bending moment is merely the moment of the 320 lb force about an axis through point B and perpendicular to the plane of the page. It is $-320(200) = -64,000$ lb-in.

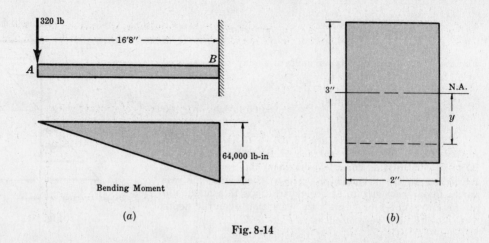

(a)   (b)

Fig. 8-14

The bending stress at a distance y from the neutral axis, which passes through the centroid of the cross-section, is $\sigma = My/I$ where y is illustrated in Fig. 8-14(b). In this expression I denotes the moment of inertia of the cross-sectional area about the neutral axis and is given by

$$I = \tfrac{1}{12}bh^3 = \tfrac{1}{12}(2)(3)^3 = 4.50 \text{ in}^4$$

Thus at the supporting wall, where the bending moment is maximum, the peak tensile stress occurs at the upper fibers of the beam and is

$$\sigma = \frac{My}{I} = \frac{(-64,000)(-1.5)}{4.50} = 21,400 \text{ lb/in}^2$$

It is evident that this stress must be tension because all points of the beam deflect downward. At the lower fibers adjacent to the wall the peak compressive stress occurs and is equal to 21,400 lb/in².

8.5. Let us reconsider Problem 8.4 for the case where the rectangular beam is replaced by a commercially available rolled steel section, designated as a 6 WF 15½. This standard manner of designation indicates that the depth of the section is 6 in., that it is a so-called *wide flange* section, and lastly that it weighs 15½ lb per ft of length. Determine the maximum tensile and compressive bending stresses.

Such a beam has the symmetric cross-section shown in Fig. 8-15 and bending takes place about the horizontal neutral axis passing through the centroid. Extensive handbooks listing properties of all available rolled steel shapes are available to designers and an abridged table is presented at the end of this chapter. From that table the moment of inertia about the neutral axis is found to be 28.1 in⁴.

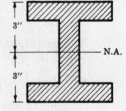

The bending stress at a distance y from the neutral axis is given by $\sigma = My/I$. At the outer fibers, $y = c$ and

$$\sigma = \frac{Mc}{I} = \frac{M}{I/c}$$

Fig. 8-15

The ratio I/c is designated as the *section modulus* and is usually denoted by the symbol Z. The units are obviously in³. From the abridged table we find Z to be 9.7 in³. Thus if one is concerned only with bending stresses occuring at the outer fibers, which is frequently the case since we are often interested only in maximum stresses, then the section modulus is a convenient quantity to work with, particularly for standard structural shapes.

The stresses in the extreme fibers at the section of the beam immediately adjacent to the wall are thus given by

$$\sigma = \frac{M}{I/c} = \frac{M}{Z} = \frac{64,000}{9.7} = 6600 \text{ lb/in}^2$$

Again, since the fibers along the top of the beam are stretching, the stress there will be tension. Along the lower face of the beam the fibers are shortening and there the stress is compressive.

8.6. A beam 8 ft in length is simply supported at each end and bears a uniformly distributed load of 400 lb per ft of length. The cross-section of the bar is rectangular, 3×6 in. Determine the magnitude and location of the maximum bending stress in the beam. Also, determine the bending stress at a point 1 in. below the upper surface of the beam at the section midway between supports.

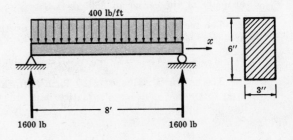

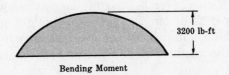

Bending Moment

Fig. 8-16

The bending moment diagram for a uniformly distributed load acting on a simply supported beam was shown in Problem 6.3 to be parabolic, varying from zero at the ends of the bar to a maximum at the center. The value of the bending moment at the center of this beam is

$$M_{x=4} = 1600(4) - 1600(2) = 3200 \text{ lb-ft} = 38,400 \text{ lb-in}$$

The bending moment diagram thus appears as in Fig. 8-16.

Since the maximum bending moment occurs at the center of the beam, then the maximum bending stress must also occur at the center of the beam, at $x = 4$ ft. At any fiber located a distance y from the neutral axis of that section, the bending stress is $\sigma = My/I = 38,400y/I$. The moment of inertia was found in Problem 7.4 to be

$$I = \tfrac{1}{12}bh^3 = \tfrac{1}{12}(3)(6)^3 = 54 \text{ in}^4$$

Thus, at the center section

$$\sigma = 38,400y/54$$

The maximum bending stress occurs at either the upper or lower extreme fibers. At the lower fibers, $y = 3$ in. and

$$\sigma = 38,400(3)/54 = 2130 \text{ lb/in}^2$$

From inspection, the fibers along the lower surface of the beam are extended in length, consequently the stress in these fibers is tensile. In the upper fibers, $y = -3$ in., a compressive stress of equal magnitude exists.

At a point 1 in. below the upper surface of the beam at this central section the bending stress is

$$\sigma = 38,400(-2)/54 = -1420 \text{ lb/in}^2$$

It is to be noted that y was taken to be negative in this calculation, since the point in question lies above the neutral axis. Consequently the bending stress is compressive as indicated by the final negative sign.

8.7. Let us reconsider the uniformly loaded beam of Problem 8.6. Determine the maximum bending stress in the beam if now the weight of the beam is considered in addition to the load of 400 lb per ft of length. The beam is steel and weighs 0.283 lb/in³.

Since the cross-section of the beam is 3×6 in., the volume of a 1 ft length of the beam is $3(6)(12) = 216$ in³ and the weight of a 1 ft length of the beam is $216(0.283) = 61.2$ lb.

For design purposes the weight of the beam is called the dead load, abbreviated D.L. This weight of the beam per ft of length may be considered to act in addition to the load of 400 lb/ft. This applied load is called the live load, abbreviated L.L. Thus the resultant load is $400 + 61.2 = 461.2$ lb/ft.

The total load over the entire beam is $461.2(8) = 3690$ lb. Each end reaction is therefore 1845 lb and the bending moment at the center of the beam is

$$M_{x=4} = 1845(4) - 1845(2) = 3690 \text{ lb-ft} = 44,300 \text{ lb-in}$$

The bending moment diagram has the same appearance as in Problem 8.6 but with a maximum ordinate at the center of 3690 lb-ft.

The maximum bending stress occurs at the outer fibers of the beam midway between supports and is given by $\sigma = My/I$ with $y = 3$ in. as before. Substituting,

$$\sigma = 44,300(3)/54 = 2460 \text{ lb/in}^2$$

The previous value of 2130 lb/in², which neglected the weight of the beam, was 13.4% lower than this value. In actual practice it is almost always necessary to take the weight of the beam into account.

8.8. A cantilever beam 10 ft long is subjected to a uniformly distributed load of 1500 lb per ft of length. The allowable working stress in either tension or compression is 20,000 lb/in². If the cross-section is to be rectangular, determine the dimensions if the height is to be twice as great as the width.

The bending moment diagram for a uniform load acting over a cantilever beam was determined in Problem 6.2. It was found to be parabolic, varying from zero at the free end of the beam to a maximum at the supporting wall. The loaded beam and the accompanying bending moment diagram are shown in Fig. 8-17. The maximum moment at the wall is given by

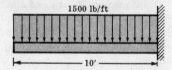
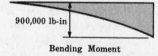

Fig. 8-17

$$M_{x=10} = -15,000(5) = -75,000 \text{ lb-ft} = -900,000 \text{ lb-in}$$

It is to be noted that this problem involves the design of a beam, whereas all previous problems in this chapter called for the analysis of stresses acting in beams of known dimensions and subject to various loadings. The only cross-section that need be considered for design purposes is the one where the bending moment is a maximum, i.e. at the supporting wall. Thus we wish to design a rectangular beam to resist a bending moment of 900,000 lb-in with a maximum bending stress of 20,000 lb/in².

Since the cross-section is to be rectangular it will have the appearance shown in Fig. 8-18 where the width is denoted by b and the height by $h = 2b$, in accordance with the specifications. The moment of inertia about the neutral axis, which passes through the centroid of the section, is given by

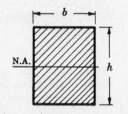

Fig. 8-18

$$I = \tfrac{1}{12}bh^3 = \tfrac{1}{12}b(2b)^3 = \tfrac{2}{3}b^4$$

At the cross-section of the beam adjacent to the supporting wall the bending stress in the beam is given by $\sigma = My/I$. The maximum bending stress in tension occurs along the upper surface of the beam, since these fibers elongate slightly, and at this surface $y = -b$ and $\sigma = 20,000$ lb/in². Then

$$\sigma = \frac{My}{I} \quad \text{or} \quad 20,000 = \frac{-900,000(-b)}{\tfrac{2}{3}b^4}$$

from which $b = 4.06$ in. and $h = 2b = 8.12$ in.

8.9. Select a suitable wide-flange section to carry the loading on the cantilever beam described in Problem 8.8. The working stress in either tension or compression is 20,000 lb/in².

The bending moment diagram is of course the same as that shown in Fig. 8-17. The maximum bending moment occurs at the supporting wall and as before is 900,000 lb-in.

For any wide-flange section, the bending stress on a fiber located a distance y from the neutral axis of the section is given by $\sigma = My/I$. It is presumed that the beam is oriented as in Fig. 8-19. The maximum bending stresses obviously occur when y takes on its maximum value, which is at the outer fibers of the beam. Let us denote this maximum value of y by c, i.e. c is half the depth of the section. Then the maximum bending stress may be written in the form

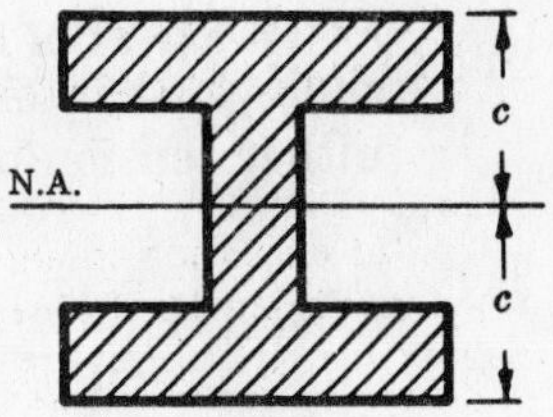

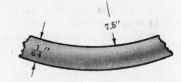

Fig. 8-19

$$\sigma_{max} \;=\; \frac{Mc}{I} \;=\; \frac{M}{I/c} \;=\; \frac{M}{Z}$$

where Z is the section modulus of the beam.

From the last equation we have $Z = M/\sigma_{max}$. Thus the required section modulus is given simply as the maximum bending moment divided by the allowable working stress. For the cantilever beam in question this becomes

$$Z \;=\; \frac{M}{\sigma_{max}} \;=\; \frac{900,000}{20,000} \;=\; 45 \text{ in}^3$$

Consequently a suitable beam will be one having a section modulus of at least 45 in³. It is of course unlikely that any standard rolled section will have exactly this section modulus and it is customary to select a section having either this Z, if possible, or a greater value. In this manner the working stress will not exceed the allowable value of 20,000 lb/in².

Reference to the abridged table of properties of wide-flange sections offered at the end of this chapter indicates that a 12 WF 36 section will be suitable for the design. It has a section modulus of 45.9 in³, which is in excess of the required value of 45 in³. This is the wide-flange section of least weight that has the necessary section modulus.

8.10. If a steel wire 1/64 in. in diameter is coiled around a pulley 15 in. in diameter, determine the maximum bending stress set up in the wire. Take $E = 30 \times 10^6$ lb/in².

Since the radius of curvature of the wire is constant, 7.5 in., it is evident from (7) of Problem 8.1, namely $M = EI/\rho$, that the bending moment M must be constant everywhere in the wire. Thus the wire acts as a beam subject to pure bending. An enlarged sketch of a portion of the wire is shown in Fig. 8-20. For any fiber in the wire at a distance y from the neutral axis, the normal strain ϵ was found in (1) of Problem 8.1 to be

$$\epsilon \;=\; \frac{y}{\rho}$$

Fig. 8-20

where ρ denotes the radius of curvature of the beam at that point.

The maximum strain occurs at the fibers where y assumes its maximum value, i.e. $\tfrac{1}{2}(1/64)$ in. from the neutral axis. The radius of curvature is approximately 7.5 in. More accurately, this radius should be measured to the neutral surface of the wire, but the value in that case would only differ from 7.5 in. by $\tfrac{1}{2}(1/64)$ in. and this quantity may reasonably be neglected.

Thus the maximum strain at the outer fibers of the wire is $\epsilon = \tfrac{1}{2}(1/64)/7.5 = 0.00104$.

The longitudinal fibers are subject to tensile stresses on one side of the wire and compressive on the other, with no other stresses acting. Hooke's law may then be used to find the stress σ:

$$\sigma \;=\; E\epsilon \;=\; (30 \times 10^6)(0.00104) \;=\; 31,200 \text{ lb/in}^2$$

This is the maximum stress in the wire.

8.11. The simply supported beam shown in Fig. 8-21(a) below is subject to a uniformly varying load having a maximum intensity of w lb per ft of length at the right end of the bar. If the beam is a 10 WF 49 section (i.e. a commercially available wide-flange section having a depth of 10 in. and weighing 49 lb/ft of length), determine the maximum load intensity w that may be applied if the working stress is 18,000 lb/in² in either tension or compression. Neglect the weight of the beam.

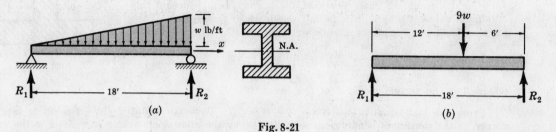

Fig. 8-21

The reactions R_1 and R_2 may readily be determined in terms of the unknown w by replacing the distributed load by its resultant. Since the average value of the distributed load is $w/2$ lb/ft acting over a length of 18 ft, the resultant is a force of magnitude $18(w/2) = 9w$ lb acting through the centroid of the triangular loading diagram, i.e., 12 ft to the right of R_1. This resultant thus appears as in Fig. 8-21(b). From statics we immediately have $R_1 = 3w$ lb and $R_2 = 6w$ lb.

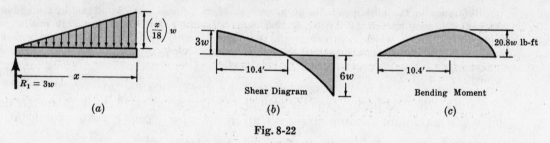

Fig. 8-22

The shearing force and bending moment diagrams for this type of loading were discussed in Problem 6.4. Let us introduce an x-axis coinciding with the beam and having its origin at the left support. Then at a distance x to the right of the left reaction, the intensity of load is found from similar-triangle relationships to be $(x/18)w$ lb/ft. This portion of the loaded beam between R_1 and the section x appears in Fig. 8-22(a). In accordance with the procedure explained in Problem 6.4 the shearing force V at the section a distance x from the left support is given by

$$V = 3w - \frac{1}{2}\left(\frac{x}{18}\right)wx = 3w - \frac{1}{36}wx^2$$

This equation holds for all values of x and from it the shear diagram is readily plotted, as shown in Fig. 8-22(b).

The point of zero shear is found by setting

$$3w - \frac{1}{36}wx^2 = 0 \qquad \text{from which} \qquad x = \sqrt{108} = 10.4 \text{ ft}$$

This is also the point where the bending moment assumes its maximum value.

The bending moment M at the section a distance x from the left support is given by

$$M = 3wx - \frac{1}{2}\left(\frac{x}{18}\right)wx\frac{x}{3} = 3wx - \frac{1}{108}wx^3$$

Again, this equation holds for all values of x and from it the bending moment diagram may be plotted as in Fig. 8-22(c). At the point of zero shear, $x = 10.4$ ft, the bending moment is found by substitution in the above equation to be

$$M_{x=10.4} = 3w\sqrt{108} - \frac{1}{108}w(\sqrt{108})^3 = 2w\sqrt{108} \text{ lb-ft} = 250w \text{ lb-in}$$

This is the maximum bending moment in the beam.

The bending stress on any fiber a distance y from the neutral axis of the beam is given by $\sigma = My/I$. The moment of inertia I of the beam is found from the table at the end of this chapter to be 272.9 in⁴. The maximum tensile stress occurs at the lower fibers of the beam where $y = 5$ in. at the section where the bending moment is a maximum. This stress is 18,000 lb/in² and thus $\sigma = My/I$ becomes

$$18,000 = \frac{(250w)(5)}{272.9} \quad \text{or} \quad w = 3930 \text{ lb/ft}$$

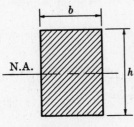

Fig. 8-23

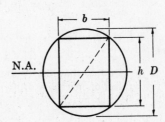

Fig. 8-24

8.12. Determine the section modulus of a beam of rectangular cross-section.

Let h denote the depth of the beam and b its width. Bending is assumed to take place about the neutral axis through the centroid of the cross-section. The moment of inertia about the neutral axis is $I = bh^3/12$.

At the outer fibers the distance to the neutral axis is $h/2$, and this is commonly denoted by c. The maximum bending stresses at these outer fibers are given by

$$\sigma_{max} = \frac{Mc}{I} = \frac{M}{I/c}$$

The ratio I/c is called the section modulus and is usually denoted by Z. Then $\sigma_{max} = M/Z$. For the beam of rectangular cross-section,

$$Z = \frac{I}{c} = \frac{bh^3/12}{h/2} = \frac{bh^2}{6}$$

The section modulus Z has units of (inches, or feet)³.

8.13. A beam of rectangular cross-section is to be cut from a circular log of diameter D. What should be the ratio of the depth of the beam to its width for maximum strength in pure bending?

A sketch of the beam cross-section appears in Fig. 8-24, where h denotes the height of the beam and b its width.

The bending is considered to take place about the horizontal neutral axis shown. The maximum bending stresses occur at the outer fibers of the rectangular section at a distance $h/2$ above or below the neutral axis. Any fiber a distance y from the neutral axis is subject to a bending stress given by $\sigma = My/I$, where I denotes the moment of inertia of the rectangular cross-section about the neutral axis, i.e. $bh^3/12$. At the outer fibers, $y = h/2$ and the maximum stresses there become

$$\sigma_{max} = \frac{M(h/2)}{bh^3/12} = \frac{M}{bh^2/6} \tag{1}$$

For any given value of the maximum bending stress, the maximum moment M that may be applied to the bar is found from the last equation to be

$$M = \sigma_{max}bh^2/6 \tag{2}$$

For the condition of maximum strength, i.e. maximum moment M, the product bh^2 must be made a maximum since σ_{max} is constant for a given material.

To maximize the quantity bh^2 we realize that it must be expressed in terms of one independent variable, say b, and we may do this from the right triangle relationship: $b^2 + h^2 = D^2$. Then

$$bh^2 = b(D^2 - b^2) = bD^2 - b^3$$

To maximize bh^2 we take the first derivative of the expression with respect to b and set it equal to zero, as follows:

$$\frac{d(bh^2)}{db} = \frac{d}{db}(bD^2 - b^3) = D^2 - 3b^2 = (b^2 + h^2) - 3b^2 = h^2 - 2b^2 = 0$$

Solving, $h/b = \sqrt{2}$. This is the desired ratio in order that the beam will carry a maximum moment M.

It is to be noted that the expression appearing in the denominator of the right side of (1), i.e. $bh^2/6$, is the section modulus of a rectangular bar. Thus the section modulus is actually the quantity to be maximized for greatest strength of the beam.

8.14. A beam is loaded by one couple at each of its ends, the magnitude of each couple being 50,000 lb-in. The beam is steel and of T-type cross-section with the dimensions indicated in Fig. 8-25. Determine the maximum tensile stress in the beam and its location, and the maximum compressive stress and its location.

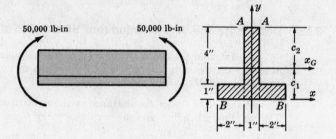

Fig. 8-25

It is first necessary to locate the centroid of the cross-sectional area since the neutral axis is known to pass through the centroid. To do this we introduce the x-y coordinate system shown and proceed as in Problem 7.10. The y-coordinate of the centroid is defined by

$$\bar{y} = \frac{\int y \, da}{A}$$

where the numerator of the right side represents the first moment of the entire area about the x-axis. The T-section may be considered to consist of the three rectangles indicated by the dashed lines and this expression becomes

$$\bar{y} = \frac{5(1)(2.5) + [2(1)(1/2)]2}{5(1) + [2(1)]2} = 1.61 \text{ in.}$$

Thus, the centroid is located 1.61 in. above the x-axis. The horizontal axis passing through this point is denoted by x_G as shown.

The moment of inertia about the x-axis is given by the sum of the moments of inertia about this same axis of each of the three component rectangles comprising the cross-section. Thus

$$I_x = \tfrac{1}{3}(2)(1)^3 + \tfrac{1}{3}(1)(5)^3 + \tfrac{1}{3}(2)(1)^3 = 43 \text{ in}^4$$

The moment of inertia about the x_G-axis may now be found by use of the parallel-axis theorem. Thus

$$I_x = I_{x_G} + A(\bar{y})^2, \quad 43 = I_{x_G} + 9(1.61)^2 \quad \text{and} \quad I_{x_G} = 19.7 \text{ in}^4$$

Evidently for the loading shown, the fibers below the x_G-axis are in tension, while the fibers above this axis are in compression. Let c_1 and c_2 denote the distances of the extreme fibers from the neutral axis (x_G) as shown. Obviously $c_1 = 1.61$ in. and $c_2 = 3.39$ in. The maximum tensile stress occurs in those fibers along B-B and is given by $\sigma = Mc_1/I$, where I denotes the moment of inertia of the entire cross-section about the neutral axis passing through the centroid of the cross-section.   Thus the maximum tensile stress is given by

$$\sigma = Mc_1/I = 50,000(1.61)/19.7 = 4100 \text{ lb/in}^2$$

The maximum compressive stress occurs in those fibers along A-A and is given by $\sigma = Mc_2/I$. To provide a consistent system of algebraic signs, it is necessary to assign a negative value to c_2 since it lies on the side of the x_G-axis opposite to that of c_1.   Hence

$$\sigma = Mc_2/I = 50,000(-3.39)/19.7 = -8600 \text{ lb/in}^2$$

The negative sign indicates that the stress is compressive.

8.15. A simply supported beam is loaded by the couple of 1000 lb-ft as shown in Fig. 8-26. The beam has a channel-type cross-section as illustrated.   Determine the maximum tensile and compressive stresses in the beam.

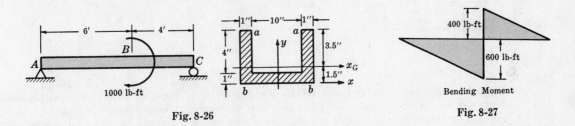

Fig. 8-26   Fig. 8-27

The bending moment diagram for this particular loading has been determined in Problem 6.11, where it was found to appear as in Fig. 8-27.

It is next necessary to locate the centroid of the cross-section since the neutral axis passes through that point. This has already been determined for this channel-type section in Problem 7.12.   There, the centroid was found to lie 1.5 in. above the x-axis and consequently the x_G-axis through the centroid lies 1.5 in. above the x-axis.   In that same problem the moment of inertia of the entire cross-section about the x_G-axis was found to be 41.6 in^4.

In this problem it is necessary to distinguish carefully between positive and negative bending moments.   One method of attack is to consider a cross-section of the beam slightly to the left of point B where the 1000 lb-ft couple is applied.   According to the bending moment diagram the moment there is -600 lb-ft and, according to the sign convention adopted in Chapter 6, since the moment is negative the beam is concave downward at that section, as shown in Fig. 8-28. Thus the upper fibers are in tension and the lower fibers in compression.   Along the upper fibers a-a the bending stress is given by $\sigma = My/I$.   Then

Fig. 8-28

$$\sigma_a = (-600)(12)(-3.5)/41.6 = 605 \text{ lb/in}^2$$

Along the lower fibers b-b the value of y in the above formula for bending stress must be taken to be positive and there we have

$$\sigma_b = (-600)(12)(+1.5)/41.6 = -260 \text{ lb/in}^2$$

It is next necessary to investigate the bending stresses at a section slightly to the right of point B.   There the bending moment is 400 lb-ft and according to the usual sign convention the beam is concave upward at that section, as shown in Fig. 8-29.   Here the upper fibers are in compression and the lower fibers in tension.   Along the upper fibers a-a the bending stress is

$$\sigma_a' = 400(12)(-3.5)/41.6 = -400 \text{ lb/in}^2$$

Along the lower fibers b-b we have

$$\sigma_b' = 400(12)(1.5)/41.6 = 170 \text{ lb/in}^2$$

Fig. 8-29

The maximum tensile and compressive stresses must now be selected from the above four values. Evidently the maximum tension is 605 lb/in² occurring in the upper fibers just to the left of point B; the maximum compression is 400 lb/in² occurring in the upper fibers also but just to the right of point B.

8.16. Consider the beam with overhanging ends loaded by the three concentrated forces shown in Fig. 8-30. The beam is simply supported and of T-type cross-section as shown. The material is gray cast iron having an allowable working stress in tension of 5000 lb/in² and in compression of 20,000 lb/in². Determine the maximum allowable value of P.

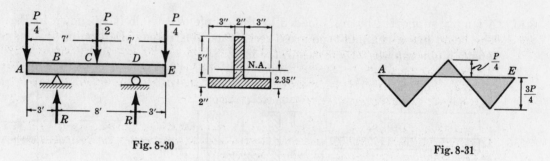

Fig. 8-30   Fig. 8-31

From symmetry each of the reactions denoted by R is equal to $P/2$. The bending moment diagram consists of a series of straight lines connecting the ordinates representing bending moments at the points A, B, C, D and E. At B the bending moment is given by the moment of the force $P/4$ acting at A about an axis through B. Thus

$$M_B = -(P/4)(3) = -3P/4 \text{ lb-ft}$$

At C the bending moment is given by the sum of the moments of the forces $P/4$ and $R = P/2$ about an axis through C. Thus

$$M_C = -(P/4)(7) + (P/2)(4) = P/4 \text{ lb-ft}$$

The bending moment at D is equal to that at B by symmetry and the moment at each of the ends A and E is zero. Hence, the bending moment diagram plots as in Fig. 8-31.

The properties of the T-section used here have already been determined in Problem 7.10. There the distance from the outer fibers of the flange to the centroid was found to be 2.35 in. and the moment of inertia about the neutral axis passing through the centroid was 101 in⁴.

It is perhaps simplest to calculate four values of P based upon the various maximum tensile and compressive stresses that may exist at each of the points B and C and then select the minimum of these values. Let us first examine point B. Since the bending moment there is negative, the beam is concave downward at that point, as shown in Fig. 8-32. Evidently the upper fibers are in tension and the lower fibers are subject to compression. We shall first calculate a value of P assuming that the allowable tensile stress of 5000 lb/in² is realized in the upper fibers. Applying the flexure formula $\sigma = My/I$ to these upper fibers, we find

$$5000 = (-3P/4)(12)(-4.65)/101 \quad \text{or} \quad P = 12,000 \text{ lb}$$

Next we shall calculate a value of P, assuming that the allowable compressive stress of 20,000 lb/in² is set up in the lower fibers. Again applying the flexure formula, we find

$$-20,000 = (-3P/4)(12)(2.35)/101 \quad \text{or} \quad P = 95,500 \text{ lb}$$

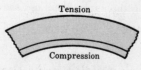

Fig. 8-32

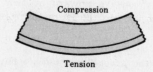

Fig. 8-33

We shall now examine point C. Since the bending moment there is positive, the beam is concave upward at that point and appears as in Fig. 8-33. Here, the upper fibers are in compression and the lower fibers are subject to tension. First we will calculate a value of P assuming that the allowable tension of 5000 lb/in² is set up in the lower fibers. From the flexure formula we find

$$5000 = (P/4)(12)(2.35)/101 \quad \text{or} \quad P = 71,800 \text{ lb}$$

Lastly, we shall assume that the allowable compression of 20,000 lb/in² is set up in the upper fibers. Applying the flexure formula, we have

$$-20,000 = (P/4)(12)(-4.65)/101 \quad \text{or} \quad P = 145,000 \text{ lb}$$

The minimum of these four values is $P = 12,000$ lb. Thus the tensile stress at the points B and D is the controlling factor in determining the maximum allowable load.

8.17. In a beam loaded by transverse forces acting perpendicular to the axis of the beam, not only are bending stresses parallel to the axis of the bar produced but shearing stresses also act over cross-sections of the beam perpendicular to the axis of the bar. Express the intensity of these shearing stresses in terms of the shearing force at the section and the properties of the cross-section.

The theory to be developed applies only to a cross-section of rectangular shape. However, the results of this analysis are commonly used to give approximate values of the shearing stress in other cross-sections having a plane of symmetry.

Let us consider an element of length dx cut from a beam as shown in Fig. 8-34. We shall denote the bending moment at the left side of the element by M and that at the right side by $M + dM$, since in general the bending moment changes slightly as we move from one section to an adjacent section of the beam. If y is measured upward from the neutral axis, then the bending stress at the left section a-a is given by

$$\sigma = \frac{My}{I}$$

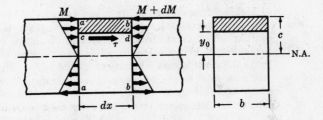

Fig. 8-34

where I denotes the moment of inertia of the entire cross-section about the neutral axis. This stress distribution is illustrated above. Similarly, the bending stress at the right section b-b is

$$\sigma' = \frac{(M + dM)y}{I}$$

Let us now consider the equilibrium of the shaded element $acdb$. The force acting on an area da of the face ac is merely the product of the intensity of the force and the area; thus

$$\sigma \, da = \frac{My}{I} da$$

The sum of all such forces over the left face ac is found by integration to be

$$\int_{y_0}^{c} \frac{My}{I} da$$

Likewise, the sum of all normal forces over the right face bd is given by

$$\int_{y_0}^{c} \frac{(M + dM)y}{I} da$$

Evidently, since these two integrals are unequal, some additional horizontal force must act on the shaded element to maintain equilibrium. Since the top face ab is assumed to be free of any externally applied horizontal forces, then the only remaining possibility is that there exists a horizontal shearing force along the lower face cd. This represents the action of the lower portion of the beam on the shaded element. Let us denote the shearing stress along this face by τ as shown. Also, let b denote the width of the beam at the position where τ acts. Then the horizontal shearing force along the face cd is $\tau b\, dx$. For equilibrium of the element $acdb$ we have

$$\Sigma F_h = \int_{y_0}^{c} \frac{My}{I}\, da - \int_{y_0}^{c} \frac{(M+dM)y}{I}\, da + \tau b\, dx = 0$$

Solving,

$$\tau = \frac{1}{Ib}\frac{dM}{dx}\int_{y_0}^{c} y\, da$$

But from Problem 6.1 we have $V = dM/dx$, where V represents the shearing force (in pounds) at the section a-a. Substituting,

$$\tau = \frac{V}{Ib}\int_{y_0}^{c} y\, da$$

The integral in this last equation represents the first moment of the shaded cross-sectional area about the neutral axis of the beam. This area is always the portion of the cross-section that is above the level at which the desired shear acts. This first moment of area is sometimes denoted by Q in which case the above formula becomes

$$\tau = \frac{VQ}{Ib}$$

The units of $\int y\, da$ or of Q are in³.

The shearing stress τ just determined acts horizontally as shown in Fig. 8-34. However, let us consider the equilibrium of a thin element $mnop$ of thickness t cut from any body and subject to a shearing stress τ_1 on its lower face, as shown in Fig. 8-35. The total horizontal force on the lower face is $\tau_1 t\, dx$. For equilibrium of forces in the horizontal direction, an equal force but acting in the opposite direction must act on the upper face, hence the shear stress intensity there too is τ_1. These two forces give rise to a couple of magnitude $\tau_1 t\, dx\, dy$. The only way in which equilibrium of the element can be maintained is for another couple to act over the vertical faces. Let the shear stress intensity on these faces be denoted by τ_2. The total force on either vertical face is $\tau_2 t\, dy$. For equilibrium of the moments about the center of the element we have

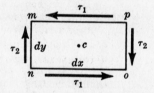

Fig. 8-35

$$\Sigma M_c = \tau_1 t\, dx\, dy - \tau_2 t\, dy\, dx = 0 \qquad \text{or} \qquad \tau_1 = \tau_2$$

Thus we have the interesting conclusion that the shearing stresses on any two perpendicular planes through a point on a body are equal. Consequently, not only are there shearing stresses τ acting horizontally at any point in the beam, but shearing stresses of an equal intensity also act vertically at that same point.

In summary, when a beam is loaded by transverse forces, both horizontal and vertical shearing stresses arise in the beam. The vertical shearing stresses are of such magnitudes that their resultant at any cross-section is exactly equal to the shearing force V at that same section.

8.18. Using the expression for shearing stress derived in Problem 8.17, determine the distribution of shearing stress in a beam of rectangular cross-section. What is the maximum shearing stress in a rectangular bar?

In Problem 8.17 the shearing stress τ at a distance y_0 from the neutral axis of the beam was found to be

$$\tau = \frac{V}{Ib}\int_{y_0}^{c} y\, da$$

where V denotes the shearing force at the cross-section and b represents the width of the beam at the position where τ is acting.

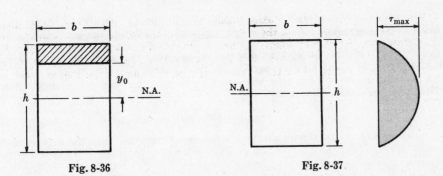

Fig. 8-36　　　　　　　　　　　　　　　　　Fig. 8-37

It is necessary to evaluate the above integral for a rectangular cross-section. Let h denote the height of the cross-section and b its width, as shown in Fig. 8-36.

The integral represents the first moment of the shaded area about the neutral axis. It is to be noted that this area extends from the level at which the desired shear stress τ acts to the extreme outer fibers of the beam. In this manner we find the shear stress τ on all fibers a distance y_0 from the neutral axis. Actually it is not necessary to integrate in such a simple case. Since the integral is known to represent the first moment of the shaded area about the neutral axis, we may calculate this first moment according to the definition given in Chapter 7. That is, the first moment of the shaded area is simply the product of the area and the perpendicular distance between the centroid of the area and the neutral axis. The area is given by $b[(h/2) - y_0]$, and the distance from the centroid of the shaded region to the neutral axis is $[(h/2) + y_0]/2$. Consequently the value of the integral representing the first moment of area is

$$\int_{y_0}^{c} y \, da = \frac{1}{2} b \left(\frac{h}{2} + y_0 \right) \left(\frac{h}{2} - y_0 \right) = \frac{1}{2} \left(\frac{h^2}{4} - y_0^2 \right) b$$

and the shearing stress τ at a distance y_0 from the neutral axis becomes

$$\tau = \frac{V}{Ib} \left[\frac{1}{2} b \left(\frac{h^2}{4} - y_0^2 \right) \right] = \frac{V}{2I} \left(\frac{h^2}{4} - y_0^2 \right)$$

From this it may be seen that the shearing stress over the cross-section varies in a parabolic manner from a maximum at the neutral axis $(y_0 = 0)$ to zero at the outer fibers of the beam $(y_0 = h/2)$. This variation may be plotted as in Fig. 8-37.

At the neutral axis, $y_0 = 0$, the maximum shearing stress is found by substitution in the above formula to be $\tau_{max} = Vh^2/8I$. But for a rectangular cross-section $I = bh^3/12$. Substituting,

$$\tau_{max} = \frac{Vh^2}{8bh^3/12} = \frac{3}{2} \frac{V}{bh}$$

Thus the maximum shearing stress in the case of a rectangular cross-section is 50 percent greater than the average shearing stress obtained by dividing the shearing force V by the cross-sectional area bh.

8.19. A beam of rectangular cross-section is simply supported at the ends and subject to the single concentrated force shown in Fig. 8-38(a). Determine the maximum shearing stress in the beam. Also, determine the shearing stress at a point one inch below the top of the beam at a section one foot to the right of the left reaction.

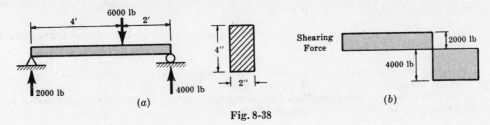

Fig. 8-38

The reactions are readily found from statics to be 2000 lb and 4000 lb as shown. The shearing force diagram for this type of loading appears in Fig. 8-38(b) above.

Inspection of the above shear diagram reveals that the maximum value of the shearing force is 4000 lb occurring at all cross-sections to the right of the 6000 lb load. The average value of the shearing stress acting over any cross-section in this region is simply the shearing force divided by the cross-sectional area, i.e.

$$\tau_{av} \; = \; 4000/2(4) \; = \; 500 \text{ lb/in}^2$$

By Problem 8.18 the maximum shearing stress is 50% greater than the average value. Hence

$$\tau_{max} \; = \; (3/2)(500) \; = \; 750 \text{ lb/in}^2$$

This maximum shearing stress occurs at all points along the neutral axis of the beam to the right of the 6000 lb load.

From the shear diagram, the shearing force acting over a section 1 ft to the right of the left reaction is 2000 lb. The shearing stress τ at any point in this section a distance y_0 from the neutral axis was shown in Problem 8.18 to be

$$\tau \; = \; \frac{V}{2I}\left(\frac{h^2}{4} - y_0^2\right)$$

At a point 1 in. below the top fibers of the beam, $y_0 = 1$ in. Also, we have $h = 4$ in. and $I = bh^3/12 = 2(4)^3/12 = 10.7$ in⁴. Substituting,

$$\tau \; = \; \frac{2000}{2(10.7)}\left(\frac{16}{4} - 1\right) \; = \; 280 \text{ lb/in}^2$$

8.20. Consider the cantilever beam subject to the concentrated load shown in Fig. 8-39. The cross-section of the beam is of T-shape. Determine the maximum shearing stress in the beam and also determine the shearing stress 1 in. from the top surface of the beam at a section adjacent to the supporting wall.

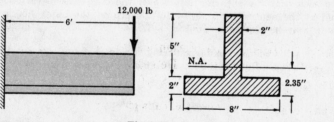
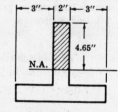

Fig. 8-39   Fig. 8-40

The shear force has a constant value of 12,000 lb at all points along the length of the beam. Because of this simple, constant value the shear diagram need not be drawn.

Further, the location of the centroid and the moment of inertia about the centroidal axis for this particular cross-section were determined in Problem 7.10. The centroid was found to be 2.35 in. above the lower surface of the beam and the moment of inertia about a horizontal axis through the centroid was found to be 101 in⁴.

The shearing stress at a distance y_0 from the neutral axis through the centroid was found in Problem 8.17 to be

$$\tau \; = \; \frac{V}{Ib}\int_{y_0}^{c} y \, da$$

Inspection of this equation reveals that the shearing stress is a maximum at the neutral axis, since at that point $y_0 = 0$ and consequently the integral assumes the largest possible value. It is not necessary to integrate, however, since the integral is known in this case to represent the the first moment of the area between the neutral axis and the outer fibers of the beam about the neutral axis. This area is represented by the shaded region in Fig. 8-40. The value of the integral could also, of course, be found by taking the first moment of the unshaded area below the neutral axis about the line, but that calculation would be somewhat more difficult.

Thus the first moment of the shaded area about the neutral axis is

$$2(4.65)(2.33) = 21.6 \text{ in}^3$$

and the shearing stress at the neutral axis, where $b = 2$ in., is found by substitution in the above general formula to be

$$\tau = \frac{12,000}{101(2)}(21.6) = 1290 \text{ lb/in}^2$$

In this formula b was taken to be 2 in., since that is the width of the beam at the point where the shearing stress is being calculated. Thus the maximum shearing stress is 1290 lb/in² and it occurs at all points on the neutral axis along the entire length of the beam, since the shearing force has a constant value along the entire length of the beam.

The shearing stress 1 in. from the top surface of the beam is again given by the formula

$$\tau = \frac{V}{Ib}\int_{y_0}^{c} y \, da$$

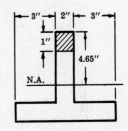

Now, the integral represents the first moment of the new shaded area shown in Fig. 8-41, about the neutral axis. Again it is not necessary to integrate to evaluate the integral, since the coordinate of the centroid of this shaded area is known. It is 4.15 in. above the neutral axis. Thus the first moment of this shaded area about the neutral axis is $2(1)(4.15) = 8.30 \text{ in}^3$, and the shearing stress 1 in. below the top fibers is

Fig. 8-41

$$\tau = \frac{12,000}{101(2)}(8.30) = 495 \text{ lb/in}^2$$

Again, b was taken to be 2 in. since that is the width of the beam at the point where the shearing stress is being evaluated. Since the shearing force is equal to 12,000 lb everywhere along the length of the beam, the shearing stress 1 in. below the top fibers is 495 lb/in² everywhere along the beam.

8.21. Consider a beam having an I-type cross-section as shown in Fig. 8-42. A shearing force V of 32,000 lb acts over the section. Determine the maximum and minimum values of the shearing stress in the vertical web of the section.

The shearing stress at any point in the cross-section is given by

$$\tau = \frac{V}{Ib}\int_{y_0}^{c} y \, da$$

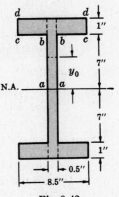

as derived in Problem 8.17. Here, y_0 represents the location of the section on which τ acts, and is measured from the neutral axis as shown. In this expression, I represents the moment of inertia of the entire cross-section about the neutral axis, which passes through the centroid of the section. I is readily calculated by dividing the section into rectangles, as indicated by the dashed lines, and we have

$$I = \tfrac{1}{12}(0.5)(16)^3 + 2[\tfrac{1}{12}(8.0)(1)^3 + 8.0(1)(7.5)^2] = 1072 \text{ in}^4$$

Inspection of the general formula for shearing stress reveals that this stress has a maximum value when $y_0 = 0$, i.e. at the neutral axis, since at that point the integral takes on its largest possible value. It is not necessary to integrate to obtain the value of $\int_{y_0}^{c} y \, da$, since this integral is shown to represent the first moment of the area between

Fig. 8-42

$y_0 = 0$ (i.e. the neutral axis) and the outer fibers of the beam. This area is shaded in Fig. 8-43. For this area we have, taking its first moment about the neutral axis,

$$\int_0^8 y\,da = 7(0.5)(3.5) + 8.5(1)(7.5) = 75.9 \text{ in}^3$$

Consequently the maximum shearing stress in the web occurs at the section *a-a* along the neutral axis and by substituting in the general formula for shearing stress is found to be

$$\tau_{max} = \frac{32,000}{1072(0.5)}(75.9) = 4540 \text{ lb/in}^2$$

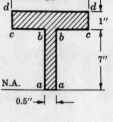

Fig. 8-43

The minimum shearing stress in the web occurs at that point in the web farthest from the neutral axis, i.e. across the section *b-b*. To calculate the shearing stress there, it is necessary to evaluate $\int_{y_0}^{c} y\,da$ for the area between *b-b* and the outer fibers of the beam. This is the shaded area shown in Fig. 8-44. Again, it is not necessary to integrate, since this integral merely represents the first moment of this shaded area about the neutral axis. It is

$$\int_7^8 y\,da = 8.5(1)(7.5) = 63.7 \text{ in}^3$$

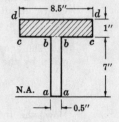

Fig. 8-44

The value of *b* is still 0.5 in., since that is the width of the beam at the position where the shearing stress is being calculated. Substituting in the general formula

$$\tau_{min} = \frac{32,000}{1072(0.5)}(63.7) = 3800 \text{ lb/in}^2$$

It is to be noted that there is not too great a difference between the maximum and minimum values of shearing stress in the web of the beam. In fact, it is customary to calculate only an approximate value of the shearing stress in the web of such an I-beam. This value is obtained by dividing the total shearing force V by the cross-sectional area of the web alone. This approximate value becomes

$$\tau_{av} = \frac{32,000}{16(0.5)} = 4000 \text{ lb/in}^2$$

A more advanced analysis of shearing stresses in an I-beam reveals that the vertical web resists nearly all of the shearing force V and that the horizontal flanges resist only a small portion of this force. The shear stress in the web of an I-beam is specified by various codes at rather low values. Thus some codes specify 10,000 lb/in², others 13,000 lb/in².

PLASTIC BENDING OF BEAMS

8.22. Consider a beam of arbitrary doubly symmetric cross-section, as in Fig. 8-45(*a*), subject to pure bending. The material is considered to be elastic-perfectly plastic, i.e. the stress-strain diagram has the appearance shown in Fig. 8-45(*b*) and stress-strain characteristics in tension and compression are identical. Determine the moment acting on the beam when all fibers a distance y_1 from the neutral axis have reached the yield point of the material.

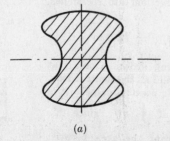 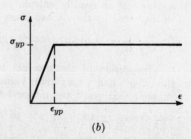

(*a*)   (*b*)

Fig. 8-45

Even though bending of the beam has caused the outer fibers to have yielded it is still realistic to assume that plane sections of the beam normal to the axis before loads are applied remain plane and normal to the axis after loading. Consequently, normal strains of the longitudinal fibers of the beam still vary linearly with the distance of the fiber from the neutral axis.

As the value of the applied moment is increased, the extreme fibers of the beam are the first to reach the yield point of the material and the normal stresses on all interior fibers vary linearly as the distance of the fiber from the neutral axis, as indicated in Fig. 8-46(a). A further increase in the value of the moment puts interior fibers at the yield point, with yielding progressing from the outer fibers inward, as indicated in Fig. 8-46(b). In the limiting case when all fibers (except those along the neutral axis) are stressed to the yield point the normal stress distribution appears as in Fig. 8-46(c). The bending moment corresponding to Fig. 8.46(c) is termed a *fully plastic moment*. For the type of stress-strain curve shown in Fig. 8-45(b), no greater moment is possible.

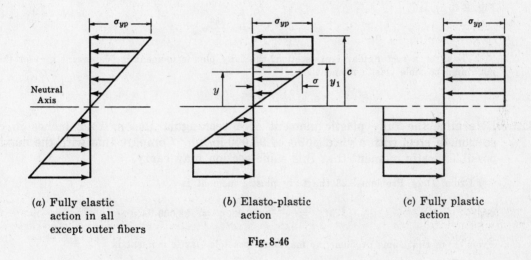

(a) Fully elastic action in all except outer fibers

(b) Elasto-plastic action

(c) Fully plastic action

Fig. 8-46

For a beam in pure bending, the sum of the normal forces over the cross-section must vanish. Hence, for the doubly symmetric section under consideration, it is evident from inspection of Fig. 8-46(b) that the neutral axis must pass through the centroid of such a section, i.e. the area above the neutral axis must be equal to the area below that axis. However, in Problem 8.27 it will be found that for a more general, nonsymmetric cross-section the location of the neutral axis after certain of the fibers have yielded is not the same as that found for purely elastic action where the neutral axis passes through the centroid of the cross-section.

From Fig. 8-46(b) we have for $y < y_1$:

$$\frac{\sigma}{y} = \frac{\sigma_{yp}}{y_1} \quad \text{or} \quad \sigma = \frac{y}{y_1}\sigma_{yp}$$

and for $y > y_1 : \sigma = \sigma_{yp} = $ constant. Thus the bending moment is

$$M = \int \sigma y \, da = 2\int_0^{y_1} \frac{y}{y_1}\sigma_{yp}y \, da + 2\int_{y_1}^c \sigma_{yp}y \, da$$

$$= \frac{2\sigma_{yp}}{y_1}\int_0^{y_1} y^2 \, da + 2\sigma_{yp}\int_{y_1}^c y \, da$$

8.23. For a beam of rectangular cross-section determine the moment acting when all fibers a distance y_1 from the neutral axis have reached the yield point of the material.

From the result of Problem 8.22 for the geometry indicated in Fig. 8-47 we have

$$M = \frac{2\sigma_{yp}}{y_1}\left(\tfrac{1}{3}by_1^3\right) + 2\sigma_{yp}b(c-y_1)\left(\frac{c+y_1}{2}\right)$$

$$= \left(bc^2 - \frac{b}{3}y_1^2\right)\sigma_{yp}$$

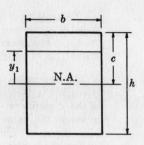

For the limiting case when $y_1 = 0$ which is indicated by Fig. 8-46(c) of Problem 8.22 the fully plastic moment of this rectangular beam is

$$M_p = bc^2\sigma_{yp} = \frac{bh^2}{4}\sigma_{yp} \qquad (1)$$

Fig. 8-47

It is to be noted that the maximum possible elastic moment, i.e. when the extreme fibers have just reached the yield point but all interior fibers are in the elastic range of action as indicated by Fig. 8-46(a), is

$$M_e = \frac{bh^2}{6}\sigma_{yp} \qquad (2)$$

Thus, for a rectangular cross-section, the fully plastic moment is 50 percent greater than the maximum possible elastic moment.

8.24. Determine the fully plastic moment of a rectangular beam, 1×2 inches in cross-section, of steel with a yield point of 38,000 lb/in². Compare this with the maximum possible elastic moment that this same section may carry.

From (1) of Problem 8.23 the fully plastic moment is

$$M_p = \frac{1(2)^2}{4}(38,000) = 38,000 \text{ lb-in}$$

From (2) of that same problem the maximum possible elastic moment is

$$M_e = \frac{1(2)^2}{6}(38,000) = 25,400 \text{ lb-in}$$

It is evident that M_p is 50 percent greater than M_e.

8.25. For a beam of rectangular cross-section determine the relation between the bending moment and the radius of curvature when all fibers at a distance y_1 from the neutral axis have reached the yield point of the material.

As in Problem 8.22 we assume that plane sections before loading remain plane and normal to the beam axis after loading. Because of this, normal strains of the longitudinal fibers vary linearly as the distance of the fibers from the neutral axis. Thus, if ϵ_{yp} denotes the strain of the fibers at a distance y_1 from the neutral axis and ϵ_c represents the outer fiber strain, we have

$$\frac{\epsilon_c}{c} = \frac{\epsilon_{yp}}{y_1} \qquad (1)$$

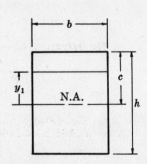

Consideration of the geometry of an originally rectangular element of length dx along the beam axis, as shown in Fig. 8-49(a), reveals that after bending it assumes the configuration indicated in Fig. 8-49(b). From that sketch we have

$$\frac{1}{R} = \frac{d\theta}{dx} = \frac{\epsilon_c}{c} \qquad (2)$$

Fig. 8-48

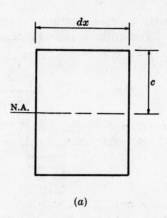

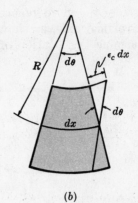

(a)   (b)

Fig. 8-49

Thus
$$\frac{d\theta}{dx} = \frac{\epsilon_{yp}}{y_1} = \frac{\sigma_{yp}}{Ey_1} \qquad (3)$$

since the fibers a distance y_1 from the neutral axis obey Hooke's law: $\sigma_{yp} = E\epsilon_{yp}$. From Problem 8.23 the moment corresponding to these strains is

$$M = \left(bc^2 y_1 - \frac{b}{3}y_1^3\right)\frac{\sigma_{yp}}{y_1} \qquad (4)$$

Thus, from (3) and (4):

$$\frac{d\theta}{dx} = \frac{M}{Eby_1(c^2 - \frac{1}{3}y_1^2)} \qquad (5)$$

Finally, from (2) and (5) we have

$$\frac{1}{R} = \frac{M}{EI\left(\dfrac{M}{M_e}\right)\sqrt{3 - \dfrac{2M}{M_e}}} \qquad (6)$$

where $M_e = bh^2\sigma_{yp}/6$ as in Problem 8.23. This is the desired relation between the bending moment M and the radius of curvature R. Equation (6) plots as shown in Fig. 8-50.

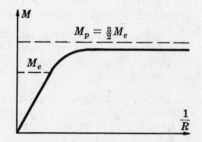

Fig. 8-50

8.26. Consider a beam of rectangular cross-section where $b = 1$ in., $h = \frac{1}{2}$ in. The material is steel for which $\sigma_{yp} = 30{,}000$ lb/in² and $E = 30 \times 10^6$ lb/in². Determine the radius of curvature corresponding to the maximum possible elastic moment and also the radius of curvature for a moment of 1850 lb-in.

From (2) of Problem 8.23 the maximum possible elastic moment is

$$M_e = \frac{1(1/2)^2}{6}(30{,}000) = 1250 \text{ lb-in}$$

The curvature corresponding to this moment is found from (6) of Problem 8.25 to be

$$\frac{1}{R} = \frac{1250}{(30 \times 10^6)\dfrac{(1)(1/2)^3}{12}\sqrt{3-2}} = 0.004 \quad \text{or} \quad R = 250 \text{ in.}$$

The value of y_1 corresponding to a moment of 1850 lb-in may be found from Problem 8.23 to be 0.05 in. The curvature corresponding to this is found from (6) of Problem 8.25 to be

$$\frac{1}{R} = \frac{1850}{(30 \times 10^6)\dfrac{(1)(1/2)^3}{12}\left(\dfrac{1850}{1250}\right)\sqrt{3-\dfrac{3700}{1250}}} = 0.02$$

or
$$R = 50 \text{ in.}$$

8.27. Consider the more general case of a beam with a cross-section symmetric only about the vertical axis, as shown in Fig. 8-51(a). For fully plastic bending [Fig. 8-51(b)] determine the location of the neutral axis.

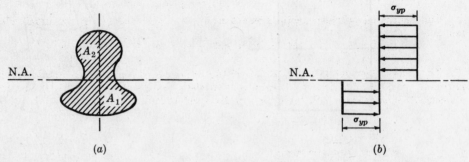

(a)
(b)

Fig. 8-51

Although the location of the neutral axis is unknown, let us denote the area of that portion of the cross-section lying below that axis by A_1 and the area of the portion above that axis by A_2. As shown by Fig. 8-51(b), all fibers in A_1 are subject to a tensile stress equal to the yield point of the material and all fibers in A_2 are subject to the same magnitude compressive stress. For horizontal equilibrium of these forces we have

$$\sigma_{yp}A_1 - \sigma_{yp}A_2 = 0 \tag{1}$$

from which
$$A_1 = A_2 = A/2 \tag{2}$$

where A is the area of the entire cross-section. Thus, for fully plastic action, the neutral axis divides the cross-section into two equal parts. This is in contrast to the situation for fully elastic action, where the neutral axis was found in Problem 8.1 to pass through the centroid of the cross-section.

Also, the sum of the moments of the tensile and compressive stresses must equal the applied moment M_p, the fully plastic moment. If $\bar{y}_1$ and $\bar{y}_2$ denote the distances from the neutral axis to the centroids of the areas A_1 and A_2 respectively, then from statics

$$\sigma_{yp}A_1\bar{y}_1 + \sigma_{yp}A_2\bar{y}_2 = M_p \tag{3}$$

From (2) this becomes
$$\sigma_{yp}\frac{A}{2}(\bar{y}_1 + \bar{y}_2) = M_p \tag{4}$$

or
$$\sigma_{yp} = \frac{M_p}{\dfrac{A}{2}(\bar{y}_1 + \bar{y}_2)} \tag{5}$$

This is frequently written in the form
$$\sigma_{yp} = \frac{M_p}{Z_p} \tag{6}$$

where $Z_p = \dfrac{A}{2}(\bar{y}_1 + \bar{y}_2)$ is termed the *plastic section modulus*.

8.28. For an 8 WF 40 wide-flange section of steel having a yield point of 38,000 lb/in², determine the fully plastic moment. Compare this with the maximum possible elastic moment that the same section can carry.

From Problem 8.27 the fully plastic moment M_p is given by

$$M_p = \sigma_{yp}Z_p$$

where Z_p is the plastic section modulus. For selected wide-flange sections Z_p is tabulated at the end of this chapter. In particular, for this section it is found to be 39.9 in^3. Thus

$$M_p = 38,000(39.9) = 1,520,000 \text{ lb-in}$$

The maximum possible elastic moment is $M_e = \sigma_{yp}Z$ where Z is the usual (elastic) section modulus. Thus

$$M_e = 38,000(35.5) = 1,350,000 \text{ lb-in}$$

The plastic moment is only 12.6 percent greater than the maximum elastic moment for this particular section. In fact, the fully plastic moment usually exceeds the maximum possible elastic moment by approximately 12 to 15 percent for most wide-flange sections.

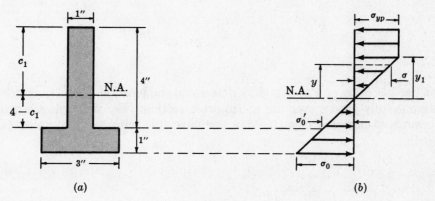

(a) (b)

Fig. 8-52

8.29. Consider the T-section in which all fibers in the vertical web at a distance y_1 from the neutral axis have reached the yield point of the material, whereas all other fibers are still in the elastic range of action. Determine the location of the neutral axis and also the moment that corresponds to this stress distribution.

The neutral axis (described by the unknown c_1) may be located by investigating the normal forces over the cross-section as shown in Fig. 8-52(b). From geometry

$$\frac{\sigma_0}{5-c_1} = \frac{\sigma_{yp}}{y_1} \quad \text{or} \quad \sigma_0 = \frac{5-c_1}{y_1}\sigma_{yp}$$

$$\frac{\sigma_0'}{4-c_1} = \frac{\sigma_{yp}}{y_1} \quad \text{or} \quad \sigma_0' = \frac{4-c_1}{y_1}\sigma_{yp}$$

For the resultant normal force to vanish

$$\Sigma F_N = (c_1 - y_1)(1)\sigma_{yp} + y_1(1)(\sigma_{yp}/2)$$

$$- \left\{ \left[\frac{5-c_1}{2y_1}(5-c_1)(3)\sigma_{yp} \right] - \left[\frac{4-c_1}{2y_1}(4-c_1)(2)\sigma_{yp} \right] \right\} = 0$$

from which we obtain the quadratic equation

$$c_1^2 - (2y_1 + 14)c_1 + (y_1^2 + 43) = 0 \tag{1}$$

which determines c_1 for any specified value of y_1. This locates the neutral axis. Note that since y_1 occurred in the denominator in the above derivation, the equation should not be used to locate the neutral axis if $y_1 = 0$. Thus, when the action is entirely elastic the neutral axis passes through the centroid of the cross-section. As plastification increases (i.e. as y_1 decreases) the neutral axis shifts to the location indicated by (1).

The moment corresponding to the stresses in Fig. 8-52(b) may be found from

$$M = \int \sigma y \, da$$

$$= \int_0^{y_1} \frac{y}{y_1}\sigma_{yp}(y)(1) \, dy + \int_{y_1}^{c_1} \sigma_{yp}(y)(1) \, dy$$

$$+ \int_0^{(5-c_1)} \left(\frac{y}{5-c_1}\right)\left(\frac{5-c_1}{y_1}\right)(\sigma_{yp})(y)(3) \, dy$$

$$- \int_0^{(4-c_1)} \left(\frac{y}{4-c_1}\right)\left(\frac{4-c_1}{y_1}\right)(\sigma_{yp})(y)(2) \, dy$$

or

$$M = \frac{\sigma_{yp}}{y_1}\left[\frac{y_1^3}{3} + \frac{y_1}{2}(c_1^2 - y_1^2) + (5-c_1)^3 - \tfrac{2}{3}(4-c_1)^3\right] \qquad (2)$$

8.30. For the T-section of Problem 8.29 determine the location of the neutral axis when the action is fully plastic over the entire cross-section. For fully plastic action determine the moment-carrying capacity and compare this with the maximum possible elastic moment.

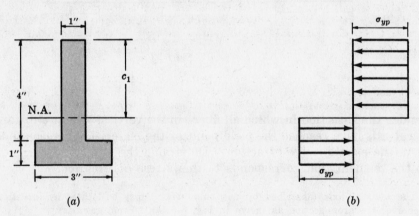

Fig. 8-53

In this case, the normal forces appear as indicated in Fig. 8-53(b). For equilibrium of normal forces over the cross-section we have

$$-\sigma_{yp}(1)(c_1) + [\sigma_{yp}(5-c_1)(3) - \sigma_{yp}(4-c_1)(2)] = 0$$

from which $c_1 = 3.5$ in. Thus, as mentioned in Problem 8.27, for fully plastic action the neutral axis divides the cross-section into two equal parts.

The moment corresponding to this fully plastic action is

$$M_p = \int \sigma y \, da$$

$$= \int_0^{c_1} \sigma_{yp}(y)(1) \, dy + \int_0^{(5-c_1)} \sigma_{yp}(y)(3) \, dy - \int_0^{(4-c_1)} \sigma_{yp}(y)(2) \, dy$$

$$= \sigma_{yp}[c_1^2 - 7c_1 + 21.5]$$

For $c_1 = 3.5$ this becomes

$$M_p = 9.25\sigma_{yp}$$

By setting $y_1 = c_1$ in (1) of Problem 8.29 the neutral axis is located for the case of the maximum possible elastic moment. This location is found to be $c_1 = 3.07$ in. (i.e. the neutral axis passes through the centroid of the cross-section). The maximum possible elastic moment is found from (2) of Problem 8.29 to be

$$M_e = 5.32\sigma_{yp}$$

The fully plastic moment exceeds this value by 74 percent.

Supplementary Problems

8.31. A beam made of titanium, type Ti-6Al-4V, has a yield point of 120,000 lb/in². The beam has 1 in × 2 in rectangular cross-section and bends about an axis parallel to the one-inch face. If the maximum bending stress is 90,000 lb/in², find the corresponding bending moment.

   Ans.   60,000 lb-in

8.32. A cantilever beam 9 ft long carries a concentrated force of 8000 lb at its free end. The material is structural steel and the maximum bending stress is not to exceed 18,000 lb/in². Determine the required diameter if the bar is to be circular.   *Ans.* 7.86 in.

8.33. A 10 WF 29 section (see Fig. 8-67 at end of this chapter for properties) is used as a cantilever beam. The beam is 5 ft long and the allowable bending stress is 18,000 lb/in². Determine the maximum allowable intensity of uniform load that may be carried along the entire length of the beam.   *Ans.* 3800 lb/ft

8.34. A steel beam 50 in. in length is simply supported at each end and carries a concentrated load of 25,000 lb acting 20 in. from one of the supports. Determine the maximum bending stresses set up in the beam if the cross-section is rectangular, 4 in. wide by 6 in. deep.   *Ans.* 12,500 lb/in²

8.35. Determine the maximum bending stresses for a bar loaded as in the preceding problem if the beam is a 10 WF 23 section.   *Ans.* 12,500 lb/in²

8.36. A strip of steel 0.04 in. thick is bent into an arc of a circle of 30 in. radius. Determine the maximum bending stress.   Take $E = 30 \times 10^6$ lb/in².   *Ans.* 20,000 lb/in²

8.37. The maximum bending moment existing in a steel beam is 750,000 lb-in. Select the most economical wide-flange section to resist this moment if the working stress both in tension and in compression is 20,000 lb/in².   *Ans.* 12 WF 32

8.38. The beam shown in Fig. 8-54 is simply supported at the ends and carries the two symmetrically placed loads of 15,000 lb each. If the working stress in either tension or compression is 18,000 lb/in², select the most economical wide-flange section to support these loads.   *Ans.* 10 WF 21

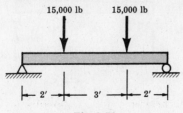

Fig. 8-54

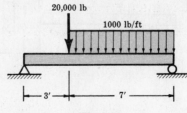

Fig. 8-55

8.39. Consider a simply supported beam carrying the concentrated and uniform loads shown in Fig. 8-55. Select a suitable wide-flange section to resist these loads based upon a working stress in either tension or compression of 20,000 lb/in².   *Ans.* 12 WF 25

8.40. The two distributed loads are carried by the simply supported beam as shown in Fig. 8-56. The cross-section of the beam is an 8 WF 28 section. Determine the magnitude and location of the maximum bending stress in the beam. *Ans.* 9000 lb/in², 5.5 ft from the right support

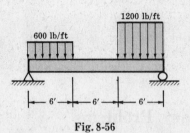

Fig. 8-56

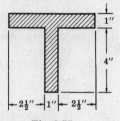

Fig. 8-57

8.41. A T-beam having the cross-section shown in Fig. 8-57 projects five feet from a wall as a cantilever beam and carries a uniformly distributed load of 600 lb/ft including its own weight. Determine the maximum tensile and compressive bending stresses. *Ans.* +6400 lb/in², −15,200 lb/in²

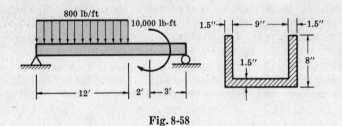

Fig. 8-58

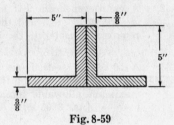

Fig. 8-59

8.42. A simply supported steel beam of channel-type cross-section is loaded by both the uniformly distributed load and the couple shown in Fig. 8-58. Determine the maximum tensile and compressive stresses set up. *Ans.* 3020 lb/in² tension, 5520 lb/in² compression

8.43. Two $5 \times 5 \times 3/8$ in. angles are welded together as shown in Fig. 8-59, and used as a beam to support loads in a vertical plane so that bending takes place about a horizontal neutral axis. Determine the maximum bending moment that may exist in the beam if the bending stress is not to exceed 20,000 lb/in² in either tension or compression. *Ans.* 97,500 lb-in

8.44. A channel-shape beam with an overhanging end is loaded as shown in Fig. 8-60. The material is gray cast iron having an allowable working stress of 5000 lb/in² in tension and 20,000 lb/in² in compression. Determine the maximum allowable value of *P*. *Ans.* 2400 lb

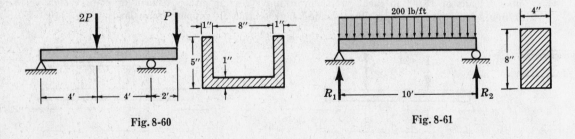

Fig. 8-60

Fig. 8-61

8.45. In Fig. 8-61 the simply supported beam of length 10 ft and cross-section 4 in. by 8 in. carries a uniform load of 200 lb/ft. Neglecting the weight of the beam, find: (*a*) the maximum normal stress in the beam, (*b*) the maximum shearing stress in the beam, (*c*) the shearing stress at a point 2 ft to the right of R_1 and 1 in. below the top surface of the beam.
Ans. (*a*) 705 lb/in², (*b*) 47 lb/in², (*c*) 12.3 lb/in²

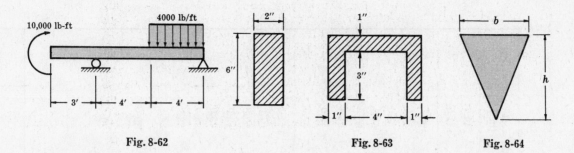

Fig. 8-62   Fig. 8-63   Fig. 8-64

8.46.   Determine (a) the maximum bending stress and (b) the maximum shearing stress in the simply supported beam shown in Fig. 8-62. *Ans.* (a) 22,000 lb/in², (b) 1660 lb/in²

8.47.   A beam has the channel-type cross-section shown in Fig. 8-63. If the maximum shearing force along the length of the beam is 6000 lb, determine the maximum shearing stress in the beam. *Ans.* 1110 lb/in²

8.48.   For a bar of solid circular cross-section, determine the amount by which the fully plastic moment exceeds the moment that just causes the yield point to be reached in the extreme fibers. *Ans.* 69.6 percent

8.49.   Consider bending of a bar of isosceles triangular cross-section (Fig. 8-64). The loads lie in the vertical plane of symmetry. Determine the ratio of the fully plastic moment to the moment that just causes yielding of the extreme fibers. *Ans.* 2.48

8.50.   For the T-section shown in Fig. 8-65, determine the location of the neutral axis for fully plastic action. *Ans.* $5\frac{1}{2}$ in. above the lowest fibers of the section

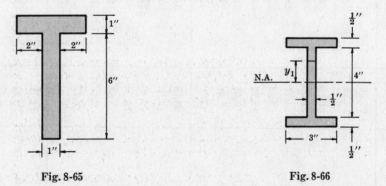

Fig. 8-65   Fig. 8-66

8.51.   A bar of solid circular cross-section of radius r is subject to bending. By what percent does the bending moment required to cause plastic action at the distance $r/2$ from the neutral axis exceed that required to just cause the yield point to be reached in the extreme fibers? *Ans.* 49.2 percent

8.52.   Determine the fully plastic moment for a 10 WF 54 steel beam if the yield point of the material is 38,000 lb-in². *Ans.* 2,546,000 lb-in

8.53.   For the section shown in Fig. 8-66 determine the value of y_1 which represents the point where elastic action terminates and plastic flow begins, when the beam is subject to a bending moment of 250,000 lb-in. Also determine the radius of curvature. Take the yield point of the material to be 30,000 lb/in². *Ans.* $y_1 = 1.59$ in., $R = 1590$ in.

Fig. 8-67. Properties of Selected Wide-Flange Sections

Designation	Weight per foot (lb/ft)	Area (in²)	I (about x-x axis) (in⁴)	Z (in³)	I (about y-y axis) (in⁴)	Z_p (plastic section modulus) (in³)
18 WF 70	70.0	20.56	1153.9	128.2	78.5	144.7
18 WF 55	55.0	16.19	889.9	98.2	42.0	111.6
12 WF 72	72.0	21.16	597.4	97.5	195.3	108.1
12 WF 58	58.0	17.06	476.1	78.1	107.4	86.5
12 WF 50	50.0	14.71	394.5	64.7	56.4	72.6
12 WF 45	45.0	13.24	350.8	58.2	50.0	64.9
12 WF 40	40.0	11.77	310.1	51.9	44.1	57.6
12 WF 36	36.0	10.59	280.8	45.9	23.7	51.4
12 WF 32	32.0	9.41	246.8	40.7	20.6	45.0
12 WF 25	25.0	7.39	183.4	30.9	14.5	35.0
10 WF 89	89.0	26.19	542.4	99.7	180.6	114.4
10 WF 54	54.0	15.88	305.7	60.4	103.9	67.0
10 WF 49	49.0	14.40	272.9	54.6	93.0	60.3
10 WF 45	45.0	13.24	248.6	49.1	53.2	55.0
10 WF 37	37.0	10.88	196.9	39.9	42.2	45.0
10 WF 29	29.0	8.53	157.3	30.8	15.2	34.7
10 WF 23	23.0	6.77	120.6	24.1	11.3	33.7
10 WF 21	21.0	6.19	106.3	21.5	9.7	24.1
8 WF 40	40.0	11.76	146.3	35.5	49.0	39.9
8 WF 35	35.0	10.30	126.5	31.1	42.5	34.7
8 WF 31	31.0	9.12	109.7	27.4	37.0	30.4
8 WF 28	28.0	8.23	97.8	24.3	21.6	27.1
8 WF 27	27.0	7.93	94.1	23.4	20.8	23.9
8 WF 24	24.0	7.06	82.5	20.8	18.2	23.1
8 WF 19	19.0	5.59	64.7	16.0	7.9	17.7
6 WF 15½	15.5	4.62	28.1	9.7	9.7	11.3

Chapter 9

Elastic Deflection of Beams: Double-Integration Method

INTRODUCTION

In Chapter 8 it was stated that lateral loads applied to a beam not only give rise to internal bending and shearing stresses in the bar, but also cause the bar to deflect in a direction perpendicular to its longitudinal axis. The stresses were examined in Chapter 8 and it is the purpose of this chapter and also Chapters 10 and 11 to examine methods for calculating the deflections.

DEFINITION OF DEFLECTION OF A BEAM

The deformation of a beam is most easily expressed in terms of the deflection of the beam from its original unloaded position. The deflection is measured from the original neutral surface to the neutral surface of the deformed beam. The configuration assumed by the deformed neutral surface is known as the elastic curve of the beam. Figure 9-1 represents the beam in its original undeformed state and Fig. 9-2 represents the beam in the deformed configuration it has assumed under the action of the loads.

Fig. 9-1   Fig. 9-2

The displacement y is defined as the deflection of the beam. Usually it will be necessary to determine the deflection y for every value of x along the beam. This relation may be written in the form of an equation which is frequently called the equation of the deflection curve (or elastic curve) of the beam.

IMPORTANCE OF BEAM DEFLECTIONS

Specifications for the design of beams frequently impose limitations upon the deflections as well as the stresses. Consequently, in addition to the calculation of stresses as outlined in Chapter 8, it is essential that the designer be able to determine deflections. For example, in many building codes the maximum allowable deflection of a beam is not to exceed 1/300 of the length of the beam. Components of aircraft usually are designed so that deflections do not exceed some preassigned value, else the aerodynamic characteristics may be altered. Thus, a well-designed beam must not only be able to carry the loads to which it will be subjected but it must not undergo undesirably large deflections. Also, the evaluation of reactions of statically indeterminate beams involves the use of various deformation relationships. These will be examined in detail in Chapter 12.

METHODS OF DETERMINING BEAM DEFLECTIONS

Numerous methods are available for the determination of beam deflections. The most commonly used are:

(a)   *double-integration method*

(b)   *moment-area method*

(c)   *method of singularity functions*

(d)   *elastic energy methods*

The first method is described in this chapter, the moment-area technique is examined in Chapter 10, the use of singularity functions is discussed in Chapter 11, and elastic energy methods are treated in Chapter 16. It is to be carefully noted that all of these methods apply *only* if all portions of the beam are acting in the *elastic range of action*.

DOUBLE-INTEGRATION METHOD

The differential equation of the deflection curve of the bent beam is

$$EI\frac{d^2y}{dx^2} \;=\; M \tag{1}$$

where x and y are the coordinates shown in Fig. 9-2. That is, y is the deflection of the beam. This equation is derived in Problem 9.1. In the equation E denotes the modulus of elasticity of the beam and I represents the moment of inertia of the beam cross-section about the neutral axis, which passes through the centroid of the cross-section. Also, M represents the bending moment at the distance x from one end of the beam. This quantity was defined in Chapter 6 to be the algebraic sum of the moments of the external forces to one side of the section at a distance x from the end about an axis through this section. Usually, M will be a function of x and it will be necessary to integrate (1) twice to obtain an algebraic equation expressing the deflection of y as a function of x.

Equation (1) is the basic differential equation that governs the elastic deflection of all beams irrespective of the type of applied loading. For applications, see Problems 9.2, 9.5, 9.6, 9.8, 9.10, 9.12, 9.14, 9.15, 9.16, 9.17 and 9.19.

THE INTEGRATION PROCEDURE

The double-integration method for calculating deflections of beams merely consists of integrating (1). The first integration yields the slope dy/dx at any point in the beam and the second integration gives the deflection y for any value of x. The bending moment M must, of course, be expressed as a function of the coordinate x before the equation can be integrated. For the cases to be studied here the integrations are extremely simple.

Since the differential equation (1) is of the second order, its solution must contain two constants of integration. These two constants must be evaluated from known conditions concerning the slope or deflection at certain points in the beam. For example, in the case of a cantilever beam the constants would be determined from the conditions of zero change of slope as well as zero deflection at the built-in end of the beam.

Frequently two or more equations are necessary to describe the bending moment in the various regions along the length of a beam. This was emphasized in Chapter 6. In such a case, (1) must be written for each region of the beam and integration of these equations yields two constants of integration for each region. These constants must then be determined so as to impose conditions of continuous deformations and slopes at the points common to adjacent regions. See Problems 9.12, 9.14, 9.16 and 9.17.

SIGN CONVENTIONS

The sign conventions for bending moment adopted in Chapter 6 will be retained here. The quantities E and I appearing in (1) are, of course, positive. Thus, from this equation, if M is positive for a certain value of x, then d^2y/dx^2 is also positive. With the above sign convention for bending moments, it is necessary to consider the coordinate x along the length of the beam to be positive to the right and the deflection y to be positive upward. This will be explained in detail in Problem 9.1. With these algebraic signs the integration of (1) may be carried out to yield the deflection y as a function of x, with the understanding that upward beam deflections are positive and downward deflections negative.

ASSUMPTIONS AND LIMITATIONS

In the derivation of (1) it is assumed that deflections caused by shearing action are negligible compared to those caused by bending action. Also, it is assumed that the deflections are small compared to the cross-sectional dimensions of the beam and that all portions of the beam are acting in the elastic range. Equation (1) is derived on the basis of the beam being straight prior to the application of loads. Beams with slight deviations from straightness prior to loading may be treated by modifying this equation as indicated in Problem 9.22.

Solved Problems

9.1. Obtain the differential equation of the deflection curve of a beam loaded by lateral forces.

In Problem 8.1 the relationship

$$M = EI/\rho \qquad (1)$$

was derived. In this expression M denotes the bending moment acting at a particular cross-section of the beam, ρ the radius of curvature to the neutral surface of the beam at this same section, E the modulus of elasticity, and I the moment of the cross-sectional area about the neutral axis passing through the centroid of the cross-section. In this book we will usually be concerned with those beams for which E and I are constant along the entire length of the beam, but in general both M and ρ will be functions of x.

Equation (1) may be written in the form

$$\frac{1}{\rho} = \frac{M}{EI} \qquad (2)$$

where the left side of equation (2) represents the curvature of the neutral surface of the beam. Since M will vary along the length of the beam, the deflection curve will be of variable curvature.

Let the heavy line in the adjacent Fig. 9-3 represent the deformed neutral surface of the bent beam. Originally the beam coincided with the x-axis prior to loading and the coordinate system that is usually found to be most convenient is shown in the sketch. The deflection y is taken to be positive in the upward direction; hence for the particular beam shown, all deflections are negative.

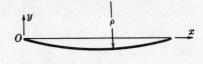

Fig. 9-3

An expression for the curvature at any point along the curve representing the deformed beam is readily available from differential calculus. The exact formula for curvature is

$$\frac{1}{\rho} = \frac{d^2y/dx^2}{[1+(dy/dx)^2]^{3/2}} \qquad (3)$$

In this expression, dy/dx represents the slope of the curve at any point; and for small beam deflections this quantity and in particular its square are small in comparison to unity and may reasonably be neglected. This assumption of small deflections simplifies the expression for curvature into

$$\frac{1}{\rho} \approx \frac{d^2y}{dx^2} \qquad (4)$$

Hence for small deflections, (2) becomes $d^2y/dx^2 = M/EI$ or

$$EI\frac{d^2y}{dx^2} = M \qquad (5)$$

This is the differential equation of the deflection curve of a beam loaded by lateral forces. In honor of its co-discoverers, it is called the Euler-Bernoulli equation of bending of a beam. In any problem it is necessary to integrate this equation to obtain an algebraic relationship between the deflection y and the coordinate x along the length of the beam. This will be carried out in the following problems.

9.2. Determine the deflection at every point of the cantilever beam subject to the single concentrated force P, as shown in Fig. 9-4.

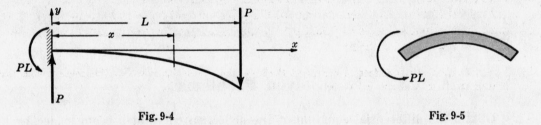

Fig. 9-4   Fig. 9-5

The x-y coordinate system shown is introduced, where the x-axis coincides with the original unbent position of the beam. The deformed beam has the appearance indicated by the heavy line. It is first necessary to find the reactions exerted by the supporting wall upon the bar; and as discussed in Problem 6.1, these are easily found from statics to be a vertical force reaction P and a moment PL as shown.

The bending moment at any cross-section a distance x from the wall is given by the sum of the moments of these two reactions about an axis through this section. Evidently the upward force P produces a positive bending moment Px, and the couple PL if acting alone would produce curvature of the bar as shown in Fig. 9-5. According to the sign convention of Chapter 6, this constitutes negative bending. Hence the bending moment M at the section x is

$$M = -PL + Px$$

The differential equation of the bent beam is

$$EI\frac{d^2y}{dx^2} = M$$

where E denotes the modulus of elasticity of the material and I represents the moment of inertia of the cross-section about the neutral axis. Substituting,

$$EI\frac{d^2y}{dx^2} = -PL + Px \qquad (1)$$

This equation is readily integrated once to yield

$$EI\frac{dy}{dx} = -PLx + \frac{Px^2}{2} + C_1 \qquad (2)$$

which represents the equation of the slope, where C_1 denotes a constant of integration. This constant may be evaluated by use of the condition that the slope dy/dx of the beam at the wall is zero since the beam is rigidly clamped there. Thus $(dy/dx)_{x=0} = 0$. Equation (2) is true for all values of x and y, and if the condition $x = 0$ is substituted we obtain $0 = 0 + 0 + C_1$ or $C_1 = 0$.

Next, integration of (2) yields

$$EIy = -PL\frac{x^2}{2} + \frac{Px^3}{6} + C_2 \qquad (3)$$

where C_2 is a second constant of integration. Again, the condition at the supporting wall will determine this constant. There, at $x = 0$, the deflection y is zero since the bar is rigidly clamped. Substituting $(y)_{x=0} = 0$ in equation (3), we find $0 = 0 + 0 + C_2$ or $C_2 = 0$.

Thus equations (2) and (3) with $C_1 = C_2 = 0$ give the slope dy/dx and deflection y at any point x in the beam. The deflection is a maximum at the right end of the beam $(x = L)$, under the load P, and from equation (3),

$$EIy_{max} = -PL^3/3$$

where the negative value denotes that this point on the deflection curve lies below the x-axis. If only the magnitude of the maximum deflection at $x = L$ is desired, it is usually denoted by Δ_{max} and we have

$$\Delta_{max} = PL^3/3EI \qquad (4)$$

9.3. The cantilever beam shown in Fig. 9-4 above is 10 ft in length and loaded by a force P of 10,000 lb. The beam is an 18 WF 47 steel section, having a moment of inertia about the neutral axis of 736.4 in⁴. Determine the maximum deflection of the beam. Take $E = 30 \times 10^6$ lb/in².

The maximum deflection occurs at the free end of the beam under the concentrated load and was found in Problem 9.2 to be

$$\Delta_{max} = \frac{PL^3}{3EI} = \frac{10,000(120)^3}{3(30 \times 10^6)(736.4)} = 0.261 \text{ in.}$$

This deflection is downward as indicated in Fig. 9-4. In the derivation of this deflection formula it was assumed that the material of the beam follows Hooke's law. Actually, from the above calculation alone there is no assurance that the material is not stressed beyond the proportional limit. If it were, then the basic beam-bending equation $EI(d^2y/dx^2) = M$ would no longer be valid and the above numerical value would be meaningless. Consequently, in every problem involving beam deflections it is to be emphasized that it is necessary to determine that the maximum bending stress in the beam is below the proportional limit of the material. This is easily done by use of the flexure formula derived in Problem 8.1. According to this formula

$$\sigma = Mc/I$$

where σ denotes the bending stress, M the bending moment, c the distance from the neutral axis to the outer fibers of the beam, and I the moment of inertia of the beam cross-section about the neutral axis. The maximum bending moment in this problem occurs at the supporting wall and is given by $M_{max} = 10,000(120) = 1,200,000$ lb-in. Substituting in the formula for bending stress, we have

$$\sigma_{max} = 1,200,000(9)/736.4 = 14,700 \text{ lb/in}^2$$

Since this value is below the proportional limit of steel, which is approximately 30,000 lb/in² for low-carbon structural steel, the use of the beam deflection equation was justifiable.

9.4. Determine the slope of the right end of the cantilever beam loaded as shown in Fig. 9-4. For the beam described in Problem 9.3, determine the value of this slope.

In Problem 9.2 the equation of the slope was found to be

$$EI\frac{dy}{dx} \;=\; -PLx + \frac{Px^2}{2}$$

At the free end, $x = L$, and

$$EI\left(\frac{dy}{dx}\right)_{x=L} \;=\; -PL^2 + \frac{PL^2}{2}$$

The slope at the free end is thus

$$\left(\frac{dy}{dx}\right)_{x=L} \;=\; -\frac{PL^2}{2EI}$$

For the beam described in Problem 9.3, this becomes

$$\left(\frac{dy}{dx}\right)_{x=L} \;=\; -\frac{10{,}000(120)^2}{2(30 \times 10^6)(736.4)} \;=\; -0.0326 \text{ radian}$$

9.5. Determine the deflection at every point of a cantilever beam subject to the uniformly distributed load of w lb per unit length shown in Fig. 9-6.

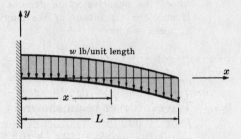

Fig. 9-6

The x-y coordinate system shown is introduced, where the x-axis coincides with the original unbent position of the beam. The deformed beam has the appearance indicated by the heavy line. The equation for the bending moment could be determined in a manner analogous to that used in Problem 9.2, but instead let us seek a slight simplification of that technique. Let us determine the bending moment at the section a distance x from the wall by considering the forces to the right of this section rather than those to the left.

The force of w lb/unit length acts over the length $L - x$ to the right of this section and hence the resultant force is $w(L - x)$ lb. This force acts at the midpoint of this length of beam to the right of x and thus its moment arm from x is $\frac{1}{2}(L - x)$. The bending moment at the section x is thus given by

$$M \;=\; -\frac{w}{2}(L - x)^2$$

the negative sign being necessary since downward loads produce negative bending.

The differential equation describing the bent beam is thus

$$EI\frac{d^2y}{dx^2} \;=\; -\frac{w}{2}(L - x)^2 \qquad\qquad (1)$$

The first integration yields

$$EI\frac{dy}{dx} \;=\; \frac{w}{2}\frac{(L - x)^3}{3} + C_1 \qquad\qquad (2)$$

where C_1 denotes a constant of integration.

This constant may be evaluated by realizing that the left end of the beam is rigidly clamped. At that point, $x = 0$, we have no change of slope and hence $(dy/dx)_{x=0} = 0$. Substituting these values in (2), we find $0 = wL^3/6 + C_1$ or $C_1 = -wL^3/6$. We thus have

$$EI\frac{dy}{dx} \;=\; \frac{w}{6}(L - x)^3 - \frac{wL^3}{6} \qquad\qquad (2')$$

The next integration yields

$$EIy \;=\; -\frac{w}{6}\frac{(L - x)^4}{4} - \frac{wL^3}{6}x + C_2 \qquad\qquad (3)$$

where C_2 represents a second constant of integration.

At the clamped end, $x = 0$, of the beam the deflection is zero and since (3) holds for all values of x and y, it is permissible to substitute this pair of values in it. Doing this, we obtain

$$0 = -wL^4/24 + C_2 \quad \text{or} \quad C_2 = wL^4/24$$

The final form of the deflection curve of the beam is thus

$$EIy = -\frac{w}{24}(L-x)^4 - \frac{wL^3}{6}x + \frac{wL^4}{24} \tag{3'}$$

The deflection is a maximum at the right end of the bar $(x = L)$ and there we have from $(3')$

$$EIy_{max} = -\frac{wL^4}{6} + \frac{wL^4}{24} = -\frac{wL^4}{8}$$

where the negative value denotes that this point on the deflection curve lies below the x-axis. The magnitude of the maximum deflection is

$$\Delta_{max} = wL^4/8EI \tag{4}$$

9.6. Obtain an expression for the deflection curve of the simply supported beam of Fig. 9-7 subject to the uniformly distributed load of w lb per unit length as shown.

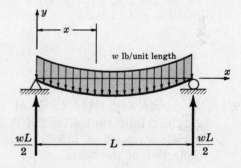

Fig. 9-7

The x-y coordinate system shown is introduced, where the x-axis coincides with the original unbent position of the beam. The deformed beam has the appearance indicated by the heavy line. The total load acting on the beam is wL lb and because of symmetry, each of the end reactions is $wL/2$ lb. Because of the symmetry of loading, it is evident that the deflected beam is symmetric about the midpoint $x = L/2$.

The equation for the bending moment at any section of a beam loaded and supported as this one is was discussed in Problem 6.3. According to the method indicated there, the portion of the uniform load to the left of the section a distance x from the left support is replaced by its resultant acting at the midpoint of the section of length x. The resultant is wx lb acting downward and hence giving rise to a negative bending moment. The reaction $wL/2$ gives rise to a positive bending moment. Consequently, for any value of x, the bending moment is

$$M = \frac{wL}{2}x - wx\frac{x}{2}$$

The differential equation of the bent beam is $EI(d^2y/dx^2) = M$. Substituting,

$$EI\frac{d^2y}{dx^2} = \frac{wL}{2}x - \frac{wx^2}{2} \tag{1}$$

Integrating,

$$EI\frac{dy}{dx} = \frac{wL}{2}\frac{x^2}{2} - \frac{w}{2}\frac{x^3}{3} + C_1 \tag{2}$$

It is to be noted that dy/dx represents the slope of the beam. Since the deflected beam is symmetric about the center of the span, i.e. about $x = L/2$, it is evident that the slope must be zero there. That is, the tangent to the deflected beam is horizontal at the midpoint of the beam. This condition enables us to determine C_1. Substituting this condition in (2) we obtain $(dy/dx)_{x=L/2} = 0$,

$$0 = \frac{wL}{4}\frac{L^2}{4} - \frac{w}{6}\frac{L^3}{8} + C_1 \quad \text{or} \quad C_1 = -\frac{wL^3}{24}$$

The slope dy/dx at any point is thus given by

$$EI\frac{dy}{dx} = \frac{wL}{4}x^2 - \frac{w}{6}x^3 - \frac{wL^3}{24} \tag{2'}$$

Integrating again, we find

$$EIy = \frac{wL}{4}\frac{x^3}{3} - \frac{w}{6}\frac{x^4}{4} - \frac{wL^3}{24}x + C_2 \tag{3}$$

This second constant of integration C_2 is readily determined by the fact that the deflection y is zero at the left support. Substituting $y_{x=0} = 0$ in (3), we find $0 = 0 - 0 - 0 + C_2$ or $C_2 = 0$.

The final form of the deflection curve of the beam is thus

$$EIy = \frac{wL}{12}x^3 - \frac{w}{24}x^4 - \frac{wL^3}{24}x \tag{3'}$$

The maximum deflection of the beam occurs at the center because of symmetry. Substituting $x = L/2$ in $(3')$, we obtain

$$EIy_{max} = -\frac{5wL^4}{384}$$

Or, without regard to algebraic sign, we have for the maximum deflection of a uniformly loaded, simply supported beam

$$\Delta_{max} = \frac{5}{384}\frac{wL^4}{EI} \tag{4}$$

9.7. A simply supported beam of length 10 ft and rectangular cross-section 1×3 in. carries a uniform load of 200 lb/ft. The beam is titanium, type Ti-5Al-2.5Sn, having a yield strength of 115,000 lb/in² and $E = 16 \times 10^6$ lb/in². Determine the maximum deflection of the beam.

From Problem 9.6 the maximum deflection is

$$\Delta_{max} = \frac{5}{384}\frac{wL^4}{EI}$$

Substituting,

$$\Delta_{max} = \frac{5}{384}\frac{(200/12)(120)^4}{(16 \times 10^6)\frac{1}{12}(1)(3)^3} = 1.25 \text{ in.}$$

Using the methods of Chapter 8, the maximum bending stress is found to be only 20,000 lb/in², well below the nonlinear range of action of the material. Thus the use of the deflection formula is justified.

9.8. Obtain an equation for the deflection curve of the simply supported beam subject to the concentrated load P applied at the center of the beam as shown in Fig. 9-8.

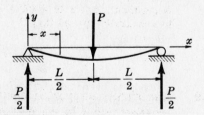

The x-y coordinate system shown is introduced. The deformed beam has the appearance indicated by the heavy line. Because of symmetry each end reaction is obviously $P/2$.

Fig. 9-8

The bending moment in the left half of the beam is

$$M = (P/2)x \quad \text{for} \quad 0 < x < L/2$$

The differential equation of the bent beam is $EI(d^2y/dx^2) = M$.   Substituting,

$$EI\frac{d^2y}{dx^2} = (P/2)x \quad \text{for} \quad 0 < x < L/2 \tag{1}$$

The first integration of this equation yields

$$EI\frac{dy}{dx} = \frac{P}{2}\frac{x^2}{2} + C_1 \tag{2}$$

The slope of the beam is represented by dy/dx.   Since the beam is loaded at its midpoint, the deflections are symmetric about the center of the beam, i.e. about the section $x = L/2$.   This condition of symmetry tells us that the slope must be zero at $x = L/2$, i.e. the tangent to the deflected beam is horizontal there.   Substituting this condition $(dy/dx)_{x=L/2} = 0$ in (2), we obtain

$$0 = \frac{P}{4}\frac{L^2}{4} + C_1 \quad \text{or} \quad C_1 = -\frac{PL^2}{16}$$

Thus the slope dy/dx at any point in the beam is given by

$$EI\frac{dy}{dx} = \frac{P}{4}x^2 - \frac{PL^2}{16} \tag{2'}$$

Integrating again, we find

$$EIy = \frac{P}{4}\frac{x^3}{3} - \frac{PL^2}{16}x + C_2 \tag{3}$$

The second constant of integration C_2 is determined by the fact that the deflection y of the beam is zero at the left support.   Thus $y_{x=0} = 0$.   Substituting in (3), we obtain $0 = 0 - 0 + C_2$ or $C_2 = 0$.

Thus the deflection curve of the left half of the beam is given by

$$EIy = \frac{P}{12}x^3 - \frac{PL^2}{16}x \tag{3'}$$

At this point it is to be carefully noted that it is *not* permissible to make use of the condition that the deflection y is zero at the right support, i.e. $y_{x=L} = 0$.   This is because the bending moment equation, $M = (P/2)x$, is valid only for values of x less than $L/2$, i.e. to the left of the applied load P.   To the right of force P the bending moment equation contains one additional term, and it would be necessary to work with the bending moment equation in the right half of the beam if the condition $y_{x=L} = 0$ were to be used.   Actually there is no need to examine deflections to the right of the load since it is known that the deflection curve of the beam is symmetric about $x = L/2$.   Briefly, in determining constants of integration one may use only those conditions on deflection or slope that pertain to the interval of the beam for which the bending moment equation was written.

Evidently the maximum deflection of the beam occurs at the center by virtue of symmetry. At this point the deflection is

$$EIy_{\max} = -PL^3/48$$

Or, without regard to algebraic sign, the maximum deflection of a simply supported beam subject to a centrally applied load P is

$$\Delta_{\max} = PL^3/48EI \tag{4}$$

9.9.   The simply supported beam described in Problem 9.8 is 14 ft long and of circular cross-section 4 in. in diameter.   If the maximum permissible deflection is 0.20 in., determine the maximum value of the load P.   The material is steel for which $E = 30 \times 10^6$ lb/in^2.

The maximum deflection, given by (4) of Problem 9.8, is $\Delta_{\max} = PL^3/48EI$.   For a circular cross-section (see Problem 7.9), $I = \pi D^4/64 = \pi 4^4/64 = 12.6$ in^4.   Also, $L = 14$ ft $= 168$ in.   Thus:

$$0.20 = \frac{P(168)^3}{48(30 \times 10^6)(12.6)} \quad \text{or} \quad P = 765 \text{ lb}$$

With this load applied at the center of the beam the reaction at each end is 383 lb and the bending moment at the center of the beam is $383(7) = 2681$ lb-ft. This is the maximum bending moment in the beam and the maximum bending stress occurs at the outer fibers at this central section. The maximum bending stress is $\sigma = Mc/I$. Then $\sigma_{max} = 2681(12)(2)/12.6 = 5100$ lb/in². This is below the proportional limit of the material; hence the use of the deflection equation was permissible.

9.10. Determine the equation of the deflection curve for a simply supported beam loaded by a couple M_1 at the right end of the bar as shown in Fig. 9-9(a).

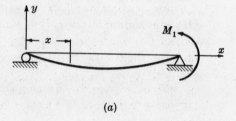

(a)

It is first necessary to determine the reactions acting on the beam. Since the applied couple M_1 can be held in equilibrium only by the action of another couple, the end reactions must be forces of equal magnitude R, but opposite in direction as indicated in Fig. 9-9(b). To find their magnitude we may write the statics equation

$$\Sigma M_0 = -M_1 + RL = 0 \quad \text{or} \quad R = M_1/L$$

The heavy line indicates the configuration of the deflected beam. The bending moment at any section a distance x from the left reaction is

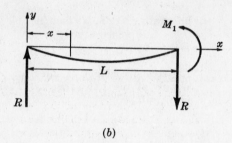

(b)

$$M = Rx = M_1 x/L$$

This equation is valid for all values of x.

Fig. 9-9

The differential equation of the deformed beam is

$$EI\frac{d^2y}{dx^2} = \frac{M_1}{L}x \tag{1}$$

Integrating once, we obtain

$$EI\frac{dy}{dx} = \frac{M_1}{L}\frac{x^2}{2} + C_1 \tag{2}$$

There is no information available concerning the slope of the beam, hence it is not possible to determine C_1 at this stage. It is to be noted that there is no symmetry to the loading, hence there is no reason to expect the slope to be zero at the midpoint of the beam. We integrate again and obtain

$$EIy = \frac{M_1}{2L}\frac{x^3}{3} + C_1 x + C_2 \tag{3}$$

At this stage we are able to determine the constants of integration C_1 and C_2. It is evident that the deflection y is zero at the left support, i.e. $y_{x=0} = 0$. Substituting these values of x and y in (3), we obtain $0 = 0 + 0 + C_2$ or $C_2 = 0$.

Also, the deflection y is zero at the right support, i.e. $y_{x=L} = 0$. Substituting these values of x and y in (3), we find $0 = \frac{M_1}{6L}L^3 + C_1 L$ or $C_1 = -\frac{M_1 L}{6}$.

The deflection curve of the beam is consequently

$$EIy = \frac{M_1 x^3}{6L} - \frac{M_1 L}{6}x \tag{3'}$$

The maximum deflection of the beam occurs at that point where the slope is zero, i.e. at that point where the tangent to the deflection curve is horizontal. The coordinate x of this point is readily found by setting the left side of (2) equal to zero.   Doing this we get

$$0 = \frac{M_1 x^2}{2L} - \frac{M_1 L}{6} \quad \text{or} \quad x = \frac{L}{\sqrt{3}}$$

The maximum deflection of the beam thus occurs at a distance $L/\sqrt{3}$ from the left reaction. The value of this deflection is found by substituting $x = L/\sqrt{3}$ in (3′).   This yields

$$EIy_{\max} = \frac{M_1}{6L}\frac{L^3}{3\sqrt{3}} - \frac{M_1 L}{6}\frac{L}{\sqrt{3}} = -\frac{M_1 L^2 \sqrt{3}}{27} \qquad (4)$$

9.11.   A simply supported beam is loaded by a couple M_1 as shown in Fig. 9-9.   The beam is 6 ft long and of square cross-section, 2 in. on a side.   If the maximum permissible deflection in the beam is 0.20 in. and the allowable bending stress is 20,000 lb/in^2, find the maximum allowable load M_1.   Take $E = 30 \times 10^6$ lb/in^2.

It is perhaps simplest to determine two values of M_1: one based upon the assumption that the deflection of 0.20 in. is realized, the other based on the assumption that the maximum bending stress in the bar is 20,000 lb/in^2.   The true value of M_1 is then the minimum of these two values.

Let us first consider that the maximum deflection in the beam is 0.20 in.   According to equation (4), Problem 9.10, we have

$$0.20 = \frac{M_1 (72)^2 \sqrt{3}}{27(30 \times 10^6)(1/12)(2)(2)^3} \quad \text{or} \quad M_1 = 24{,}000 \text{ lb-in}$$

We shall now assume that the allowable bending stress of 20,000 lb/in^2 is set up in the outer fibers of the beam at the section of maximum bending moment. The bending moment diagram is shown at the right.   From this it is seen that the maximum bending moment in the beam is M_1.   Using the usual flexure formula, $\sigma = Mc/I$, we have at the outer fibers of the bar at the right end, i.e. at the section of maximum bending moment,

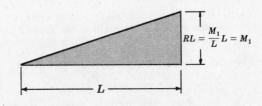

$RL = \dfrac{M_1}{L}L = M_1$

Fig. 9-10

$$20{,}000 = \frac{M_1(1)}{\frac{1}{12}(2)(2)^3} \quad \text{or} \quad M_1 = 26{,}700 \text{ lb-in}$$

Thus the maximum allowable load is $M_1 = 24{,}000$ lb-in.

9.12.   Determine the deflection curve of a simply supported beam subject to the concentrated force P applied as shown in Fig. 9-11.

The x-y coordinate system is introduced as shown.   The heavy line indicates the configuration of the deformed beam.   From statics the reactions are easily found to be $R_1 = Pb/L$ and $R_2 = Pa/L$.

This problem presents one feature that distinguishes it from the other problems solved thus far in this chapter.   Namely, it is essential to consider two different equations describing the bending moment in the beam.   One equation is valid to the left of the load P, the other holds to the right of this force.   The integration of each equation gives rise to two constants of integration and thus there are four constants of integration to be determined.

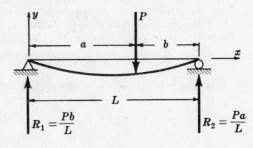

Fig. 9-11

All problems met thus far have offered only two constants.

In the region to the left of the force P we have the bending moment $M = (Pb/L)x$ for $0 < x < a$. The differential equation of the bent beam thus becomes

$$EI\frac{d^2y}{dx^2} = \frac{Pb}{L}x \qquad \text{for} \quad 0 < x < a \tag{1}$$

The first integration yields

$$EI\frac{dy}{dx} = \frac{Pb}{L}\frac{x^2}{2} + C_1 \tag{2}$$

No numerical information is available about the slope dy/dx at any point in this region. Since the load is not applied at the center of the beam there is no reason to believe that the slope is zero at $x = L/2$. However, for the slope of the beam under the point of application of the force P we can write

$$EI\left(\frac{dy}{dx}\right)_{x=a} = \frac{Pba^2}{2L} + C_1 \tag{3}$$

The next integration of (2) yields

$$EIy = \frac{Pb}{2L}\frac{x^3}{3} + C_1x + C_2 \tag{4}$$

At the left support, $y = 0$ when $x = 0$. Substituting these values in (4) we immediately find $C_2 = 0$. It is to be noted that it is not permissible to use the condition $y = 0$ at $x = L$ in (4) since (1) is not valid in that region. We have for the deflection under the point of application of the force P

$$EIy_{x=a} = \frac{Pba^3}{6L} + C_1a \tag{5}$$

In the region to the right of the force P the bending moment equation is $M = (Pb/L)x - P(x-a)$ for $a < x < L$. Thus

$$EI\frac{d^2y}{dx^2} = \frac{Pb}{L}x - P(x-a) \qquad \text{for} \quad a < x < L \tag{6}$$

The first integration of this equation yields

$$EI\frac{dy}{dx} = \frac{Pb}{L}\frac{x^2}{2} - \frac{P(x-a)^2}{2} + C_3 \tag{7}$$

Although nothing definite may be said about the slope in this portion of the beam, we have for the slope under the point of application of the force P

$$EI\left(\frac{dy}{dx}\right)_{x=a} = \frac{Pba^2}{2L} + C_3 \tag{8}$$

Under the concentrated load P the slope as given by (3) must be equal to that given by (8). Consequently the right sides of these two equations must be equal and we have

$$\frac{Pba^2}{2L} + C_1 = \frac{Pba^2}{2L} + C_3 \qquad \text{or} \qquad C_1 = C_3$$

Equation (7) may now be integrated to give

$$EIy = \frac{Pb}{2L}\frac{x^3}{3} - \frac{P(x-a)^3}{6} + C_3x + C_4 \tag{9}$$

We may write for the deflection under the concentrated load

$$EIy_{x=a} = \frac{Pba^3}{6L} + C_3a + C_4 \tag{10}$$

The deflection at $x = a$ given by (5) must equal that given by (10). Thus the right sides of these two equations are equal and we have

$$\frac{Pba^3}{6L} + C_1 a = \frac{Pba^3}{6L} + C_3 a + C_4$$

Since $C_1 = C_3$, we have $C_4 = 0$.

The condition that $y = 0$ when $x = L$ may now be substituted in (9), yielding

$$0 = \frac{PbL^2}{6} - \frac{Pb^3}{6} + C_3 L \qquad \text{or} \qquad C_3 = \frac{Pb}{6L}(b^2 - L^2)$$

In this manner all four constants of integration are determined. These values may now be substituted in equations (4) and (9) to give

$$EIy = \frac{Pb}{6L}[x^3 - (L^2 - b^2)x] \qquad \text{for} \quad 0 < x < a \qquad (4')$$

$$EIy = \frac{Pb}{6L}\left[x^3 - \frac{L}{b}(x - a)^3 - (L^2 - b^2)x\right] \qquad \text{for} \quad a < x < L \qquad (9')$$

These two equations are necessary to describe the deflection curve of the bent beam. Each equation is valid only in the region indicated.

9.13. Consider the simply supported beam described in Problem 9.12. If the cross-section is rectangular, 2×4 in., and $P = 4000$ lb with $a = 4$ ft, $b = 2$ ft, determine the maximum deflection of the beam. The beam is steel, for which $E = 30 \times 10^6$ lb/in².

Since $a > b$ it is evident that the maximum deflection must occur to the left of the load P. It occurs at that point where the slope of the beam is zero.

Differentiating equation (4') of Problem 9.12, we find that the slope in this region is given by

$$EI\frac{dy}{dx} = \frac{Pb}{6L}[3x^2 - (L^2 - b^2)]$$

Setting the slope equal to zero, we find $x = \sqrt{(L^2 - b^2)/3}$ for the point where the deflection is maximum. The deflection at this point is found by substituting this value of x in (4'):

$$EIy_{\max} = -\frac{Pb\sqrt{3}}{27L}(L^2 - b^2)^{3/2}$$

For the rectangular section $I = 2(4^3)/12 = 10.7$ in⁴. Also, $P = 4000$ lb, $b = 24$ in., $L = 72$ in., and $E = 30 \times 10^6$ lb/in². Substituting,

$$y_{\max} = -\frac{4000(24)\sqrt{3}}{30 \times 10^6(10.7)(27)(72)}(72^2 - 24^2)^{3/2} = -0.0834 \text{ in.}$$

The negative sign indicates that this point on the bent beam lies below the x-axis.

From $\sigma = Mc/I$ the maximum bending stress, which occurs under the load P, is 12,000 lb/in². This is below the proportional limit of steel, so use of the above deflection equations is valid.

9.14. Determine the equation of the deflection curve for the cantilever beam loaded by a uniformly distributed load of w lb per unit length over the portion of the beam shown in Fig. 9-12.

It is first necessary to determine the reactions exerted by the supporting wall upon the beam. These are easily found from statics to be a vertical force of magnitude wa lb

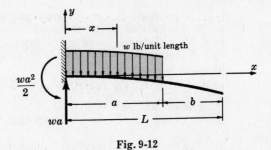

Fig. 9-12

together with a couple of magnitude $wa^2/2$. Again, two equations are required to describe the bending moment along the length of the bar.

For any point under the uniform load the bending moment at a distance x from the wall is given by

$$M = wax - \frac{wa^2}{2} - \frac{wx^2}{2}$$

In obtaining this equation the portion of the uniform load to the left of the section x was replaced by its resultant of wx lb acting downward at a distance $x/2$ from the wall. Also, according to the sign convention adopted in Chapter 6 the couple $wa^2/2$ produces negative bending. The differential equation of the loaded portion of the beam becomes

$$EI\frac{d^2y}{dx^2} = wax - \frac{wa^2}{2} - \frac{wx^2}{2} \qquad \text{for} \quad 0 < x < a \tag{1}$$

Integrating the first time we obtain

$$EI\frac{dy}{dx} = wa\frac{x^2}{2} - \frac{wa^2}{2}x - \frac{w}{2}\frac{x^3}{3} + C_1 \tag{2}$$

Since the bar is clamped at the left end, $x = 0$, we know that the slope dy/dx must be zero there. Substituting these values in (2), we find $C_1 = 0$. Integrating again, we find

$$EIy = \frac{wa}{2}\frac{x^3}{3} - \frac{wa^2}{2}\frac{x^2}{2} - \frac{w}{6}\frac{x^4}{4} + C_2 \tag{3}$$

The deflection y of the beam is zero at the wall where $x = 0$. Substituting in (3), we obtain $C_2 = 0$. Thus the equation of the bent beam in the loaded region is

$$EIy = \frac{wa}{6}x^3 - \frac{wa^2}{4}x^2 - \frac{w}{24}x^4 \tag{4}$$

From (4) the deflection y at $x = a$ is given by

$$EIy_{x=a} = -wa^4/8 \tag{5}$$

Also, from (2) the slope dy/dx at $x = a$ is given by

$$EI(dy/dx)_{x=a} = -wa^3/6 \tag{6}$$

At any section in the unloaded region of the beam, i.e. $a < x < L$, the bending moment is zero. This is most easily seen by considering the moments of the forces to the right of such a section about an axis through this section and perpendicular to the plane of the page. Since there are no loads to the right of the section, the bending moment is zero everywhere in this region. Thus in this region we have

$$EI(d^2y/dx^2) = 0 \qquad \text{for} \quad a < x < L$$

Integrating once we obtain

$$EI(dy/dx) = C_3 \tag{7}$$

The constant C_3 may be evaluated by realizing that the slope dy/dx at $x = a$ is the same whether calculated from the loaded or unloaded region of the beam. For the loaded region the slope at $x = a$ was found in (6); for the unloaded region the slope according to (7) is a constant C_3. Equating the right sides of these two equations we have $C_3 = -wa^3/6$. The slope in the unloaded region is thus

$$EI(dy/dx) = -wa^3/6 \tag{7'}$$

Integrating this we obtain

$$EIy = (-wa^3/6)x + C_4 \tag{8}$$

The constant C_4 may be evaluated by realizing that at $x = a$ the deflection y as given by (5) must be equal to the deflection as given by (8) for the unloaded region. Equating the right sides of these two equations at the common point $x = a$, we have $C_4 = wa^4/24$.

Thus two equations are necessary to describe the deflection curves of the loaded and unloaded regions of the beam. They are

$$EIy = \frac{wa}{6}x^3 - \frac{wa^2}{4}x^2 - \frac{w}{24}x^4 \qquad \text{for} \quad 0 < x < a \qquad (4')$$

$$EIy = -\frac{wa^3}{6}x + \frac{wa^4}{24} \qquad \text{for} \quad a < x < L \qquad (8')$$

Inspection of (7') reveals that the slope of the beam is constant in the unloaded region. Thus the deflected beam is straight in that region.

9.15. Determine the equation of the deflection curve for a cantilever beam loaded by a uniformly distributed load of w lb per unit length, as well as by a concentrated force P at the free end. See Fig. 9-13.

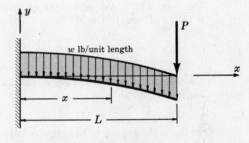

Fig. 9-13

The deformed beam has the configuration indicated by the heavy line. The x-y coordinate system is introduced as shown. One logical approach to this problem is to determine the reactions at the wall, then write the differential equation of the bent beam, integrate this equation twice and determine the constant of integration from the conditions of zero slope and zero deflection at the wall.

Actually this procedure has already been carried out in Problem 9.2 for the case in which only the concentrated load acts on the beam, and in Problem 9.5 when only the uniformly distributed load is acting. For the concentrated force alone the deflection y was found in (3) of Problem 9.2 to be

$$EIy = -PL\frac{x^2}{2} + \frac{Px^3}{6} \qquad (1)$$

For the uniformly distributed load alone the deflection y was found in (3') of Problem 9.5 to be

$$EIy = -\frac{w}{24}(L-x)^4 - \frac{wL^3}{6}x + \frac{wL^4}{24} \qquad (2)$$

It is possible to obtain the resultant effect of these two loads when they act simultaneously merely by adding together the effects of each as they act separately. This is called the *method of superposition*. It is useful in determining deflections of beams subject to a combination of loads, such as we have here. Essentially it consists in utilizing the results of simpler beam-deflection problems to build up the solutions of more complicated problems. Thus it is not an independent method of determining beam deflections.

According to this method the deflection at any point of a beam subject to a combination of loads can be obtained as the sum of the deflections produced at this point by each of the loads acting separately. The final deflection equation resulting from the combination of loads is then obtained by adding the deflection equations for each load.

For the present beam the final deflection equation is given by adding equations (1) and (2):

$$EIy = -PL\frac{x^2}{2} + \frac{Px^3}{6} - \frac{w}{24}(L-x)^4 - \frac{wL^3}{6}x + \frac{wL^4}{24} \qquad (3)$$

The slope dy/dx at any point in the beam is merely found by differentiating both sides of (3) with respect to x.

The method of superposition is valid in all cases where there is a linear relationship between each separate load and the separate deflection which it produces.

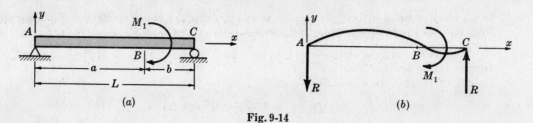

Fig. 9-14

9.16. Determine the deflection curve of the simply supported beam loaded by the couple M_1 as shown in Fig. 9-14(a).

The reactions and bending moment equations for this type of loading have been discussed in Problem 6.11. As demonstrated there, the reactions must constitute a couple as shown in Fig. 9-14(b). From statics we have

$$\Sigma M_A = M_1 - RL = 0 \quad \text{or} \quad R = M_1/L$$

The configuration of the bent beam is indicated by the heavy line. The x-axis coincides with the original unbent position of the bar. The bending moment in the region to the left of the load M_1 is evidently

$$M = -Rx \quad \text{for} \quad 0 < x < a \tag{1}$$

while in the region to the right of M_1 the bending moment is given by

$$M = -Rx + M_1 \quad \text{for} \quad a < x < L \tag{2}$$

The couple M_1 produces a positive bending moment, since if it alone acted upon the region BC it would produce bending as shown in Fig. 9-15. According-ing to the sign convention of Chapter 6 this constitutes positive bending, hence in (2) M_1 bears a positive sign.

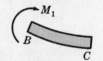

Fig. 9-15

The differential equation of the region of the bent beam to the left of the couple M_1 is

$$EI(d^2y/dx^2) = -Rx \quad \text{for} \quad 0 < x < a \tag{3}$$

Integrating once we obtain

$$EI(dy/dx) = -(Rx^2/2) + C_1 \tag{4}$$

Since we have no definite information concerning the slope in this region, we are unable to evaluate C_1 immediately. However, we may say that the slope of the beam under the point of application of the couple M_1 is given by

$$EI(dy/dx)_{x=a} = -(Ra^2/2) + C_1 \tag{5}$$

Integration of (4) yields

$$EIy = -\frac{R}{2}\frac{x^3}{3} + C_1x + C_2 \tag{6}$$

It is evident that the deflection y is zero at the left support where $x = 0$. Substituting this value, $y_{x=0} = 0$, in (6) we obtain $0 = 0 + 0 + C_2$ or $C_2 = 0$.

The differential equation of the region of the bent beam to the right of M_1 is

$$EI(d^2y/dx^2) = -Rx + M_1 \quad \text{for} \quad a < x < L \tag{7}$$

Integrating the first time we find

$$EI(dy/dx) = -(Rx^2/2) + M_1x + C_3 \tag{8}$$

Again no definite information is available concerning the slope in this region, but we may say that the slope under the point of application of M_1 is given by

$$EI(dy/dx)_{x=a} = -(Ra^2/2) + M_1a + C_3 \tag{9}$$

However, the slope of the beam under the point of application of M_1 has only one value, its value as represented by the right sides of equations (5) and (9). Equating the right sides of these two equations,

$$-(Ra^2/2) + C_1 = -(Ra^2/2) + M_1 a + C_3 \qquad \text{or} \qquad C_1 = M_1 a + C_3 \qquad (10)$$

The second integration of (8) yields

$$EIy = -\frac{R}{2}\frac{x^3}{3} + M_1\frac{x^2}{2} + C_3 x + C_4 \qquad (11)$$

It is evident that the deflection y is zero at the right support where $x = L$. Substituting this value, $y_{x=L} = 0$, in (11) we obtain

$$0 = -\frac{RL^3}{6} + M_1\frac{L^2}{2} + C_3 L + C_4 \qquad (12)$$

One additional relationship is needed to determine all of the constants of integration. This relation is that the deflection of the beam under the point of application of M_1 is the same no matter if it is calculated from the equation for the left region or the right region of the beam. It is to be emphasized that there is no reason for assuming the deflection to be zero under the point of application of the couple. Substituting $x = a$ in (6) and (11) and equating the right sides, we obtain

$$-\frac{Ra^3}{6} + C_1 a = -\frac{Ra^3}{6} + M_1\frac{a^2}{2} + C_3 a + C_4 \qquad \text{or} \qquad C_1 a = M_1\frac{a^2}{2} + C_3 a + C_4 \qquad (13)$$

Solving (10), (12) and (13) simultaneously, we find

$$C_1 = -\frac{M_1 L}{3} + M_1 a - \frac{M_1 a^2}{2L}, \qquad C_3 = -\frac{M_1 L}{3} - \frac{M_1 a^2}{2L}, \qquad C_4 = \frac{M_1 a^2}{2}$$

Substituting these values in (6) and (11), we obtain the two equations required to describe the deflection curve of the bent beam:

$$EIy = -\frac{M_1 x^3}{6L} - \frac{M_1 L x}{3} + M_1 a x - \frac{M_1 a^2 x}{2L} \qquad \text{for} \quad 0 < x < a \qquad (14)$$

$$EIy = -\frac{M_1 x^3}{6L} + \frac{M_1 x^2}{2} - \frac{M_1 L x}{3} - \frac{M_1 a^2 x}{2L} + \frac{M_1 a^2}{2} \qquad \text{for} \quad a < x < L \qquad (15)$$

To summarize, two equations were required to define the bending moment along the entire length of the beam. Two second-order differential equations then had to be integrated and two constants of integration arose from the solution of each of these two equations. Thus we had four constants of integration and it was necessary to use four boundary conditions to determine them. These conditions were:

(a) $y = 0$ when $x = 0$.

(b) $y = 0$ when $x = L$.

(c)   When $x = a$, the deflections given by equations (6) and (11) are equal.

(d)   When $x = a$, the slopes given by equations (4) and (8) are equal.

9.17. Determine the deflection curve of an overhanging beam subject to a uniform load of w lb per unit length and supported as shown in Fig. 9-16.

We replace the distributed load by its resultant of wL lb acting at the midpoint of the length L. Taking moments about the right reaction, we have

$$\Sigma M_C = R_1 b - \frac{wL^2}{2} = 0 \qquad \text{or} \qquad R_1 = \frac{wL^2}{2b}$$

Summing forces vertically, we find

$$\Sigma F_v = \frac{wL^2}{2b} + R_2 - wL = 0$$

or

$$R_2 = wL - \frac{wL^2}{2b}$$

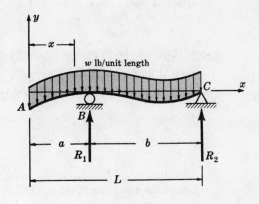

Fig. 9-16

The bending moment equation in the left overhanging region is $M = -wx^2/2$ for $0 < x < a$. Consequently the differential equation of the bent beam in that region is

$$EI(d^2y/dx^2) = -wx^2/2 \quad \text{for} \quad 0 < x < a \tag{1}$$

Two successive integrations yield

$$EI\frac{dy}{dx} = -\frac{w}{2}\frac{x^3}{3} + C_1 \tag{2}$$

$$EIy = -\frac{w}{6}\frac{x^4}{4} + C_1 x + C_2 \tag{3}$$

The bending moment equation in the region between supports is $M = -wx^2/2 + R_1(x - a)$. The differential equation of the bent beam in that region is thus

$$EI\frac{d^2y}{dx^2} = -\frac{wx^2}{2} + \frac{wL^2}{2b}(x - a) \quad \text{for} \quad a < x < L \tag{4}$$

Two integrations of this equation yield

$$EI\frac{dy}{dx} = -\frac{w}{2}\frac{x^3}{3} + \frac{wL^2}{2b}\frac{(x-a)^2}{2} + C_3 \tag{5}$$

$$EIy = -\frac{w}{6}\frac{x^4}{4} + \frac{wL^2}{4b}\frac{(x-a)^3}{3} + C_3 x + C_4 \tag{6}$$

Since we started with two second-order differential equations, (1) and (4), and two constants of integration arose from each, we have four constants C_1, C_2, C_3 and C_4 to evaluate from known conditions concerning slopes and deflections. These conditions are:

(a)   When $x = a$, $y = 0$ in the overhanging region.

(b)   When $x = a$, $y = 0$ in the region between supports.

(c)   When $x = L$, $y = 0$ in the region between supports.

(d)   When $x = a$, the slope given by (2) must be equal to that given by (5); consequently the right sides of these equations must be equal when $x = a$.

Substituting condition (a) in (3), we obtain

$$0 = -wa^4/24 + C_1 a + C_2 \tag{7}$$

Substituting condition (b) in (6), we find

$$0 = -wa^4/24 + C_3 a + C_4 \tag{8}$$

Substituting condition (c) in (6), we get

$$0 = -wL^4/24 + wL^2b^2/12 + C_3 L + C_4 \tag{9}$$

Lastly, equating slopes at the left reaction by substituting $x = a$ in the right sides of equations (2) and (5), we obtain

$$-wa^3/6 + C_1 = -wa^3/6 + C_3 \tag{10}$$

Note that there is no reason for assuming the slope to be zero at the left support, $x = a$.

These last four equations (7), (8), (9), (10) may now be solved for the four unknown constants C_1, C_2, C_3, C_4. The solution is found to be

$$C_1 = C_3 = \frac{w(L^4 - a^4)}{24b} - \frac{wL^2b}{12} \tag{11}$$

$$C_2 = C_4 = \frac{wa^4}{24} - \frac{w(L^4 - a^4)a}{24b} + \frac{wL^2ab}{12} \tag{12}$$

The two equations describing the deflection curve of the bent bar are found by substituting these values of the constants in (3) and (6). These equations may be written in the final forms

$$EIy = -\frac{wx^4}{24} + \frac{w(L^4 - a^4)x}{24b} - \frac{wL^2bx}{12} + \frac{wa^4}{24} - \frac{w(L^4 - a^4)a}{24b} + \frac{wL^2ab}{12} \quad \text{for} \quad 0 < x < a \tag{3'}$$

$$EIy \;=\; -\frac{wx^4}{24} + \frac{wL^2(x-a)^3}{12b} + \frac{w(L^4-a^4)x}{24b} - \frac{wL^2bx}{12} + \frac{wa^4}{24} - \frac{w(L^4-a^4)a}{24b} + \frac{wL^2ab}{12} \quad \text{for } a < x < L$$

$$(6')$$

9.18. For the overhanging beam treated in Problem 9.17 consider the uniform load to be 120 lb/ft, $a = 3$ ft and $b = 12$ ft. The bar has a 3×4 in. rectangular cross-section. Determine the maximum deflection of the beam. Take $E = 30 \times 10^6$ lb/in².

An approximate representation of the deflected beam is given in Fig. 9-16. The point where the maximum deflection occurs is not immediately evident. It may be at the extreme left end of the beam, where $x = 0$, or it may be at some intermediate point between the supports. If the maximum does occur between supports, it is unlikely that it occurs midway between them since there is no symmetry to the system. In the event that it occurs between supports, the location of the point may be determined by finding the point where the slope of the beam is zero. The slope anywhere in the region between supports is given by (5) of Problem 9.17, and if we set the slope dy/dx in that equation equal to zero and use the value of C_3 given by (11) we find

$$0 \;=\; -\frac{wx^3}{6} + \frac{wL^2(x-a)^2}{4b} + \frac{w(L^4-a^4)}{24b} - \frac{wL^2b}{12}$$

Substituting $w = 120$ lb/ft $= 10$ lb/in, $a = 3$ ft $= 36$ in., $b = 12$ ft $= 144$ in. and $L = 180$ in., we find

$$0 \;=\; -\frac{10x^3}{6} + \frac{10(180)^2(x-36)^2}{4(144)} + \frac{10[(180)^4-(36)^4]}{24(144)} - \frac{10(180)^2(144)}{12}$$

Solving by trial and error, $x = 110.4$ in. $= 9.20$ ft. This locates the point where the slope is zero.

The deflection at $x = 110.4$ in. may be found by substituting this value in (6') of Problem 9.17. This leads to the relation

$$(30 \times 10^6)\,\frac{1}{12}(3)(4^3)y_{x\,=\,110.4\,\text{in.}} \;=\; -\frac{10(110.4)^4}{24} + \frac{10(180)^2(110.4-36)^3}{12(144)} + \frac{10[(180)^4-(36)^4]110.4}{24(144)}$$

$$- \frac{10(180)^2(144)(110.4)}{12} + \frac{10(36)^4}{24}$$

$$- \frac{10[(180)^4-(36)^4](36)}{24(144)} + \frac{10(180)^2(36)(144)}{12}$$

Solving, $y_{x\,=\,110.4\,\text{in.}} = -0.10$ in.

The calculus technique of equating the first derivative dy/dx to zero for the purpose of determining the location of a point where the value of a function is maximum fails to detect any maximum deflection that may exist at such a point as $x = 0$. Hence it is necessary to investigate the deflection at that point. Substituting $x = 0$ in (3') of Problem 9.17, we find

$$(30 \times 10^6)\,\frac{1}{12}(3)(4^3)y_{x=0} \;=\; \frac{10(36)^4}{24} - \frac{10[(180)^4-(36)^4](36)}{24(144)} + \frac{10(180)^2(36)(144)}{12}$$

Solving, $y_{x=0} = +0.065$ in.

Thus the assumed form of the deflection curve shown in Fig. 9-16 is incorrect in the overhanging region for this particular beam. Actually, the beam bends upward in this region. For other values of a and b it would be possible for the beam to bend into the configuration shown there.

The maximum deflection of the beam is consequently 0.10 in. downward at a point 9.20 ft from the left end.

The maximum bending moment for this beam is found to be 1900 lb-ft. The maximum bending stress is given by

$$\sigma_{\text{max}} \;=\; \frac{Mc}{I} \;=\; \frac{1900(12)(2)}{\frac{1}{12}(3)(4^3)} \;=\; 2840 \text{ lb/in}^2$$

The use of the above deflection equations is thus justified.

It should be noted that the section where the bending stress is a maximum is not the section where the deflection is a maximum.

9.19. A cantilever beam when viewed from the top [Fig. 9-17(a)] has a triangular configuration. The thickness of the beam is constant, as shown in the side view [Fig. 9-17(b)]. Determine the deflection due to a concentrated force P at the tip.

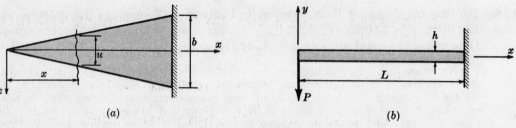

(a)   (b)

Fig. 9-17

From similar triangles, $u = bx/L$. Using the Euler-Bernoulli equation from Problem 9.1 we have, since I is variable,

$$E[\tfrac{1}{12}(bx/L)h^3]\frac{d^2y}{dx^2} = -Px \quad \text{or} \quad \frac{d^2y}{dx^2} = -\frac{12LP}{Ebh^3}$$

Integrating,

$$\frac{dy}{dx} = -\frac{12LP}{Ebh^3}x + C_1$$

But when $x = L$, $dy/dx = 0$, from which $C_1 = 12PL^2/Ebh^3$. Integrating again,

$$y = -\frac{12LP}{Ebh^3}\frac{x^2}{2} + \frac{12PL^2}{Ebh^3}x + C_2$$

When $x = L$, $y = 0$, from which $C_2 = -6PL^3/Ebh^3$.

The equation of the deflection curve is thus

$$y = -\frac{6PL}{Ebh^3}x^2 + \frac{12PL^2}{Ebh^3}x - \frac{6PL^3}{Ebh^3}$$

from which $y_{x=0} = -6PL^3/Ebh^3$.

9.20. Consider the bending of a cantilever beam which remains in contact with a rigid cylindrical surface as it deflects. The tangent to the cantilever is horizontal at point A in Fig. 9-18. Determine the deflection of the tip B due to the load P.

If the curvature of the cantilever at A is less than the curvature of the rigid cylindrical surface, then the cantilever touches the surface only at point A and the deflection is exactly as found in Problem 9.2. From Problem 9.1, the curvature of the beam at A is given by

$$1/\rho = M/EI = PL/EI$$

and thus this curvature must be less than the curvature of the rigid surface, which is $1/R$.

If however $1/R = PL/EI$ then the beam comes into contact with the surface to the right of point A. We shall denote by P^* the limiting value

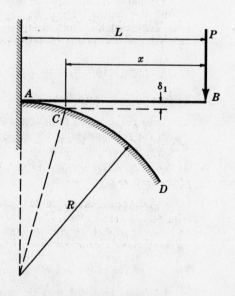

Fig. 9-18

of the load given by $P^* = EI/RL$. For $P > P^*$ some region AC of the beam will be in contact with the surface and at point C the curvature of the rigid surface $1/R$ is equal to the curvature of the beam, i.e. $Px/EI = 1/R$ from which $x = EI/PR$.

The deflection at the tip B may now be found as the sum of

(a)   the deflection of C from the tangent at A, which is given by δ_1 in the diagram and is found from the relation

$$(R + \delta_1)^2 = R^2 + (L - x)^2$$

to be approximately

$$\delta_1 = (L - x)^2/2R$$

(b)   the deflection of the portion of the beam of length x acting as a simple cantilever, given by

$$\delta_2 = Px^3/3EI = (EI)^2/3P^2R^3$$

(c)   the deflection owing to the rotation at point C, given by

$$\delta_3 = \frac{x(L - x)}{R} = \frac{EI}{PR^2}\left(L - \frac{EI}{PR}\right)$$

The desired deflection at the tip is thus

$$\delta = \delta_1 + \delta_2 + \delta_3 = \frac{L^2}{2R} - \frac{(EI)^2}{6P^2R^3}$$

9.21.   A thermostat consists of two strips of different materials of equal thickness bonded together at their interface. Frequently this configuration takes the form of a cantilever beam, as in Fig. 9-19. If E_1 and E_2 denote the Young's moduli and α_1 and α_2 denote the coefficients of linear expansion of the two materials, each of thickness h, determine the deflection of the end of the cantilever assembly due to a temperature rise T.

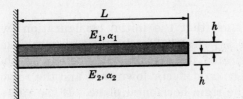

Fig. 9-19

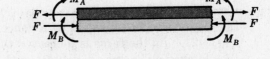

Fig. 9-20

Let b represent the width of the assembly. As in Problem 8.1, we shall assume that a plane section prior to deformation remains plane after deformation. The resultant normal forces F acting over each strip must be numerically equal since no external forces are applied along the length of the beam. Thus a cross-section at any station along the length has Fig. 9-20 as its free-body representation.

The normal strain in the lower fibers of the top strip is found as the sum of (a) the strain due to the normal load, F/E_1bh, (b) the strain due to bending, which is $M_A(h/2)/E_1I$ from Problem 8.1 and (c) the strain due to the temperature rise, which is α_1T as mentioned in Chapter 1. The sum of these strains must be the same as the strain in the upper fibers of the lower strip. Thus

$$\frac{F}{E_1bh} + \frac{M_A(h/2)}{E_1I} + \alpha_1T = \frac{-F}{E_2bh} - \frac{M_B(h/2)}{E_2I} + \alpha_2T \qquad (1)$$

The curvatures at this interface must also be equal. Thus, from Problem 9.1,

$$1/R_1 = M_A/E_1 I \quad \text{and} \quad 1/R_2 = M_B/E_1 I \tag{2}$$

and since $R_1 = R_2$, we have

$$M_A = (E_1/E_2)M_B \tag{3}$$

From statics it is evident that

$$M_A + M_B = Fh \tag{4}$$

from which

$$M_B = \frac{Fh}{1 + (E_1/E_2)} \qquad M_A = \frac{Fh}{1 + (E_2/E_1)} \tag{5}$$

Substituting (5) in (1) we find

$$F = \frac{(\alpha_2 - \alpha_1)TbhE_1E_2(E_1 + E_2)}{E_1^2 + E_2^2 + 14E_1E_2} \tag{6}$$

and from (5) we get

$$M_A = \frac{(\alpha_2 - \alpha_1)Tbh^2 E_1^2 E_2}{E_1^2 + E_2^2 + 14E_1E_2} \tag{7}$$

We may now use the result obtained in Problem 9.20 for the deflection δ of a point on a cylindrical surface (which represents the interface, since in pure bending the assembly deforms into a circular configuration according to Problem 9.1) and express the deflection δ of the end of the assembly as

$$\delta = L^2/2R \tag{8}$$

Substituting from Eq. (2),

$$\delta = M_A L^2/2E_1 I$$

From (7) we then get

$$\delta = \frac{6(\alpha_2 - \alpha_1)TE_1E_2L^2}{h(E_1^2 + E_2^2 + 14E_1E_2)}$$

9.22. A beam has a slight initial curvature such that the initial configuration (which is stress free) is described by the relation $y_0 = Kx^3$. The beam is rigidly clamped at the origin and is subjected to a concentrated force at its extreme end as shown in Fig. 9-21. As the force is increased the beam deflects downward and the region near the clamped end comes in contact with the rigid horizontal plane. If the value of the applied force is P, determine the length of the beam in contact with the horizontal plane and the vertical distance of the extreme end from the plane.

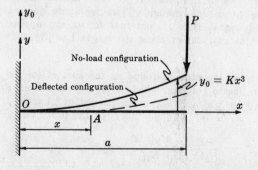

Fig. 9-21

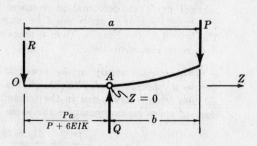

Fig. 9-22

The initial curvature may be determined from the expression $y_0 = Kx^3$ so that the bending moment arising from straightening the portion of the beam near the support is readily found to be $EI(d^2y_0/dx^2) = 6EIKx$, where x is the length of beam in contact with the horizontal plane. If this expression for moment is equated to the moment of the applied load about the point of contact, i.e. $P(a-x)$, we have

$$6EIKx = P(a-x) \qquad \text{whence} \qquad x = Pa/(P+6EIK)$$

Since the beam is considered to be weightless, there is no normal force between the beam and the rigid horizontal plane between the clamp at O and the extreme point of contact at A. The beam is flat between O and A. A free-body diagram of the deformed beam thus appears as in Fig. 9-22. A simple statics equation for equilibrium of moments about point A indicates that the clamp exerts a downward force equal to $6EIK$. For vertical equilibrium there is a concentrated force reaction $Q = P + 6EIK$ acting on the beam at the extreme point of contact A.

We now seek the equation of the deflection curve in the region to the right of point A. In Problem 9.1 equation (5) indicated that for an initially straight beam the bending moment M is proportional to the curvature, d^2y/dx^2. However, in the present problem it is necessary to modify (5) to say that the bending moment M is proportional to the *change of curvature* since the beam is not initially straight. Thus, the Euler-Bernoulli equation for the portion of the beam to the right of point A is:

$$EI\left(\frac{d^2y_0}{dx^2} - \frac{d^2y}{dZ^2}\right) = P(b-Z)$$

where a new coordinate Z has been introduced. This coordinate runs along the x-axis but has its origin at point A. It is important to note that, as the beam deflects, the curvature decreases from its original value, hence the quantity in parentheses on the left side of the equation is positive. Accordingly, the right side must be written as positive. This does not contradict our previous sign convention of downward forces giving negative moments since it was applied to *initially straight* beams. If we substitute $EI(d^2y_0/dx^2) = 6EIKx$, the last equation becomes

$$EI\frac{d^2y}{dZ^2} = 6EIK\left[\frac{Pa}{P+6EIK} + Z\right] - Pb + PZ$$

Integrating twice and imposing the boundary conditions that $y = dy/dZ = 0$ at $Z = 0$, we obtain the desired deflection

$$EIy_{Z=b} = \frac{36(EIKa)^3}{(P+6EIK)^2}$$

Supplementary Problems

9.23. The cantilever beam loaded as shown in Problem 9.2 is made of titanium, type Ti-6Al-4V. The load P is 5000 lb, $L = 12$ ft, the moment of inertia of the cross-section is 250 in^4, and $E = 16.5 \times 10^6$ lb/in^2. Determine the maximum deflection of the beam. *Ans.* -1.20 in.

9.24. Consider the uniformly loaded cantilever beam discussed in Problem 9.5. The total load is 6000 lb, the length of the beam is 10 ft and the moment of inertia of the cross-section is 200 in^4. Determine the deflection and slope at the free end of the beam. Take $E = 30 \times 10^6$ lb/in^2. *Ans.* -0.216 in., -0.0024 rad

9.25. Consider the simply supported beam loaded as shown in Problem 9.12. The length of the beam is 20 ft, $a = 15$ ft, the load $P = 1000$ lb, and $I = 150$ in^4. Determine the deflection at the center of the beam. Take $E = 30 \times 10^6$ lb/in^2. *Ans.* -0.044 in.

9.26. Refer to Fig. 9-23. Determine the deflection at every point of the cantilever beam subject to the single moment M_1 shown.   *Ans.* $EIy = -M_1 x^2/2$

9.27. The cantilever beam described in Problem 9.26 is of circular cross-section, 5 in. in diameter. The length of the beam is 10 ft and the applied moment is 5000 lb-ft. Determine the maximum deflection of the beam. Take $E = 30 \times 10^6$ lb/in². *Ans.* -0.469 in.

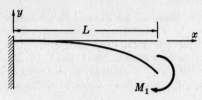

Fig. 9-23

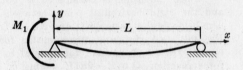

Fig. 9-24

9.28. Refer to Fig. 9-24. Determine the equation of the deflection curve for a simply supported beam loaded by a couple M_1 at the left end of the bar as shown.

Ans. $EIy = -\dfrac{M_1}{6L}x^3 + \dfrac{M_1}{2}x^2 - \dfrac{M_1 L}{3}x$

9.29. Determine the equation of the deflection curve for a simply supported beam loaded by a uniformly distributed load of w lb/unit length as well as a concentrated force P applied at the midpoint as shown in Fig. 9-25.

Ans. $EIy = \dfrac{wLx^3}{12} - \dfrac{wx^4}{24} - \dfrac{wL^3x}{24} + \dfrac{Px^3}{12} - \dfrac{PL^2x}{16}$ for $0 < x < \dfrac{L}{2}$

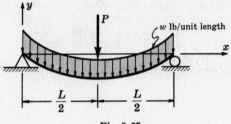

Fig. 9-25

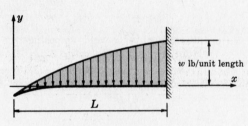

Fig. 9-26

9.30. Refer to Fig. 9-26. Find the equation of the deflection curve for the cantilever beam subject to the uniformly varying load shown.

Ans. $EIy = -\dfrac{wx^5}{120L} + \dfrac{wL^3x}{24} - \dfrac{wL^4}{30}$

9.31. The cross-section of the cantilever beam loaded as shown in Fig. 9-26 is rectangular, 2×3 in. The bar, 3 ft long, is aluminum for which $E = 10 \times 10^6$ lb/in². Determine the permissible maximum intensity of loading if the maximum deflection is not to exceed 0.15 in. and the maximum stress is not to exceed 8000 lb/in². *Ans.* $w = 1333$ lb/ft

9.32. Refer to Fig. 9-27. Determine the equation of the deflection curve for the simply supported beam supporting the load of uniformly varying intensity.

Ans. $EIy = \dfrac{wL}{2}\left(-\dfrac{x^5}{60L^2} + \dfrac{x^3}{18} - \dfrac{7L^2x}{180}\right)$

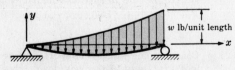

Fig. 9-27

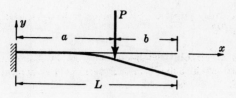

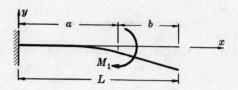

Fig. 9-28   Fig. 9-29

9.33. Determine the equation of the deflection curve for the cantilever beam loaded by the concentrated force P as shown in Fig. 9-28.

Ans. $EIy = -\frac{P}{6}(a-x)^3 - \frac{Pa^2}{2}x + \frac{Pa^3}{6}$ for $0 < x < a$; $EIy = -\frac{Pa^2}{2}x + \frac{Pa^3}{6}$ for $a < x < L$

9.34. For the cantilever beam of Fig. 9-28, take $P = 1000$ lb, $a = 6$ ft, and $b = 4$ ft. The beam is of equilateral triangular cross-section, 6 in. on a side, with a vertical axis of symmetry. Determine the maximum deflection of the beam. Take $E = 30 \times 10^6$ lb/in². *Ans.* −0.355 in.

9.35. Refer to Fig. 9-29. Find the equation of the deflection curve for the cantilever beam loaded by the couple M_1. *Ans.* $EIy = -M_1 x^2/2$ for $0 < x < a$; $EIy = -M_1 ax + (M_1 a^2/2)$ for $a < x < L$

9.36. Refer to Fig. 9-30. A simply supported beam is subject to the two symmetrically placed loads shown. Determine the deflection curve of the bent beam.

Ans. $EIy = \frac{Px^3}{6} + \left(\frac{Pa^2}{2} - \frac{PaL}{2}\right)x$ for $0 < x < a$

$EIy = \frac{Pax^2}{2} - \frac{PaLx}{2} + \frac{Pa^3}{6}$ for $a < x < a+b$

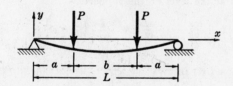

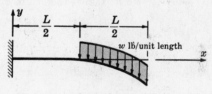

Fig. 9-30   Fig. 9-31

9.37. Refer to Fig. 9-31. Find the equation of the deflection curve for the cantilever beam loaded over half its length by a uniformly distributed load of w lb per unit length as shown. Using this equation, determine the maximum deflection.

Ans. $EIy = \frac{wLx^3}{12} - \frac{3wL^2x^2}{16}$ for $0 < x < \frac{L}{2}$

$EIy = -\frac{w(L-x)^4}{24} - \frac{7wL^3x}{48} + \frac{15wL^4}{384}$ for $\frac{L}{2} < x < L$ $\Delta_{max} = \frac{41}{384}\frac{wL^4}{EI}$

9.38. Consider a simply supported beam subjected to a triangular-type load distributed over the central portion of the beam as indicated in Fig. 9-32. Determine the equation of the deflection curve.

Ans. $EIy = \frac{23wL^3x}{1536} - \frac{wLx^3}{48}$ for $0 < x < \frac{L}{4}$

$EIy = \frac{w}{30L}\left(x - \frac{L}{4}\right)^5 - \frac{wLx^3}{48} + \frac{23wL^3x}{1536}$ for $\frac{L}{4} < x < \frac{L}{2}$

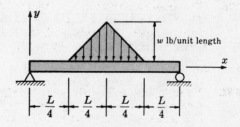

Fig. 9-32

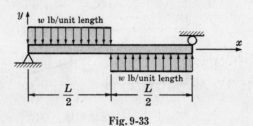

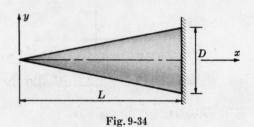

Fig. 9-33   Fig. 9-34

9.39. The simply supported beam is subjected to the uniformly distributed loads shown in Fig. 9-33. Determine the equation of the deflection curve of the beam.

$$\text{Ans.}\quad EIy = \frac{wLx^3}{24} - \frac{wx^4}{24} - \frac{wL^3x}{192} \quad \text{for} \quad 0 < x < \frac{L}{2}$$

$$EIy = -\frac{wLx^3}{8} + \frac{wx^4}{24} + \frac{wL^2x^2}{8} - \frac{3wL^3x}{64} + \frac{wL^4}{192} \quad \text{for} \quad \frac{L}{2} < x < L$$

9.40. Consider the cantilever beam of triangular configuration shown in Fig. 9-17 of Problem 9.19. Determine the deflection of the tip due to a uniformly distributed load w per unit area.

$$\text{Ans.}\quad wL^4/2Eh^3$$

9.41. Figure 9-34 shows a cantilever beam in the form of a circular cone whose length L is large compared to the base diameter D. If the only force acting is its own weight, which is γ per unit volume, determine the equation of the deflection curve.

$$\text{Ans.}\quad y = -\frac{2\gamma L^2}{45ED^2}(x^3 + 2L^3 - 3L^2x)$$

9.42. A simply supported overhanging beam supports a uniform load (Fig. 9-35). Find the equation of the deflection curve, taking the origin of coordinates at the level of the supports.

$$\text{Ans.}\quad EIy = -\frac{wx^4}{24} + \frac{wL^3x}{48} - \frac{wLx}{4}\left(\frac{L}{2} - a\right)^2 + \frac{wa^4}{24} - \frac{waL^3}{48} + \frac{wLa}{4}\left(\frac{L}{2} - a\right)^2$$
$$\text{for} \quad 0 < x < a$$

$$EIy = -\frac{wx^4}{24} + \frac{wL(x-a)^3}{12} + \frac{wL^3x}{48} - \frac{wLx}{4}\left(\frac{L}{2} - a\right)^2 + \frac{wa^4}{24} - \frac{waL^3}{48} + \frac{wLa}{4}\left(\frac{L}{2} - a\right)^2$$
$$\text{for} \quad a < x < a + b$$

9.43. The symmetrically supported beam shown in Fig. 9-35 is 30 ft long and the distance between supports is 20 ft. The moment of inertia of the cross-section is 400 in⁴ and the uniform load is 800 lb/ft. Find the deflection at the center of the beam. Assume $E = 30 \times 10^6$ lb/in². *Ans.* −0.166 in.

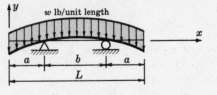

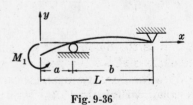

Fig. 9-35   Fig. 9-36

9.44. A beam with an overhanging end is loaded by a couple M_1 (Fig. 9-36). Determine the equation of the deflection curve. Take the origin of coordinates at the level of the supports.

$$\text{Ans.}\quad EIy = -\frac{M_1x^2}{2} + M_1ax + \frac{M_1(L-a)x}{3} - \frac{M_1a^2}{2} - \frac{M_1a(L-a)}{3} \quad \text{for} \quad 0 < x < a$$

$$EIy = -\frac{M_1(L-x)^3}{6(L-a)} - \frac{M_1x(L-a)}{6} + \frac{M_1L(L-a)}{6} \quad \text{for} \quad a < x < L$$

Chapter 10

Elastic Deflection of Beams: Moment-Area Method

INTRODUCTION

In Chapter 9 it was mentioned that several methods are available for the determination of elastic deflections of beams. That chapter was devoted to an exposition of the double-integration method. In the present chapter a second method, the *moment-area method*, will be investigated in detail. It may be considered to constitute an alternative procedure to the double-integration process.

STATEMENT OF THE PROBLEM

A given system of loads acts upon a beam. The dimensions of the beam and the modulus of elasticity are known. It is desired to determine the deflection of any point in the bent beam from its original position.

FIRST MOMENT-AREA THEOREM

In Fig. 10-1, *AB* represents a portion of the deflection curve of a bent beam. The shaded diagram immediately below *AB* represents the corresponding portion of the bending moment diagram. The construction of this diagram for many different types of loadings was examined in detail in Chapter 6. Tangents to the deflection curve are drawn at each of the points *A* and *B* as indicated.

The *first moment-area theorem* states that: The angle between the tangents at *A* and *B* is equal to the area of the bending moment diagram between these two points, divided by the product *EI*.

If θ denotes the angle between the tangents as shown in the diagram, then this theorem may be stated in equation form as follows:

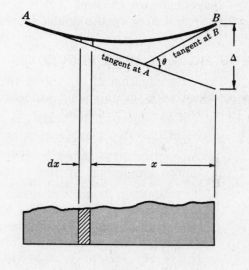

Fig. 10-1

$$\theta = \int_A^B \frac{M\,dx}{EI}$$

where *E* is the modulus of elasticity of the beam, *I* is the moment of inertia of the beam cross-section about the neutral axis which passes through the centroid of the cross-section and *M* is the bending moment at the distance *x* from the point *B* as shown. This theorem is derived in Problem 10.1. For applications, see Problems 10.5 and 10.11.

SECOND MOMENT-AREA THEOREM

Let us consider the vertical distance between the point *B* on the deflection curve shown in Fig. 10-1 and the tangent to this curve drawn at point *A*. This vertical distance is denoted by Δ in the diagram.

The *second moment-area theorem* states that: The vertical distance of point B on a deflection curve from the tangent drawn to the curve at A is equal to the moment with respect to the vertical through B of the area of the bending moment diagram between A and B, divided by the product EI.

This theorem may be stated in equation form as follows:

$$\Delta = \int_A^B \frac{Mx\,dx}{EI}$$

This theorem is derived in Problem 10.2. For applications, see Problems 10.4, 10.6–10.10 and 10.12–10.17.

SIGN CONVENTIONS

In the use of the first theorem, areas corresponding to a positive bending moment diagram are considered positive, those arising from a negative moment diagram are taken as negative. With reference to Fig. 10-1, a positive net area implies that the tangent at B makes a positive or counterclockwise angle with the tangent drawn at A. In the use of the second theorem, the moments of areas of positive bending moment diagrams are considered to be positive and such positive products of areas and moment arms give rise to positive deflections. Positive deflections are taken to be those where point B lies *above* the tangent drawn at point A. It is to be noted that this sign convention bears no relation to that used in the double-integration method.

THE MOMENT-AREA PROCEDURE

The determination of the deflection of a specified point on a loaded beam is made in accordance with the following procedure.

1. The reactions of the beam are determined. In the particular case of a cantilever beam this step may frequently be omitted.

2. An approximate deflection curve is drawn. This curve must be consistent with the known conditions at the supports, such as zero slope or zero deflection.

3. The bending moment diagram is drawn for the beam. The procedure for this was discussed in Chapter 6. Frequently it is convenient to construct the moment diagram by *parts*, as discussed in Problems 6.5 and 6.6. Actually, the M/EI diagram must be used in connection with either of the above theorems, but for beams of constant cross-section, the M/EI digram has the same shape as the ordinary bending moment diagram, except that each ordinate is divided by the product EI. Accordingly it is permissible in the case of beams of *constant* cross-section to work directly with the bending moment diagram and then divide the computed areas or moment-areas by EI. Or, equivalently, the angles or deflections may be multiplied by EI when areas or moment-areas of the ordinary moment diagram are used. This is actually the procedure commonly followed and is used in all illustrative problems of this chapter except Problem 10.17 which involves a beam of variable cross-section.

4. Convenient points A and B are selected and a tangent is drawn to the assumed deflection curve at one of these points, say A.

5. The deflection of point B from the tangent at A is then calculated by the second moment-area theorem.

In certain simple cases, particularly those involving cantilever beams, this deflection of B from the tangent at A may actually be the desired deflection. But often it will be necessary to apply the second moment-area theorem to another point in the beam and then examine the geometric relationship between these two calculated deflections in order to obtain the desired deflection. No general statements regarding this phase of the procedure may be made. Specific examples of this technique may be found in Problems 10.13–10.15.

COMPARISON OF MOMENT-AREA AND DOUBLE-INTEGRATION METHODS

If the deflection of only a single point of a beam is desired, the moment-area method is usually more convenient than the double-integration method. On the other hand, if the equation of the deflection curve of the entire beam is desired the double-integration method (or possibly the method of singularity functions discussed in Chapter 11) is usually superior to the moment-area method. Occasionally there is little preference of one method over the other and often the preference is entirely a personal one.

ASSUMPTIONS AND LIMITATIONS

As explained in Problems 10.1 and 10.2, the moment-area method may be derived from the equation relating bending moment at a point in a beam and the curvature of the neutral surface at that same point. This same relation was used in deriving the double-integration procedure. Hence both methods are based upon the same fundamental relationship and thus both are subject to the same limitations. These are mentioned in the corresponding section in Chapter 9.

Solved Problems

10.1. Derive the first moment-area theorem.

In Fig. 10-2, let AB represent a portion of the deflection curve of a bent beam. Let us consider an element of the beam of length ds. The radius of curvature of this element is denoted by ρ, and the bending moment in the beam at this point is denoted by M. From Problem 8.1, we have the relationship given in equation (7)

$$M = \frac{EI}{\rho} \qquad (1)$$

where E represents the modulus of elasticity of the material and I denotes the moment of inertia of the cross-sectional area of the beam about its neutral axis.

The plot immediately below AB represents the bending moment diagram corresponding to the length AB of the beam. Construction of this diagram was discussed in Chapter 6.

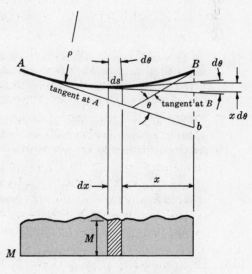

Fig. 10-2

The element of length ds is considered to subtend an angle $d\theta$, measured with respect to the center of curvature of the element ds, as shown. It is evident that $ds = \rho \, d\theta$; hence $\rho = ds/d\theta$. Substituting in (1),

$$d\theta = \frac{M}{EI} ds$$

Since we are concerned only with very small lateral deflections of beams, it will be satisfactory to replace ds by its horizontal projection dx.   Thus

$$d\theta = \frac{M\,dx}{EI}$$

This angle $d\theta$ may also be thought of as the angle between tangents to the deflection curve at the ends of the element of length ds, as shown above. This is true because the sides of these two angles are perpendicular. The angle θ between tangents to the deflection curve at the points A and B may now be found by summing all such angles $d\theta$, that is,

$$\theta = \int d\theta = \int_A^B \frac{M\,dx}{EI}$$

This is called the *first moment-area theorem*. In words this theorem states: The angle between the tangents at two points A and B of the deflection curve of a beam is equal to the area under the bending moment diagram between these two points, divided by EI.

For a sign convention, we shall take positive areas to be those arising from positive moment diagrams. A positive net area will be taken to denote that the right hand tangent at B makes a counter-clockwise angle with the left hand tangent at A.

10.2. Derive the second moment-area theorem.

Let us again refer to Fig. 10-2. It is desired to calculate the vertical distance of point B on the deflection curve from the tangent drawn to this curve at a point A. This distance is represented by the line segment Bb shown there. The contribution to this length Bb made by the bending of the element of length ds is the vertical element $x\,d\theta$ shown there. However, in Problem 10.1 it was shown that $d\theta = M\,dx/EI$. Hence

$$x\,d\theta = \frac{Mx\,dx}{EI}$$

With reference to the figure, the right side of this equation represents the moment of the shaded area $M\,dx$ about a vertical line through B, divided by EI.   Integration yields

$$Bb = \int_A^B \frac{Mx\,dx}{EI}$$

In words this equation states that if A and B are points on the deflection curve of a beam, the vertical distance of B from the tangent drawn to the curve at A is equal to the moment with respect to the vertical through B of the area of the bending moment diagram between A and B, divided by EI. This is called the *second moment-area theorem*.

For a sign convention, we shall take moments of areas of positive moment diagrams to be positive and such positive moment-areas will produce positive deflections. We shall further define positive deflections to be those where the point B lies above the tangent drawn at A.

It is important to note that this theorem indicates relative deflections, i.e. the deflection of point B with respect to the tangent drawn to the curve at A. The true or absolute deflection of point B may be zero, as in the case of a point directly over one support of a beam, yet in such a case there may be a nonzero relative deflection with respect to the tangent at A.

10.3. Determine the areas and locate the centroids of the figures commonly occurring in bending moment diagrams drawn by parts.

We need be concerned primarily with only three geometric figures: the rectangle, the triangle, the parabola. For the rectangle, the area is of course equal to the product of the lengths of two adjacent sides, and the centroid lies at the geometric center of the rectangle.   The area of a triangle is half the product of its base and its altitude, and the centroid of a triangle was located in Problem 7.1, page 109.

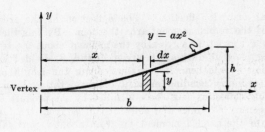

Fig. 10-3

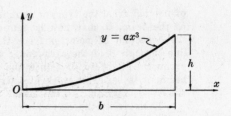

Fig. 10-4

Let us now consider the case of the parabola shown in the adjoining figure. Note carefully that the parabola is placed so that its vertex lies at the origin of the coordinate system.

To determine the area, we consider first the area of the small shaded element of width dx and altitude y. Evidently its area is $y\,dx$. To find the area under the entire parabola shown, we must sum the areas of all such elements by means of an integral. Thus

$$A = \int y\,dx = \int_0^b ax^2\,dx = \tfrac{1}{3}a[x^3]_0^b = \tfrac{1}{3}ab^3$$

But when $x = b$, $y = h$; hence $a = h/b^2$ and thus $A = bh/3$.

To locate the x-coordinate of the centroid of this parabolic area, we employ the definition of the centroid given in Chapter 7, i.e.

$$\bar{x} = \frac{\int x\,da}{A}$$

Then
$$\bar{x} = \frac{\int x(y\,dx)}{bh/3} = \frac{\int_0^b x(ax^2)\,dx}{bh/3} = \frac{a[x^4/4]_0^b}{bh/3} = \frac{(h/b^2)(b^4/4)}{bh/3} = \frac{3}{4}b$$

For the so-called cubic parabola, described by the equation $y = ax^3$ and having the shape illustrated in Fig. 10-4, a comparable analysis yields the area as $A = bh/4$ and the centroid is found to lie a distance $4b/5$ from the origin. These will be the only geometric figures occurring in the moment-area treatments of beams subject to couples, concentrated forces, uniformly distributed loads and linearly varying loads.

Thus the moments of areas under the bending moment diagrams, used in the second moment-area theorem, may be found in many cases by use of the above expressions for areas and centroids. The first moment of the area is equal to the product of its centroidal distance from the moment axis and its area.

10.4. The cantilever beam of Fig. 10-5(a) is subject to the concentrated force P applied at the free end. Determine the deflection under the point of application of the load.

In the case of a cantilever beam, the reactions at the wall need not be determined although their determination is, of course, extremely simple. It is known that the slope and deflection at the clamped end A are each zero by definition of a cantilever beam. Hence the heavy curved line represents a realistic deflection curve.

The bending moment diagram is most easily drawn by working from right to left across the beam and is shown in Fig. 10-5(b).

Next, a tangent to the deflection curve is drawn at point A. In the case of a cantilever beam, this tangent coincides with the original unbent position of the bar and is represented by the straight line shown. Hence, in this particular case, the deflection

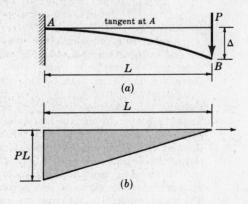

Fig. 10-5

of point B from the tangent at A is the actual desired deflection. The deflection of the free end of the beam, B, may now be found by use of the second moment-area theorem. By this theorem, the deflection of point B from the tangent drawn at A is given by the moment about the vertical line through B of the area under the bending moment diagram between A and B, divided by the product EI. Actually, since the cross-section of the beam is constant along the length of the beam, it is easier to work directly with the ordinary bending moment diagram rather than with the M/EI diagram. In that event, the resulting deflection must be multiplied by the product EI.

Thus, from the second moment-area theorem, EI times the deflection of B from the tangent at A, denoted by Δ, is given by the moment of the shaded moment diagram about a vertical line through B. This moment of area may be calculated by multiplying the area by the distance of the centroid of the area from the vertical line through B. The area of the triangular moment diagram is $(L/2)(-PL)$, where the negative sign is used because the bending moment is negative. The centroid of the moment diagram lies at a distance $2L/3$ from the right end. Hence the moment-area theorem becomes

$$EI\Delta = (L/2)(-PL)(2L/3) = -PL^3/3 \quad \text{or} \quad \Delta = -PL^3/3EI$$

The negative sign implies that the final position of B lies below the tangent drawn at A.

10.5. Determine the slope at the right end of the cantilever beam discussed in Problem 10.4.

The heavy curved line representing the deflected beam has been sketched in Fig. 10-5. Also, the moment diagram was presented there. These are reproduced in Fig. 10-6. In Fig. 10-6(a) tangents are drawn to the deflection curve of the bent beam, at the clamped end A and the free end B. These are designated as the tangents at A and B respectively. According to the first moment-area theorem the angle θ between these two tangents is equal to the area under the bending moment diagram between A and B, divided by EI. Thus from this theorem EI times the angle θ is given by the area under the shaded moment diagram, and we have

$$EI\theta = (L/2)(-PL)$$

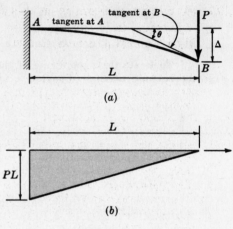

(a)

(b)

Fig. 10-6

where the negative sign accompanying PL is used because the bending moment is negative. Solving,

$$\theta = -PL^2/2EI$$

The negative sign denotes that the right hand tangent at B makes a clockwise angle with the left hand tangent at A. This is in accordance with the sign convention discussed in Problem 10.1. The angle θ is of course in radians.

10.6. A cantilever beam is subject to the uniformly distributed load acting over the entire length of the beam as shown in Fig. 10-7(a). Determine the deflection of the free end of the beam.

Again, as in Problem 10.4, the reactions at the end of the beam need not be found. It is first necessary to sketch an approximate deflection curve for the bent beam. Since the beam is clamped at the left end, it is evident that the slope and also the deflection at that end are each zero. Thus the heavy curved line represents a deflection curve that is in agreement with the known conditions of zero slope and zero deflection at the left end of the beam.

The bending moment diagram is best drawn by working from right to left across the beam. The construction of this diagram was discussed in Problem 6.2. As mentioned in that problem,

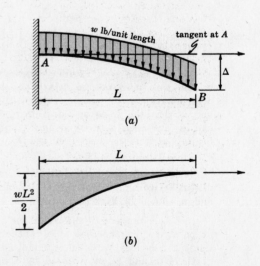

(a)

(b)

Fig. 10-7

the maximum bending moment occurs at the supporting wall and has the value $wL^2/2$, where w represents the intensity of the uniform load per unit length of the beam. Also in that problem it was shown that the bending moment diagram is a parabola.

Since the tangent at point A coincides with the original position of the beam, the deflection of point B from the tangent at A represents the desired deflection and may be found by use of the second moment-area theorem. Thus we must calculate the moment about the vertical line through B of the area under the bending moment diagram between A and B, divided by EI. Actually it is easier to work with the bending moment diagram shown above and then multiply the resulting deflection by EI. Using the expressions for area and location of the centroid of a parabolic area as found in Problem 10.3 and noting that the altitude of the parabola must be taken as negative since the bending moment is negative, we immediately have

$$EI\Delta = \frac{1}{3}L\left(-\frac{wL^2}{2}\right)\left(\frac{3L}{4}\right) = -\frac{wL^4}{8} \quad \text{or} \quad \Delta = -\frac{wL^4}{8EI}$$

The negative sign indicates that the final position of B lies below the tangent drawn at A.

10.7. A cantilever beam is subject to the uniformly distributed load extending from the midpoint of the beam to the extreme end as shown in Fig. 10-8(a). Determine the deflection of the free end of the beam.

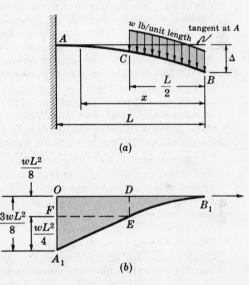

(a)

(b)

Fig. 10-8

The reactions at the left end of the beam need not be calculated. The heavy curved line represents a reasonable approximation to the deflection curve of the beam. This curve is in agreement with the known conditions of zero slope and zero deflection at the left end of the beam. The bending moment diagram can be obtained by working from right to left across the beam. Under the uniform load, the moment must be parabolic, as discussed in Problem 6.2, page 85. In constructing the portion of the moment diagram between A and C, the uniform load extending from C to B may be replaced by its resultant which is $wL/2$ lb acting downward. At any point between A and C, located a distance x from the end B, the bending moment is given by the moment of the resultant of the distributed load about an axis through this same point and perpendicular to the plane of the page. Thus the bending moment anywhere between A and C is given by $-\frac{1}{2}wL(x - L/4)$. This is a linear function, and hence the bending moment diagram plots as a straight line between A and C. Thus the bending moment diagram consists of a parabolic region DB_1E together with a trapezoidal region $ODEA_1$, as shown above.

The tangent at point A coincides with the original position of the beam. The deflection of point B from the tangent at A represents the desired deflection and may be found by use of the second moment-area theorem. To apply this theorem we must calculate the moment about the vertical line through B of the area under the bending moment diagram between A and B and divide the result by EI. To calculate the moment of the area OB_1A_1 about a vertical line through B it is simplest to divide the shaded area into a parabolic region DB_1E, a rectangular region $ODEF$, and a triangular region FEA_1. Using the results of Problem 10.3 and again noting that the altitudes of each of these three regions must be taken as negative, we have

$$EI\Delta = \frac{1}{3}\left(\frac{L}{2}\right)\left(\frac{-wL^2}{8}\right)\left(\frac{3}{4}\frac{L}{2}\right) + \frac{L}{2}\left(\frac{-wL^2}{8}\right)\left(\frac{L}{2} + \frac{1}{2}\frac{L}{2}\right) + \frac{1}{2}\left(\frac{L}{2}\right)\left(\frac{-wL^2}{4}\right)\left(\frac{L}{2} + \frac{2}{3}\frac{L}{2}\right)$$

or

$$\Delta = \frac{-41wL^4}{384EI}$$

The negative sign indicates that the final position of B lies below the tangent drawn at A.

10.8. In Fig. 10-9(a) a cantilever beam is loaded by the moment M_1 as well as the uniformly distributed load extending over half of its length. Determine the deflection of the free end of the beam.

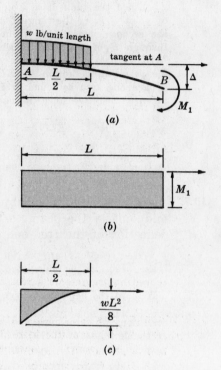

(a)

(b)

(c)

Fig. 10-9

The reactions at the left end of the beam need not be calculated. The heavy curved line represents the approximate shape of the deflection curve of the beam. This curve is in agreement with the conditions of zero slope and zero deflection at the left end of the beam. In this case, since there are two loads acting on the beam, it is perhaps simplest to construct the bending moment diagram by *parts* as discussed in Problems 6.5 and 6.6, pages 89-91. One bending moment diagram will be constructed to represent the bending moment due to M_1 alone without regard to the uniform load, and a second moment diagram will be drawn for the uniform load alone without regard to the moment M_1. In the construction of each of these diagrams, it is most convenient to work from the right end of the beam toward the left end.

The first bending moment diagram [Fig. 10-9 (b)], due to M_1 alone, is evidently a rectangle, since the bending moment due to M_1 is the same at all points along the beam. The moment M_1 causes the beam to bend into a configuration that is concave downward, which, according to the sign convention of Chapter 6, constitutes a negative bending moment.

The second bending moment diagram [Fig. 10-9(c)], due to the uniform load, is parabolic, as discussed in Problem 10.6, except that the parabola here corresponds only to the portion of the beam that is subject to the uniform load, i.e. the left half of the beam.

A tangent to the deflection curve is now drawn at point A. The free end of the beam is designated as point B. This tangent coincides with the original position of the beam and is represented by the straight line shown. Thus the deflection of point B from the tangent drawn at A represents the desired deflection of the free end of the beam. According to the second moment-area theorem, this deflection of B from the tangent at A is given by the moment about the vertical line through B of the area under the entire bending moment diagram between A and B, divided by EI.

The moment of the area of the entire bending moment diagram between A and B about the vertical line through B is most easily found by adding the moments of the rectangular and parabolic areas about this vertical line. For each of these areas, the moment about the line through B is given by the product of the area and the distance from the centroid of the area to B. These are the same areas and centroidal distances used in previous problems. Thus the second moment-area theorem gives

$$EI\Delta = (-M_1)(L)\left(\frac{L}{2}\right) + \frac{1}{3}\left(\frac{L}{2}\right)\left(\frac{-wL^2}{8}\right)\left(\frac{L}{2} + \frac{3}{4}\frac{L}{2}\right) \quad \text{or} \quad \Delta = -\frac{M_1L^2}{2EI} - \frac{7wL^4}{384EI}$$

The negative signs indicate that the final position of B lies below the tangent drawn at A.

10.9. The simply supported beam of Fig. 10-10(a) is loaded by the two symmetrically placed forces shown. Find the maximum deflection of the beam.

The reactions at the ends of the beam are each equal to P by symmetry. The heavy line represents the deflection curve of the beam and this curve must be symmetric about the midpoint of the beam because the loads are symmetrically applied. The tangent to the deflection curve at the midpoint of the beam, denoted by A, is horizontal. The right end of the beam is designated as point B. Because of symmetry, the maximum deflection of the beam must occur at its midpoint.

However, the diagram indicates that the deflection at the midpoint is equal to the deflection of point B from the tangent at A, both deflections being represented by Δ. It is simple to calculate this second deflection by use of moment-areas.

The bending moment diagram is perhaps best drawn in *parts*, working from right to left along the beam. In this procedure, discussed in Problems 6.5 and 6.6, the moment diagram due to each of the forces P applied individually is drawn with the result that the final moment diagram appears as shown above.

By the second moment-area theorem, the deflection of B from the tangent at A is equal to the moment with respect to the vertical line through B of the area of the bending moment diagram between A and B, divided by EI. This area consists of the two triangles deh and efg whose centroids were located in Problem 10.3. From the second moment-area theorem:

$$EI\Delta = \frac{1}{2}\left(\frac{L}{2}\right)\left(\frac{PL}{2}\right)\left(\frac{2}{3}\frac{L}{2}\right) + \frac{1}{2}\left(\frac{L-2a}{2}\right)$$
$$\times \left[-\left(\frac{PL}{2}-Pa\right)\right]\left[a+\frac{2}{3}\left(\frac{L-2a}{2}\right)\right]$$

$$= \frac{PL^2a}{8} - \frac{Pa^3}{6}$$

or $$\Delta = \frac{PL^3}{24EI}\left(\frac{3a}{L}-\frac{4a^3}{L^3}\right)$$

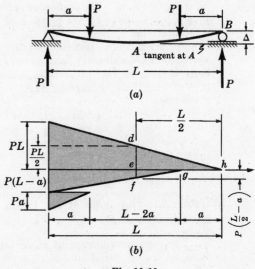

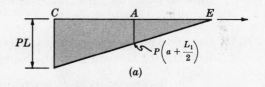

Fig. 10-10

10.10. The overhanging beam in Fig. 10-11 is loaded by two concentrated forces. Find the deflection at the midpoint.

Each of the reactions is equal to P lb by symmetry. The deflection curve of the beam, shown by the heavy line, is symmetric about the midpoint of the beam and consequently the tangent to the deflection curve is horizontal at the midpoint. The midpoint of the beam is designated as point A and the point directly over the right support is B. From the diagram it is evident that the deflection of the midpoint is given by Δ, which may also be considered to be the deflection of B from the tangent drawn at A. Again, it is to be noted that the moment-area technique indicates relative deflections, in this case the deflection of B relative to the tangent drawn at A. Of course, the absolute deflection of B is zero. However, the use of the moment-area theorem enables us to find the desired deflection of the midpoint by the somewhat indirect method of determining the deflection of B relative to the tangent at A.

Let us draw the moment diagram by parts. Working from right to left across the beam we first find a triangular diagram due to the downward applied load P acting at E, as shown in Fig. 10-12(a). The bending moment due to the reaction P at B does not come into being until we have passed to the left of B. The moment diagram for this load only is shown in Fig. 10-12(b). Finally, after we pass to the left of D, another portion of the bending moment diagram appears as shown in Fig. 10-12(c).

The entire bending moment diagram is composed of these three triangles and is shown

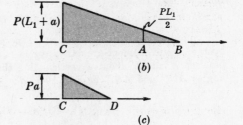

Fig. 10-11

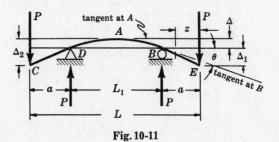

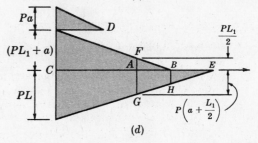

Fig. 10-12

in Fig. 10-12(d). The deflection of point B from the tangent drawn to the deflection curve at point A is given by the moment of the area under the moment diagram between A and B about a vertical line through B, divided by EI. This area under the moment diagram consists of a triangle AFB and a trapezoid $ABHG$ in Fig. 10-12(d). The trapezoid may be treated most conveniently by dividing it into a rectangle of altitude Pa and a triangle of altitude $PL_1/2$. The area of each of these figures together with the location of each centroid of area was discussed in Problem 10.3. According to the second moment-area theorem the desired deflection Δ is thus given by

$$EI\Delta = \frac{1}{2}\left(\frac{L_1}{2}\right)\left(\frac{PL_1}{2}\right)\left(\frac{2}{3}\frac{L_1}{2}\right) + \left(\frac{L_1}{2}\right)(-Pa)\left(\frac{L_1}{4}\right) + \frac{1}{2}\left(\frac{L_1}{2}\right)\left(\frac{-PL_1}{2}\right)\left(\frac{2}{3}\frac{L_1}{2}\right)$$

or
$$\Delta = -\frac{PaL_1^2}{8EI}$$

10.11. Determine the slope of the beam described in Problem 10.10 at a point directly over the support at B.

For the determination of slopes, the first moment-area theorem is useful. Again, let us consider the two points A and B as described in Problem 10.10. As stated above, the tangent to the deflection curve of the beam is horizontal at the point A because of symmetry of both loading and support. If the change of slope of the deflection curve can be evaluated between A and B, then evidently this change will also be equal to the slope itself at B, since the slope at A is zero. The first moment-area theorem tells us that the angle between the tangents at A and B is equal to the area of the bending moment diagram between these two points divided by EI. Evidently the angle that the tangent at A makes with the horizontal is zero, and we shall denote the angle that the tangent at B makes with the horizontal by θ. Then, from the moment-area theorem, we have merely to evaluate the area under the moment diagram between A and B. This area, as before, consists of the triangle AFB together with the trapezoid $ABHG$. From moment-areas we thus have

$$EI\theta = \frac{1}{2}\left(\frac{L_1}{2}\right)\left(\frac{PL_1}{2}\right) + \left(\frac{L_1}{2}\right)(-Pa) + \frac{1}{2}\left(\frac{L_1}{2}\right)\left(\frac{-PL_1}{2}\right) \quad \text{or} \quad \theta = -\frac{PaL_1}{8EI}$$

The negative sign indicates that the tangent at B makes a clockwise angle with the tangent drawn at A. This agrees with the sign convention adopted in Problem 10.1.

10.12. Consider again the overhanging beam described in Problem 10.10. Determine the deflection of the end E of the beam with respect to its original position.

The desired deflection is designated as Δ_1 in Fig. 10-11. From that figure, the following relationship exists:
$$\Delta_1 = \Delta_2 - \Delta$$

The quantity Δ_2 is the deflection of point E from the tangent to the deflection curve at A and by the second moment-area theorem is equal to the moment of the area under the moment diagram between A and E about a vertical line through E, divided by EI. This area consists of the triangles AFB and AGE in Fig. 10-12(d). Hence,

$$EI\Delta_2 = \frac{1}{2}\left(\frac{L_1}{2}\right)\left(\frac{PL_1}{2}\right)\left[a + \frac{2}{3}\frac{L_1}{2}\right] + \frac{1}{2}\left(a + \frac{L_1}{2}\right)\left[-P\left(a + \frac{L_1}{2}\right)\right]\left[\frac{2}{3}\left(a + \frac{L_1}{2}\right)\right]$$

$$= -\frac{PaL_1^2}{8} - \frac{Pa^2L_1}{2} - \frac{Pa^3}{3}$$

The quantity Δ was found in Problem 10.10 to be $\Delta = -PaL_1^2/8EI$.

Finally, substituting into the first equation above, the required end deflection Δ_1 is

$$\Delta_1 = -\frac{PaL_1^2}{8EI} - \frac{Pa^2L_1}{2EI} - \frac{Pa^3}{3EI} + \frac{PaL_1^2}{8EI} = -\frac{Pa^2L_1}{2EI} - \frac{Pa^3}{3EI}$$

10.13. A simply supported beam is loaded by the concentrated load shown in Fig. 10-13. Determine the deflection at the midpoint of the beam.

From statics the end reactions have the values indicated. The approximate form of the deflection curve is indicated by the heavy line. Evidently there is no symmetry of the deflection curve about the midpoint of the beam, and this feature renders this problem somewhat more difficult than the preceding ones. The left end of the beam is designated as point A, the midpoint as B, the point of application of the force P as C, and the right end as D.

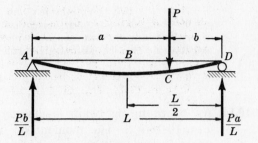

Fig. 10-13

The moment diagram by parts, working from left to right along the beam, has the components indicated in Fig. 10-14(a), (b), and the entire bending moment diagram appears as in Fig. 10-14(c). A tangent to the deflection curve is drawn at point A as shown in Fig. 10-15 and the deflection of D from the tangent at A determined by the second moment-area theorem. It is to be remembered that the theorem yields relative deflections, i.e. the deflection of D relative to the tangent at A. The absolute deflection of D is of course zero. This relative deflection is found by calculating the moment of the area under the moment diagram between A and D about a vertical line through D, divided by EI. This leads to

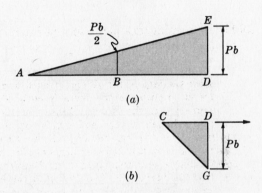

$$EI(\overline{Dd}) = \frac{1}{2}(L)(Pb)\left(\frac{L}{3}\right) + \frac{1}{2}(b)(-Pb)\left(\frac{b}{3}\right)$$

$$= \frac{PbL^2}{6} - \frac{Pb^3}{6}$$

As stated previously, B represents the midpoint of the beam. Evidently from similar triangles, the line segment $\overline{Bf}$ shown in Fig. 10-15 is exactly half the length of $\overline{Dd}$ and thus we may write

$$EI(\overline{Bf}) = \frac{PbL^2}{12} - \frac{Pb^3}{12}$$

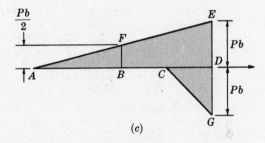

Fig. 10-14

Next, it is possible to calculate the deflection of the midpoint of the beam from the tangent drawn at A. This deflection is represented by the line segment $\overline{ef}$ in Fig. 10-15. According to the second moment-area theorem this is given by the moment of the area under the bending moment diagram between A and B about a vertical line through B, divided by EI. This portion of the moment diagram is represented by the triangle ABF. Thus we have

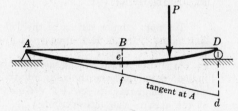

Fig. 10-15

$$EI(\overline{ef}) = \frac{1}{2}\left(\frac{L}{2}\right)\left(\frac{Pb}{2}\right)\left(\frac{1}{3}\frac{L}{2}\right) = \frac{PbL^2}{48}$$

From the above representation of the deflection curve of the beam it is apparent that the desired midpoint deflection is represented by the line segment $\overline{Be}$. This may be found from the relationship

$$\overline{Be} = \overline{Bf} - \overline{ef}$$

Substituting the above values in the right side of the equation, we find for the desired deflection of the midpoint

$$EI(\overline{Be}) = \frac{PbL^2}{12} - \frac{Pb^3}{12} - \frac{PbL^2}{48} \quad \text{or} \quad \overline{Be} = \frac{PbL^2}{48EI}\left(3 - \frac{4b^2}{L^2}\right)$$

Note that this is not the maximum deflection of the beam except in the special case when $a = b = L/2$. Also, it is assumed that the load P lies to the right of the midpoint of the beam, otherwise the triangular moment diagram CDG would extend to the left of the midpoint of the beam and it would then be necessary to take a portion of it into account in calculating the deflection ef.

10.14. An overhanging beam loaded by the concentrated force of 10,000 lb is shown in Fig. 10-16(a). The moment of inertia of the cross-section of the beam about its neutral axis is 1000 in⁴ and $E = 30 \times 10^6$ lb/in². Determine the deflection under the point of application of the 10,000 lb force.

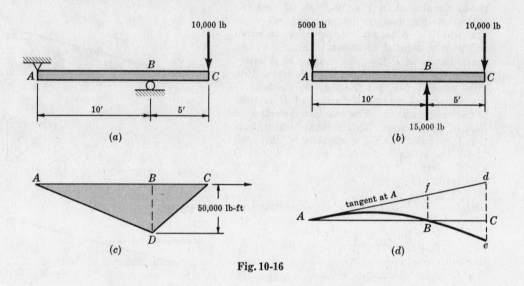

Fig. 10-16

From statics the reactions are easily found to be a downward force of 5000 lb acting at A and an upward force of 15,000 lb applied at B. The free-body diagram of the beam is shown in Fig. 10-16(b).

The bending moment diagram for this problem may be drawn in any one of several equally convenient ways. Let us draw it in the conventional manner, in which case it appears as in Fig. 10-16(c).

The approximate nature of the deflection curve is represented by the heavy line in Fig. 10-16(d). A tangent to the deflection curve is drawn at the point A, as indicated.

According to the second moment-area theorem the deflection of point C, whose final position is indicated by e, from the tangent at A is given by the moment of the area under the bending moment diagram between A and C about a vertical line through C, divided by EI. Thus the moment of the area of the entire triangle ADC about a vertical line through C must be calculated. This gives for the deflection of C with respect to the tangent at A,

$$EI(\overline{de}) = \tfrac{1}{2}(10)(-50,000)(5 + \tfrac{1}{3} \cdot 10) + \tfrac{1}{2}(5)(-50,000)(\tfrac{2}{3} \cdot 5) = -2,496,000$$

It is to be carefully noted that the units of the right side of this equation are lb-ft³.

Next, it is necessary to calculate the deflection of point B from the tangent drawn at A. This is represented by the line segment $\overline{fB}$ in Fig. 10-16(d). Again, it is to be remembered that the moment-area theorem indicates relative deflections, in this case the deflection of B relative to the tangent at A. Actually, of course, the true or absolute deflection of point B is zero. This relative deflection may be found by taking the moment of the area of the triangle ABD about a vertical line through B and then dividing by EI. This gives

$$EI(\overline{fB}) = \tfrac{1}{2}(10)(-50,000)(\tfrac{1}{3} \cdot 10) = -833,000 \text{ lb-ft}^3$$

From a consideration of the similar triangles AfB and AdC in Fig. 10-16(d), we may write $\overline{fB}/10 = \overline{dC}/15$. Hence from the above value of $EI(\overline{fB})$ we have

$$EI(\overline{dC}) = (15/10)(-833,000) = -1,250,000 \text{ lb-ft}^3$$

Evidently the desired deflection of the point C, represented by the line segment $\overline{Ce}$, is given by

$$\overline{Ce} = \overline{de} - \overline{dC}$$

Substituting the above values in the right side of this equation, we find for the desired deflection of the point C,

$$EI(\overline{Ce}) = -2,496,000 - (-1,250,000) = -1,246,000 \text{ lb-ft}^3$$

The negative sign indicates that the final position of point C, designated as e, lies below the tangent drawn at A. Let us now substitute $I = 1000 \text{ in}^4$ and $E = 30 \times 10^6 \text{ lb/in}^2$ in the last equation. This gives the desired deflection

$$\overline{Ce} = \frac{-1,246,000(1728)}{(30 \times 10^6)(1000)} = -0.0720 \text{ in.}$$

Note the factor 1728 introduced to convert to consistent units, since the quantity $-1,246,000$ is in lb-ft^3 units whereas E and I are expressed in lb/in^2 and in^4 respectively.

10.15. The overhanging beam in Fig. 10-17(a) is loaded by the uniformly distributed load shown. The bar has a 3×4 in. cross-section. Find the deflection of point A. Take $E = 30 \times 10^6$ lb/in^2.

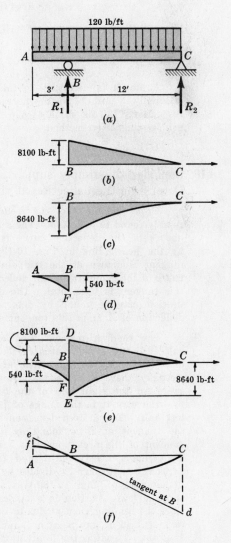

Fig. 10-17

From statics the reactions are readily found to be $R_1 = 1125$ lb and $R_2 = 675$ lb. The bending moment diagram for this beam is best adapted to the moment-area technique if it is drawn by parts working from both ends toward the support at point B. We first work from point C to the left, considering first the bending moment due to the reaction R_2 only, which appears in Fig. 10-17(b); and then the bending moment due to the uniform load of 120 lb/ft only, which appears as the parabola in Fig. 10-17(c).

We next work from point A to the right and obtain the parabolic bending moment diagram due to the uniform load in the region AB, as shown in Fig. 10-17(d).

The complete bending moment diagram drawn by parts thus appears as shown in Fig. 10-17(e).

The approximate nature of the deflection curve is represented by the heavy line in Fig. 10-17(f). A tangent to the deflection curve is drawn at the point B as indicated.

The deflection represented by $\overline{Af}$ is desired. Let us begin by finding the deflection of A (whose final position is represented by f) from the tangent drawn at B. This is readily given by the second moment-area theorem to be the moment of the area under the bending moment diagram between A and B about a vertical line through A, divided by EI. This area under the moment diagram between A and B is represented by the parabola ABF. Applying the second theorem, we find

$$EI(\overline{ef}) = \tfrac{1}{3}(3)(-540)(\tfrac{3}{4} \cdot 3) = -1216 \text{ lb-ft}^3$$

The negative sign indicates that the final position of point A (represented by f) lies below the tangent drawn at B.

We shall next calculate the deflection of C from the tangent drawn at B. This distance, represented by $\overline{Cd}$, is readily given by the second moment-area theorem to be the moment about the vertical line through C of the area under the bending moment diagram between B and C, divided by EI. Thus,

$$EI(\overline{Cd}) = \tfrac{1}{2}(12)(8100)(\tfrac{2}{3} \cdot 12) + \tfrac{1}{3}(12)(-8640)(\tfrac{3}{4} \cdot 12) = 77{,}900 \text{ lb-ft}^3$$

Again, it is to be remembered that this is a relative deflection, since the absolute deflection of point C is zero.

From a consideration of the similar triangles BCd and ABe, we have $\overline{eA}/3 = \overline{Cd}/12$. Hence from the above value of $EI(\overline{Cd})$ we have

$$EI(\overline{eA}) = (3/12)(77{,}900) = 19{,}500 \text{ lb-ft}^3$$

From Fig. 10-17(f) it is evident that the desired deflection $\overline{Af}$ is given by

$$\overline{Af} = \overline{eA} - \overline{ef}$$

Substituting the above values in the right side of the equation, we find

$$EI(\overline{Af}) = 19{,}500 - 1216 = 18{,}284 \text{ lb-ft}^3$$

Now substitute $E = 30 \times 10^6 \text{ lb/in}^2$ and $I = 3(4^3)/12 = 16 \text{ in}^4$ in the last equation and obtain

$$\overline{Af} = \frac{18{,}284(1728)}{(30 \times 10^6)(16)} = 0.065 \text{ in.}$$

Again, it is necessary to introduce the factor 1728 since the right side of the equation was originally in units of lb-ft^3 whereas the values of E and I were stated in units of lb/in^2 and in^4 respectively. This value of deflection agrees with that found in Problem 9.18, page 173, by the double-integration method.

10.16. Consider the simply supported beam subject to the linearly varying load indicated. Determine the deflection at the midpoint of the beam.

From statics the end reactions are readily found to be $w_0L/4$. The approximate nature of the deflection curve is indicated by the heavy line in Fig. 10-18(a) and a tangent is drawn at the midpoint B. Because of the symmetry of the deflection curve this tangent is horizontal. The desired deflection may be found by determining the deflection of A from this tangent at B.

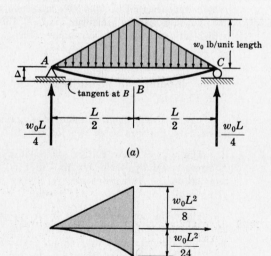

Fig. 10-18

The bending moment diagram is best suited to the moment-area method if it is drawn by parts working from left to right. Actually, there is no need to draw this diagram for the right half of the beam, since by symmetry it is the image of that in the left half. The desired deflection of A from the tangent at B is found by taking the moment of the area under the bending moment diagram between A and B about a vertical line through A, divided by EI. The upper part of Fig. 10-18(b) is a triangle, and the lower portion is a cubic parabola, since the bending moment due to the linearly varying load at any point a distance x from the point A is $M = -w_0x^3/3L$. The area under this diagram, as well as the location of its centroid, was determined in Problem 10.3. Utilizing these results we have

$$EI\Delta \;=\; \frac{1}{2}\left(\frac{L}{2}\right)\left(\frac{w_0L^2}{8}\right)\left(\frac{2}{3}\frac{L}{2}\right) + \frac{1}{4}\left(\frac{L}{2}\right)\left(-\frac{w_0L^2}{24}\right)\left(\frac{4}{5}\frac{L}{2}\right) \quad\text{or}\quad \Delta \;=\; \frac{w_0L^4}{120EI}$$

10.17. The simply supported beam shown in Fig. 10-19(a) has flexural rigidity EI in the central portion of length $L/2$. The beam is infinitely rigid in each of the end quarters. Determine the central deflection due to the concentrated central force.

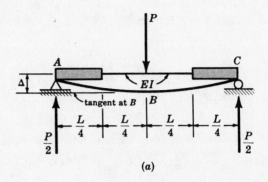

(a)

The bending moment diagram follows from statics and is indicated in Fig. 10-19(b). To convert this into an M/EI diagram, it is necessary to divide all ordinates by the appropriate value of EI. Since this value is infinite in the end quarters, the M/EI diagram is zero in those regions and thus the true M/EI diagram appears as in Fig. 10-19(c).

The approximate nature of the deflection curve is indicated by the heavy line and the tangent at the midpoint B is horizontal by virtue of symmetry. The desired deflection is determined as the deflection of A from this tangent at B. By taking the moment of the area under the M/EI diagram between A and B about a vertical line through A this is found to be

$$\Delta \;=\; \left(\frac{L}{4}\right)\left(\frac{PL}{8EI}\right)\left(\frac{3}{8}L\right)$$

$$\quad + \frac{1}{2}\left(\frac{L}{4}\right)\left(\frac{PL}{8EI}\right)\left[\frac{L}{4} + \frac{2}{3}\left(\frac{L}{4}\right)\right]$$

$$=\; \frac{7PL^3}{384EI}$$

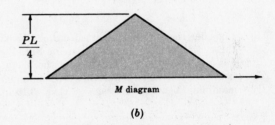

M diagram

(b)

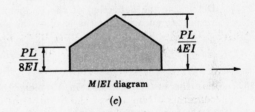

M/EI diagram

(c)

Fig. 10-19

Supplementary Problems

10.18. A cantilever beam of length L is subject to a concentrated force P applied at a distance a from the fixed end of the beam. Determine the deflection at the free end of the beam. *Ans.* $Pa^2(3L-a)/6EI$

10.19. The cantilever beam described in Problem 10.18 is 10 ft long and a load of 1000 lb is applied 6 ft from the support. The cross-section of the beam is 2×6 in. The material is steel for which $E = 30 \times 10^6$ lb/in². Determine the maximum deflection of the beam. *Ans.* 0.230 in.

10.20. A cantilever beam of length L is loaded by a couple of magnitude M_1 at its free end. Determine the deflection of the free end of the beam. *Ans.* $M_1L^2/2EI$

10.21. Consider the cantilever beam loaded as in Problem 10.6. Determine the slope of the beam at the end B.   Ans. $\theta_B = -wL^3/6EI$

10.22. A cantilever beam of length L is loaded by a couple of magnitude M_1 applied at the midpoint of the bar. Determine the deflection of the free end of the bar.   Ans. $3M_1L^2/8EI$

10.23. A cantilever beam of length L is loaded by a uniform load of w lb/ft extending from the clamped end to the midpoint of the length of the bar. Determine the maximum deflection of the beam. Ans. $7wL^4/384EI$

10.24. A simply supported beam is loaded by the uniformly distributed loads shown in Fig. 10-20. Determine the maximum deflection of the beam.

Ans. $-\dfrac{wa^4}{24EI} + \dfrac{wa^2L^2}{16EI}$

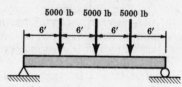

Fig. 10-20

10.25. Consider the simply supported hickory beam subject to the three concentrated forces shown in Fig. 10-21. The beam has a 6×12 in. cross-section and the modulus of elasticity is 2.2×10^6 lb/in². Determine the maximum deflection and maximum bending stress.   Ans. 3.10 in., 5000 lb/in²

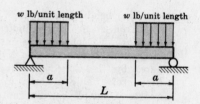

Fig. 10-21

10.26. A simply supported beam with overhanging ends is loaded by the uniformly distributed loads shown in Fig. 10-22. Determine the deflection of the midpoint of the beam with respect to an origin at the level of the supports.

Ans. $wa^2(L-2a)^2/16EI$ (above level of supports)

10.27. For the beam described in Problem 10.26, determine the deflection of one end of the beam with respect to an origin at the level of the supports.

Ans. $\dfrac{wa^3L}{4EI} - \dfrac{3wa^4}{8EI}$ (below level of supports)

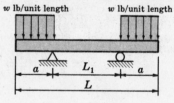

Fig. 10-22

10.28. Consider the simply supported beam described in Problem 9.44, page 180. By use of the moment-area method, determine the deflection of the beam under the point of application of the applied moment M_1. Compare this result with that obtained by use of the double-integration method.

Ans. $\dfrac{M_1a^2}{2EI} + \dfrac{M_1a(L-a)}{3EI}$

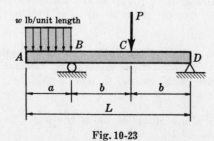

Fig. 10-23

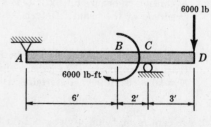

Fig. 10-24

10.29. The overhanging beam of Fig. 10-23 is loaded by the uniformly distributed load as well as the concentrated force shown. Determine the deflection of point A on the beam.

Ans. $-\dfrac{wa^3b}{3EI} + \dfrac{Pab^2}{4EI} - \dfrac{wa^4}{8EI}$ (below level of supports)

10.30. Consider the simply supported overhanging beam shown in Fig. 10-24. The loading consists of a moment of 6000 lb-ft and a force of 6000 lb applied as shown. The moment of inertia of the cross-sectional area of the beam is 42 in⁴ and $E = 30 \times 10^6$ lb/in². Determine the deflection of point B where the moment is applied.   *Ans.* 0.104 in.

10.31. Determine the slope at the end of the beam in Problem 10.16.   *Ans.* $5w_0L^3/192EI$

10.32. Consider the simply supported beam subject to the load indicated in Fig. 10-25. Determine the slope at the end of the beam.   *Ans.* $7w_0L^3/2880EI$

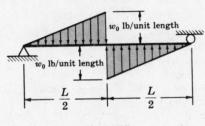

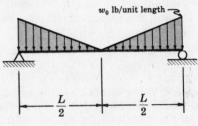

Fig. 10-25   **Fig. 10-26**

10.33. Determine the central deflection of the beam loaded as indicated in Fig. 10-26.
Ans. $3w_0L^4/640EI$

10.34. Determine the central deflection of the simply supported beam in Fig. 10-27. The end quarters have flexural rigidity EI, the central half has flexural rigidity $2EI$.   *Ans.* $3PL^3/256EI$

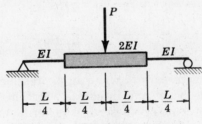

Fig. 10-27

Chapter 11

Elastic Deflection of Beams: Method of Singularity Functions

In the previous two chapters we have found the elastic deflections of transversely loaded beams through integration of the second-order Euler-Bernoulli equation and also by the moment-area method. The first method is direct but may become extremely lengthy. The second method is best suited to finding particular values of beam deflections and usually offers little advantage if the equation of the entire deflection curve is desired.

A third approach is based upon use of the singularity functions introduced in Chapter 6. This method is direct and may be applied in a straightforward fashion to a beam subject to any combination of concentrated forces, moments, and/or distributed loads. One must only remember the definition of the singularity functions given in Problem 6.7, i.e. the quantity $\langle x - a \rangle$ vanishes if $x < a$ but is equal to $x - a$ if $x > a$.

If at any point along the length of a beam we let w denote the intensity of loading, V the transverse shearing force, M the bending moment, and y the deflection, then from Problem 6.1 we have

$$w = \frac{dV}{dx} \tag{a}$$

$$V = \frac{dM}{dx} \tag{b}$$

From Problem 9.1, the Euler-Bernoulli equation describing the deflected beam is

$$M = EI \frac{d^2 y}{dx^2} \tag{c}$$

Thus,

$$V = EI \frac{d^3 y}{dx^3} \tag{d}$$

$$w = EI \frac{d^4 y}{dx^4} \tag{e}$$

Consequently, it is only necessary to write an expression for the load $w(x)$ for all values of x along the length of the beam and integrate equation (e) four times to find the deflection $y(x)$. The load must of course be described in terms of the singularity functions.

The attractive feature of the method of singularity functions is that *one* equation describes the entire beam deflection curve, in contrast to a separate equation for each region between loads, as was the case of the direct application of the Euler-Bernoulli method. This one equation of course contains the singularity functions.

As shown in Problems 11.1 through 11.8, this method is relatively short and easy to apply. It avoids the problem of evaluation of pairs of constants of integration for each region of the beam between forces. Instead, only a single pair of constants arises and they are evaluated from the boundary conditions of support of the beam.

Solved Problems

11.1. Using singularity functions, determine the deflection curve of the simply supported beam subject to a concentrated force (see Problem 6.8, page 94).

Let us work with general dimensions rather than numbers. It is easily seen from statics that the left reaction is Pb/L and the right reaction is Pa/L. In Problem 6.8, the bending moment along the entire length of the beam was found to be represented by the relation

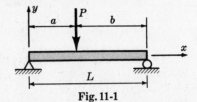

Fig. 11-1

$$M(x) = \frac{Pb}{L}\langle x\rangle^1 - P\langle x-a\rangle^1 + \frac{Pa}{L}\langle x-L\rangle^1 \qquad (1)$$

where the pointed brackets have the meaning given in Problem 6.7. From the differential equation of the bent beam as found in Problem 9.1, we immediately have

$$EI\frac{d^2y}{dx^2} = M = \frac{Pb}{L}\langle x\rangle^1 - P\langle x-a\rangle^1 + \frac{Pa}{L}\langle x-L\rangle^1 \qquad (2)$$

The first integration yields

$$EI\frac{dy}{dx} = \frac{Pb}{2L}\langle x\rangle^2 - \frac{P}{2}\langle x-a\rangle^2 + \frac{Pa}{2L}\langle x-L\rangle^2 + C_1 \qquad (3)$$

where C_1 is a constant of integration. The next integration gives

$$EIy = \frac{Pb}{6L}\langle x\rangle^3 - \frac{P}{6}\langle x-a\rangle^3 + \frac{Pa}{6L}\langle x-L\rangle^3 + C_1 x + C_2 \qquad (4)$$

where C_2 is a second constant of integration. These two constants are found from the conditions that $y = 0$ at $x = 0$ and $x = L$. From (4) we find

$$C_1 = -\frac{PbL}{6} + \frac{Pb^3}{6L}$$

$$C_2 = 0 \qquad (5)$$

The desired deflection curve is thus

$$EIy = \frac{Pb}{6L}\langle x\rangle^3 - \frac{P}{6}\langle x-a\rangle^3 + \frac{Pa}{6L}\langle x-L\rangle^3 + \left(-\frac{PbL}{6} + \frac{Pb^3}{6L}\right)x \qquad (6)$$

The third term on the right side of this equation of course vanishes for the finite beam of length L and arises only because we have considered the hypothetical extended beam of infinite length as mentioned in Problem 6.8. The term could be dropped in (6).

11.2. Using singularity functions, determine the deflection curve of a simply supported beam subject to a concentrated moment.

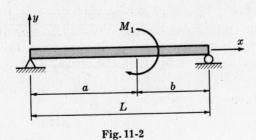

Fig. 11-2

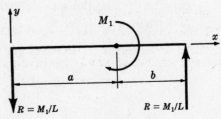

Fig. 11-3

As indicated in Problem 9.16, the end reactions are readily found from statics to be $R = M_1/L$ acting as shown in Fig. 11-3.

Following Problem 6.11, we may write

$$w(x) = -\frac{M_1}{L}\langle x\rangle^{-1} + M_1\langle x-a\rangle^{-2} + \frac{M_1}{L}\langle x-L\rangle^{-1} \tag{1}$$

Integrating:

$$V(x) = -\frac{M_1}{L}\langle x\rangle^0 + M_1\langle x-a\rangle^{-1} + \frac{M_1}{L}\langle x-L\rangle^0 \tag{2}$$

Integrating again:

$$M(x) = -\frac{M_1}{L}\langle x\rangle^1 + M_1\langle x-a\rangle^0 + \frac{M_1}{L}\langle x-L\rangle^1 \tag{3}$$

Next, we use the differential equation of the bent beam to write

$$EI\frac{d^2y}{dx^2} = M = -\frac{M_1}{L}\langle x\rangle^1 + M_1\langle x-a\rangle^0 + \frac{M_1}{L}\langle x-L\rangle^1 \tag{4}$$

Integrating:

$$EI\frac{dy}{dx} = -\frac{M_1}{2L}\langle x\rangle^2 + M_1\langle x-a\rangle^1 + \frac{M_1}{2L}\langle x-L\rangle^2 + C_1 \tag{5}$$

The last integration yields

$$EIy = -\frac{M_1}{6L}\langle x\rangle^3 + \frac{M_1}{2}\langle x-a\rangle^2 + \frac{M_1}{6L}\langle x-L\rangle^3 + C_1 x + C_2 \tag{6}$$

For boundary conditions, we have $y = 0$ at $x = 0$ and $x = L$. Using these boundary conditions to determine C_1 and C_2 from (6), we find

$$C_1 = \frac{M_1 L}{6} - \frac{M_1 b^2}{2L} \qquad C_2 = 0$$

The deflection curve is thus

$$EIy = -\frac{M_1}{6L}\langle x\rangle^3 + \frac{M_1}{2}\langle x-a\rangle^2 + \left(\frac{M_1 L}{6} - \frac{M_1 b^2}{2L}\right)x \tag{7}$$

where the third term on the right side of (6) has been dropped since it vanishes for the beam of finite length L and arises only because we originally considered the hypothetical beam of infinite length. It is evident that the single equation (7) above is equivalent to the two equations found in Problem 9.16 where this same problem was solved without the aid of the singularity functions.

11.3. Consider a simply supported beam subject to a uniform load distributed over a portion of its length as indicated in Fig. 11-4. Use singularity functions to determine the deflection curve of the beam.

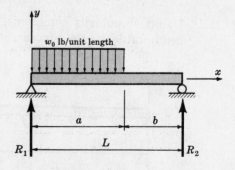

From statics the reactions are found to be

$$R_1 = \frac{w_0}{2L}(L^2 - b^2)$$

$$R_2 = w_0 a - \frac{w_0}{2L}(L^2 - b^2)$$

Fig. 11-4

It is not mandatory to establish these reactions at the start of the analysis; they could be determined as in Problem 6.8, page 94.

The load intensity function is

$$w(x) = R_1\langle x\rangle^{-1} - w_0\langle x\rangle^0 + w_0\langle x-a\rangle^0 + R_2\langle x-L\rangle^{-1} \qquad (1)$$

Note that the third term on the right is required to cancel the distributed load represented by the second term for all values of x greater than $x = a$. Integrating:

$$V(x) = R_1\langle x\rangle^0 - w_0\langle x\rangle^1 + w_0\langle x-a\rangle^1 + R_2\langle x-L\rangle^0 \qquad (2)$$

$$M(x) = R_1\langle x\rangle^1 - \frac{w_0}{2}\langle x\rangle^2 + \frac{w_0}{2}\langle x-a\rangle^2 + R_2\langle x-L\rangle^1 \qquad (3)$$

Thus

$$EI\frac{d^2y}{dx^2} = M = R_1\langle x\rangle^1 - \frac{w_0}{2}\langle x\rangle^2 + \frac{w_0}{2}\langle x-a\rangle^2 + R_2\langle x-L\rangle^1 \qquad (4)$$

Integrating:

$$EI\frac{dy}{dx} = \frac{R_1}{2}\langle x\rangle^2 - \frac{w_0}{6}\langle x\rangle^3 + \frac{w_0}{6}\langle x-a\rangle^3 + \frac{R_2}{2}\langle x-L\rangle^2 + C_1 \qquad (5)$$

Lastly:

$$EIy = \frac{R_1}{6}\langle x\rangle^3 - \frac{w_0}{24}\langle x\rangle^4 + \frac{w_0}{24}\langle x-a\rangle^4 + \frac{R_2}{6}\langle x-L\rangle^3 + C_1x + C_2 \qquad (6)$$

To determine C_1 and C_2, we impose the boundary conditions that $y = 0$ at $x = 0$ and $x = L$. From (6) we thus find

$$C_1 = \frac{w_0L^3}{24} - \frac{w_0b^4}{24L} - \frac{w_0L}{12}(L^2 - b^2)$$

$$C_2 = 0$$

The deflection curve is accordingly

$$EIy = \frac{w_0}{12L}(L^2-b^2)\langle x\rangle^3 - \frac{w_0}{24}\langle x\rangle^4 + \frac{w_0}{24}\langle x-a\rangle^4 + \left[-\frac{w_0L^3}{24} - \frac{w_0b^4}{24L} + \frac{w_0Lb^2}{12}\right]x \qquad (7)$$

where the term involving R_2 in (6) has been omitted for the reasons given previously.

11.4. Consider the overhanging beam shown in Fig. 11-5. Determine the equation of the deflection curve using singularity functions.

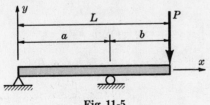

Fig. 11-5

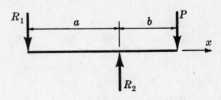

Fig. 11-6

From statics the reactions are first found to be $R_1 = Pb/a$ and $R_2 = P[1 + (b/a)]$ acting as indicated in Fig. 11-6. The load intensity function is

$$w(x) = -R_1\langle x\rangle^{-1} + R_2\langle x-a\rangle^{-1} - P\langle x-L\rangle^{-1} \qquad (1)$$

Integrating:

$$V(x) = -R_1\langle x\rangle^0 + R_2\langle x-a\rangle^0 - P\langle x-L\rangle^0 \qquad (2)$$

$$M(x) = -R_1\langle x\rangle^1 + R_2\langle x-a\rangle^1 - P\langle x-L\rangle^1 \tag{3}$$

Thus

$$EI\frac{d^2y}{dx^2} = M = -R_1\langle x\rangle^1 + R_2\langle x-a\rangle^1 - P\langle x-L\rangle^1 \tag{4}$$

from which

$$EI\frac{dy}{dx} = -\frac{R_1}{2}\langle x\rangle^2 + \frac{R_2}{2}\langle x-a\rangle^2 - \frac{P}{2}\langle x-L\rangle^2 + C_1 \tag{5}$$

$$EIy = -\frac{R_1}{6}\langle x\rangle^3 + \frac{R_2}{6}\langle x-a\rangle^3 - \frac{P}{6}\langle x-L\rangle^3 + C_1x + C_2 \tag{6}$$

The boundary conditions are $y = 0$ at $x = 0$ and $x = a$. From these conditions, C_1 and C_2 are found from (6) to be

$$C_1 = \frac{Pab}{6} \qquad C_2 = 0$$

The deflection curve is thus

$$EIy = -\frac{Pb}{6a}\langle x\rangle^3 + \frac{P}{6}\left(1+\frac{b}{a}\right)\langle x-a\rangle^3 + \frac{Pabx}{6} \tag{7}$$

11.5. Consider a simply supported beam subject to a uniform load acting over a portion of the beam as indicated in Fig. 11-7. Use singularity functions to determine the equation of the deflection curve.

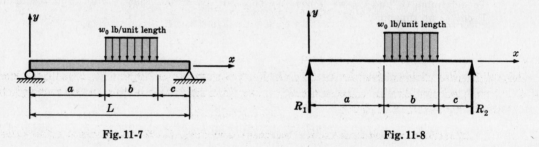

Fig. 11-7   **Fig. 11-8**

The reactions as indicated in Fig. 11-8 are found to be

$$R_1 = \frac{w_0 b}{L}\left(\frac{b}{2}+c\right) \quad \text{and} \quad R_2 = \frac{w_0 b}{L}\left(\frac{b}{2}+a\right)$$

The load intensity function is

$$w(x) = R_1\langle x\rangle^{-1} - w_0\langle x-a\rangle^0 + w_0\langle x-a-b\rangle^0 + R_2\langle x-L\rangle^{-1} \tag{1}$$

Integrating:

$$V(x) = R_1\langle x\rangle^0 - w_0\langle x-a\rangle^1 + w_0\langle x-a-b\rangle^1 + R_2\langle x-L\rangle^0 \tag{2}$$

$$M(x) = R_1\langle x\rangle^1 - \frac{w_0}{2}\langle x-a\rangle^2 + \frac{w_0}{2}\langle x-a-b\rangle^2 + R_2\langle x-L\rangle^1 \tag{3}$$

Thus

$$EI\frac{d^2y}{dx^2} = M = R_1\langle x\rangle^1 - \frac{w_0}{2}\langle x-a\rangle^2 + \frac{w_0}{2}\langle x-a-b\rangle^2 + R_2\langle x-L\rangle^1 \tag{4}$$

Integrating:

$$EI\frac{dy}{dx} = \frac{R_1}{2}\langle x\rangle^2 - \frac{w_0}{6}\langle x-a\rangle^3 + \frac{w_0}{6}\langle x-a-b\rangle^3 + \frac{R_2}{2}\langle x-L\rangle^2 + C_1 \tag{5}$$

$$EIy = \frac{R_1}{6}\langle x\rangle^3 - \frac{w_0}{24}\langle x-a\rangle^4 + \frac{w_0}{24}\langle x-a-b\rangle^4 + \frac{R_2}{6}\langle x-L\rangle^3 + C_1x + C_2 \tag{6}$$

The boundary conditions are: $y = 0$ at $x = 0$ and $x = L$. Applying these conditions to (6) yields

$$C_1 = \frac{w_0}{24L}[(L-a)^4 - (L-c)^4] - \frac{w_0 bL}{6}\left(\frac{b}{2}+c\right)$$

$$C_2 = 0$$

The desired deflection curve is thus

$$EIy = \frac{w_0 b}{6L}\left(\frac{b}{2}+c\right)\langle x\rangle^3 - \frac{w_0}{24}\langle x-a\rangle^4 + \frac{w_0}{24}\langle x-a-b\rangle^4$$

$$+ \left\{\frac{w_0}{24L}[(L-a)^4 - (L-c)^4] - \frac{w_0 bL}{6}\left(\frac{b}{2}+c\right)\right\} x \tag{7}$$

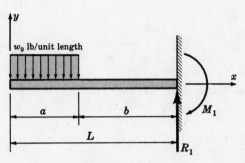

Fig. 11-9

11.6. Consider the cantilever beam of Fig. 11-9, which is loaded by a uniform load over part of its length. Determine the equation of the deflection curve through the use of singularity functions.

It is not necessary to determine the reactions at the wall. The load intensity function is

$$w(x) = -w_0\langle x\rangle^0 + w_0\langle x-a\rangle^0 + R_1\langle x-L\rangle^{-1} + M_1\langle x-L\rangle^{-2} \tag{1}$$

Integrating:

$$V(x) = -w_0\langle x\rangle^1 + w_0\langle x-a\rangle^1 + R_1\langle x-L\rangle^0 + M_1\langle x-L\rangle^{-1} \tag{2}$$

$$M(x) = -\frac{w_0}{2}\langle x\rangle^2 + \frac{w_0}{2}\langle x-a\rangle^2 + R_1\langle x-L\rangle^1 + M_1\langle x-L\rangle^0 \tag{3}$$

Thus

$$EI\frac{d^2y}{dx^2} = M = -\frac{w_0}{2}\langle x\rangle^2 + \frac{w_0}{2}\langle x-a\rangle^2 + R_1\langle x-L\rangle^1 + M_1\langle x-L\rangle^0 \tag{4}$$

Integrating:

$$EI\frac{dy}{dx} = -\frac{w_0}{6}\langle x\rangle^3 + \frac{w_0}{6}\langle x-a\rangle^3 + \frac{R_1}{2}\langle x-L\rangle^2 + M_1\langle x-L\rangle^1 + C_1 \tag{5}$$

Since $dy/dx = 0$ when $x = L$, equation (5) yields $C_1 = w_0(L^3 - b^3)/6$. Integrating again:

$$EIy = -\frac{w_0}{24}\langle x\rangle^4 + \frac{w_0}{24}\langle x-a\rangle^4 + \frac{R_1}{6}\langle x-L\rangle^3 + \frac{M_1}{2}\langle x-L\rangle^2 + \frac{w_0}{6}(L^3-b^3)x + C_2 \tag{6}$$

Since $y = 0$ when $x = L$, equation (6) yields $C_2 = \frac{w_0}{24}(L^4-b^4) - \frac{w_0 L}{6}(L^3-b^3)$.

The desired equation of the deflection curve is thus

$$EIy = -\frac{w_0}{24}\langle x\rangle^4 + \frac{w_0}{24}\langle x-a\rangle^4 + \frac{w_0}{6}(L^3-b^3)x + \frac{w_0}{24}(L^4-b^4) - \frac{w_0L}{6}(L^3-b^3) \quad (7)$$

The reactive shear R_1 is found from (2) to be $R_1 = w_0a$ and the reactive moment M_1 is found from (3) to be $M_1 = (w_0/2)(L^2-b^2)$.

11.7. Determine the equation of the deflection curve of the simply supported beam shown in Fig. 11-10(a). Use singularity functions.

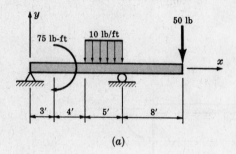 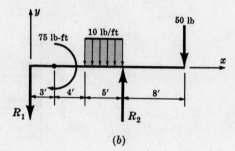

(a)   (b)

Fig. 11-10

The free-body diagram is shown in Fig. 11-10(b). From statics the reactions are readily found to be $R_1 = 29$ lb, $R_2 = 129$ lb. The load intensity function is

$$w(x) = -29\langle x\rangle^{-1} + 75\langle x-3\rangle^{-2} - 10\langle x-7\rangle^0 + 10\langle x-12\rangle^0 + 129\langle x-12\rangle^{-1} - 50\langle x-20\rangle^{-1} \quad (1)$$

Integrating:

$$V(x) = -29\langle x\rangle^0 + 75\langle x-3\rangle^{-1} - 10\langle x-7\rangle^1 + 10\langle x-12\rangle^1 + 129\langle x-12\rangle^0 - 50\langle x-20\rangle^0 \quad (2)$$

$$M(x) = -29\langle x\rangle^1 + 75\langle x-3\rangle^0 - 5\langle x-7\rangle^2 + 5\langle x-12\rangle^2 + 129\langle x-12\rangle^1 - 50\langle x-20\rangle^1 \quad (3)$$

Thus

$$EI\frac{d^2y}{dx^2} = M = -29\langle x\rangle^1 + 75\langle x-3\rangle^0 - 5\langle x-7\rangle^2 + 5\langle x-12\rangle^2$$
$$+ 129\langle x-12\rangle^1 - 50\langle x-20\rangle^1 \quad (4)$$

Integrating:

$$EI\frac{dy}{dx} = -\frac{29}{2}\langle x\rangle^2 + 75\langle x-3\rangle^1 - \frac{5}{3}\langle x-7\rangle^3 + \frac{5}{3}\langle x-12\rangle^3$$
$$+ \frac{129}{2}\langle x-12\rangle^2 - 25\langle x-20\rangle^2 + C_1 \quad (5)$$

$$EIy = -\frac{29}{6}\langle x\rangle^3 + \frac{75}{2}\langle x-3\rangle^2 - \frac{5}{12}\langle x-7\rangle^4 + \frac{5}{12}\langle x-12\rangle^4$$
$$+ \frac{129}{6}\langle x-12\rangle^3 - \frac{25}{3}\langle x-20\rangle^3 + C_1x + C_2 \quad (6)$$

The boundary conditions are $y=0$ at $x=0$, $x=12$. Using these conditions in (6) to determine C_1 and C_2, we find $C_1 = 465$, $C_2 = 0$.

The desired deflection curve is thus

$$EIy = -\frac{29}{6}\langle x\rangle^3 + \frac{75}{2}\langle x-3\rangle^2 - \frac{5}{12}\langle x-7\rangle^4 + \frac{5}{12}\langle x-12\rangle^4$$
$$+ \frac{43}{2}\langle x-12\rangle^3 - \frac{25}{3}\langle x-20\rangle^3 + 465x \quad (7)$$

Care must of course be taken to express EI in consistent units, which, since we have used units of feet in measuring distances along the beam, would be lb-ft².

11.8. Use singularity functions to determine the equation of the deflection curve of the simply supported beam subject to a uniformly varying load as in Fig. 11-11(a). What is the central deflection of the beam?

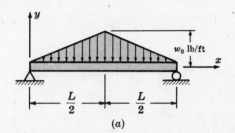   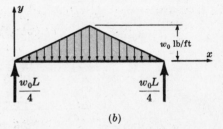

(a)   (b)

Fig. 11-11

The free-body diagram with the reactions found from statics is shown in Fig. 11-11(b). The load intensity function is

$$w(x) = \frac{w_0 L}{4}\langle x \rangle^{-1} - \frac{2w_0}{L}\langle x \rangle^1 + 2\left(\frac{2w_0}{L}\right)\left\langle x - \frac{L}{2}\right\rangle^1 + \frac{w_0 L}{4}\langle x - L\rangle^{-1} \qquad (1)$$

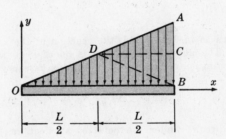

Fig. 11-12

The second term on the right side of (1) represents a uniformly varying load extending completely across the beam as indicated by the triangle OAB in Fig. 11-12. To remove the portion of this loading represented by triangle ABD, we add the third term on the right side, which leaves the true load represented by triangle ODB.

Integrating:

$$V(x) = +\frac{w_0 L}{4}\langle x \rangle^0 - \frac{w_0}{L}\langle x \rangle^2 + \frac{2w_0}{L}\left\langle x - \frac{L}{2}\right\rangle^2 + \frac{w_0 L}{4}\langle x - L\rangle^0 \qquad (2)$$

$$M(x) = +\frac{w_0 L}{4}\langle x \rangle^1 - \frac{w_0}{3L}\langle x \rangle^3 + \frac{2w_0}{3L}\left\langle x - \frac{L}{2}\right\rangle^3 + \frac{w_0 L}{4}\langle x - L\rangle^1 \qquad (3)$$

Thus

$$EI\frac{d^2y}{dx^2} = M = +\frac{w_0 L}{4}\langle x \rangle^1 - \frac{w_0}{3L}\langle x \rangle^3 + \frac{2w_0}{3L}\left\langle x - \frac{L}{2}\right\rangle^3 + \frac{w_0 L}{4}\langle x - L\rangle^1 \qquad (4)$$

from which

$$EI\frac{dy}{dx} = +\frac{w_0 L}{8}\langle x \rangle^2 - \frac{w_0}{12L}\langle x \rangle^4 + \frac{w_0}{6L}\left\langle x - \frac{L}{2}\right\rangle^3 + \frac{w_0 L}{8}\langle x - L\rangle^2 + C_1 \qquad (5)$$

From symmetry we have as a boundary condition $dy/dx = 0$ at $x = L/2$. From (5) we find that $C_1 = -5w_0 L^3/192$. Integrating again we get the desired deflection curve.

$$EIy = \frac{w_0 L}{24}\langle x\rangle^3 - \frac{w_0}{60L}\langle x\rangle^5 + \frac{w_0}{24L}\left\langle x - \frac{L}{2}\right\rangle^4 + \frac{w_0 L}{24}\langle x - L\rangle^3 - \frac{5}{192}w_0 L^3 x + C_2 \qquad (6)$$

Since $y = 0$ at $x = 0$ it follows that $C_2 = 0$. The central deflection is found from (6) to be

$$y = -\frac{w_0 L^4}{120EI} \qquad (7)$$

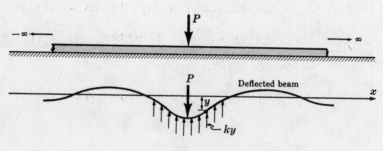

Fig. 11-13

11.9. Determine the equation of the deflection curve of an infinitely long beam resting on an elastic foundation, i.e. one which at each point resists deflection with a force proportional to the deflection at that point. The beam is subject to a single concentrated force.

The beam appears as in Fig. 11-13, the reaction of the foundation being ky at a point where the deflection is y. The governing equation is (e) of page 198. The w in that equation represents load intensity acting on the beam, which in this case is $-ky$. Thus equation (e) becomes

$$EI \frac{d^4 y}{dx^4} + ky = 0 \qquad (a)$$

at all points except directly under the point of application of the concentrated force P. If we assume a solution of (a) of the form $y = e^{mx}$ then we obtain the auxiliary equation

$$m^4 + \frac{k}{EI} = 0$$

which has the four roots $m = \pm\beta(1 \pm i)$, where $i = \sqrt{-1}$ and $\beta = \sqrt[4]{k/4EI}$.

The general solution of (a) thus becomes

$$y = e^{-\beta x}(A \cos \beta x + B \sin \beta x) + e^{\beta x}(C \cos \beta x + D \sin \beta x) \qquad (b)$$

From this we immediately have

$$dy/dx = e^{-\beta x}(-A\beta \sin \beta x + B\beta \cos \beta x) - \beta e^{-\beta x}(A \cos \beta x + B \sin \beta x)$$
$$+ e^{\beta x}(-C\beta \sin \beta x + D\beta \cos \beta x) + \beta e^{\beta x}(C \cos \beta x + D \sin \beta x) \qquad (c)$$

Evidently the deflections of the beam will be symmetric about the point of application of the concentrated force. In considering the portion of the beam to the right of that force it is obvious that the deflection and slope must tend to zero for very large values of x, hence in (b) we must set $C = D = 0$. In fact, if we took any other values the deflection and slope would increase beyond bound as x becomes infinitely large.

Since the deflection curve must have a horizontal tangent at $x = 0$ by virtue of symmetry, it is evident from (c) that $A = B$. Thus, (b) becomes:

$$y = Ae^{-\beta x}(\cos \beta x + \sin \beta x) \qquad (d)$$

From this we immediately have

$$dy/dx = -2\beta Ae^{-\beta x}\sin \beta x \qquad (e)$$

$$d^2y/dx^2 = 2\beta^2 Ae^{-\beta x}(\sin \beta x - \cos \beta x) \qquad (f)$$

$$d^3y/dx^3 = 4\beta^3 Ae^{-\beta x}\cos \beta x \qquad (g)$$

Lastly, the shearing force in the beam at a point immediately to the right of the force P is $-P/2$ where the negative sign is introduced because of the sign convention for shearing forces introduced in Chapter 6. Employing equation (d), page 198, together with (g) above, we have

$$-\frac{P}{2} = 4EI\beta^3 A \qquad \text{from which} \qquad A = \frac{-P}{8\beta^3 EI}$$

The equation of the deflection curve is thus

$$y = -\frac{P}{8\beta^3 EI}(\cos \beta x + \sin \beta x) \qquad (h)$$

Supplementary Problems

In Problems 11.10 through 11.16 use singularity functions to determine the equation of the deflection curve of the beam.

11.10.

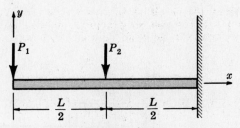

Fig. 11-14

Ans. $EIy = -\dfrac{P_1}{6}\langle x\rangle^3 - \dfrac{P_2}{6}\left\langle x - \dfrac{L}{2}\right\rangle^3 + \left(\dfrac{P_1 L^2}{2} + \dfrac{P_2 L^2}{8}\right)x - \dfrac{P_1 L^3}{3} - \dfrac{5P_2 L^3}{48}$

11.11.

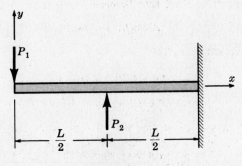

Fig. 11-15

Ans. $EIy = -\dfrac{P_1}{6}\langle x\rangle^3 + \dfrac{P_2}{6}\left\langle x - \dfrac{L}{2}\right\rangle^3 + \left(\dfrac{P_1 L^2}{2} - \dfrac{P_2 L^2}{8}\right)x - \dfrac{P_1 L^3}{3} + \dfrac{5P_2 L^3}{48}$

11.12.

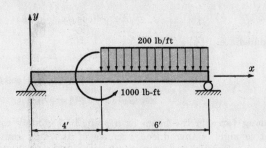

Fig. 11-16

Ans. $EIy = \dfrac{230}{3}\langle x\rangle^3 - 500\langle x-4\rangle^2 - \dfrac{100}{12}\langle x-4\rangle^4 + \dfrac{370}{3}\langle x-10\rangle^3 - 6930x$

11.13.

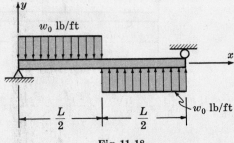

Fig. 11-17

Ans. $EIy = -\dfrac{w_0 a}{24}\langle x\rangle^3 - \dfrac{w_0}{24}\langle x\rangle^4 + \dfrac{w_0}{24}\langle x-a\rangle^4 + \dfrac{w_0 a^2}{2}\langle x-a\rangle^2 + \dfrac{9}{24}w_0 a\langle x-2a\rangle^3$

$\qquad\qquad -\dfrac{w_0 a}{6}\langle x-3a\rangle^3 + \dfrac{11}{48}w_0 a^3 x$

11.14.

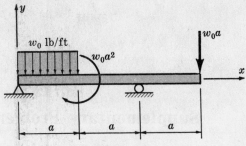

Fig. 11-18

Ans. $EIy = \dfrac{w_0 L}{24}\langle x\rangle^3 - \dfrac{w_0}{24}\langle x\rangle^4 + \dfrac{w_0}{12}\left\langle x-\dfrac{L}{2}\right\rangle^4 - \dfrac{w_0 L^3}{192}x$

11.15.

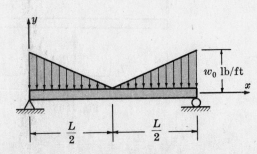

Fig. 11-19

Ans. $EIy = \dfrac{w_0 L}{24}\langle x\rangle^3 - \dfrac{w_0}{24}\langle x\rangle^4 + \dfrac{w_0}{60L}\langle x\rangle^5 - \dfrac{w_0}{10L}\left\langle x - \dfrac{L}{2}\right\rangle^5 - \dfrac{3}{192}w_0 L^3 x$

11.16.

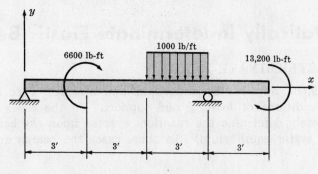

Fig. 11-20

Ans. $EIy = -\dfrac{850}{3}\langle x\rangle^3 + 3300\langle x-3\rangle^2 - \dfrac{500}{12}\langle x-6\rangle^4 + \dfrac{500}{12}\langle x-9\rangle^4$

$+ \dfrac{2350}{3}\langle x-9\rangle^3 + 10{,}175x$

Chapter 12

Statically Indeterminate Elastic Beams

STATICALLY DETERMINATE BEAMS

In Chapters 9, 10, and 11 the deflections and stresses were determined for beams having various conditions of loading and support. In the cases treated it was always possible to completely determine the reactions exerted upon the beam merely by applying the equations of static equilibrium. In these cases the beams are said to be *statically determinate*.

STATICALLY INDETERMINATE BEAMS

In this chapter we shall consider those beams where the number of unknown reactions exceeds the number of equilibrium equations available for the system. In such a case it is necessary to supplement the equilibrium equations with additional equations stemming from the deformations of the beam. In these cases the beams are said to be *statically indeterminate*.

TYPES OF STATICALLY INDETERMINATE BEAMS

Several common types of statically indeterminate beams are illustrated below. Although a wide variety of such structures exists in practice the following four diagrams will illustrate the nature of an indeterminate system. For the beams shown below the reactions of each constitute a parallel force system and hence there are two equations of static equilibrium available. Thus the determination of the reactions in each of these cases necessitates the use of additional equations arising from the deformation of the beam.

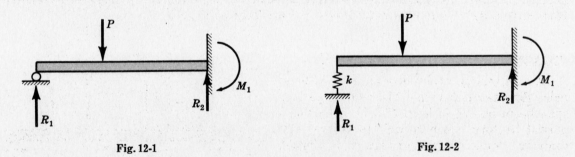

Fig. 12-1   Fig. 12-2

In the case (Fig. 12-1) of a beam fixed at one end and supported at the other, sometimes termed a supported cantilever, we have as unknown reactions R_1, R_2, and M_1. The two statics equations must be supplemented by one equation based upon deformations. For applications, see Problems 12.1 and 12.5.

In Fig. 12-2 the beam is fixed at one end and has a flexible springlike support at the other. In the case of a simple linear spring the flexible support exerts a force proportional to the beam deflection at that point. The unknown reactions are again R_1, R_2, and M_1. The two statics equations must be supplemented by one equation stemming from deformations. For applications see Problems 12.4 and 12.11.

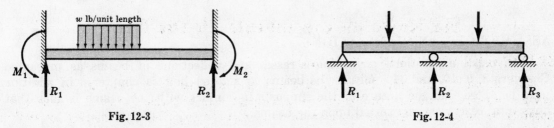

Fig. 12-3   Fig. 12-4

As shown in Fig. 12-3, a beam fixed or clamped at both ends has the unknown reactions R_1, R_2, M_1, and M_2. The two statics equations must be supplemented by two equations arising from the deformations. For applications, see Problems 12.6, 12.7, 12.8, and 12.10.

In Fig. 12-4 the beam is supported on three supports at the same level. The unknown reactions are R_1, R_2, and R_3. The two statics equations must be supplemented by one equation based upon deformations. A beam of this type that rests on more than two supports is called a *continuous beam*. For applications, see Problems 12.17 through 12.19.

NATURE OF EQUATIONS ARISING FROM THE BEAM DEFORMATIONS

In the first and third of the above illustrations it is convenient to employ the moment-area theorems discussed in Chapter 10 to obtain the supplementary equations. In problems involving flexible supports, such as that indicated in the second illustration, the method of singularity functions is preferable.

The continuous beam illustrated in Fig. 12-4 is usually investigated in a somewhat different manner by use of the *three-moment theorem*. The derivation of this theorem from simple deformation principles is given in Problem 12.14.

THREE-MOMENT THEOREM

A continuous beam is one that rests on more than two supports. The continuous two-span beam shown in Fig. 12-5 is subject to a partial uniform load as well as several concentrated forces. It is convenient to consider the bending moments at the various supports as the unknowns (rather than the reactions themselves) and write deformation equations in terms of these bending moments. This leads to the three-moment theorem:

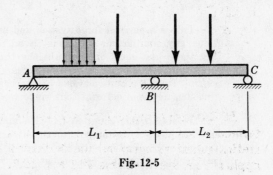

Fig. 12-5

$$M_A L_1 + 2M_B(L_1 + L_2) + M_C L_2 = -\frac{6A_1 \bar{a}_1}{L_1} - \frac{6A_2 \bar{b}_2}{L_2}$$

In this equation M_A, M_B, and M_C designate bending moments at the supports A, B, and C respectively; L_1 and L_2 denote the span lengths; A_1 and A_2 represent the areas of the moment diagrams drawn on the temporary assumption that each span of the beam is *simply supported*; and $\bar{a}_1$ and $\bar{b}_2$ designate the distances of the centroids of each of these moment diagrams from A and C respectively. The theorem in this form is applicable to all continuous beams having all supports at the same level. For applications see Problems 12.15 through 12.19.

ASSUMPTIONS AND LIMITATIONS

The usual asumptions governing stresses and deflections of beams as presented in Chapters 8, 9, 10, and 11 apply to the beams considered in this chapter. In addition, it is to be noted that the nature of the supports in contact with the beams is such that no horizontal reactions are exerted upon the beams.

Solved Problems

12.1. Consider a beam that is supported at the left end, clamped at the right end and subject to the concentrated load as shown in Fig. 12-6. Determine the reactions R_1, R_2, and M_1.

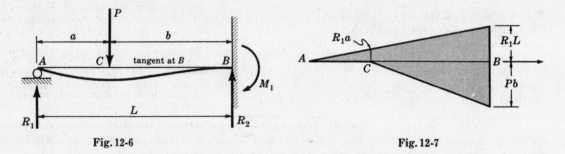

Fig. 12-6   Fig. 12-7

From statics we have

$$\Sigma M_B = R_1 L + M_1 - Pb = 0 \tag{1}$$

$$\Sigma F_v = R_1 + R_2 - P = 0 \tag{2}$$

This is a parallel force system and hence there are only two equations of equilibrium available. Thus any equilibrium equations in addition to the above two would not be independent equations. Yet these two equations contain the three unknowns R_1, R_2, and M_1. Thus we have a statically indeterminate system and it is necessary to supplement the statics equations with relationships arising from the deformations of the beam. In this case it is necessary to employ only one equation arising from deformations since we will then have three equations containing three unknowns.

To obtain this equation let us examine the deflected beam, indicated by the heavy curved line in Fig. 12-6. If a tangent to the deformed beam is drawn at B, i.e. the clamped end, then this tangent will coincide with the original unbent position of the beam. The deflection of the support end A from this tangent at B is zero. Thus we may apply the second moment-area theorem developed in Problem 10-2. The moment diagram when drawn by parts is shown in Fig. 12-7. From the second moment-area theorem, realizing that the deflection of A from the tangent at B is zero and that this deflection is given by the moment of the area under the above moment diagram between A and B about a vertical line through A, we have

$$\tfrac{1}{2}R_1L(L)(\tfrac{2}{3}L) + \tfrac{1}{2}(-Pb)(b)(a + \tfrac{2}{3}b) = 0 \quad \text{or} \quad R_1 = \frac{3Pb^2}{2L^3}(a + \tfrac{2}{3}b) = \frac{Pb^2}{2L^3}(2L + a) \tag{3}$$

Substituting for R_1 in (2), we find

$$R_2 = \frac{Pa}{2L^3}(3L^2 - a^2) \tag{4}$$

Substituting these values in (1),

$$M_1 = \frac{Pa}{2L^2}(L^2 - a^2) \tag{5}$$

The unknown reactions are thus completely determined.

12.2. The beam discussed in Problem 12.1 is an 8 WF 24 section. The load P is 5000 lb, $L = 20$ ft, and $a = 10$ ft. Determine the reactions and the maximum bending stresses in the beam.

Substituting in (3) of Problem 12.1,

$$R_1 = \frac{5000(10)^2}{2(20)^3}(40 + 10) = 1560 \text{ lb}$$

From (4) we have

$$R_2 = \frac{5000(10)}{2(20)^3}[3(400) - 100] = 3440 \text{ lb}$$

Lastly, from (5) we find

$$M_1 = \frac{5000(10)}{2(20)^2}(400 - 100) = 18,750 \text{ lb-ft}$$

It is to be noted that these expressions are valid only if the proportional limit of the material is not exceeded at any point in the beam. This is because the moment-area theorem was derived upon the assumption that this was true, and this theorem was used in Problem 12.1 to determine the reactions. Hence it is necessary to investigate the maximum bending stress in the beam. The only points that need be investigated are at the clamped end and those directly under the concentrated load. From the table at the end of Chapter 8 we have $I = 82.5$ in^4 for this section.

At the clamped end the moment M_1 gives rise to a maximum bending stress in the outer fibers of

$$\sigma = \frac{Mc}{I} = \frac{18,750(12)(4)}{82.5} = 10,900 \text{ lb/in}^2$$

Under the concentrated load the bending moment is $1560(10) = 15,600$ lb-ft. At the outer fibers the maximum bending is

$$\sigma = \frac{15,600(12)(4)}{82.5} = 9100 \text{ lb/in}^2$$

The first value is the peak bending stress. Since it is less than the proportional limit of steel the use of the above expressions for reactions is justified.

12.3. For the beam described in Problem 12.2, determine the deflection under the point of application of the 5000 lb force.

This deflection is readily determined by use of the second moment-area theorem together with the bending moment diagram shown in Fig. 12-7 and the reactions as found in Problem 12.2. According to the second moment-area theorem the deflection of point C (under the load) from the tangent drawn at B is equal to the moment of the area under the M/EI diagram between C and B about a vertical line through C. The area consists of a rectangle together with two triangles. The desired deflection Δ_C is then given by

$$EI\Delta_C = 10(15,600)(5) + \tfrac{1}{2}(10)(15,600)(\tfrac{2}{3} \cdot 10) + \tfrac{1}{2}(10)(-50,000)(\tfrac{2}{3} \cdot 10) = -366,500$$

or

$$\Delta_C = -\frac{(366,500)(1728)}{(30 \times 10^6)(82.5)} = -0.255 \text{ in.}$$

It is to be noted that the factor of 1728 is introduced to convert to consistent units, since the quantity $-366,500$ is in lb-ft^3 whereas E and I are expressed in lb/in^2 and in^4 respectively.

12.4. The beam shown in Fig. 12-8 is clamped at the left end and spring-supported at the right end. When the beam is unloaded the spring is free of force. When a vertical force of 1 ton is applied at point C when there is *no* spring there, the deflection at C is two inches. The spring constant k is 1 ton/in. Determine the deflection at point C when a load of two tons is applied at the midpoint B and the spring *is* attached to the beam.

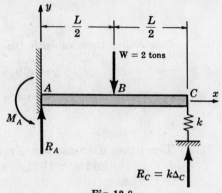

Fig. 12-8

Let us employ the method of singularity functions so that with the coordinate system indicated the load intensity function is

$$w(x) = -M_A\langle x\rangle^{-2} + R_A\langle x\rangle^{-1} - W\left\langle x - \frac{L}{2}\right\rangle^{-1} + R_C\langle x - L\rangle^{-1}$$

Integrating:

$$V(x) = -M_A\langle x\rangle^{-1} + R_A\langle x\rangle^0 - W\left\langle x - \frac{L}{2}\right\rangle^0 + R_C\langle x - L\rangle^0$$

$$M(x) = -M_A\langle x\rangle^0 + R_A\langle x\rangle^1 - W\left\langle x - \frac{L}{2}\right\rangle^1 + R_C\langle x - L\rangle^{+1}$$

Thus

$$EI\frac{d^2y}{dx^2} = -M_A\langle x\rangle^0 + R_A\langle x\rangle^1 - W\left\langle x - \frac{L}{2}\right\rangle^1 + R_C\langle x - L\rangle^{+1}$$

Integrating:

$$EI\frac{dy}{dx} = -M_A\langle x\rangle^1 + \frac{R_A}{2}\langle x\rangle^2 - \frac{W}{2}\left\langle x - \frac{L}{2}\right\rangle^2 + \frac{R_C}{2}\langle x - L\rangle^2 + C_1$$

But when $x = 0$, $dy/dx = 0$, hence $C_1 = 0$. Integrating again:

$$EIy = -\frac{M_A}{2}\langle x\rangle^2 + \frac{R_A}{6}\langle x\rangle^3 - \frac{W}{6}\left\langle x - \frac{L}{2}\right\rangle^3 + \frac{R_C}{6}\langle x - L\rangle^3 + C_2$$

But when $x = 0$, $y = 0$, hence $C_2 = 0$. When $x = L$, the deflection is denoted by Δ_C; hence from the last equation we have

$$EI\Delta_C = -\frac{M_A L^2}{2} + \frac{R_A L^3}{6} - \frac{W}{6}\left(\frac{L}{2}\right)^3 \qquad (a)$$

By the nature of the spring reaction at point C, $R_C = -k\Delta_C$ where the negative sign is introduced because a downward deflection (which is negative) corresponds to an upward force R_C acting on the beam. This, together with the two statics relations

$$R_A + R_C - W = 0 \qquad\qquad WL/2 + M_A - R_A L = 0$$

may be substituted into the equation (a) to yield

$$-\frac{EI(W - R_A)}{k} = \frac{R_A L^3}{6} - \frac{WL^3}{48} - \frac{L^2}{2}\left(R_A L - \frac{WL}{2}\right)$$

which becomes

$$R_A\left(\frac{EI}{k} + \frac{L^3}{3}\right) = \frac{EIW}{k} + \frac{11WL^3}{48} \qquad (b)$$

We also know that when no spring supports the tip, a load of 1 ton at C causes a two inch deflection there. Thus, from Problem 10.4, we have

$$2 \text{ in.} = \frac{(1 \text{ ton})L^3}{3EI} \quad \text{from which} \quad \frac{EI}{L^3} = \frac{1}{6} \text{ ton/in}$$

This value, together with the value of the spring constant $k = 1$ ton/in when substituted in (b), yields $R_A = 19/12$ ton. From the statics equation, $R_C = 2 - 19/12 = 5/12$ ton and since $k = 1$ ton/in, the desired deflection of point C is $\Delta_C = 5/12$ in.

12.5. Consider the overhanging beam shown in Fig. 12-9. Determine the value of the supporting force at B.

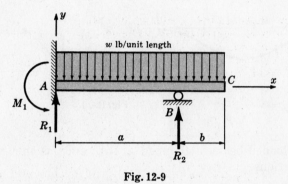

Fig. 12-9

From statics we have

$$\Sigma M_A = M_1 + R_2 a - w(a+b)^2/2 = 0 \tag{1}$$

$$\Sigma F_v = R_1 + R_2 - w(a+b) = 0 \tag{2}$$

We shall use the method of singularity functions. The load intensity function is

$$w(x) = -M_1\langle x\rangle^{-2} + R_1\langle x\rangle^{-1} - w\langle x\rangle^0 + R_2\langle x-a\rangle^{-1}$$

Integrating:

$$V(x) = -M_1\langle x\rangle^{-1} + R_1\langle x\rangle^0 - w\langle x\rangle^1 + R_2\langle x-a\rangle^0$$

$$M(x) = -M_1\langle x\rangle^0 + R_1\langle x\rangle^1 - \frac{w}{2}\langle x\rangle^2 + R_2\langle x-a\rangle^1 = EI\frac{d^2y}{dx^2}$$

Integrating again:

$$EI\frac{dy}{dx} = -M_1\langle x\rangle^1 + \frac{R_1}{2}\langle x\rangle^2 - \frac{w}{6}\langle x\rangle^3 + \frac{R_2}{2}\langle x-a\rangle^2 + C_1$$

But when $x = 0$, $dy/dx = 0$, hence $C_1 = 0$. Integrating again:

$$EIy = -\frac{M_1}{2}\langle x\rangle^2 + \frac{R_1}{6}\langle x\rangle^3 - \frac{w}{24}\langle x\rangle^4 + \frac{R_2}{6}\langle x-a\rangle^3 + C_2 \tag{3}$$

But when $x = 0$, $y = 0$, so that $C_2 = 0$.

Since the support at point B is unyielding, y must vanish in (3) when $x = a$. Substituting, we find

$$0 = -\frac{M_1 a^2}{2} + \frac{R_1 a^3}{6} - \frac{wa^4}{24} \quad \text{from which} \quad M_1 = R_1\frac{a}{3} - \frac{wa^2}{12}$$

Solving this in conjunction with the statics equations, we find

$$R_1 = \frac{5}{8}wa - \frac{3wb^2}{4a} \qquad R_2 = \frac{3}{8}wa + wb + \frac{3wb^2}{4a}$$

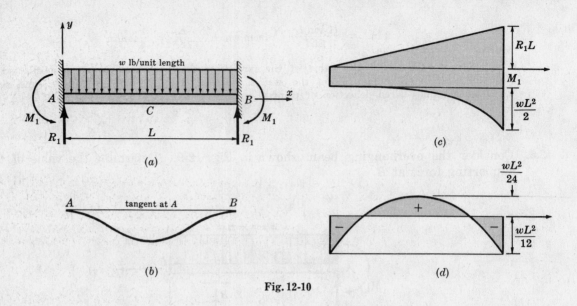

Fig. 12-10

12.6. A uniformly loaded beam is clamped at both ends as shown in Fig. 12-10(a). Determine the reactions.

From symmetry, the force reactions at each end must be equal and each is denoted by R_1. Also, the moment reactions must be equal and each of these is represented by M_1. From statics we have

$$\Sigma F_v = 2R_1 - wL = 0 \quad\quad \text{or} \quad\quad R_1 = wL/2 \quad\quad\quad (1)$$

Although there were initially two equations of static equilibrium available for such a parallel force system, we have essentially utilized one of these equations when we made use of the symmetry considerations. Thus to determine M_1 we must now examine the deformations of the system.

The deflected beam has the symmetrical appearance shown in Fig. 12-10(b). The tangent at A as well as the tangent at B each coincide with the original, undeflected position of the beam. The bending moment diagram is drawn by parts in Fig. 12-10(c). Applying the second moment-area theorem to the deflection of B from the tangent at A (which is zero) we have

$$\tfrac{1}{2}L(R_1 L)\left(\frac{L}{3}\right) + L(-M_1)\left(\frac{L}{2}\right) + \tfrac{1}{3}L\left(-\frac{wL^2}{2}\right)\left(\frac{L}{4}\right) = 0$$

Substituting R_1 from (1) and solving, we have

$$M_1 = wL^2/12 \quad\quad\quad (2)$$

The bending moment at the center of the span is now readily found to be $wL^2/24$. It is frequently convenient to present the bending moment diagram in a composite form rather than by parts. This may be done by superposing on the parabolic diagram due to a uniform load acting on a simply supported beam the rectangular moment diagram corresponding to the end moments. According to the sign convention of Chapter 6 these end moments are negative. The resultant moment diagram then corresponds to the shaded areas shown in Fig. 12-10(d).

12.7. Solve Problem 12.6 by use of singularity functions.

The load intensity function is

$$w(x) = -M_1\langle x\rangle^{-2} + R_1\langle x\rangle^{-1} - w\langle x\rangle^0 + R_1\langle x - L\rangle^{-1} + M_1\langle x - L\rangle^{-2}$$

Integrating:

$$V(x) = -M_1\langle x\rangle^{-1} + R_1\langle x\rangle^0 - w\langle x\rangle^1 + R_1\langle x - L\rangle^0 + M_1\langle x - L\rangle^{-1}$$

$$M(x) = -M_1\langle x\rangle^0 + R_1\langle x\rangle^1 - \frac{w}{2}\langle x\rangle^2 + R_1\langle x - L\rangle^1 + M_1\langle x - L\rangle^0 = EI\frac{d^2y}{dx^2}$$

Integrating again:

$$EI\frac{dy}{dx} = -M_1\langle x\rangle^1 + \frac{R_1}{2}\langle x\rangle^2 - \frac{w}{6}\langle x\rangle^3 + \frac{R_1}{2}\langle x-L\rangle^2 + M_1\langle x-L\rangle^1 + C_1 \qquad (a)$$

But when $x = 0$, $dy/dx = 0$, hence $C_1 = 0$. Integrating again:

$$EIy = -\frac{M_1}{2}\langle x\rangle^2 + \frac{R_1}{6}\langle x\rangle^3 - \frac{w}{24}\langle x\rangle^4 + \frac{R_1}{6}\langle x-L\rangle^3 + \frac{M_1}{2}\langle x-L\rangle^2 + C_2 \qquad (b)$$

But when $x = 0$, $y = 0$, so that $C_2 = 0$.

If we now let $x = L$ in (b) we obtain

$$0 = -\frac{M_1 L^2}{2} + \frac{R_1 L^3}{6} - \frac{wL^4}{24}$$

If the value of $R_1 = wL/2$ as found from statics in Problem 12.6 is now substituted into (c) we find $M_1 = wL^2/12$ as before. The same result could be obtained by substituting $x = L$ in (a).

12.8. The clamped-end beam is loaded by the couple M_0 applied as shown in Fig. 12-11(a). Determine all reactions.

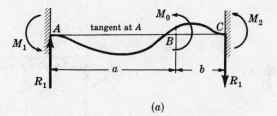

(a)

The deformed configuration is indicated by the heavy curved line. For vertical equilibrium the force reactions at each end must be equal and each is denoted by R_1. Also, from statics we have

$$\Sigma M_A = M_1 + M_2 + M_0 - R_1(a+b) = 0 \qquad (1)$$

This equation contains R_1, M_1, and M_2 as unknowns. Since there are no more statics equations available, the problem is statically indeterminate, and we must supplement (1) with two additional equations coming from the deformations of the system.

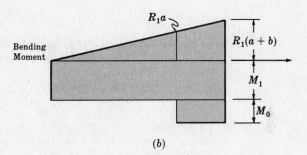

(b)

Fig. 12-11

The bending moment diagram drawn by parts (working from left to right) is shown in Fig. 12-11(b).

The first equation is found from the relation that the tangent drawn to the deformed beam at A remains horizontal. Hence the deflection of the right end of the beam, C, from the tangent at A is zero. Employing the second moment-area theorem between points A and C we have

$$\tfrac{1}{2}(a+b)[R_1(a+b)]\left(\frac{a+b}{3}\right) + (a+b)(-M_1)\left(\frac{a+b}{2}\right) + b(-M_0)\left(\frac{b}{2}\right) = 0 \qquad (2)$$

The second equation is perhaps found most simply by realizing that since both ends of the beam are clamped the angle between the tangents at A and C is zero. Actually, the tangents to the ends of the deformed beam coincide with the original straight configuration of the beam. The first moment-area theorem may now be employed between points A and C. It states that the angle between the tangents at A and C is equal to the area of the bending moment diagram between A and C, divided by EI. Thus

$$\tfrac{1}{2}(a+b)[R_1(a+b)] + (a+b)(-M_1) + b(-M_0) = 0 \qquad (3)$$

Solving (1), (2) and (3), we find

$$R_1 = \frac{6M_0 ab}{(a+b)^3} \qquad M_1 = \frac{M_0(2ab - b^2)}{(a+b)^2} \qquad M_2 = \frac{M_0(2ab - a^2)}{(a+b)^2}$$

It is to be carefully noted that there is no reason to assume that the vertical deflection of point B, the point of application of the couple M_0, is zero.

12.9. Determine the deflection of point B, the point of application of the couple M_0, in Problem 12.8.

This is readily determined since the desired deflection is equal to the deflection of point B from the tangent drawn at A. This tangent remains horizontal during deformation of the beam. From the second moment-area theorem the deflection of B from the tangent at A is given by the moment about the vertical line through B of the area under the bending moment diagram between A and B, divided by EI. Referring to the moment diagram, Fig. 12-11(b), we have

$$EI\Delta_B = \tfrac{1}{2}a(R_1a)(a/3) + a(-M_1)(a/2)$$

Substituting the values of R_1 and M_1 found in Problem 12.8, we have $\Delta_B = \dfrac{M_0a^2b^2(b-a)}{2(a+b)^3EI}.$

12.10. A clamped-end beam is subject to a concentrated force as in Fig. 12-12. Determine the reactions.

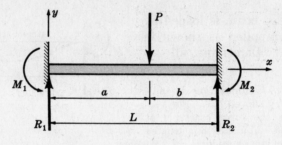

Fig. 12-12

The method of singularity functions will be used. The load intensity function is

$$w(x) = -M_1\langle x\rangle^{-2} + R_1\langle x\rangle^{-1} - P\langle x-a\rangle^{-1} + R_2\langle x-L\rangle^{-1} + M_2\langle x-L\rangle^{-2}$$

Integrating:

$$V(x) = -M_1\langle x\rangle^{-1} + R_1\langle x\rangle^0 - P\langle x-a\rangle^0 + R_2\langle x-L\rangle^0 + M_2\langle x-L\rangle^{-1}$$

$$M(x) = -M_1\langle x\rangle^0 + R_1\langle x\rangle^1 - P\langle x-a\rangle^1 + R_2\langle x-L\rangle^1 + M_2\langle x-L\rangle^0 = EI\frac{d^2y}{dx^2}$$

Integrating again:

$$EI\frac{dy}{dx} = -M_1\langle x\rangle^1 + \frac{R_1}{2}\langle x\rangle^2 - \frac{P}{2}\langle x-a\rangle^2 + \frac{R_2}{2}\langle x-L\rangle^2 + M_2\langle x-L\rangle^1 + C_1 \qquad (a)$$

But when $x = 0$, $dy/dx = 0$, hence $C_1 = 0$. Integrating again:

$$EIy = -\frac{M_1}{2}\langle x\rangle^2 + \frac{R_1}{6}\langle x\rangle^3 - \frac{P}{6}\langle x-a\rangle^3 + \frac{R_2}{6}\langle x-L\rangle^3 + \frac{M_2}{2}\langle x-L\rangle^2 + C_2 \qquad (b)$$

But when $x = 0$, $y = 0$, so that $C_2 = 0$.

The deflection and slope are also zero at the right end of the bar so that we may substitute $x = L$ in (a) and (b) to obtain

$$0 = -M_1L + \frac{R_1L^2}{2} - \frac{Pb^2}{2} \qquad (c)$$

$$0 = -\frac{M_1L^2}{2} + \frac{R_1L^3}{6} - \frac{Pb^3}{6} \qquad (d)$$

Solving (c) and (d) simultaneously we find

$$M_1 = \frac{Pab^2}{L^2} \qquad R_1 = \frac{Pb^2}{L^2} + \frac{2Pab^2}{L^3}$$

From statics, the sum of the vertical forces must vanish; hence

$$\frac{Pb^2}{L^2} + \frac{2Pab^2}{L^3} + R_2 - P = 0$$

Solving:

$$R_2 = \frac{Pa^2}{L^2} + \frac{2Pa^2b}{L^3}$$

For static equilibrium of moments about any point, such as the right end of the bar, we immediately find $M_2 = Pa^2b/L^2$.

12.11. The beam shown in Fig. 12-13(a) is simply supported at the left and right ends and spring-supported at the center. Determine the spring constant so that the bending moment will be zero at the point where the spring supports the beam.

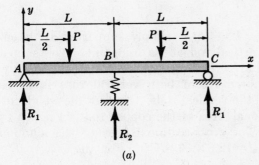

Fig. 12-13

Although this problem at first may appear to be indeterminate, the fact that the bending moment at B is prescribed to be zero actually makes the system statically determinate. Thus, if the left half of the system is isolated as indicated in Fig. 12-13(b) and the equation expressing equilibrium of moments about point B is written

$$\Sigma M_B = R_1 L - P\frac{L}{2} = 0$$

one finds $R_1 = P/2$. Then, from a consideration of the equilibrium of the entire system we have

$$2R_1 + R_2 - 2P = 0 \quad \text{or} \quad R_2 = P$$

where $R_2 = P$ is the force exerted by the spring upon the beam.

We now employ the method of singularity functions to determine the deflections of the entire beam. The load intensity function is

$$w(x) = \frac{P}{2}\langle x \rangle^{-1} - P\left\langle x - \frac{L}{2}\right\rangle^{-1} + P\langle x - L\rangle^{-1} - P\left\langle x - \frac{3L}{2}\right\rangle^{-1} + \frac{P}{2}\langle x - 2L\rangle^{-1}$$

Integrating:

$$V(x) = \frac{P}{2}\langle x \rangle^{0} - P\left\langle x - \frac{L}{2}\right\rangle^{0} + P\langle x - L\rangle^{0} - P\left\langle x - \frac{3L}{2}\right\rangle^{0} + \frac{P}{2}\langle x - 2L\rangle^{0}$$

$$M(x) = \frac{P}{2}\langle x \rangle^{1} - P\left\langle x - \frac{L}{2}\right\rangle^{1} + P\langle x - L\rangle^{1} - P\left\langle x - \frac{3L}{2}\right\rangle^{1} + \frac{P}{2}\langle x - 2L\rangle^{1} = EI\frac{d^2y}{dx^2}$$

$$EI\frac{dy}{dx} = \frac{P}{4}\langle x \rangle^{2} - \frac{P}{2}\left\langle x - \frac{L}{2}\right\rangle^{2} + \frac{P}{2}\langle x - L\rangle^{2} - \frac{P}{2}\left\langle x - \frac{3L}{2}\right\rangle^{2} + \frac{P}{4}\langle x - 2L\rangle^{2} + C_1$$

From symmetry, when $x = L$, $dy/dx = 0$. Hence

$$0 = \frac{P}{4}(L^2) - \frac{P}{2}\left(\frac{L}{2}\right)^2 + C_1 \quad \text{or} \quad C_1 = -\frac{PL^2}{8}$$

Integrating again:

$$EIy \ = \ \frac{P}{12}\langle x \rangle^3 - \frac{P}{6}\left\langle x - \frac{L}{2} \right\rangle^3 - \frac{P}{6}\langle x - L \rangle^3 - \frac{P}{6}\left\langle x - \frac{3L}{2} \right\rangle^2$$

$$+ \ \frac{P}{12}\langle x - 2L \rangle^3 - \frac{PL^2}{8}x + C_2$$

But when $x = 0$, $y = 0$, hence $C_2 = 0$. From this last equation the deflection at midpoint B is found by substituting $x = L$:

$$EIy_{x=L} \ = \ \frac{PL^3}{16}$$

But the spring exerts a force $R_2 = ky_{x=L}$. Thus

$$P \ = \ k\left(\frac{PL^3}{16EI}\right) \quad \text{or} \quad k \ = \ \frac{16EI}{L^3}$$

12.12. The horizontal beam shown in Fig. 12-14(a) is simply supported at the ends and is connected to a composite elastic vertical rod at its midpoint. The supports of the beam and the top of the copper rod are originally at the same elevation, at which time the beam is horizontal. The temperature of both vertical rods is then decreased 100 F°. Find the stress in each of the vertical rods. Neglect the weight of the beam and of the rods. The cross-sectional area of the copper rod is 1 in², $E_{cu} = 15 \times 10^6$ lb/in², and $\alpha_{cu} = 9.3 \times 10^{-6}$/F°. The cross-sectional area of the aluminum rod is 2 in², $E_{al} = 10 \times 10^6$ lb/in², and $\alpha_{al} = 12.8 \times 10^{-6}$/F°. For the beam, $E = 1.5 \times 10^6$ lb/in² and $I = 1000$ in⁴.

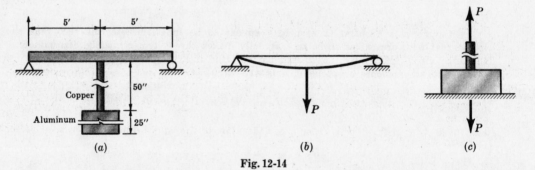

(a)   (b)   (c)

Fig. 12-14

A free-body diagram of the horizontal beam appears as in Fig. 12-14(b). Here, P denotes the force exerted upon the beam by the copper rod. Since this force is initially unknown there are three forces acting upon the beam, but only two equations of equilibrium for a parallel force system; hence the problem is statically indeterminate. It will thus be necessary to consider the deformations of the system.

A free-body diagram of the two vertical rods appears as in Fig. 12-14(c). The simplest procedure is temporarily to cut the connection between the beam and the copper rod, and then allow the vertical rods to contract freely because of the decrease in temperature. If the horizontal beam offers no restraint, the copper rod will contract an amount

$$\Delta_{cu} \ = \ (9.3 \times 10^{-6})(50)(100) \ = \ 0.0465 \text{ in.}$$

and the aluminum rod will contract by an amount

$$\Delta_{al} \ = \ (12.8 \times 10^{-6})(25)(100) \ = \ 0.0320 \text{ in.}$$

However, the beam exerts a tensile force P upon the copper rod and the same force acts in the aluminum rod as shown in Fig. 12-14(c) above. These axial forces elongate the vertical rods and this elongation (see Problem 1.1) is

$$\frac{P(50)}{1(15 \times 10^6)} + \frac{P(25)}{2(10 \times 10^6)}$$

The downward force P exerted by the copper rod upon the horizontal beam causes a vertical deflection of the beam. In Problem 9.8 this central deflection was found to be $\Delta = PL^3/48EI$.

Actually, of course, the connection between the copper rod and the horizontal beam is not cut in the true problem and we realize that the resultant shortening of the vertical rods is exactly equal to the downward vertical deflection of the midpoint of the beam. This change of length of the vertical rods is caused partially by the decrease in temperature and partially by the axial force acting in the rods. For the shortening of the rods to be equal to the deflection of the beam we must have

$$[0.0465 + 0.0320] \; - \; \left[\frac{P(50)}{1(15 \times 10^6)} + \frac{P(25)}{2(10 \times 10^6)} \right] \; = \; \frac{P(120)^3}{48(1.5 \times 10^6)(1000)}$$

Solving, $P = 2740$ lb. Then $\sigma_{cu} = 2740/1 = 2740$ lb/in^2 and $\sigma_{al} = 2740/2 = 1370$ lb/in^2.

12.13. The beam shown in Fig. 12-15(a) is clamped at both ends so as to completely restrain any angular rotation at these ends. The right end B is then displaced downward an amount Δ relative to the left end A. Determine the reactions developed because of this displacement.

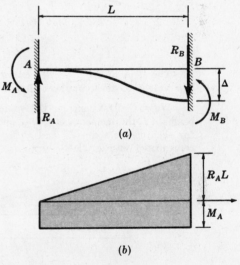

Let us represent the reactions developed because of the displacement by the forces and moments indicated. The moment diagram by parts, working from left to right, has the appearance shown in Fig. 12-15(b). The second moment-area theorem may be used to represent the deflection of B from the tangent at A as follows:

$$-EI\Delta \;=\; \tfrac{1}{2}L(R_A L)\left(\frac{L}{3}\right) + (-M_A)(L)\left(\frac{L}{2}\right)$$

or
$$-EI\Delta \;=\; \frac{R_A L^3}{6} - \frac{M_A L^2}{2} \qquad (a)$$

Next, the second moment-area theorem may be used to represent the deflection of A from the tangent at B as follows:

$$EI\Delta \;=\; \tfrac{1}{2}L(R_A L)\left(\frac{2L}{3}\right) + (-M_A)(L)\left(\frac{L}{2}\right)$$

or
$$EI\Delta \;=\; \frac{R_A L^3}{3} - \frac{M_A L^2}{2} \qquad (b)$$

(a)

(b)

Fig. 12-15

The statics equations are simply

$$\Sigma F_V \;=\; R_A - R_B \;=\; 0 \qquad (c)$$

$$\Sigma M_B \;=\; R_A L - M_A - M_B \;=\; 0 \qquad (d)$$

Solving (a), (b), (c), and (d) simultaneously:

$$M_A \;=\; M_B \;=\; \frac{6EI\Delta}{L^2} \qquad R_A \;=\; R_B \;=\; \frac{12EI\Delta}{L^3}$$

12.14. Derive the three-moment theorem for continuous beams.

A continuous beam is one that rests on more than two supports. The particular continuous beam shown in Fig. 12-16 consists of two spans and is subject to a series of uniform and concen-

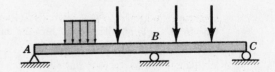

Fig. 12-16

trated loads as shown. It will be assumed that the nature of the supports is such that no horizontal reactions are present.

Continuous beams are statically indeterminate and thus it is necessary to supplement the available statics equations with equations arising from the deformations of the system. One possible way of deriving these equations is to treat the vertical forces at the various supports as the unknowns. However, an even simpler technique is to consider the bending moments at the supports as the unknowns. Deformation equations are then written, the bending moments determined, and lastly the values of the various vertical force reactions determined. This last procedure will be employed here.

Figure 12-17 displays free-body diagrams of any two adjacent spans of a continuous beam subject to any arbitrary loading. As shown in these diagrams M_A, M_B, and M_C represent the bending moments at the supports A, B, and C respectively. Although the directions of these moments are of course functions of the loads, we have assumed the moments to be positive in the sense of the definition offered in Chapter 6. Hence the directions indicated below are for positive moments.

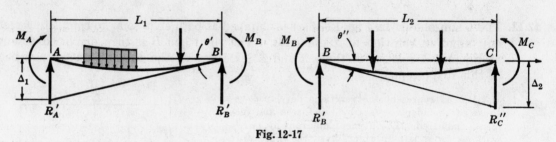

Fig. 12-17

The slope of the deflection curve must be continuous over the central support; hence $\theta' = -\theta''$. The values of these angles will now be found by the moment-area method.

We shall first consider the various known loadings to act on corresponding simply supported beams, i.e. the moments M_A, M_B, and M_C are temporarily omitted. The bending moment diagrams for each of the spans L_1 and L_2 may then be determined by the techniques of Chapter 6 and are represented in Fig. 12-18.

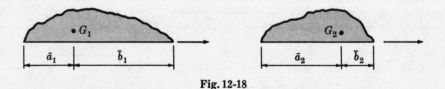

Fig. 12-18

In these diagrams G_1 and G_2 denote the centroids of the areas of the moment diagrams; and $\bar{a}_1$, $\bar{b}_1$, $\bar{a}_2$, and $\bar{b}_2$ have the meanings shown. It is to be carefully noted that these moment diagrams are determined on the assumption that each of the spans is simply supported. We shall let A_1 and A_2 denote the areas of these moment diagrams for the left and right spans respectively.

The deflection of A from the tangent at B is found from the second moment-area theorem to be $\Delta_1 = \dfrac{A_1 \bar{a}_1}{EI}$ and consequently the slope θ' is $\theta' = \dfrac{A_1 \bar{a}_1}{L_1 EI}$.

It is next necessary to consider the effects of M_A and M_B on the slope θ' in the left span. From a consideration of Problem 9.10, the rotation of the left span at point B produced by these moments is

$$\frac{M_B L_1}{3EI} + \frac{M_A L_1}{6EI}$$

The total angle of rotation is thus the sum of these, or

$$\theta' = \frac{A_1 \bar{a}_1}{L_1 EI} + \frac{M_B L_1}{3EI} + \frac{M_A L_1}{6EI}$$

Similarly, for the right span we have

$$\theta'' = \frac{A_2 \bar{b}_2}{L_2 EI} + \frac{M_B L_2}{3EI} + \frac{M_C L_2}{6EI}$$

Substituting in the relationship $\theta' = -\theta''$, we have

$$M_A L_1 + 2M_B(L_1 + L_2) + M_C L_2 = -\frac{6A_1 \bar{a}_1}{L_1} - \frac{6A_2 \bar{b}_2}{L_2}$$

This is the three-moment theorem and in this general form is applicable to any type of loading. Application of this equation to a continuous beam, together with the statics equations, enables one to solve for the various reactions. The above equation is sometimes called *Clapeyron's equation*.

12.15. Discuss the special form of the three-moment theorem for uniformly distributed loads acting on two adjacent spans.

We shall refer to the diagrams of Problem 12.14. Let w_1 represent the intensity of uniform load acting on the left span, and w_2 the intensity acting on the right span. If each of these loadings is considered to act on simply supported spans of lengths L_1 and L_2 respectively, then the bending moment diagrams are parabolic as shown in Fig. 12-19. In Problem 6.3 the maximum ordinates were found to have the values indicated.

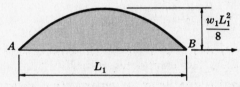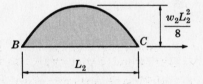

Fig. 12-19

In Problem 10.3 it was shown that the area under such a parabola is 2/3 the area of the surrounding rectangle. Hence, since A_1 represents the area of the left bending moment diagram, we have

$$A_1 = \tfrac{2}{3} L_1 \frac{w_1 L_1^2}{8}$$

Similarly,

$$A_2 = \tfrac{2}{3} L_2 \frac{w_2 L_2^2}{8}$$

The centroids of these areas are located at the midpoints of each of these spans.

Substituting these values in the right-hand side of the general form of the three-moment theorem found in Problem 12.14, we obtain

$$M_A L_1 + 2M_B(L_1 + L_2) + M_C L_2 = -\frac{w_1 L_1^3}{4} - \frac{w_2 L_2^3}{4}$$

This is the three-moment equation for uniformly distributed loads.

12.16. Determine the special form of the three-moment theorem for a single concentrated force acting on each of two adjacent spans.

Again we shall refer to the diagrams of Problem 12.14. Let P_1 denote the concentrated force acting on the left span, and a_1 the distance of this load from the left support A. Also, P_2 denotes the concentrated force acting on the right span at a distance b_2 from the right support C. If each of these loadings is considered to act on simply supported spans of length L_1 and L_2 respectively, then the bending moment diagrams are triangular as shown in Fig. 12-20.

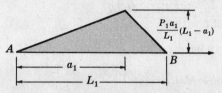

Fig. 12-20

In the general form of the three-moment theorem presented in Problem 12.14 the quantity $A_1\bar{a}_1$ appears. For the left span this designates the moment of the area under the above bending moment diagram about a vertical line through the left end A. Referring to the above moment diagram for the left span L_1, we may calculate this most easily by dividing the triangle into two triangles, one having a base equal to a_1, the other having a base equal to $L_1 - a_1$. The product of the area of each of these triangles times the distance from its centroid to the vertical line through A yields the required product $A_1\bar{a}_1$. Thus

$$A_1\bar{a}_1 = \tfrac{1}{2}a_1\frac{P_1a_1}{L_1}(L_1-a_1)(\tfrac{2}{3}a_1) + \tfrac{1}{2}(L_1-a_1)\left[\frac{P_1a_1}{L_1}(L_1-a_1)\right][a_1 + \tfrac{1}{3}(L_1-a_1)]$$

From this, after simplification, we obtain

$$\frac{6A_1\bar{a}_1}{L_1} = \frac{P_1a_1}{L_1}(L_1^2 - a_1^2)$$

Similarly, for the right span L_2 we find

$$\frac{6A_2\bar{b}_2}{L_2} = \frac{P_2b_2}{L_2}(L_2^2 - b_2^2)$$

For a number of concentrated loads the three-moment theorem thus becomes

$$M_AL_1 + 2M_B(L_1+L_2) + M_CL_2 = -\Sigma\frac{P_1a_1}{L_1}(L_1^2-a_1^2) - \Sigma\frac{P_2b_2}{L_2}(L_2^2-b_2^2)$$

where the summation sign is used so as to include the effects of all concentrated loads.

12.17. The two-span continuous beam shown in Fig. 12-21(a) below supports a uniform load of w lb per unit length. Determine the various reactions.

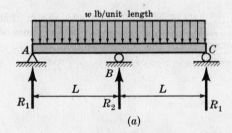

w lb/unit length

(a)

The three-moment theorem for uniform loads as given in Problem 12.15 is applicable. Here the load is constant across the entire beam and we have $w_1 = w_2 = w$. Also, $L_1 = L_2 = L$. The ends A and C are simply supported and hence we have $M_A = M_C = 0$. Substituting in the three-moment theorem,

$$M_AL_1 + 2M_B(L_1+L_2) + M_CL_2$$
$$= -\frac{w_1L_1^3}{4} - \frac{w_2L_2^3}{4}$$

we obtain

$$0 + 2M_B(2L) + 0 = -\frac{wL^3}{4} - \frac{wL^3}{4}$$

or

$$M_B = -\frac{wL^2}{8}$$

Perhaps the simplest method for determining the reactions is to write the expression for the bending moment at B (which has just been determined) in terms of the moments of the forces to the left of B. Thus

$$R_1L - wL\frac{L}{2} = -\frac{wL^2}{8} \quad \text{or} \quad R_1 = \tfrac{3}{8}wL$$

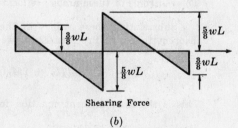

Shearing Force

(b)

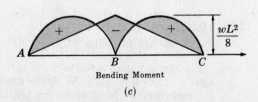

Bending Moment

(c)

Fig. 12-21

The end forces designated as R_1 are equal by symmetry. From statics,

$$2(\tfrac{3}{8}wL) + R_2 - 2wL = 0 \quad \text{or} \quad R_2 = \tfrac{5}{4}wL$$

The shear diagram may now be constructed by the usual methods of Chapter 6 and has the appearance shown in Fig. 12-21(b).

The bending moment diagram may also be plotted by the techniques of Chapter 6. However, a somewhat simpler procedure for continuous beams is to plot the moment diagram for the load acting on each span, on the basis that the span is simply supported. For this loading this diagram is of course parabolic for the uniform load. Then, the moment diagram due to the moments at the supports is constructed. This moment is zero at either end of the beam and equal to $-wL^2/8$ at the point B. From a consideration of Problem 9.10 it is evident that the variation of moment from either end of the beam to the midpoint (due to the moment $-wL^2/8$ only) is linear, i.e. straight lines must connect the value of the moment at the midpoint with the values at the ends, which are zero. Also, the moment diagrams due to the uniform load are positive, whereas those due to M_B are negative. By superposition of these diagrams the final form appears as shown by the shaded areas in Fig. 12-21(c).

12.18. The two-span continuous beam shown in Fig. 12-22(a) supports the centrally applied forces. Determine the various reactions.

The three-moment theorem for concentrated loads as given in Problem 12-16 is applicable. Here we have $P_1 = P_2 = P$, $L_1 = L_2 = L$, and $a_1 = b_2 = L/2$. The ends A and C are simply supported and hence we have $M_A = M_C = 0$. Substituting in the three-moment theorem,

$$M_A L_1 + 2M_B(L_1 + L_2) + M_C L_2 = -\sum \frac{P_1 a_1}{L_1}(L_1^2 - a_1^2) - \sum \frac{P_2 b_2}{L_2}(L_2^2 - b_2^2)$$

we obtain

$$0 + 2M_B(2L) + 0 = -\frac{2P(L/2)}{L}(L^2 - L^2/4)$$

or
$$M_B = -\frac{3}{16}PL$$

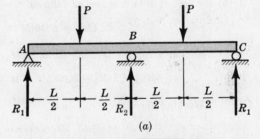

We may now express this bending moment at B in terms of moments of the forces to the left of B as follows:

$$R_1 L - \frac{PL}{2} = -\frac{3PL}{16} \quad \text{or} \quad R_1 = \frac{5}{16}P$$

The end forces designated as R_1 are equal by symmetry. From statics,

$$2\left(\frac{5P}{16}\right) + R_2 - 2P = 0 \quad \text{or} \quad R_2 = \frac{11}{8}P$$

The shear diagram thus appears as in Fig. 12-22(b).

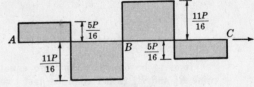

Shearing Force

(b)

The bending moment diagram may be plotted by the technique mentioned in Problem 12.17. The moment diagrams for the loads on each span on the basis that the span is simply supported appear as triangles of altitude $PL/4$. The moment diagram due to the moments over the supports is next constructed. It varies linearly from zero at either end of the beam to a value of $-3PL/16$ at the midpoint B. Again, the moment diagrams due to the concen-

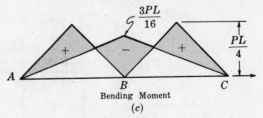

Bending Moment

(c)

Fig. 12-22

trated loads are positive whereas those due to M_B are negative. Superposition of these diagrams yields the final form of the moment diagram as shown by the shaded areas in Fig. 12-22(c).

12.19. In Fig. 12-23 the three-span continuous beam is subject to the uniform load as well as the two concentrated loads shown. Determine the four reactions.

We shall denote the moments acting at the supports from left to right by M_1, M_2, M_3, and M_4 respectively. We immediately have $M_1 = M_4 = 0$, since these ends are simply supported.

We shall first apply the three-moment theorem to the left and central spans. This will evidently give rise to an equation containing M_2 and M_3 as unknowns. Since these two spans are subject to a concentrated force and a uniform load respectively, the special forms of the three-moment theorem developed in Problems 12.15 and 12.16 are applicable. We must take $L_1 = 12$ ft, $L_2 = 24$ ft. We thus have

$$0 + 2M_2(12 + 24) + M_3(24) = -\frac{8000(6)}{12}[(12)^2 - (6)^2] - \frac{1000(24)^3}{4}$$

Simplifying,

$$3M_2 + M_3 = -18{,}000 - 144{,}000 \qquad (a)$$

Next we shall apply the three-moment theorem to the central and right spans. This gives rise to another equation containing M_2 and M_3. It is to be carefully noted that we must now take $L_1 = 24$ ft, $L_2 = 16$ ft. This equation is

$$M_2(24) + 2M_3(24 + 16) + 0 = -\frac{1000(24)^3}{4} - \frac{6000(8)}{16}[(16)^2 - (8)^2]$$

Simplifying,

$$M_2 + 3.33M_3 = -144{,}000 - 24{,}000 \qquad (b)$$

Solving equations (a) and (b) simultaneously, we find $M_2 = -41{,}300$ lb-ft and $M_3 = -38{,}000$ lb-ft.

We may now express the bending moment M_2 at support 2 in terms of moments of the forces to the left of this reaction as follows:

$$12R_1 - 8000(6) = -41{,}300 \qquad \text{or} \qquad R_1 = 560 \text{ lb}$$

Similarly for the bending moment M_3 at support 3,

$$36(560) + 24R_2 - 8000(30) - 1000(24)(12) = -38{,}000 \qquad \text{or} \qquad R_2 = 19{,}600 \text{ lb}$$

Working from the right end,

$$16R_4 - 6000(8) = -38{,}000 \qquad \text{or} \qquad R_4 = 640 \text{ lb}$$

Lastly,

$$40(640) + 24R_3 - 1000(24)(12) - 6000(32) = -41{,}300 \qquad \text{or} \qquad R_3 = 17{,}200 \text{ lb}$$

It is to be observed that two equations of static equilibrium could have been used to determine R_3 and R_4 rather than the last two equations if desired. However, the above procedure has the advantage that the statics equations are still available for checking the accuracy of the above numerical work.

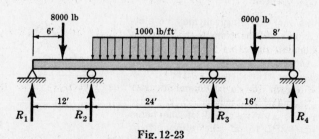

Fig. 12-23

Supplementary Problems

12.20. A clamped-end beam is supported at the right end, clamped at the left and carries the two concentrated forces shown in Fig. 12-24. Determine the reaction at the wall and the reaction at the right end of the beam.

Ans. $4P/3$ acting upward at left end, $PL/3$ acting counterclockwise at left end,
  $2P/3$ acting upward at right end.

12.21. Determine the deflection under the point of application of the force P located a distance $L/3$ from the right end of the beam described in Problem 12.20.   *Ans.* $7PL^3/486EI$

12.22. The beam of Problem 12.20 is of titanium Ti-4Al-3Mo-IV (STA) with a tensile ultimate strength of 175,000 lb/in² at room temperature. If the cross-section is 2 in. by 5 in. and a safety factor of 1.4 is employed, determine the maximum allowable value of each load P.   *Ans.* 17,400 lb

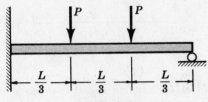

Fig. 12-24

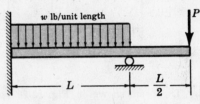

Fig. 12-25

12.23. A clamped-end beam is supported at an intermediate point and loaded as shown in Fig. 12-25. Determine the various reactions.

Ans. $\frac{5}{8}wL - \frac{3}{4}P$ upward at left end, $\frac{1}{8}wL^2 - \frac{1}{4}PL$ counterclockwise at left end,
  $\frac{3}{8}wL + \frac{7}{4}P$ upward at support.

12.24 For the beam shown in Problem 12.23, determine the deflection of the right end of the beam (under the point of application of the force P).   *Ans.* $(wL^4/96EI) - (5PL^3/48EI)$

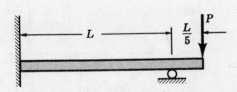

Fig. 12-26

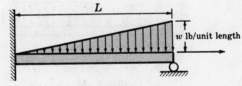

Fig. 12-27

12.25 The clamped-end beam of Fig. 12-26 is supported ·at an intermediate point and loaded as shown. Determine the various reactions.

Ans. $3P/10$ acting downward at left end, $PL/10$ acting clockwise at left end,
  $13P/10$ acting upward at support.

12.26. For the beam shown in Fig. 12-26 determine the deflection under the point of application of force P.   *Ans.* $19PL^3/1500EI$

12.27. A clamped-end beam is supported at the right end, clamped at the left and carries the load of uniformly varying intensity as indicated in Fig. 12-27. Determine the moment exerted by the support on the beam.   *Ans.* $7wL^2/120$

12.28. Determine the deflection of the right end C of the overhanging beam discussed in Problem 12.5.

$$Ans. \quad \Delta_C = \frac{wa^3b}{48EI} - \frac{wa^2b^2}{2EI} - \frac{wab^3}{8EI} - \frac{wb^4}{6EI}$$

12.29. Determine the central deflection of the clamped-end beam of Problem 12.6.   *Ans.* $wL^4/384EI$

12.30. The beam shown in Fig. 12-28 supports a concentrated load as well as a partial uniform load. Determine the reaction at the right end of the beam.

$$Ans. \quad \frac{81}{128}P + \frac{7}{128}wL$$

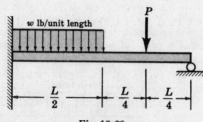

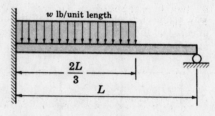

Fig. 12-28   Fig. 12-29

12.31. The beam shown in Fig. 12-29 supports a uniform load extending over two-thirds of its length. Determine the reaction at the right end of the beam.   *Ans.* $10wL/81$

12.32. A beam is clamped at both ends and supports a uniform load over its right half as shown in Fig. 12-30.   Determine all reactions.

Ans. $3wL/32$ acting upward at left end, $5wL^2/192$ acting counterclockwise at left end,
            $13wL/32$ acting upward at right end, $11wL^2/192$ acting clockwise at right end

12.33. Determine the central deflection of the beam described in Problem 12.32.   *Ans.* $wL^4/768EI$

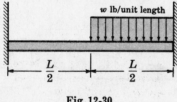

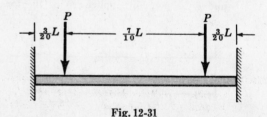

Fig. 12-30   Fig. 12-31

12.34. A beam is clamped at both ends and supports two symmetrically located concentrated forces as shown in Fig. 12-31.   Determine the various reactions.

Ans.   an upward force equal to P together with a counterclockwise moment equal to $51PL/400$ at
            the left end; symmetric reactions at the right end

12.35. Determine the central deflection of the beam described in Problem 12.34.   *Ans.* $9PL^3/4000EI$

12.36. The beam shown in Fig. 12-32 is clamped at the left end, supported at the right, and is loaded by a couple M_0.   Determine the reaction at the right support.   *Ans.* $3M_0a(a + 2b)/2(a + b)^3$

12.37. For the beam shown in Fig. 12-32, determine the deflection under the point of application of the applied moment M_0.   *Ans.* $M_0a^2b(a^2 - 2b^2)/4(a + b)^3EI$

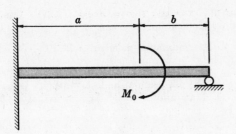

Fig. 12-32

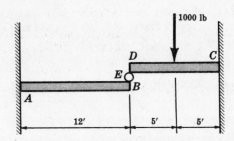

Fig. 12-33

12.38. In Fig. 12-33 AB and CD are cantilever beams with a roller E between their end points. A load of 1000 lb is applied as shown. Both beams are made of steel for which $E = 30 \times 10^6$ lb/in². For beam AB, $I = 50$ in⁴; for CD, $I = 80$ in⁴. Find the reaction at E.   *Ans.* 83 lb

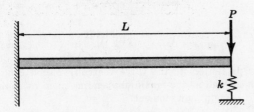

Fig. 12-34

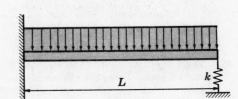

Fig. 12-35

12.39. The cantilever beam of Fig. 12-34 is supported by a spring with spring constant k and loaded by the concentrated force as indicated. Determine the reaction exerted by the spring upon the beam.

Ans. $$\frac{P}{\left(1 + \dfrac{3EI}{kL^3}\right)}$$

12.40. The cantilever beam of Fig. 12-35 is supported by a spring with spring constant k and subject to the uniformly distributed load shown. Determine the reaction exerted by the spring upon the beam.

Ans. $$\frac{w_0 L}{8\left(\dfrac{1}{3} + \dfrac{EI}{kL^3}\right)}$$

12.41. The beam shown in Fig. 12-36 is clamped at the left end and the right end is rotated through an angle ϕ as indicated. Determine the moment reactions developed at each end of the beam because of this rotation.   *Ans.* $2\phi EI/L$ at left end, $4\phi EI/L$ at right end

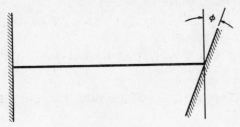

Fig. 12-36

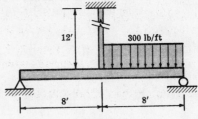

Fig. 12-37

12.42. A 16 ft beam carries a uniform load over the right half of its span and is supported at the center of the span by a vertical rod as shown in Fig. 12-37. The rod is steel, 12 ft in length, 0.5 in² in cross-sectional area, and $E_s = 30 \times 10^6$ lb/in². The beam is wood 4 in. by 8 in. in cross-section and $E_w = 1.5 \times 10^6$ lb/in². Determine the stress in the vertical steel rod.   *Ans.* 2960 lb/in²

12.43.   A three-span continuous beam is subject to the uniform load shown in Fig. 12-38. Determine the various reactions as well as the maximum bending moment in the beam.

   Ans.   reactions: $4wL/10$, $11wL/10$, $11wL/10$, $4wL/10$ maximum moment: $wL^2/10$

12.44.   The continuous beam shown in Fig. 12-38 is a 12 WF 32 section, and $L = 12$ ft. Determine the allowable load per unit length so as not to exceed an allowable bending stress of 18,000 lb/in².

   Ans.   428 lb/ft

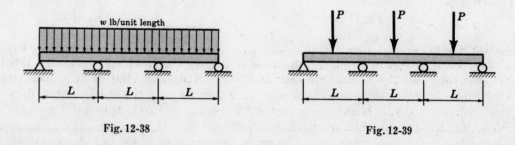

Fig. 12-38   Fig. 12-39

12.45.   The three-span continuous beam in Fig. 12-39 is subject to the three centrally placed concentrated loads shown. Determine the various reactions as well as the maximum bending moment in the beam.
   Ans.   reactions: $7P/20$, $23P/20$, $23P/20$, $7P/20$ maximum moment: $7PL/40$

12.46.   A two-span continuous beam is subject to the single concentrated force shown in Fig. 12-40. Determine the various reactions.

   Ans. $\dfrac{Pb}{L} - \dfrac{Pa}{4L^3}(L^2 - a^2)$ upward, $\dfrac{Pa}{L} + \dfrac{Pa}{2L^3}(L^2 - a^2)$ upward, $\dfrac{Pa}{4L^3}(L^2 - a^2)$ downward

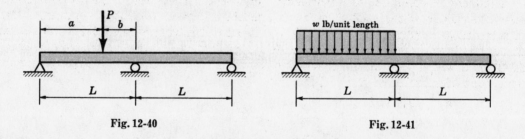

Fig. 12-40   Fig. 12-41

12.47.   A two-span continuous beam is subject to the uniform load shown in Fig. 12-41. Determine the various reactions. Also, determine the maximum bending moment in the beam.

   Ans.   reactions: $7wL/16$ upward, $5wL/8$ upward, $wL/16$ downward
          maximum moment: $49wL^2/512$

Chapter 13

Special Topics in Elastic Beam Theory

SHEAR CENTER

The simple flexure formula $\sigma = \dfrac{My}{I}$ determined in Problem 8.1 is valid only if the transverse loads which give rise to bending act in a plane of symmetry of the beam cross-section. In this type of loading there is obviously no torsion of the beam. However, in more general cases the beam cross-section will have no axes of symmetry and the problem of where to apply transverse loads so that the action is entirely bending with no torsion arises. Every elastic beam cross-section possesses a point through which transverse forces may be applied so as to produce *bending only* with *no torsion* of the beam. This point is called the *shear center*. In general, determination of the shear center location is extremely difficult and requires use of the theory of elasticity. However, in this chapter we will be concerned only with beams of *thin-walled open* cross-section having a *single axis of symmetry*, with the loads acting in a plane perpendicular to this axis of symmetry. (The contrasting case of closed thin-walled tubes is treated in Problem 5.15, page 74.) We will locate the shear center of the open cross-section on the axis of symmetry of the beam. For applications, see Problems 13.1 through 13.4.

UNSYMMETRIC BENDING

Frequently beams are of unsymmetric cross-section, or even if the cross-section is symmetric the plane of the applied loads may not be one of the planes of symmetry. In either of these cases the expression $\sigma = My/I$ derived in Problem 8.1 is not valid for determination of the bending stress. It is convenient to resolve the bending moment into components along the y and z axes of the cross-section as indicated by the double-headed vector representations of these moments in Fig. 13-1.

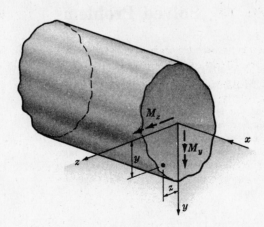

Fig. 13-1

The bending stress at a point located by the coordinates y, z is shown in Problem 13.5 to be

$$\sigma = \frac{(M_z I_y + M_y I_{yz})y + (-M_y I_z - M_z I_{yz})z}{I_y I_z - I_{yz}^2}$$

where I_y and I_z denote the moments of inertia about the y and z axes respectively and I_{yz} is the product of inertia. These quantities are determined by the methods of Chapter 7. There exists a *neutral axis* and those longitudinal fibers lying on the neutral axis are not subject to any normal stress. However, the neutral axis is usually not perpendicular to the plane of the applied loads nor does it coincide with either of the principal axes. For applications, see Problems 13.6 and 13.7.

CURVED BEAMS

Occasionally initially curved beams are encountered in machine design and other areas. Here we consider only those elastic beams for which the plane of curvature is also a plane of symmetry of every cross-section and the bending loads act in this plane of symmetry. Unlike the case of the initially straight beam the neutral axis no longer passes through the centroid of the cross-section but instead shifts toward the center of curvature of the beam by a distance denoted by $\bar{y}$. The bending stress distribution over the cross-section is hyperbolic in nature and in Problem 13.8 it is shown that these stresses are given by

$$\sigma = \frac{My}{A\bar{y}(r+y)}$$

where M is the bending moment, A the cross-sectional area, r is the radius of curvature of the neutral axis, and y denotes the distance of any fiber from the neutral axis. The peak stress always occurs at the outer fibers on the concave side of the beam. For applications see Problems 13.9 and 13.10.

Solved Problems

SHEAR CENTER

13.1. Determine the shear center of half of a thin-walled cylindrical section oriented as shown in Fig. 13-2 and subject to a vertical load.

Since the beam action is one of bending only with no torsion it follows that normal stresses are distributed over the cross-section in accordance with the flexure formula $\sigma = My/I$. Consequently, according to Problem 8.17, page 139, horizontal shearing stresses acting perpendicular to the plane of the cross-section are generated and are determined by the relation

$$\tau = \frac{V}{Ib}\int_{y_0}^{c} y\,da$$

As indicated in Problem 8.17, the presence of these horizontal shearing stresses necessitates the presence of equal intensity shear stresses acting over the vertical

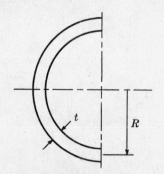

Fig. 13-2

cross-section.   In Fig. 13-3(a) these shear stresses have been shown as acting tangential to the
center line of the cross-section and further, for a thin-walled section, it is customary to assume
a uniform distribution of the shear stresses across the thickness t, as mentioned in Problem 5.15.
Lastly, it is assumed that shearing stresses perpendicular to the circular center-line of the section
are negligible.   In Fig. 13-3(a), V denotes the resultant of the distributed shearing stresses and it,
of course, acts vertically, since the horizontal components of the various stress vectors above and
below the axis of symmetry annul one another.

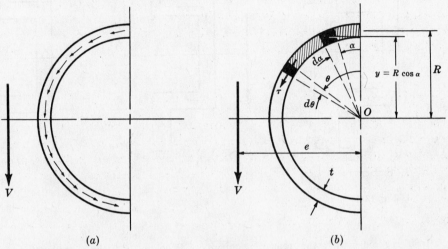

(a) (b)

Fig. 13-3

Let us examine the shearing stress τ at an arbitrary point denoted by the angle θ as indicated in Fig. 13-3(b).   Determination of this stress from the relation

$$\tau \;=\; \frac{V}{Ib} \int_{y_0}^{c} y \, da \tag{a}$$

necessitates evaluation of I as well as the integral, which, as explained in Problem 8.17, represents the first moment of the shaded area about the axis of symmetry.   This is accomplished by
introducing an auxiliary variable α $(0 < \alpha < \theta)$ as shown in Fig. 13-3(b) so that

$$\int_{y_0}^{c} y \, da \;=\; \int_{0}^{\theta} (R \cos \alpha) t (R \, d\alpha) \;=\; R^2 t \sin \theta \tag{b}$$

Next, the moment of inertia of the entire cross-section about the axis of symmetry is given by

$$I \;=\; \int y^2 \, da \;=\; \int_{0}^{\pi} (R \cos \theta)^2 t R \, d\theta \;=\; \frac{\pi R^3 t}{2} \tag{c}$$

The shearing stress at any point represented by θ is now found from (a), (b) and (c) to be

$$\tau \;=\; \frac{V}{(\pi R^3 t/2)t} [R^2 t \sin \theta] \;=\; \frac{2V}{\pi R t} \sin \theta \tag{d}$$

The moment of these distributed shearing stresses about any point, say O, must be equal to the
moment of the resultant V about that same point.   Thus since τ acts over an area $t(R d\theta)$ we have

$$\int_{\theta=0}^{\theta=\pi} \left(\frac{2V}{\pi R t} \sin \theta \right) (Rt \, d\theta) R \;=\; Ve$$

Thus

$$e \;=\; \frac{4R}{\pi}$$

gives the location of the shear center.

13.2. Determine the shear center of the "hat"-type thin-walled section indicated in Fig. 13.4. The thickness t is constant throughout the beam.

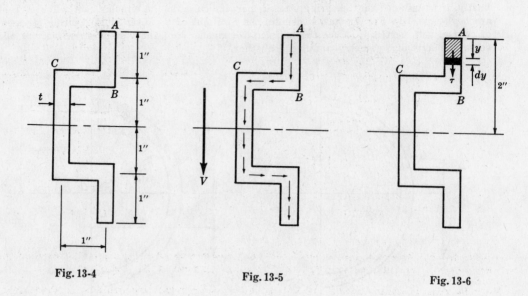

Fig. 13-4   Fig. 13-5   Fig. 13-6

In accordance with the reasoning given in Problem 13.1 the distribution of shear stresses over the cross-section appears as in Fig. 13-5. The resultant of the distributed shearing stresses, denoted V, acts vertically because the net horizontal effect of the shearing stresses in the two horizontal portions of the "hat" is zero. Let us first examine the shearing stress in the upper vertical member AB. At a distance y below the extreme point A, as shown in Fig. 13-6, the shearing stress is given by

$$\tau = \frac{V}{It} \int_{y_0}^{c} y \, da \qquad (a)$$

The integral represents the first moment of the shaded area about the axis of symmetry and may be readily evaluated as the product of the area, i.e. yt, and the distance from the centroid of the area to the axis of symmetry, i.e. $2 - \frac{y}{2}$. The shear stress at y is thus

$$\tau = \frac{V}{It}\left(2 - \frac{y}{2}\right) yt \qquad (b)$$

where it is to be remembered that V and I pertain to the shear force acting over the *entire* cross-section and the moment of inertia of the *entire* cross-section respectively. The resultant shear force V_1 acting over the vertical region AB, as indicated in Fig. 13-7, is found by integration to be

$$V_1 = \int_{y=0}^{y=1} \tau t \, dy = \frac{Vt}{I} \int_0^1 \left(2y - \frac{y^2}{2}\right) dy = \frac{5}{6} \frac{Vt}{I} \qquad (c)$$

Let us next examine the shearing stress in the upper horizontal member BC. At a distance x from point B, as indicated in Fig. 13-8, the shearing stress is given by equation (a), where now the integral represents the first moment of the shaded area in Fig. 13-8 about the axis of symmetry. By inspection the integral has the value $(1)(t)(1.5) + (x)(t)(1)$ and shear stress at x is thus

$$\tau = \frac{V}{It}[1.5t + xt] \qquad (d)$$

where V and I again pertain to the resultant shear over the *entire* cross-section and the moment of inertia of the *entire* cross-section respectively. The resultant shear force V_2, as indicated in Fig. 13-7, is found to be

$$V_2 = \int_{x=0}^{x=1} \tau t \, dx = \frac{Vt}{I} \int_0^1 (1.5 + x) \, dx = \frac{3Vt}{2I} \qquad (e)$$

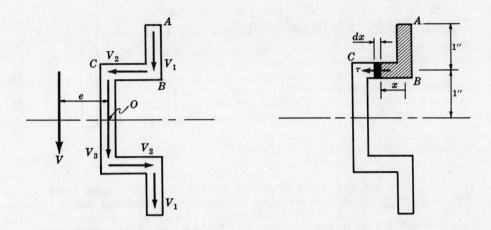

Fig. 13-7   Fig. 13-8

Since the entire section is thin-walled it is customary to use only nominal dimensions and thus neglect any slight duplication of areas at the intersections of the various members.

Because of symmetry the forces on the lower members are identical to those just found. The sum of the moments of these forces about any point, such as O in Fig. 13-7, must equal the moment of the resultant V about that same point.   Thus, we have $-2V_1(1) + 2V_2(1) = Ve$ or

$$e = \frac{4t}{3I} \qquad (f)$$

Finally, I may be calculated by the methods of Chapter 7 to be

$$I = \frac{1}{12}(t)(4)^3 + 2[(1)(t)(1)^2] = \frac{22t}{3} \qquad (g)$$

The shear center from (f) thus becomes

$$e = \frac{4t}{3(22t/3)} = \frac{4}{22} = 0.182 \text{ in.} \qquad (h)$$

Note that by choosing the moment center at O it is not necessary to determine V_3.

13.3. Determine the shear center of a thin-walled rectangular section in which there is a narrow longitudinal slit (see Fig. 13-9).   The thickness t is constant.

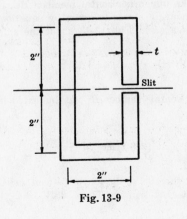

Fig. 13-9

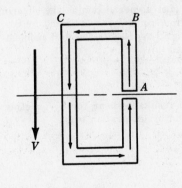

Fig. 13-10

236   SPECIAL TOPICS IN ELASTIC BEAM THEORY   [CHAP. 13

Observe that this section corresponds to the "hat" section of Problem 13.2 except that the outstanding flanges of the "hat" are turned toward the axis of symmetry here. The distribution of shear stresses appears as indicated in Fig. 13-10 and the vertical force V denotes the resultant of these distributed shearing stresses. Let us first examine the shearing stress in the vertical member AB. At a distance z above the axis of symmetry (assuming the slit to be of negligible thickness) the shearing stress is again given by

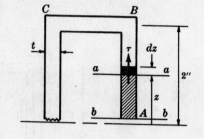

Fig. 13-11

$$\tau = \frac{V}{Ib} \int_{y_0}^{c} y \, da \qquad (a)$$

where it is of utmost importance to observe that the integral represents the first moment of the area lying *between* the section a-a where the shear stress is desired and the extreme fibers b-b of the section. This is true even though fibers b-b lie *closer* to the axis of symmetry than a-a. This statement follows from the derivation of the above equation as given in Problem 8.17. The integral is evaluated as the product of the area, i.e. zt, and the distance from the centroid of the area to the axis of symmetry, i.e. $z/2$. The shear stress at z is thus

$$\tau = \frac{V}{It} \left[zt \frac{z}{2} \right] = \frac{Vz^2}{2I} \qquad (b)$$

The resultant shear force V_1 acting over the vertical region AB, indicated in Fig. 13-12, is found by integration to be

$$V_1 = \int_{z=0}^{z=2} \tau t \, dz = \int_0^2 \frac{Vz^2}{2I} t \, dz = \frac{4Vt}{3I} \qquad (c)$$

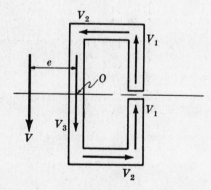

Fig. 13-12

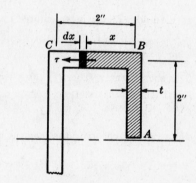

Fig. 13-13

Let us next examine the shearing stress in the upper horizontal member BC. At a distance x from point B, as indicated in Fig. 13-13, the shearing stress is given by equation (a) where the integral represents the first moment of the shaded area in Fig. 13-13 about the axis of symmetry. From (a),

$$\tau = \frac{V}{It} [(x)(t)(2) + (2)(t)(1)] = \frac{2V}{I}(x+1) \qquad (d)$$

The resultant shear force V_2 acting over the horizontal member BC, as indicated in Fig 13-12, is found by integration to be

$$V_2 = \int_{x=0}^{x=2} \tau t \, dx = \int_0^2 \frac{2Vt}{I}(x+1) \, dx = \frac{8Vt}{I} \qquad (e)$$

From Fig. 13-12 the sum of the moments of the forces V_1, V_2, and V_3 about any point, such as O, must equal the moment of the resultant about that point. Thus $2(2V_1) + 2(2V_2) = Ve$.

Substituting from (c) and (e):

$$\frac{16Vt}{3I} + \frac{32Vt}{I} = Ve \qquad (f)$$

$$e = \frac{112t}{3I} \qquad (g)$$

The moment of inertia is given by

$$I = 2[\tfrac{1}{12}(t)(4)^3] + 2[2t(2)^2] = 26.7t$$

Thus
$$e = \frac{112t}{3(26.7t)} = 1.40 \text{ in.}$$

which locates the shear center.

13.4. Determine the shear center of the thin-walled section indicated in Fig. 13-14. The thickness t is constant.

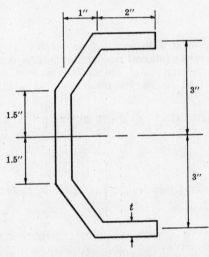

Fig. 13-14

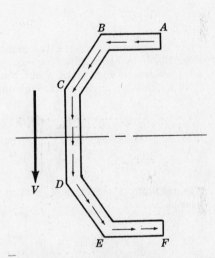

Fig. 13-15

The distribution of shear stresses appears as in Fig. 13-15 where the vertical force V denotes the resultant of these distributed shearing stresses. Let us first determine the shearing stress in the horizontal member AB. At a distance x from point A, as indicated in Fig. 13-16 the shearing stress is found to be

$$\tau = \frac{V}{Ib}\int_{y_0}^{c} y\,da \qquad (a)$$

or
$$\tau = \frac{V}{It}[(x)(t)(3)] = \frac{3Vx}{I} \qquad (b)$$

The resultant shear force V_1 acting over AB, as indicated in Fig. 13-17, is found by integration to be

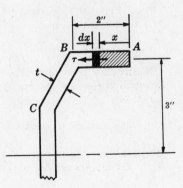

Fig. 13-16

$$V_1 = \int_{x=0}^{x=2} \tau t\,dx = \int_0^2 \frac{3Vxt}{I}dx = \frac{6Vt}{I} \qquad (c)$$

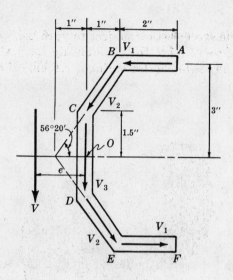

Fig. 13-17

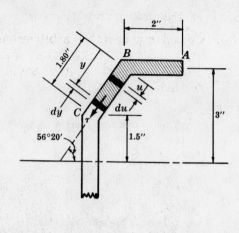

Fig. 13-18

The shearing stress in the inclined member BC at a distance y from point B, as indicated in Fig. 13-18, is again given by equation (a) where the integral represents the first moment of the shaded area in Fig. 13-18 about the axis of symmetry. For the inclined portion of that area it is simplest to integrate through introduction of an auxiliary variable u as indicated. Thus

$$\tau = \frac{V}{It}\left[(2)(t)(3) + \int_{u=0}^{u=y}[1.5 + (1.80 - u)\sin 56°20']t\,du\right]$$

$$= \frac{V}{I}(6 + 3y - 0.416y^2) \tag{d}$$

The resultant shear force V_2 acting over the inclined member BC in Fig. 13-17 is found by integration to be

$$V_2 = \int_{y=0}^{y=1.80} \tau t\,dy$$

$$= \int_0^{1.80} \frac{Vt}{I}(6 + 3y - 0.416y^3)\,dy = \frac{14.85Vt}{I} \tag{e}$$

From Fig. 13-17 the sum of the moments of the forces V_1, V_2, and V_3 about any point, such as O, must equal the moment of the resultant about that point. Thus

$$2(3V_1) + 2(V_2 \sin 56°20')(1) = Ve$$

Substituting from (c) and (e):

$$e = \frac{60.8t}{I} \tag{f}$$

The moment of inertia is given by

$$I = \tfrac{1}{12}(t)(3)^3 + 2[(2)(t)(3)^2] + 2\int_{u=0}^{u=1.80}[1.5 + (1.80 - u)\sin 56°20']^2 t\,du$$

$$= 57.2t$$

We then have

$$e = \frac{60.8t}{57.2t} = 1.06 \text{ in.}$$

which locates the shear center.

UNSYMMETRIC BENDING

13.5. Consider a beam of arbitrary unsymmetric cross-section subject to pure bending as indicated in Fig. 13-19(a). Derive an expression for the relationship between the bending moment and the bending stress at any point in this section. Assume Hooke's law holds.

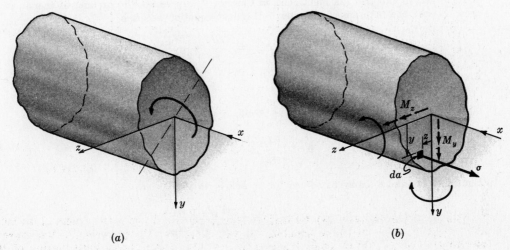

(a)   (b)

Fig. 13-19

It is convenient to resolve the moment M, which acts in a plane oblique to the y- and z-axes, into moment components about those axes. These components are designated as M_y and M_z and their positive directions are indicated by the double-headed vectors in Fig. 13-19(b).

As in Problem 8.1 it is reasonable to assume that cross-sections that were plane prior to bending remain plane after application of the loads. However, in the general case being considered here there is one radius of curvature ρ_z in the x-y plane and another ρ_y in the x-z plane. Thus, for a longitudinal fiber of area da as indicated in Fig. 13-19(b) the normal strain, analogous to (1) of Problem 8.1, is given by

$$\epsilon = \frac{y}{\rho_z} + \frac{z}{\rho_y} \tag{1}$$

Since Hooke's law holds we immediately have

$$\sigma = \frac{Ey}{\rho_z} + \frac{Ez}{\rho_y} \tag{2}$$

and this longitudinal, or bending, stress is indicated in the figure.

The resultant longitudinal force acting over the cross-section is zero (for the case of pure bending) and this condition may be expressed as:

$$\int_A \sigma \, da = 0 \quad \text{or} \quad \int_A \left(\frac{Ey}{\rho_z} + \frac{Ez}{\rho_y} \right) da = 0$$

where the integration is extended over the cross-sectional area A. Since ρ_y and ρ_z are constant over the cross-section, we have

$$\frac{E}{\rho_z} \int_A y \, da + \frac{E}{\rho_y} \int_A z \, da = 0 \tag{3}$$

This equation is satisfied if the integrals vanish. This implies taking the origin of the y-z coordinate system to coincide with the centroid of the cross-section.

From Fig. 13-19(b) it is evident that

$$M_z = \int_A \sigma y \, da = \int_A \left(\frac{Ey^2}{\rho_z} + \frac{Eyz}{\rho_y}\right) da$$

$$= \frac{E}{\rho_z}\int_A y^2 \, da + \frac{E}{\rho_y}\int_A yz \, da$$

where the first integral represents the moment of inertia of the cross-sectional area about the z-axis and the second integral (as mentioned in Chapter 7) represents the product of inertia of the same area. Using the notation of Chapter 7 this last equation becomes

$$M_z = \frac{EI_z}{\rho_z} + \frac{EI_{yz}}{\rho_y} \tag{4}$$

Also from Fig. 13-19(b) we have

$$M_y = -\int_A \sigma z \, da = -\int_A \left(\frac{Eyz}{\rho_z} + \frac{Ez^2}{\rho_y}\right) da$$

$$= -\frac{EI_{yz}}{\rho_z} - \frac{EI_y}{\rho_y} \tag{5}$$

Equations (4) and (5) may be solved for ρ_y and ρ_z to yield

$$\frac{1}{\rho_y} = \frac{-M_y I_z - M_z I_{yz}}{E(I_y I_z - I_{yz}^2)} \tag{6}$$

$$\frac{1}{\rho_z} = \frac{M_z I_y + M_y I_{yz}}{E(I_y I_z - I_{yz}^2)} \tag{7}$$

Substituting (6) and (7) in (2) yields the bending stress

$$\sigma = \frac{(M_z I_y + M_y I_{yz})y + (-M_y I_z - M_z I_{yz})z}{I_y I_z - I_{yz}^2} \tag{8}$$

Equation (8) is termed the *generalized flexure formula* and holds for an elastic beam of arbitrary cross-section with bending loads in an arbitrary plane. For the special case $M_y = I_{yz} = 0$ (implying that the y-z axes are principal axes and that bending takes place only about the z-axis) (8) reduces to $\sigma = M_z y/I_z$ which is equivalent to (9) of Problem 8.1.

The equation of the neutral axis is readily found by setting the stress from (8) equal to zero, since by definition the fibers along the neutral axis are free of longitudinal stress. Thus

$$\frac{y}{z} = \frac{M_y I_z + M_z I_{yz}}{M_z I_y + M_y I_{yz}} = \tan \alpha \tag{9}$$

where α denotes the angle of inclination of the neutral axis as indicated in Fig. 13-20. In general the neutral axis is *not* perpendicular to the plane of the applied moments nor does it coincide with either of the principal axes.

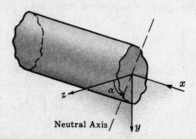

Fig. 13-20

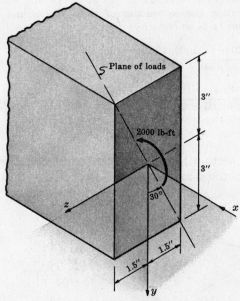

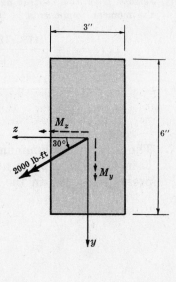

Fig. 13-21   Fig. 13-22

13.6. The rectangular beam of Fig. 13-21 is subject to loads that create a bending moment of 2000 lb-ft acting in a plane oriented at 30° to the y-axis. Determine the peak tensile and compressive stresses in the beam.

The vector representation of the 2000 lb-ft moment is indicated by the solid double-headed vector in Fig. 13-22, together with its moment components (dashed vectors) in the y- and z-directions. This convenient vector representation enables us to find the components as

$$M_y = 2000 \sin 30° = 1000 \text{ lb-ft} \qquad M_z = 2000 \cos 30° = 1732 \text{ lb-ft}$$

From Problem 7.4, page 110, we have

$$I_y = \tfrac{1}{12}(6)(3)^3 = 13.5 \text{ in}^4 \qquad I_z = \tfrac{1}{12}(3)(6)^3 = 54 \text{ in}^4$$

Also, since the y-z axes are axes of symmetry, they are principal axes of the cross-section and, from Chapter 7, the product of inertia with respect to these axes vanishes: $I_{yz} = 0$.

The angle of inclination of the neutral axis (which passes through the centroid) is given by (9) of Problem 13.5 to be

$$\tan \alpha = \frac{M_y I_z + M_z I_{yz}}{M_z I_y + M_y I_{yz}}$$

$$= \frac{(1000)(54) + (1732)(0)}{(1732)(13.5) + 1000(0)} = 2.31$$

$$\alpha = 66°40'$$

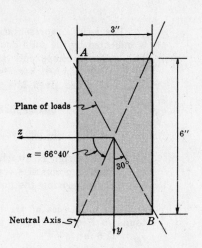

Fig. 13-23

As mentioned in Problem 13.5 there is no reason to expect the neutral axis, as indicated in Fig. 13-23, to be normal to the plane of the loads.

In Problem 13.5 it was assumed that plane sections remain plane during bending. The originally plane section rotates about the neutral axis indicated in Fig. 13-23 and since both strains as well as stresses vary as the distance from the neutral axis it is evident that the peak tensile stress occurs at point B and the peak compressive stress occurs at A, i.e. *at those points most remote from the neutral axis.* Substituting the coordinates of these points and the values of the moment components in (8) of Problem 13.5, we obtain

$$\sigma_B = \frac{[(1732)(12)(13.5) + 0](3) + [-(1000)(12)(54) - 0](-1.5)}{(13.5)(54) - 0} = 2480 \text{ lb/in}^2$$

$$\sigma_A = \frac{[(1732)(12)(13.5) + 0](-3) + [-(1000)(12)(54) - 0](1.5)}{(13.5)(54) - 0} = -2480 \text{ lb/in}^2$$

13.7. The structural aluminum 6 Z 5.42 section is subject to a bending moment of 50,000 lb-in lying in a plane at 26°33′ to the y-axis. Determine the peak tensile and compressive stresses in the beam.

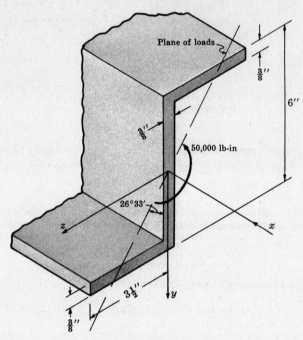

Fig. 13-24

The vector representation of the 50,000 lb-in moment is indicated by the solid double-headed vector in Fig. 13-25. The moment components in the y- and z-directions are indicated by dashed vectors in that same diagram. The components are given by

$$M_y = -50,000 \sin 26°33′ = -22,350 \text{ lb-in}$$

$$M_z = 50,000 \cos 26°33′ = 44,700 \text{ lb-in}$$

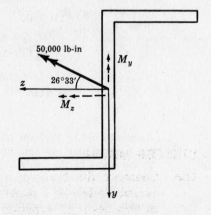

The negative sign is attached to M_y because the direction of M_y in this problem is opposite to that assumed in Problem 13.5 in deriving the generalized flexure formula.

The properties of the cross-section have already been found in Problem 7.20 to be $I_y = 9.08 \text{ in}^4$, $I_z = 25.27 \text{ in}^4$, $I_{yz} = 11.6 \text{ in}^4$.

From Problem 13.5 the angle of inclination of the neutral

Fig. 13-25

axis (which passes through the centroid) is given by

$$\tan \alpha = \frac{M_y I_z + M_z I_{yz}}{M_z I_y + M_y I_{yz}}$$

$$= \frac{(-22,350)(25.75) + (44,700)(11.6)}{(44,700)(9.08) + (-22,350)(11.6)} = -0.317$$

$$\alpha = -17°35'$$

The negative value of the angle indicates the neutral axis orientation shown in Fig. 13-26, i.e. opposite in direction to the positive direction indicated in Fig. 13-20 of Problem 13.5.

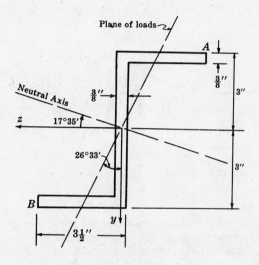

Fig. 13-26

The peak stresses occur at those points *most remote from the neutral axis*. From Fig. 13-26 it is evident that the peak compression occurs at point A having coordinates $y = -3$ in., $z = -3\frac{5}{16}$ in. The peak tension occurs at point B with coordinates $y = 3$ in., $z = 3\frac{5}{16}$ in. The bending stresses at these points are found from (8) of Problem 13.5 to be

$$\sigma_A = \frac{[(44,700)(9.08) + (-22,350)(11.6)](-3) + [-(-22,350)(25.27) - (44,700)(11.6)][-3\frac{5}{16}]}{(9.08)(25.27) - (11.6)^2}$$

$$= -6150 \text{ lb/in}^2$$

$$\sigma_B = \frac{[(44,700)(9.08) + (-22,350)(11.6)](3) + [-(-22,350)(25.27) - (44,700)(11.6)][3\frac{5}{16}]}{(9.08)(25.27) - (11.6)^2}$$

$$= 6150 \text{ lb/in}^2$$

CURVED BEAMS

13.8. Consider the bending of an initially curved elastic beam for which the plane of curvature is also a plane of symmetry of every cross-section. The bending loads act in this plane of symmetry. Derive an expression for the relationship between the bending moment and the bending stress at any point in the cross-section. Assume Hooke's law holds.

The beam is illustrated in Fig. 13-27 where R denotes the distance from the center of curvature C to the axis through the centroid of the cross-section. The bending moment M is taken to be positive in the direction indicated, i.e. when it tends to *increase* the curvature (*decrease* the radius of curvature).

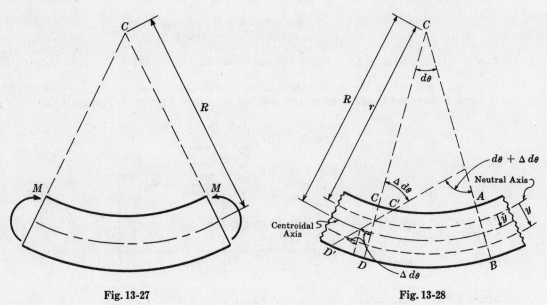

Fig. 13-27   Fig. 13-28

Let us examine the behavior of a part of the beam corresponding to a central angle $d\theta$ before deformation. After deformation, this angle changes to $d\theta + \Delta \, d\theta$ as shown in Fig. 13-28. Just as in the case of the initially straight beam studied in Problem 8.1 we will assume that plane cross-sections originally perpendicular to the geometric axis of the beam remain plane after bending. Thus, the normal section CD prior to loading moves to $C'D'$ after loading. For convenience we shall assume that AB remains fixed in space but this in no way influences the results we will obtain. It will still be assumed that there exists one axis, the neutral axis, for which the longitudinal fibers do not change length, and thus the section CD may be considered to rotate about this neutral axis as indicated in Fig. 13-28. However, there is no reason to believe that the neutral axis coincides with the centroid of the cross-section as it did for the initially straight beam in Problem 8.1. In the present problem involving the curved beam Fig. 13-28 indicates that the *total elongation* of a longitudinal fiber varies as the distance y of the fiber from the neutral axis. The coordinate y is measured positive away from the center of curvature. However, the lengths of these fibers prior to loading are obviously different, hence the *unit* elongations, i.e. normal strains, are *not* proportional to the distances from the neutral axis. This point constitutes the fundamental difference between behavior of a curved beam and behavior of the initially straight beam discussed in Problem 8.1. Since Hooke's law is assumed to hold for this curved beam it follows that stresses on these fibers are *not* proportional to the distances from the neutral axis.

Let us consider the elongation of the fiber at a distance y from the neutral axis. From Fig. 13-28 this is $y(\Delta \, d\theta)$. Dividing this elongation by the original length of the fiber, $(r+y) \, d\theta$, yields the normal strain as

$$\epsilon \;=\; \frac{y(\Delta \, d\theta)}{(r+y) \, d\theta} \tag{a}$$

where r denotes the radius of curvature of the neutral axis. Since Hooke's law holds, the normal stress is

$$\sigma \;=\; \frac{Ey(\Delta \, d\theta)}{(r+y) \, d\theta} \tag{b}$$

The neutral axis may now be located by requiring the resultant normal force over the cross-section to vanish. Thus

$$\int_A \sigma \, da \;=\; \int_A \frac{Ey(\Delta \, d\theta) \, da}{(r+y) \, d\theta} \;=\; \frac{E(\Delta \, d\theta)}{d\theta} \int_A \frac{y \, da}{(r+y)} \;=\; 0 \tag{c}$$

where the integration is over the entire cross-section area A. If $u = r + y$ (i.e. the distance of any fiber from the center of curvature C) then (c) becomes

$$\int_A \frac{(u-r)\,da}{u} = 0 \quad \text{or} \quad r = \frac{A}{\int_A \frac{da}{u}} \qquad (d)$$

where the integral in the denominator represents a mathematical property of the cross-sectional area and is analogous to the moment of inertia that arises in the case of bending of an initially straight beam.

The sum of the moments of the normal forces on the fibers must equal the bending moment:

$$M = \int_A \sigma y\,da = \int_A \frac{Ey^2(\Delta\,d\theta)\,da}{(r+y)\,d\theta} = \frac{E(\Delta\,d\theta)}{d\theta}\int_A \frac{y^2\,da}{r+y}$$

Simplifying:

$$\int_A \frac{y^2\,da}{r+y} = \int_A y\,da - r\int_A \frac{y\,da}{r+y}$$

The first integral represents the first or static moment of the cross-sectional area about the neutral axis, and the second according to (c) vanishes. Thus

$$M = \frac{E(\Delta\,d\theta)}{d\theta}[A\bar{y}] \qquad (e)$$

where $\bar{y}$ denotes the distance from the neutral axis to the centroidal axis. Combining (b) and (e) we find the normal stress on any fiber to be

$$\sigma = \frac{My}{A\bar{y}(r+y)} \qquad (f)$$

From (f) it is evident that the stress distribution across the depth of the curved beam is *hyperbolic*. The maximum stress always occurs at the outer fibers on the concave side of the beam. Further, the neutral axis always lies between the centroidal axis and the center of curvature.

13.9. A curved beam is of rectangular cross-section 3 in. by 6 in. Determine the peak tensile and compressive stresses if the beam is loaded by a moment of 120,000 lb/in².

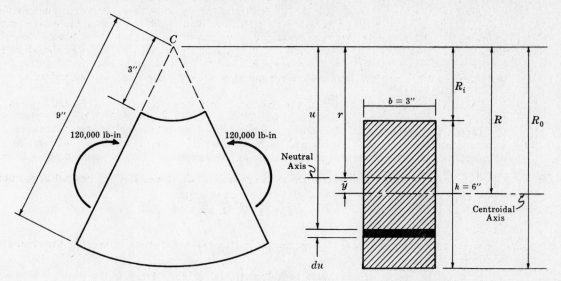

Fig. 13-29

It is first necessary to employ (d) of Problem 13.8 to locate the neutral axis. For the special case of a rectangular cross-section with the coordinates indicated in Fig. 13-29 we find

$$r = \frac{bh}{\displaystyle\int_A \frac{b\,du}{u}} = \frac{bh}{b[\ln u]_{R_i}^{R_0}} = \frac{h}{\ln\left(\dfrac{R_0}{R_i}\right)} \qquad (a)$$

For the present geometry

$$r = \frac{6}{\ln\left(\frac{9}{3}\right)} = 5.46 \text{ in.}$$

indicating that the neutral axis lies between the centroidal axis and the center of curvature, the distance between the centroidal axis and the neutral axis being 0.54 in.

The stresses in the extreme fibers on the concave side are found from (f) of Problem 13.8 to be

$$\sigma_{R_i} = \frac{120,000(-2.46)}{(3)(6)(0.54)(5.46 - 2.46)} = -10,000 \text{ lb/in}^2$$

The stresses in the extreme fibers on the convex side are

$$\sigma_{R_0} = \frac{120,000(3.54)}{(3)(6)(0.54)(5.46 + 3.54)} = 4860 \text{ lb/in}^2$$

The hyperbolic stress distribution over the cross-section appears as in Fig. 13-30.

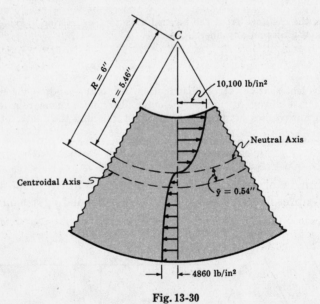

Fig. 13-30

In (a) the evaluation of the denominator presented no particular problem. However, in a thinner beam the ratio of R_0 to R_i may be much closer to unity, in which case extreme precision of calculation is required because of the nature of the logarithmic function near unity. In such a case it is best to proceed as follows:

$$\ln\left(\frac{R_0}{R_i}\right) = \ln\left(\frac{R + h/2}{R - h/2}\right) = \ln\left(\frac{1 + h/2R}{1 - h/2R}\right)$$

$$= \frac{h}{R}\left[1 + \frac{1}{3}\left(\frac{h}{2R}\right)^2 + \frac{1}{5}\left(\frac{h}{2R}\right)^4 + \cdots\right] \approx \frac{h}{R}\left[1 + \frac{1}{12}\frac{h^2}{R^2}\right] \qquad (b)$$

where R is the radius of curvature to the centroidal axis. Since $\bar{y}$ is the distance between the neutral axis and the centroidal axis, we have from (a) and (b)

$$\bar{y} \;=\; R - r \;\approx\; \frac{h^2}{12R} \qquad \text{(for a rectangular section } only\text{)} \tag{c}$$

13.10. A thick ring is of circular cross-section and is loaded by two opposed diametral loads as shown in Fig. 13-31. Determine the peak compressive stress.

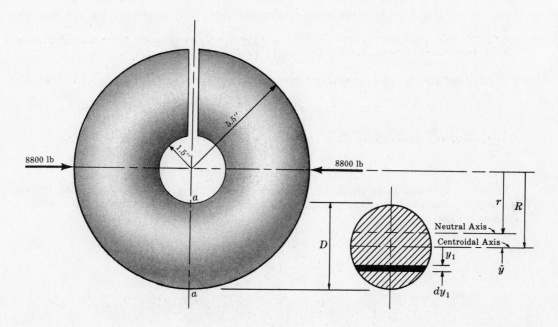

Fig. 13-31

To locate the neutral axis of the section at *a-a* we examine equation (*d*) of Problem 13.8. For a circular cross-section, as well as for many others, direct use of this equation leads to certain computational difficulties. It is often more convenient to determine $\bar{y}$, the distance from the neutral axis to the centroidal axis, directly. Let us introduce a new variable y_1 which represents the distance of a fiber from the centroidal axis. From Fig. 13-28 of Problem 13.8, $y_1 = y - \bar{y}$. Then from (*c*) of Problem 13.8 we have

$$0 \;=\; \int_A \frac{y\,da}{r+y} \;=\; \int_A \frac{(y_1+\bar{y})\,da}{R+y_1} \;=\; \left\{\int_A \frac{y_1\,da}{R+y_1}\right\} + \bar{y}\left\{\int_A \frac{da}{R+y_1}\right\} \tag{a}$$

The first bracketed integral may be regarded as being analogous to the first moment of area and may be written as

$$\int_A \frac{y_1\,da}{R+y_1} \;=\; kA \tag{b}$$

where the quantity k must be determined for each cross-section. The second bracketed integral may be put in the form

$$\int_A \frac{da}{R+y_1} \;=\; \frac{1}{R}\int_A \left[1 - \frac{y_1}{R+y_1}\right]da \;=\; \frac{A}{R} - k\frac{A}{R} \tag{c}$$

Substituting (*b*) and (*c*) into (*a*) we find that

$$\bar{y} \;=\; \frac{kR}{1-k} \tag{d}$$

For the circular cross-section shown in Fig. 13-31

$$da = 2\sqrt{(D/2)^2 - (y_1)^2}\, dy_1$$

Substituting this into (c) we find

$$\int_{-D/2}^{D/2} \frac{2\sqrt{(D/2)^2 - (y_1)^2}\, dy_1}{R + y_1} = \frac{A}{R}(1-k)$$

Integrating and expanding the radical in power series and retaining only the first two terms we obtain:

$$k = \frac{1}{4}\left(\frac{D}{2R}\right)^2 + \frac{1}{8}\left(\frac{D}{2R}\right)^4 \tag{e}$$

For the ring under consideration this becomes

$$k = \tfrac{1}{4}(\tfrac{4}{7})^2 + \tfrac{1}{8}(\tfrac{4}{7})^4 = 0.0947$$

From (d) we locate the neutral axis by

$$\bar{y} = \frac{0.0947(3.5)}{1 - 0.0947} = 0.366 \text{ in.}$$

The peak compressive stress occurs at the extreme fibers on the concave side at section a-a and is found as the sum of the bending stress given by (f) of Problem 13.8 and the direct, i.e. P/A, stress. Thus

$$\sigma_{\max} = \frac{(8800)(3.5)(-1.634)}{\pi(2)^2(0.366)(1.5)} - \frac{8800}{\pi(2)^2} = -7980 \text{ lb/in}^2$$

Supplementary Problems

13.11. Locate the shear center of a thin-walled circular section with a longitudinal slit (Fig. 13-32).
Ans. $e = 2R$

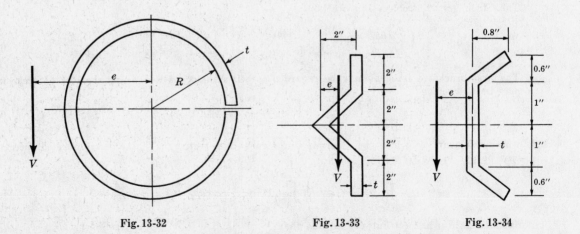

Fig. 13-32   Fig. 13-33   Fig. 13-34

13.12. Determine the shear center of the thin-walled "hat" section shown in Fig. 13-33.
Ans. $e = 0.51$ in.

13.13. Detemine the shear center of the thin-walled section indicated in Fig. 13-34.

   Ans. $e = 0.32$ in.

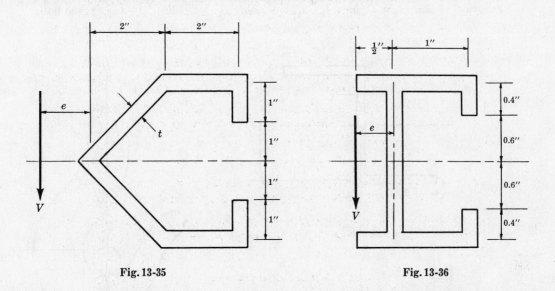

Fig. 13-35   Fig. 13-36

13.14. Determine the shear center of the thin-walled section in Fig. 13-35.   *Ans.* $e = 1.31$ in.

13.15. Determine the shear center of the thin-walled section in Fig. 13-36.   *Ans.* $e = 0.366$ in.

13.16. A wooden cantilever beam 12 ft long with cross-section 4 in. wide by 8 in. deep is subjected to a uniform load of 30 lb/foot of length. The load acts in a plane oriented at 20° to the vertical axis of symmetry of the section (see Fig. 13-37). Determine the maximum tension in the beam.

   Ans.   990 lb/in²

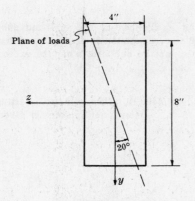

Fig. 13-37

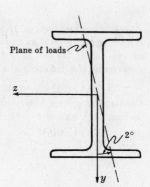

Fig. 13-38

13.17. A structural steel I-beam 10 in. deep and weighing 35 lb/ft is subjected to a bending moment lying in a plane oriented at 2° to the vertical axis of symmetry of the beam (see Fig. 13-38). Determine the percentage increase in elastic tensile stress over the stress that would exist if the moment acted in the vertical plane of symmetry. For this section $I_z = 145.8$ in⁴ and $I_y = 8.5$ in⁴.

   Ans.   29.6 percent

13.18. A structural steel Z-section with the dimensions shown in Fig. 13-39 is used as a beam 16 ft in length. It is subjected to a uniformly distributed vertical load of 50 lb/ft together with a centrally applied horizontal force of 120 lb. The end conditions correspond to a simple support in both the horizontal and vertical directions. Determine the maximum tensile stress in the beam. For this section $I_y = 4.2$ in^4, $I_z = 6.3$ in^4 and $I_{yz} = 4.46$ in^4. *Ans.* 34,800 lb/in^2

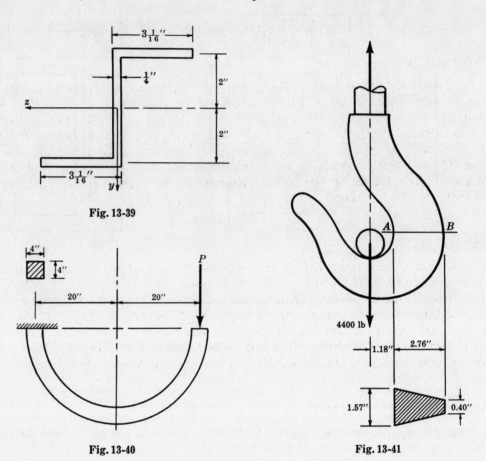

Fig. 13-39

Fig. 13-40

Fig. 13-41

13.19. A semicircular bar is of square cross-section and is clamped at one end and subject to a load P at the other end, as indicated in Fig. 13-40. The cross-section is 4 in. on a side and the radius of the bar is 20 in. If the maximum tensile stress at the support is not to exceed 28,000 lb/in^2, determine the maximum allowable value of the load P. *Ans.* 6460 lb

13.20. Consider a crane hook subject to a vertical load of 4400 lb. The contour is trapezoidal and at section AB has the dimensions shown in Fig. 13-41. Determine the tensile stress at point A. *Ans.* 13,100 lb/in^2

Chapter 14

Plastic Deformations of Beams

INTRODUCTION

In certain situations in structural design it is acceptable to permit a modest amount of permanent deformation of the structural element. If this is the case, then it is possible to permit loads greater than indicated by elastic theory, which permits no stress greater than the yield point of the material to develop at any point. This results in more efficient use of the material and is called *plastic design*. Fundamentally, this more efficient design is possible because of the ability of certain materials, such as structural steel, to undergo relatively large plastic deformations after the yield point has been reached. This is illustrated by the horizontal region of the stress-strain diagram shown in Fig. 1-5, page 3.

PLASTIC HINGE

As the transverse loads on a beam increase, yielding begins at the outer fibers at some critical station along the length of the beam and progresses rather rapidly toward the central fibers at this station. When finally all the fibers on one side of the neutral axis are in a state of tension corresponding to the yield point of the material and all those on the other side are in a state of compression, again at the yield point, then a flowing or *hinging* action occurs at that station and the bending moment transmitted across the *plastic hinge* remains constant. In this book a plastic hinge is denoted by a small, open circle.

FULLY PLASTIC MOMENT

The bending moment developed at a plastic hinge is termed a *fully plastic moment*. This concept was discussed in Chapter 8.

LOCATION OF PLASTIC HINGES

In general, plastic hinges form at points of maximum moment. For beams subject to concentrated forces and moments, the peak bending moment must always occur under one of these loadings or at some reaction and thus the plastic hinges must develop first at these points. In the case of distributed loads, the location of the plastic hinges is considerably more difficult to determine and often several possible points must be investigated. This is discussed in Problems 14.7 and 14.13.

COLLAPSE MECHANISM

When enough plastic hinges have formed in a structure to develop its full plastic load-carrying capacity, then portions of the structure (such as a beam or frame) between hinges may displace without any further increase of load, i.e., the portions between hinges behave as a *mechanism*. Essentially, the hinges allow a kinematic freedom of motion.

Under these conditions the shape of the deformed body may be characterized as a straight line between any pair of hinges. Typical representations of collapse mechanisms are shown in Problems 14.1, 14.4, and 14.6 through 14.12.

LIMIT LOAD

The external load sufficient to cause the structure to behave as a mechanism is termed the *limit load* or *collapse load*. Any design based upon the concept of development of a mechanism is termed *limit design*. All problems in this chapter illustrate computation of the limit load.

Solved Problems

14.1. Determine the limit load P of the simply supported beam shown in Fig. 14-1.

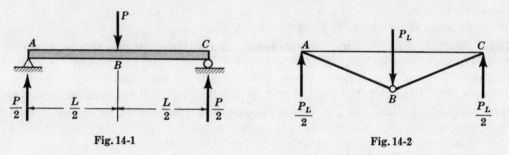

Fig. 14-1   Fig. 14-2

The maximum value of the bending moment in this beam occurs under the concentrated force P. The end reactions are each $P/2$ by symmetry irrespective of whether the beam is in the elastic or plastic state. The maximum bending moment (at the center) is thus $(P/2)(L/2)$ or $PL/4$. When this bending moment reaches a value corresponding to fully plastic action of the section of the beam at B, which we term M_p, a plastic hinge forms at B and the beam continues to deflect without further increase of load P. The collapse mechanism has the appearance of Fig. 14-2.

The value of the load P corresponding to this condition is termed the *limit load* P_L. The end reactions are each $P_L/2$ under this limit load so that $(P_L/2)(L/2) = M_p$ from which the limit load is $P_L = 4M_p/L$.

Dividing P_L by some suitable safety factor gives an allowable working load. This procedure is termed *limit design*.

14.2. The beam in Problem 14.1 is of rectangular cross-section 2×3 in. It is titanium, type Ti-8Mn, with a yield-point stress of 110,000 lb/in². If the length of the beam is 6 ft, determine the central force P necessary to develop the plastic hinge at B.

From Problem 8.23, the fully plastic moment for a rectangular cross-section is given by

$$M_p = \sigma_{yp} \frac{bh^2}{4}$$

Substituting:

$$M_p = 110,000 \frac{2(3)^2}{4} = 495,000 \text{ lb-in}$$

Using the result of Problem 14.1:

$$P_L = \frac{4M_p}{L} = \frac{4(495,000)}{6(12)} = 27,500 \text{ lb}$$

This is the limit load of the beam.

From Problem 8.23, the maximum elastic moment that this beam could withstand is given by

$$M_e = \sigma_{yp} \frac{bh^2}{6} = 330,000 \text{ lb-in}$$

from which the maximum allowable load P_e based upon elastic design is

$$P_e = \frac{4M_e}{L} = 18,200 \text{ lb}$$

For a bar of rectangular cross-section we thus have

$$\frac{M_p}{M_e} = \frac{P_L}{P_e} = \frac{3}{2}$$

That is, use of limit design as the criterion permits a load 50 percent greater than elastic analysis, where the yield-point stress is reached only in the outer fibers of the beam and all interior fibers are in the elastic range of action.

14.3. Determine the limit load of the simply supported beam subject to a uniformly distributed load.

According to the methods developed in Chapter 6, the peak bending moment occurs at the midpoint of the length of the beam and is given by $wL^2/8$. For fully plastic action at the midpoint, this moment is denoted by M_p. Thus, when the plastic hinge forms at the center, the uniform load has the value w_L (limit load) so that

$$\frac{w_L L^2}{8} = M_p \quad \text{or} \quad w_L = \frac{8M_p}{L^2}$$

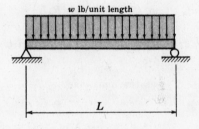

w lb/unit length

L

Fig. 14-3

The collapse mechanism has the same appearance as indicated in Fig. 14-2.

14.4. The beam shown in Fig. 14-4 is clamped at the left end, simply supported at the right, and subject to the concentrated load indicated. Determine the magnitude of the limit load P_L corresponding to plastic collapse.

This statically indeterminate beam cannot collapse plastically through formation of a single plastic hinge at B because the region AB is constrained to very small lateral deflections until another hinge forms somewhere along its length. It has been demonstrated in Chapter 6 that significant bending moments in a beam subject to concentrated forces always occur either at the points of application of these forces or where the reactions are applied. In the present case, this would imply the formation of another plastic hinge at A. With hinges at A and B, we have a so-called *kinematically admissible mechanism* of collapse. The order in which the plastic hinges are formed is of no consequence. The collapse mechanism appears in Fig. 14.5.

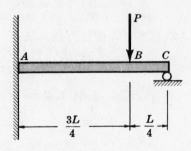

A P B C

$\dfrac{3L}{4}$ $\dfrac{L}{4}$

Fig. 14-4

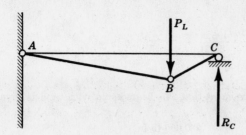

Fig. 14-5

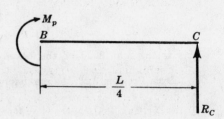

Fig. 14-6

The free-body diagram of the right portion of the beam, extending from C to a point just to the right of the applied load P when that force is the limit load P_L, is shown in Fig. 14-6, in which M_p denotes the fully plastic moment at B. From statics:

$$M_p - \frac{R_c L}{4} = 0 \quad \text{or} \quad R_c = \frac{4M_p}{L} \tag{1}$$

Next, from the free-body diagram of the entire beam, with plastic hinges at A and B, we have

$$\Sigma F_v = R_A + \frac{4M_p}{L} - P_L = 0$$

Hence

$$R_A = P_L - \frac{4M_p}{L} \tag{2}$$

$$\Sigma M_c = R_A L - M_p - P_L \left(\frac{L}{4}\right) = 0 \tag{3}$$

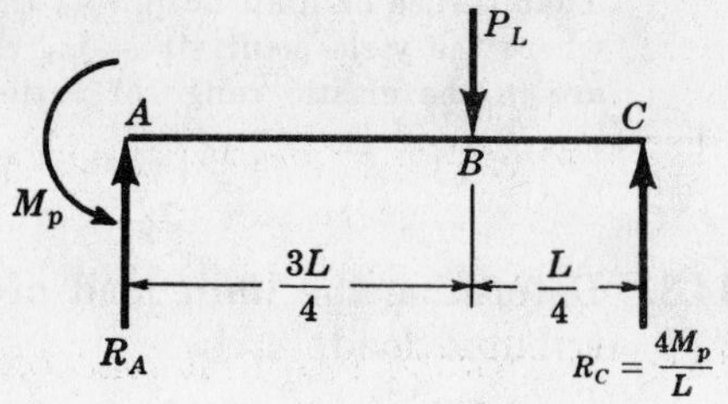

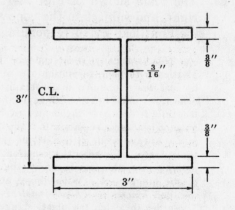

Fig. 14-7

Substituting R_A from (2) in (3) yields

$$P_L = \frac{20}{3} \frac{M_p}{L}$$

as the limit load.

14.5. The beam described in Problem 14.4 is a wide-flange section having the dimensions indicated in Fig. 14-8. For this section, determine the limit load P_L. The material is structural steel with a yield point of 38,000 lb/in² and the length of the beam is 5 ft.

As mentioned in Problem 8.27, for fully plastic action, the neutral axis divides the cross-sectional area into two parts of equal area. Here, because of the symmetry the neutral axis coincides with the centerline (C.L.) and the centroidal distances from that line are

$$\bar{y}_1 = \bar{y}_2 = \frac{(3)(\frac{3}{8})(\frac{3}{2} - \frac{3}{16}) + (1\frac{1}{8})(\frac{3}{16})(\frac{9}{16})}{(3)(\frac{3}{8}) + (1\frac{1}{8})(\frac{3}{16})} = 1.20 \text{ in.}$$

The fully plastic moment is thus

$$M_p = \sigma_{yp} \frac{A}{2} (\bar{y}_1 + \bar{y}_2)$$

$$= 38,000 \left(\frac{9}{8} + \frac{9}{8} \cdot \frac{3}{16}\right)(1.20 + 1.20)$$

$$= 122,000 \text{ lb-in}$$

Fig. 14-8

The limit load from Problem 14.4 is

$$P_L = \frac{20}{3} \frac{(122{,}000)}{(60)} = 13{,}600 \text{ lb}$$

It is of interest to carry out an elastic analysis of this same beam. In this case the outer fibers are taken to be stressed to the yield point and of course the stresses vary linearly over the depth, being zero at the neutral axis. The moment of inertia of the cross-section is found by the methods of Chapter 7 to be

$$I = \tfrac{1}{12}(3)(3)^3 - \tfrac{1}{12}(2\tfrac{13}{16})(2\tfrac{1}{4})^3 = 4.09 \text{ in}^4$$

and the outer fiber stresses are found from

$$\sigma_{yp} = \frac{M_e c}{I} \quad \text{or} \quad 38{,}000 = \frac{M_e(1.5)}{4.09}$$

and thus the maximum elastic moment M_e that the section can support is $M_e = 102{,}000$ lb-in. From Problem 12.1 the bending moment at point A is found to be $0.116PL$ while that at point B is $0.159PL$. Using the latter value we can find the maximum load that the beam can support for entirely elastic action to be

$$0.159 P_e L = 102{,}000 \quad \text{or} \quad P_e = 11{,}000 \text{ lb}$$

The load P_L corresponding to plastic collapse exceeds this value by 24 percent.

14.6. Determine the limit load of a clamped-end beam carrying a uniformly distributed load (Fig. 14-9).

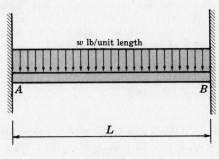

Fig. 14-9

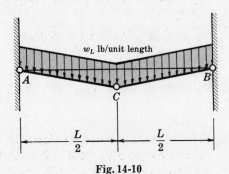

Fig. 14-10

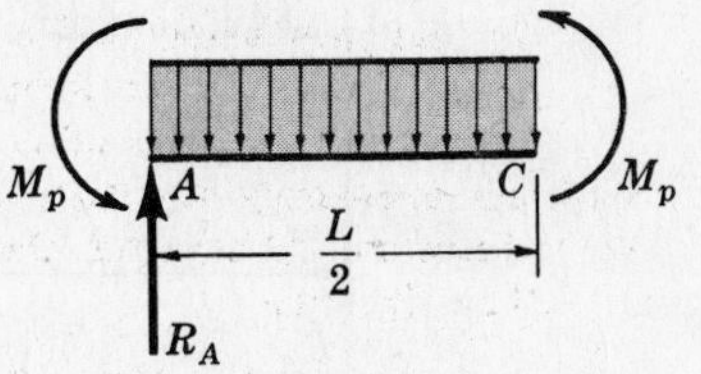

The collapse mechanism appears in Fig. 14-10 where plastic hinges have formed at points A, B, and C. By virtue of symmetry the shear is zero at the midpoint C; hence we may draw the free-body diagram of the left half of the beam as in Fig. 14-11. From statics:

$$\Sigma M_A = 2M_p - w_L\left(\frac{L}{2}\right)\left(\frac{L}{4}\right) = 0$$

Fig. 14-11

The limit load is thus $w_L = 16M_p/L^2$. From Problem 12.6 the permissible load based upon the outer fibers being at the yield point and all interior fibers acting in the elastic range of action is $w_e = 12M_e/L^2$ so that in this case the ratio of limit load w_L to maximum elastic load w_e is $\tfrac{4}{3}M_p/M_e$. However, the ratio M_p/M_e itself may be significant. For a rectangular cross-section it has the value 3/2, as indicated in Problem 14.2. For such a rectangular bar we then have

$$\frac{w_L}{w_e} = \frac{4}{3}\frac{M_p}{M_e} = \frac{4}{3}\left(\frac{3}{2}\right) = 2$$

indicating that in this particular case, limit design permits application of twice the load permitted by elastic analysis. This rather large variation between the permissible loads is due partially to the indeterminate nature of this beam. The variation found in Problem 14.2 for a determinate beam involved only the ratio M_p/M_e and was not nearly as great. However, it should be noted that there are exceptional cases where the limit load and maximum elastic load coincide even for an indeterminate system.

14.7. The beam shown in Fig. 14-12 is clamped at the left end, simply supported at the right, and subject to a uniformly distributed load. Determine the magnitude of this load corresponding to plastic collapse of the beam.

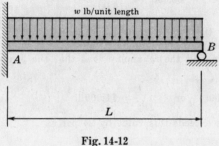

Fig. 14-12

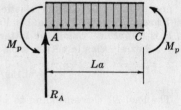

Fig. 14-13

This problem is somewhat analogous to Problem 14.4 because the beam cannot collapse plastically through formation of a single plastic hinge but instead, two hinges must form. One of these is obviously at the clamped end A but the location of the other is not immediately apparent. It of course occurs at the position of relative maximum moment (excluding point A) but that point is not known. However, since the shear is known to be zero at the point of maximum moment, we may draw the free-body diagram of the left region of the beam of length La and regard a as an unknown. It thus appears as in Fig. 14-13 where M_p denotes the fully plastic moment at each of the two sections.

From statics:

$$\Sigma F_v = R_A - w_L La = 0 \tag{1}$$

$$\Sigma M_A = 2M_p - \frac{w_L L^2 a^2}{2} = 0 \tag{2}$$

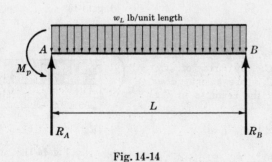

Fig. 14-14

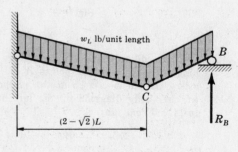

Fig. 14-15

Next, let us consider the free-body diagram of the entire beam. From statics:

$$\Sigma M_B = -R_A L + \frac{w_L L^2}{2} + M_p = 0 \tag{3}$$

Solving (1), (2), and (3) simultaneously we arrive at the single equation

$$a^2 - 4a + 2 = 0$$

for determination of the point of relative maximum moment. Solving; we obtain $a = 2 - \sqrt{2}$, the other root of the quadratic being of no physical significance.

Substituting this value in (2) we find

$$w_L = (6 + 4\sqrt{2})\frac{M_p}{L^2}$$

as the limit load. The collapse mechanism appears in Fig. 14-15.

14.8. The clamped-end beam is subject to a concentrated force as shown in Fig. 14-16. Determine the magnitude of this load corresponding to plastic collapse of the beam.

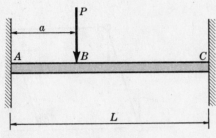

Fig. 14-16

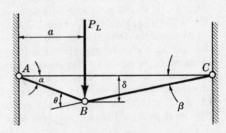

Fig. 14-17

The only logical collapse mechanism is that of Fig. 14-17, where plastic hinges form at A, B, and C. From the geometry of triangle ABC we have

$$\alpha + \beta = \theta \tag{1}$$

or

$$\frac{\delta}{a} + \frac{\delta}{L - a} = \theta \tag{2}$$

since the deflection δ is still small compared to L even though plastic collapse has occurred. Solving (2) we obtain

$$\delta = \theta a \left(1 - \frac{a}{L}\right) \tag{3}$$

and from geometry we have

$$\alpha = \theta \left(1 - \frac{a}{L}\right) \qquad \beta = \frac{\theta a}{L} \tag{4}$$

This problem could be solved by use of statics equations as employed in Problems 14.4 and 14.6. However, let us introduce another technique which will be well suited to even more complex problems. This involves a consideration of the work done by the load P_L after plastic collapse has occurred. If we assume that the elastic deflection is very small compared to the plastic deflection, then the work done by the load P_L during plastic collapse is $P_L\delta$. It is to be carefully noted that the load assumes the value P_L at the start of the collapse through the deflection δ and maintains this constant value throughout the collapse process. During the collapse, the beam develops the fully plastic moment M_p at each of the hinge points A, B, and C. The total energy dissipated at these hinges is provided by and is equal to the work done by the load P_L.

The work done by the plastic hinge at A is given by $M_p\alpha$, at B it is given by $M_p\theta$ and at C by $M_p\beta$. Thus, equating work done by P_L to the net work done by these three plastic moments, and using (4) we have

$$P_L\delta = M_p\theta\left(1 - \frac{a}{L}\right) + M_p\theta + M_p\left(\frac{\theta a}{L}\right) \tag{5}$$

Substituting δ from (3) we have as the collapse load:

$$P_L = \frac{2M_pL}{a(L-a)}$$

14.9. Figure 14-18(a) shows a square frame with both bases clamped subject to the single horizontal load P. Determine the magnitude of this load corresponding to plastic collapse of the frame.

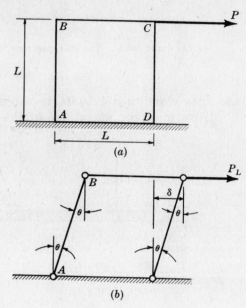

The plastic collapse mechanism is as indicated in Fig. 14-18(b) where plastic hinges have formed at each of the points $A, B, C,$ and D. As in Problem 14.8 it will be assumed that the elastic deflection is very small compared to the plastic deflection so that the work done by the load P_L during plastic collapse is $P_L\delta$. This work is equal to the energy dissipated at the plastic hinges at $A, B, C,$ and D. From the diagram, the work done by the plastic hinges at $A, B, C,$ and D is $M_p\theta$ at each hinge. Thus, the work-energy balance requires that

$$P_L\delta = 4M_p\theta$$

But $\theta = \delta/L$, so that the collapse load is

Fig. 14-18

$$P_L = \frac{4M_p}{L}$$

14.10. Consider the rectangular frame with both bases clamped subject to the two equal loads shown in Fig. 14-19. Determine the magnitude of the loads corresponding to plastic collapse of the frame.

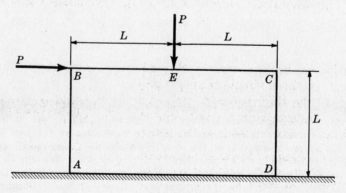

Fig. 14-19

In this situation there are three possible plastic collapse mechanisms. These are shown in Fig. 14-20 where Cases I and II correspond to individual actions of the applied loads and Case III is a composite mechanism formed as a combination of I and II so as to eliminate a plastic hinge at point B. We shall determine the collapse loads of each of these three cases and then select the minimum of the three loads as the correct one.

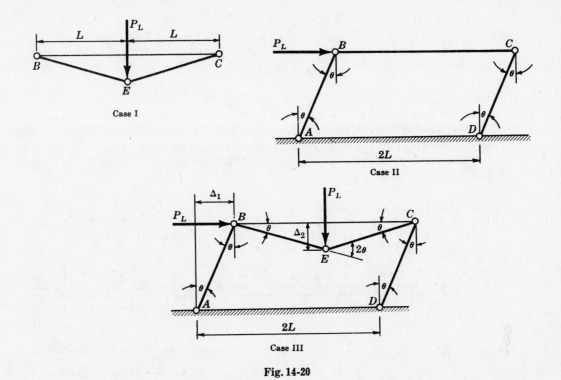

Fig. 14-20

Case I has already been treated in Problem 14.8, so that we immediately have $P_{L1} = 4M_p/L$.

Case II was treated in Problem 14.9, so for it we have $P_{L2} = 4M_p/L$.

For Case III there are plastic hinges at A, E, C, and D, with B constituting a rigid joint. Work-energy balance requires that

$$P_{L3}\Delta_1 + P_{L3}\Delta_2 = [M_p\theta]_A + [M_p(2\theta)]_E + [M_p(2\theta)]_C + [M_p\theta]_D$$

or

$$P_{L3}(L\theta) + P_{L3}(L\theta) = 6M_p\theta$$

from which $P_{L3} = 3M_p/L$.

Thus, the collapse load is $P_L = P_{L3} = 3M_p/L$ and collapse occurs as indicated by the sketch for Case III.

14.11. The continuous beam shown in Fig. 14-21(a) rests on three simple supports and is subject to the single concentrated load indicated. Determine the magnitude of this load for plastic collapse of the beam.

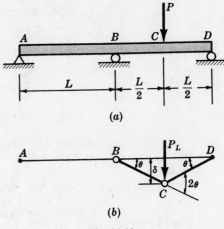

Fig. 14-21

The plastic collapse of such a beam usually occurs in only one of the spans and, in this case, collapse could occur by formation of a mechanism as indicated in Fig. 14-21(b), where plastic hinges form at points B and C.

The work done by the load P_L during plastic collapse is $P_L\delta$. The fully plastic moment M_p develops at each of the hinge points B and C. Work-energy balance requires that

$$P_L\delta = [M_p\theta]_B + [M_p(2\theta)]_C$$

or
$$P_L\left(\frac{L}{2}\theta\right) \;=\; 3M_p\theta$$

from which the collapse load is $P_L = 6M_p/L$.

14.12. A two-span continuous steel beam supports the concentrated forces indicated in Fig. 14-22(a). The beam is of rectangular cross-section, 2 in. wide by 4 in. high, with the yield point of the steel being 38,000 lb/in². Determine the value of P to cause plastic collapse.

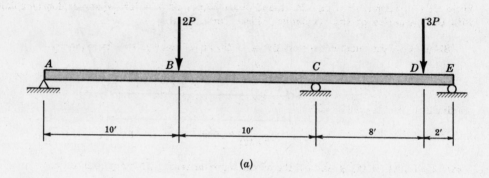

(a)

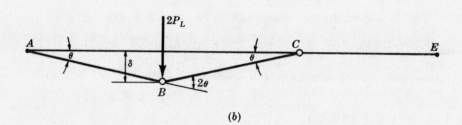

(b)

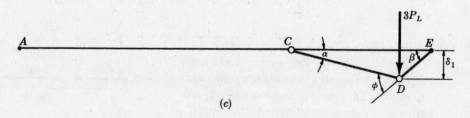

(c)

Fig. 14-22

Let us first assume that collapse occurs in the span AC with the formation of the mechanism indicated in Fig. 14-22(b). Fully plastic moments develop at B and C and the work-energy balance requires that

$$2P_L(10\theta) \;=\; [M_p(2\theta)]_B + [M_p\theta]_C \qquad \text{or} \qquad P_L \;=\; 3M_p/20$$

Next, consider the possibility of collapse in the span CE with the formation of the mechanism shown in Fig. 14-22(c). From the geometry of triangle CDE we have

$$\phi \;=\; \alpha + \beta$$

But since α is small compared to the span CE, this becomes

$$\frac{\delta_1}{8} + \frac{\delta_1}{2} = \phi$$

where δ_1 must of course be in consistent units (i.e. feet). Thus

$$\delta_1 = \tfrac{8}{5}\phi$$

and from geometry

$$\alpha = \tfrac{1}{5}\phi \qquad \beta = \tfrac{4}{5}\phi$$

In this case fully plastic moments develop at C and D and work-energy balance requires that

$$3P_L(8\alpha) = [M_p\phi]_D + [M_p\alpha]_C \quad \text{or} \quad P_L = M_p/4$$

Since this is larger than the P_L found for collapse of the left span, evidently collapse occurs with the formation of the mechanism shown for span AC.

Since the fully plastic moment for a rectangular cross-section is given by

$$M_p = \sigma_{yp}\left(\frac{bh^2}{4}\right)$$

we find the collapse load to be

$$P_L = \frac{3}{20(12)}(38,000)\frac{(2)(4)^2}{4} = 3800 \text{ lb}$$

where the factor of 12 appears in the denominator to render the units consistent.

14.13. A simply supported beam of 2×3 in. rectangular cross-section has a yield-point stress of 38,000 lb/in^2 and carries the loads indicated in Fig. 14-23(a). Use the limit design criterion to determine the maximum load P.

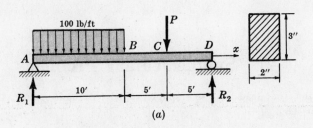

(a)

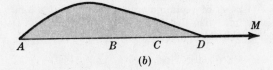

(b)

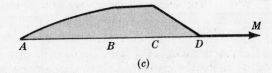

(c)

Fig. 14-23

From statics the reactions are $R_1 = 750 + (P/4)$ and $R_2 = 250 + (3P/4)$.

Fully plastic action of this beam corresponds to a moment of

$$M_p = \sigma_{yp} \left(\frac{bh^2}{4}\right) = 38,000 \frac{(2)(3)^2}{4} = 171,000 \text{ lb-in}$$

In any problem involving several loads, the location of the first plastic hinge to form is usually not apparent. Here, two possibilities exist. In the first [Fig. 14-23(b)], the maximum moment would occur between points A and B. If this is the correct form of the moment diagram then the shear must vanish at some point for which $x < 10$. Thus, since

$$V = 750 + \frac{P}{4} - 100x \qquad \text{for } 0 < x < 10$$

we must find P from the equation

$$0 = 750 + \frac{P}{4} - 100x \qquad \text{or} \qquad x = 7.5 + \frac{P}{400}$$

Since $x < 10$ in this consideration, this implies $P < 1000$ lb. For $P = 1000$ lb, $R_1 = 1000$ lb and the bending moment at B is given by

$$M_B = 1000(10) - 100(10)(5) = 5000 \text{ lb-ft} = 60,000 \text{ lb-in}$$

Although the moment slightly to the left of B may be somewhat greater than this value, it is apparent that the fully plastic moment cannot first form between A and B for any value of P.

For the second possibility [Fig. 14-23(c)], the maximum moment occurs at point C. The presence of a plastic hinge at C corresponds to a load P given by

$$\left(250 + \frac{3P}{4}\right)(5)(12) = 171,000 \qquad \text{or} \qquad P = 3460 \text{ lb}$$

In this case the moment at B must be less than that at C, since the moment diagram must have a common tangent to the two branches meeting at B. Hence there is no need to investigate the moment at B. Thus $P = 3460$ lb is the peak load that may be applied according to the limit design criterion.

Supplementary Problems

14.14. Determine the limit load P of the simply supported beam of Fig. 14-24.
Ans. $P_L = 4.5M_p/L$

14.15. The beam of Fig. 14-24 is of rectangular cross-section, 1×2 in. It is Hy-80 steel with a yield strength of 80,000 lb/in². The length of the beam is 4 ft. Determine the limit load when the loading is applied at the third point as indicated. Ans. $P_L = 7500$ lb

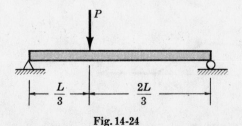

Fig. 14-24

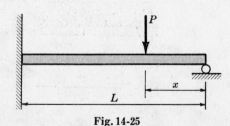

Fig. 14-25

14.16. The beam of Problem 14.3 is 5 ft long and of square cross-section 2×2 in. It is structural steel with a yield stress of 36,000 lb/in². Determine the limit load.   *Ans.* $w_L = 160$ lb/in

14.17. Determine the magnitude of the limit load P_L for the beam clamped at one end and simply supported at the other (Fig. 14-25).

Ans. $P_L = M_p \dfrac{L + x}{(Lx - x^2)}$

14.18. In Problem 14.17 determine x so that P_L is a minimum.

Ans. $x = 0.41L$, $(P_L)_{\min} = 5.64 M_p/L$

14.19. Determine the magnitude of each of the loads P in Fig. 14-26 for plastic collapse of the square frame having both bases clamped.   *Ans.* $P_L = 4M_p/L$

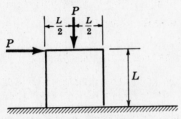

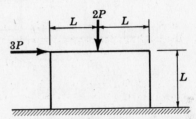

Fig. 14-26   Fig. 14-27

14.20. Determine the magnitude of P in Fig. 14-27 for plastic collapse of the rectangular frame having both bases clamped.   *Ans.* $P_L = 1.2M_p/L$

14.21. Determine the magnitude of P for plastic collapse of the rectangular frame having both bases pinned (Fig. 14-28).   *Ans.* $P_L = 4M_p/3L$

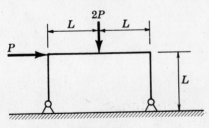

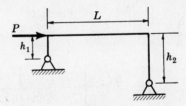

Fig. 14-28   Fig. 14-29

14.22. Determine the magnitude of the force P for plastic collapse of the unsymmetric frame having both bases pinned (Fig. 14-29).   *Ans.* $P_L = M_p(h_1 + h_2)/h_1 h_2$

In Problems 14.23 through 14.29 determine the magnitude of the load for plastic collapse of the system.

14.23.

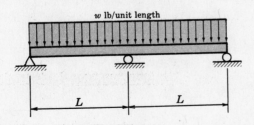

Ans. $w_L = (6 + 4\sqrt{2}) \dfrac{M_p}{L^2}$

Fig. 14-30

14.24.

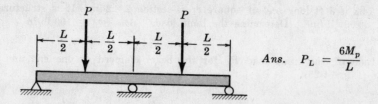

Fig. 14-31

$Ans. \quad P_L = \dfrac{6M_p}{L}$

14.25.

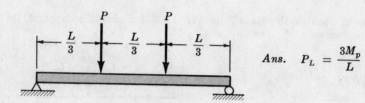

Fig. 14-32

$Ans. \quad P_L = \dfrac{3M_p}{L}$

14.26.

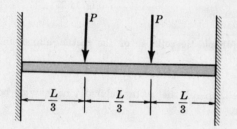

Fig. 14-33

$Ans. \quad P_L = \dfrac{6M_p}{L}$

14.27.

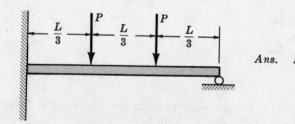

Fig. 14-34

$Ans. \quad P_L = \dfrac{4M_p}{L}$

14.28.

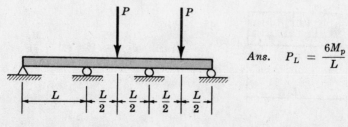

Fig. 14-35

$Ans. \quad P_L = \dfrac{6M_p}{L}$

14.29.

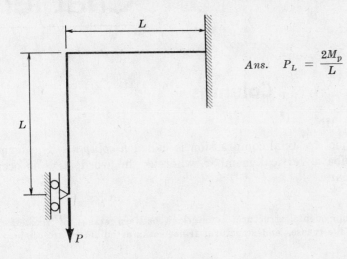

$Ans.$ $P_L = \dfrac{2M_p}{L}$

Fig. 14-36

Chapter 15

Columns

DEFINITION OF A COLUMN

A long slender bar subject to axial compression is called a *column*. The term *column* is frequently used to describe a vertical member, whereas the word *strut* is occasionally used in regard to inclined bars.

Examples.

Many aircraft structural components, structural connections between stages of boosters for space vehicles, certain members in bridge trusses and structural frameworks of buildings are common examples of columns.

TYPE OF FAILURE OF A COLUMN

Failure of a column occurs by buckling, i.e. by lateral deflection of the bar. In comparison it is to be noted that failure of a short compression member occurs by yielding of the material. Buckling, and hence failure, of a column may occur even though the maximum stress in the bar is less than the yield point of the material.

DEFINITION OF THE CRITICAL LOAD OF A COLUMN

The critical load of a slender bar subject to axial compression is that value of the axial force that is just sufficient to keep the bar in a slightly deflected configuration. Figure 15-1 shows a pin-ended bar in a buckled configuration due to the critical load P.

Fig. 15-1

SLENDERNESS RATIO OF A COLUMN

The ratio of the length of the column to the minimum radius of gyration of the cross-sectional area is termed the slenderness ratio of the bar. This ratio is of course dimensionless. The method of determining the radius of gyration of an area was discussed in Chapter 7.

If the column is free to rotate at each end, then buckling takes place about that axis for which the radius of gyration is a minimum.

CRITICAL LOAD OF A LONG SLENDER COLUMN

If a long slender bar of constant cross-section is pinned at each end and subject to axial compression, the load P_{cr} that will cause buckling is given by

$$P_{cr} = \frac{\pi^2 EI}{L^2}$$

where E denotes the modulus of elasticity, I the minimum moment of inertia of the cross-sectional area about an axis through the centroid, and L the length of the bar. The derivation of this formula is presented in Problem 15.1.

This formula was first obtained by the Swiss mathematician Leonhard Euler (1707-1783) and the load P_{cr} is called the *Euler buckling load*. As discussed in Problem 15.2, this expression is not immediately applicable if the corresponding axial stress, found from the expression $\sigma_{cr} = P_{cr}/A$, where A represents the cross-sectional area of the bar, exceeds the proportional limit of the material. For example, for a steel bar having a proportional limit of 30,000 lb/in², the above formula is valid only for columns whose slenderness ratio exceeds 100. The value of P_{cr} represented by this formula is a failure load; consequently, a safety factor must be introduced to obtain a design load. Applications of this expression may be found in Problems 15.7–15.9.

DESIGN OF ECCENTRICALLY LOADED COLUMNS

The derivation of the expression leading to the Euler buckling load assumes that the column is loaded perfectly concentrically. If the axial force P is applied with an eccentricity e the peak compressive stress in the bar occurs at the outer fibers at the midpoint and is given by

$$\sigma_{max} = \frac{P}{A}\left[1 + \frac{ec}{r^2}\sec\left(\frac{L}{2r}\sqrt{\frac{P}{AE}}\right)\right]$$

where c is the distance from the neutral axis to the outer fibers, r is the radius of gyration, L the length of the column and A the cross-sectional area. This is the *secant formula* for columns. It is discussed in detail in Problem 15.14.

INELASTIC COLUMN BUCKLING

The expression for the Euler buckling load may be extended into the inelastic range of action by replacing Young's modulus by the tangent modulus E_t. The resulting *tangent-modulus formula* is then

$$P_{cr} = \frac{\pi^2 E_t I}{L^2}$$

This is discussed in Problem 15.15.

DESIGN FORMULAS FOR COLUMNS HAVING INTERMEDIATE SLENDERNESS RATIOS

The design of compression members having large values of the slenderness ratio proceeds according to the Euler formula presented above, together with an appropriate safety factor. For the design of shorter compression members it is customary to employ any one of the many empirical formulas giving a relationship between the critical stress and the slenderness ratio of the bar. Actually, the formulas usually present an expression for the working stress as a function of the slenderness ratio, i.e. a safety factor has already been incorporated into the expression. Only two of the many existing empirical relations will be considered in this book.

The first, a so-called straight-line formula, originated in the Chicago Building Code and states that the allowable working stress in a column is given by

$$\sigma_w = 16,000 - 70(L/r)$$

where L/r represents the slenderness ratio of the bar. These specifications state that this expression is to be used only in the range $30 < L/r < 120$ for so-called main members and as high as $L/r = 150$ for so-called secondary members, i.e. bars used as lateral bracing between roof trusses, or bars to reduce the slenderness ratio of a long column by bracing it at some intermediate point. This formula is discussed in detail in Problem 15.16.

The second relation, to be found in the specifications of the American Institute of Steel Construction (A.I.S.C.), is a so-called parabolic formula and states that the allowable working stress in a column is given by

$$\sigma_w = 17,000 - 0.485(L/r)^2$$

provided L/r is less than 120. This expression is discussed in detail in Problem 15.17. For applications, see Problems 15.18 and 15.19.

The effect of each of these two expressions is to reduce the working stress in a column for increasing values of the slenderness ratio.

BEAM-COLUMN

Bars subjected to axial compression while simultaneously supporting lateral loads are termed *beam-columns*. An example is presented in Problem 15.21.

THE ELASTICA

The shape of the deflection curve of a buckled bar as determined by solution of the exact differential equation. An example is discussed in Problem 15.22.

BUCKLING DUE TO "FOLLOWER" FORCES

If a bar is loaded by a force which always acts tangentially to the deformed bar, the system is *nonconservative* and the approach utilized in deriving the Euler buckling load is no longer valid. In such a case a dynamic analysis is required to determine load-deflection relations. This is illustrated in Problem 15.23.

Solved Problems

15.1. Determine the critical load for a long slender pin-ended bar loaded by an axial compressive force at each end. The line of action of the forces passes through the centroid of the cross-section of the bar.

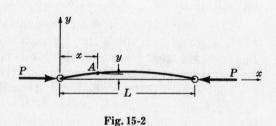

Fig. 15-2

The critical load is defined to be that axial force that is just sufficient to hold the bar in a slightly deformed configuration. Under the action of the load P the bar has the deflected shape shown in Fig. 15-2.

It is of course necessary that one end of the bar be able to move axially with respect to the other end in order that the lateral deflection may take place. The differential equation of the deflection curve is the same as that presented in Chapter 9, namely

$$EI \frac{d^2y}{dx^2} = M \tag{1}$$

Here the bending moment at the point A having coordinates (x, y) is merely the moment of the force P applied at the left end of the bar about an axis through the point A and perpendicular to the plane of the page. It is to be carefully noted that this force produces curvature of the bar that is concave downward, which, according to the sign convention of Chapter 6, constitutes negative bending. Hence the bending moment is $M = -Py$. Thus we have

$$EI \frac{d^2y}{dx^2} = -Py \tag{2}$$

If we set

$$P/EI = k^2 \tag{3}$$

(2) becomes

$$\frac{d^2y}{dx^2} + k^2y = 0 \tag{4}$$

This equation is readily solved by any one of several standard techniques discussed in works on differential equations. However, the solution is almost immediately apparent. We need merely find a function which when differentiated twice and added to itself (times a constant) is equal to zero. Evidently either $\sin kx$ or $\cos kx$ possesses this property. In fact, a combination of these terms in the form

$$y = C \sin kx + D \cos kx \tag{5}$$

may also be taken to be a solution of (4). This may be readily checked by substitution of y as given by (5) into (4).

Having obtained y in the form given in (5), it is next necessary to determine C and D. At the left end of the bar, $y = 0$ when $x = 0$. Substituting these values in (5) we obtain

$$0 = 0 + D \quad \text{or} \quad D = 0$$

At the right end of the bar, $y = 0$ when $x = L$. Substituting these values in (5) with $D = 0$ we obtain

$$0 = C \sin kL$$

Evidently either $C = 0$ or $\sin kL = 0$. But if $C = 0$ then y is everywhere zero and we have only the trivial case of a straight bar which is the configuration prior to the occurrence of buckling. Since we are not interested in this solution, then we must take

$$\sin kL = 0 \tag{6}$$

For this to be true, we must have

$$kL = n\pi \text{ radians } (n = 1, 2, 3, \ldots) \tag{7}$$

Substituting $k^2 = P/EI$ in (7) we find

$$\sqrt{\frac{P}{EI}} L = n\pi \quad \text{or} \quad P = \frac{n^2\pi^2 EI}{L^2} \tag{8}$$

The smallest value of this load P evidently occurs when $n = 1$. Then we have the so-called first mode of buckling where the critical load is given by

$$P_{cr} = \frac{\pi^2 EI}{L^2} \tag{9}$$

This is called Euler's buckling load for a pin-ended column. The deflection shape corresponding to this load is

$$y = C \sin \left(\sqrt{\frac{P}{EI}} \, x \right) \tag{10}$$

Substituting in this equation from (9) we obtain

$$y = C \sin \frac{\pi x}{L} \tag{11}$$

Thus the deflected shape is a sine curve. Because of the approximations introduced in the derivation of (1) it is not possible to obtain the amplitude of the buckled shape, denoted by C in (11).

As may be seen from (9), buckling of the bar will take place about that axis in the cross-section for which I assumes a minimum value.

15.2. Determine the axial stress in the column considered in Problem 15.1.

In the derivation of the equation $EI(d^2y/dx^2) = M$ used to determine the critical load in Problem 15.1, it was assumed that there is a linear relationship between stress and strain (see Problem 9.1). Thus the critical load indicated by (9) of Problem 15.1 is correct only if the proportional limit of the material has not been exceeded.

The axial stress in the bar immediately prior to the instant when the bar assumes its buckled configuration is given by

$$\sigma_{cr} = P_{cr}/A \qquad (1)$$

where A represents the cross-sectional area of the bar. Substituting for P_{cr} its value as given by (9) of Problem 15.1, we find

$$\sigma_{cr} = \pi^2 EI/AL^2 \qquad (2)$$

But from Chapter 7 we know that we may write

$$I = Ar^2 \qquad (3)$$

where r represents the radius of gyration of the cross-sectional area. Substituting this value in (2), we find

$$\sigma_{cr} = \pi^2 EAr^2/AL^2 = \pi^2 E(r/L)^2 \qquad (4)$$

or

$$\sigma_{cr} = \frac{\pi^2 E}{(L/r)^2} \qquad (5)$$

The ratio L/r is called the *slenderness ratio* of the column.

Let us consider a steel column having a proportional limit of 30,000 lb/in² and $E = 30 \times 10^6$ lb/in². The stress of 30,000 lb/in² marks the upper limit of stress for which (5) may be used. To find the value of L/r corresponding to these constants, we substitute in (5) and obtain

$$30,000 = \frac{\pi^2(30 \times 10^6)}{(L/r)^2} \qquad \text{or} \qquad L/r \approx 100$$

Thus for this material the buckling load as given by (9) of Problem 15.1 and the axial stress as given by (5) are valid only for those columns having $L/r \gtrsim 100$. For those columns having $L/r < 100$, the compressive stress exceeds the proportional limit before elastic buckling takes place and the above equations are not valid.

Equation (5) may be plotted as shown in Fig. 15-3. For the particular values of proportional limit and modulus of elasticity assumed above, the portion of the curve to the left of $L/r = 100$ is not valid. Thus for this material, point A marks the upper limit of applicability of the curve.

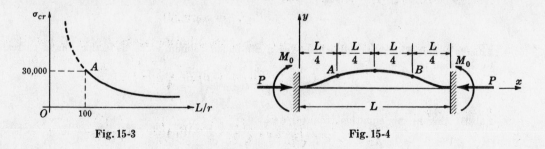

Fig. 15-3   Fig. 15-4

15.3. Determine the critical load for a long slender bar clamped at each end and loaded by an axial compressive force at each end.

The critical load is that axial compressive force P that is just sufficient to keep the bar in a slightly deformed configuration, as shown in Fig. 15-4. The moments M_0 at each end of the bar represent the actions of the supports on the column; these moments prevent any angular rotation of the bar at either end.

Inspection of the above deflection curve for the buckled column indicates that the central portion of the bar between points A and B corresponds to the deflection curve for the pin-ended column discussed in Problem 15.1. Thus for the fixed-end column, the length $L/2$ corresponds to the entire length L for the pin-ended bar. Hence the critical load for a clamped-end bar may be found from (9), Problem 15.1, by replacing L by $L/2$. This yields

$$P_{cr} = \frac{\pi^2 EI}{(L/2)^2} = \frac{4\pi^2 EI}{L^2}$$

Again, it is assumed that the maximum stress in the column does not exceed the proportional limit of the material.

The above formula, derived here on an intuitive basis, could be derived more rigorously by solving the usual differential equation for a bent bar. This is carried out in detail in Problem 15.4.

15.4. Determine the critical load for the long slender clamped-end bar described in Problem 15.3 by direct solution of the governing differential equation.

Let us introduce the x-y coordinate system shown in Fig. 15-4 and let (x, y) represent the coordinates of an arbitrary point on the bar. The bending moment at this point is found as the sum of the moments of the forces to the left of this section about an axis through this point and perpendicular to the plane of the page. Hence at this point we have $M = -Py + M_0$. The differential equation for the bending of the bar is then $EI(d^2y/dx^2) = -Py + M_0$, or

$$\frac{d^2y}{dx^2} + \frac{P}{EI}y = \frac{M_0}{EI} \tag{1}$$

As discussed in texts on differential equations, the solution to (1) consists of two parts. The first part is merely the solution of the so-called homogeneous equation obtained by setting the right hand side of (1) equal to zero. We must then solve the equation

$$\frac{d^2y}{dx^2} + \frac{P}{EI}y = 0 \tag{2}$$

But the solution to this equation has already been found in Problem 15.1 to be

$$y = A_1 \cos\left(\sqrt{\frac{P}{EI}}\,x\right) + B_1 \sin\left(\sqrt{\frac{P}{EI}}\,x\right) \tag{3}$$

The second part of the solution of (1) is given by a so-called particular solution, i.e. any function satisfying (1). Evidently one such function is given by

$$y = M_0/P \ (= \text{constant}) \tag{4}$$

The general solution of (1) is given by the sum of the solutions represented by (3) and (4), or

$$y = A_1 \cos\left(\sqrt{\frac{P}{EI}}\,x\right) + B_1 \sin\left(\sqrt{\frac{P}{EI}}\,x\right) + \frac{M_0}{P} \tag{5}$$

Consequently

$$\frac{dy}{dx} = -A_1\sqrt{\frac{P}{EI}}\,\sin\left(\sqrt{\frac{P}{EI}}\,x\right) + B_1\sqrt{\frac{P}{EI}}\,\cos\left(\sqrt{\frac{P}{EI}}\,x\right) \tag{6}$$

At the left end of the bar we have $y = 0$ when $x = 0$. Substituting these values in (5) we find $0 = A_1 + M_0/P$. Also, at the left end of the bar we have $dy/dx = 0$ when $x = 0$; substituting in (6) we obtain $0 = 0 + B_1\sqrt{P/EI}$ or $B_1 = 0$.

At the right end of the bar we have $dy/dx = 0$ when $x = L$; substituting in (6), with $B_1 = 0$, we find

$$0 = -A_1 \sqrt{\frac{P}{EI}} \sin \left(\sqrt{\frac{P}{EI}} L \right)$$

But $A_1 = -M_0/P$ and since this ratio is not zero, then $\sin (\sqrt{P/EI}\, L) = 0$. This occurs only when $\sqrt{P/EI}\, L = n\pi$ where $n = 1, 2, 3, \ldots$. Consequently

$$P_{cr} = \frac{n^2\pi^2 EI}{L^2} \tag{7}$$

For the so-called first mode of buckling illustrated in Problem 15.3 the deflection curve of the bent bar has a horizontal tangent at $x = L/2$, i.e. $dy/dx = 0$ there. Equation (6) now takes the form

$$\frac{dy}{dx} = \frac{M_0}{P} \left(\frac{n\pi}{L} \right) \sin \frac{n\pi x}{L} \tag{6'}$$

and since $dy/dx = 0$ at $x = L/2$, we find

$$0 = \frac{M_0}{P} \left(\frac{n\pi}{L} \right) \sin \frac{n\pi}{2}$$

The only manner in which this equation may be satisfied is for n to assume even values, i.e. $n = 2, 4, 6, \ldots$.

Thus for the smallest possible value of $n = 2$, equation (7) becomes

$$P_{cr} = \frac{4\pi^2 EI}{L^2}$$

This is the critical load for a clamped-end bar subject to axial compression. The result found by the less rigorous treatment of Problem 15.3 is thus confirmed.

15.5. Determine the critical load for a long slender bar clamped at one end, free at the other, and loaded by an axial compressive force applied at the free end.

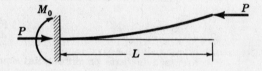

Fig. 15-5

The critical load is that axial compressive force P that is just sufficient to keep the bar in a slightly deformed configuration, as shown in Fig. 15-5. The moment M_0 represents the effect of the support in preventing any angular rotation of the left end of the bar.

Inspection of the above deflection curve for the buckled column indicates that the entire bar corresponds to one-half of the deflected pin-ended bar discussed in Problem 15.1. Thus for the column under consideration, the length L corresponds to $L/2$ for the pin-ended column. Hence the critical load for the present column may be found from equation (9), Problem 15.1, by replacing L by $2L$. This yields

$$P_{cr} = \frac{\pi^2 EI}{(2L)^2} = \frac{\pi^2 EI}{4L^2}$$

15.6. Determine the slenderness ratio for a timber column 8×10 in. in cross-section and 25 ft long.

As mentioned in Problem 15.1, buckling of such a bar will take place about that axis in the cross-section for which the moment of inertia assumes a minimum value. This moment of inertia for a rectangular area about an axis through its centroid is

$$I = bh^3/12 = 10(8^3)/12 = 426 \text{ in}^4$$

The cross-sectional area is 80 in²; hence the minimum radius of gyration is

$$r = \sqrt{I/A} = \sqrt{426/80} = 2.31 \text{ in.}$$

The slenderness ratio L is thus $\dfrac{L}{r} = \dfrac{25(12)}{2.31} = 130$.

15.7. A steel bar of rectangular cross-section 1.5 in. by 2 in. and pinned at each end is subject to axial compression. If the proportional limit of the material is 33,000 lb/in², and $E = 30 \times 10^6$ lb/in², determine the minimum length for which Euler's equation may be used to determine the buckling load.

The minimum moment of inertia is $I = \frac{1}{12}bh^3 = \frac{1}{12}(2)(1.5)^3 = 0.562$ in⁴.

Hence the least radius of gyration is $r = \sqrt{\dfrac{I}{A}} = \sqrt{\dfrac{0.562}{(1.5)(2)}} = 0.434$ in.

The axial stress for such an axially loaded bar was found in Problem 15.2 to be

$$\sigma_{cr} = \frac{\pi^2 E}{(L/r)^2}$$

The minimum length for which Euler's equation may be applied is found by placing the critical stress in the above formula equal to 33,000 lb/in². Doing this, we obtain

$$33,000 = \frac{\pi^2(30 \times 10^6)}{(L/0.434)^2} \quad \text{or} \quad L = 41.0 \text{ in.}$$

15.8. Consider again a rectangular steel bar 1.5 in. by 2 in. in cross-section, pinned at each end and subject to axial compression. The bar is 70 in. long and $E = 30 \times 10^6$ lb/in². Determine the buckling load using Euler's formula.

The minimum moment of inertia of this cross-section was found in Problem 15.7 to be 0.562 in⁴. Applying the expression for buckling load given in (9) of Problem 15.1, we find

$$P_{cr} = \frac{\pi^2 EI}{L^2} = \frac{\pi^2(30 \times 10^6)(0.562)}{(70)^2} = 34,000 \text{ lb}$$

The axial stress corresponding to this load is $\sigma_{cr} = \dfrac{P_{cr}}{A} = \dfrac{34,000}{(1.5)(2)} = 11,300$ lb/in².

15.9. Determine the critical load for a 10 WF 21 section acting as a pin-ended column. The bar is 12 ft long and $E = 30 \times 10^6$ lb/in². Use Euler's theory.

From the table of Fig. 8-67, page 154, we find the minimum moment of inertia to be 9.7 in⁴. This is the value that must be used in the expression for the buckling load. Thus

$$P_{cr} = \pi^2 EI/L^2 = \pi^2(30 \times 10^6)(9.7)/(144)^2 = 138,000 \text{ lb}$$

15.10. Consider the frame shown in Fig. 15-6, in which the horizontal beam BC of flexural rigidity EI_1 and length L_1 is supported by two columns each of flexural rigidity EI and length L. The columns are pinned at their lower ends, and the entire frame is free to displace laterally at the top. Determine the buckling load P.

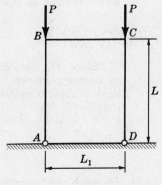

Fig. 15-6

Due to the vertical symmetry of the structure, each vertical member may be considered to be a column pinned at the lower end and elastically restrained at the upper end. Thus, in the buckled configuration the column AB has the appearance shown in Fig. 15-7. Its governing equation is

$$EI(d^2y/dx^2) = -Py$$

If $k^2 = P/EI$, then from Problem 15.1 we have

$$y = C \sin kx + D \cos kx$$

But when $x = 0$, $y = 0$, from which $D = 0$. For the column AB, the angular rotation at the upper end B is readily found to be

$$\left[\frac{dy}{dx}\right]_{x=L} = Ck \cos kL$$

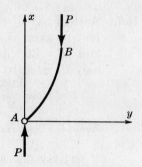

As shown in Fig. 15-8, the beam BC is subject to the action of the two end couples, each equal to $P[y]_{x=L}$, which together with vertical reactions create moment equilibrium.

Fig. 15-7

$$M_1 = P[y]_{x=L}$$

Fig. 15-8

It is possible to utilize the results obtained in Problems 9.10 and 9.28 to immediately write the equation of the deflection curve of the beam BC by superposition of those two solutions:

$$EI_1 y_1 = -\frac{M_1 x_1^3}{6L_1} + \frac{M_1 x_1^2}{2} - \frac{M_1 L_1 x_1}{3} - \frac{M_1 x_1^3}{6L_1} + \frac{M_1 L_1 x_1}{6}$$

From this

$$EI\frac{dy_1}{dx_1} = -\frac{M_1 x_1^2}{L_1} + M_1 x_1 - \frac{M_1 L_1}{6}$$

The slope at the left end B is thus

$$EI_1 \left[\frac{dy_1}{dx_1}\right]_{x_1=0} = -\frac{M_1 L_1}{6} = -\frac{PL_1[y]_{x=L}}{6}$$

But this end rotation must be numerically equal to that of the vertical column AB. Thus:

$$Ck \cos kL = \frac{PL_1[y]_{x=L}}{6EI_1} = \frac{PL_1 C \sin kL}{6EI_1}$$

which reduces to the following equation for kL:

$$kL \tan kL = \frac{6I_1 L}{IL_1}$$

Since $P = (kL)^2 EI/L^2$, the buckling load can be found by solving the above equation for any particular geometry. For example, if the vertical columns and the horizontal beam are of identical flexural rigidities, the equation for kL becomes $kL \tan kL = 6$. Such a transcendental equation is readily solved by trial-and-error to yield the smallest root as $kL = 1.35$, from which $P_{cr} = 1.82EI/L^2$. This is for the case $L_1 = L$.

15.11. Develop an approximate method of determining buckling loads of axially compressed bars utilizing energy methods.

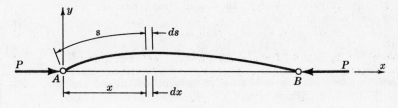

Fig. 15-9

Consider the pin-end axially compressed bar in Fig. 15-9. The bar is loaded by the axial forces P and it is assumed that end A does not move horizontally, but end B is free to approach A as the axial force progressively increases in magnitude. Let us denote by δ this horizontal displacement of B, δ being measured from the position of B just before buckling to the final position of B after buckling has occurred. It is evident that δ is equal to the difference between the arc length of the deflected bar and the chord AB, which is the horizontal projection of the arc. To determine δ we note that the difference between the length of an element of arc ds and an element of the chord dx is

$$ds - dx = \sqrt{(dx)^2 + (dy)^2} - dx = dx\sqrt{1 + \left(\frac{dy}{dx}\right)^2} - dx$$

$$= dx\left[1 + \frac{1}{2}\left(\frac{dy}{dx}\right)^2 - \cdots\right] - dx \approx \frac{1}{2}\left(\frac{dy}{dx}\right)^2$$

Integrating:

$$\delta = \frac{1}{2}\int_0^L \left(\frac{dy}{dx}\right)^2 dx \tag{a}$$

The axial force of course starts from a value of zero and gradually increases to a value just slightly less than the buckling load. It is not necessary to consider the elastic shortening of the bar during this phase of the process. Then, the right end of the bar moves to the left an amount δ as the bar goes into its laterally deflected configuration. Since the force P essentially remains constant during this axial shortening process, it does an amount of work given by

$$\frac{P}{2}\int_0^L \left(\frac{dy}{dx}\right)^2 dx \tag{b}$$

This work is stored within the bar as internal bending energy, found in Problem 16.3, page 295, to be

$$\int_0^L \frac{M^2\, dx}{2EI} \tag{c}$$

Thus

$$\int_0^L \frac{M^2\, dx}{2EI} = \frac{P}{2}\int_0^L \left(\frac{dy}{dx}\right)^2 dx \tag{d}$$

The procedure for obtaining an approximate solution for the buckling load is to assume a physically plausible deflection configuration compatible with the boundary conditions of the bar. This configuration, with y expressed as some function of x but with unknown amplitude, is substituted in equation (d) and the buckling load P determined.

Should one postulate the exact shape of the buckled configuration, one obtains the exact solution for the buckling load, agreeing with the result found by solving the governing differential equation. If any shape other than the exact one is employed, the buckling load obtained is always greater than the exact value. Unfortunately, it is not possible to estimate the magnitude of this discrepancy. However, in almost all cases reasonable shapes of the deflection configuration lead to results of satisfactory engineering accuracy.

15.12. Determine the buckling load for the axially compressed bar of variable cross-section shown in Fig. 15-10. The lower end of the bar is clamped, the upper end free to deflect laterally.

A plausible representation of the buckled configuration, indicated by the dashed curve, is

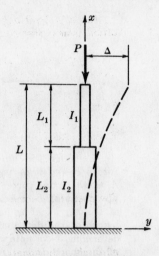

$$y = \Delta\left(1 - \cos\frac{\pi x}{2L}\right) \qquad (a)$$

This satisfies the boundary conditions of zero deflection and zero rotation at the base. Employing (d) of Problem 15.11 we obtain

$$\int_0^{L_2} \frac{M^2\,dx}{2EI_2} + \int_{L_2}^{L} \frac{M^2\,dx}{2EI_1} = \frac{P}{2}\int_0^{L}\left(\frac{dy}{dx}\right)^2 dx \qquad (b)$$

Substituting (a) into (b) and using $M = P(\Delta - y)$ we obtain

$$\frac{P^2\Delta^2}{2EI_2}\int_0^{L_2}\cos^2\frac{\pi x}{2L}\,dx + \frac{P^2\Delta^2}{2EI_1}\int_{L_2}^{L}\cos^2\frac{\pi x}{2L}\,dx = \frac{P\pi^2\Delta^2}{16L} \qquad (c)$$

Note that the amplitude factor Δ is common to all terms and thus cannot be determined. This is true in all such energy treatments. From (c) we find the approximate buckling load to be

Fig. 15-10

$$P_{cr} = \frac{\pi^2 EI_2}{4L^2}\ \frac{1}{\left[\dfrac{L_2}{L} + \dfrac{L_1}{L}\dfrac{I_2}{I_1} - \dfrac{1}{\pi}\left(\dfrac{I_2}{I_1}-1\right)\sin\dfrac{\pi L_2}{L}\right]}$$

For the case $I_1 = I_2$ this reduces to the result found in Problem 15.5.

15.13. A long thin bar of length L and rigidity EI is pinned at end A, and at the end B rotation is resisted by a restoring moment of magnitude λ per radian of rotation at that end. Derive the equation for the axial buckling load P. Neither A nor B can displace laterally, but A is free to approach B.

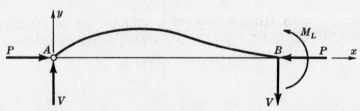

Fig. 15-11

The buckled bar is shown in Fig. 15-11, where M_L represents the restoring moment. The differential equation of the buckled bar is

$$EI\frac{d^2y}{dx^2} = Vx - Py$$

or

$$\frac{d^2y}{dx^2} + \frac{P}{EI}y = \frac{V}{EI}x$$

Let $\alpha^2 = P/EI$. Then

$$\frac{d^2y}{dx^2} + \alpha^2 y = \frac{V}{EI}x$$

The general solution of this equation is easily found to be

$$y = A\sin\alpha x + B\cos\alpha x + \frac{V}{P}x \qquad (a)$$

As the first boundary condition, when $x = 0$, $y = 0$, hence $B = 0$. As the second boundary condition, when $x = L$, $y = 0$, hence from (a) we obtain

$$0 = A \sin \alpha L + \frac{VL}{P} \quad \text{or} \quad \frac{V}{P} = -\frac{A}{L} \sin \alpha L$$

Thus

$$y = A \left[\sin \alpha x - \frac{x}{L} \sin \alpha L \right] \tag{b}$$

From (b) the slope at $x = L$ is found to be

$$\left[\frac{dy}{dx} \right]_{x=L} = A \left[\alpha \cos \alpha L - \frac{1}{L} \sin \alpha L \right] \tag{c}$$

The restoring moment at end B is thus

$$M_L = A\lambda \left[\alpha \cos \alpha L - \frac{1}{L} \sin \alpha L \right] \tag{d}$$

Also, since in general $M = EI(d^2y/dx^2)$, from (b) we have

$$M_L = -A\alpha^2 EI \sin \alpha L \tag{e}$$

Equating expressions (d) and (e) after carefully noting that as M_L increases dy/dx at that point decreases (necessitating the insertion of a negative sign), we have

$$-A\alpha^2 EI \sin \alpha L = -\left[A\lambda\alpha \cos \alpha L - \frac{A\lambda}{L} \sin \alpha L \right] \tag{f}$$

Simplifying, the equation for determination of the buckling load P becomes

$$\frac{PL}{\lambda} - \alpha L \cot \alpha L + 1 = 0 \tag{g}$$

This equation would have to be solved numerically for specific values of EI, L, and λ.

15.14. Consider an initially straight, pin-end column subject to an axial compressive force applied with known eccentricity e (see Fig. 15-12). Determine the maximum compressive stress in the column.

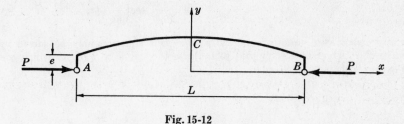

Fig. 15-12

The differential equation of the bar in its deflected configuration is

$$EI \frac{d^2y}{dx^2} = -Py$$

which has the standard solution

$$y = C_1 \sin \left(\sqrt{\frac{P}{EI}} x \right) + C_2 \cos \left(\sqrt{\frac{P}{EI}} x \right)$$

Since $y = e$ at each of the ends $x = -L/2$ and $x = L/2$, the values of the two constants of integration are readily found to be

$$C_1 = 0 \qquad C_2 = \frac{e}{\cos\left(\sqrt{\dfrac{P}{EI}}\dfrac{L}{2}\right)}$$

Thus, the deflection curve of the bent bar is

$$y = \frac{e}{\cos\left(\sqrt{\dfrac{P}{EI}}\dfrac{L}{2}\right)}\cos\left(\sqrt{\frac{P}{EI}}\,x\right)$$

The maximum value of deflection occurs at $x = 0$, by symmetry, and is

$$y_{max} = e\sec\left(\sqrt{\frac{P}{EI}}\frac{L}{2}\right)$$

Introducing the value of the critical load P_{cr} as given by (9) of Problem 15.1 this becomes

$$y_{max} = e\sec\left(\frac{\pi}{2}\sqrt{\frac{P}{P_{cr}}}\right)$$

Evidently the maximum deflection, which occurs at the center of the bar, becomes very great as the load P approaches the critical value. The phenomenon is one of gradually increasing lateral deflections, not buckling. The maximum compressive stress occurs on the concave side of the bar at C and is given by

$$\sigma_{max} = \frac{P}{A} + \frac{M_{max}c}{I} = \frac{P}{A} + \frac{Pec}{I}\sec\left(\frac{\pi}{2}\sqrt{\frac{P}{P_{cr}}}\right)$$

where c denotes the distance from the neutral axis to the outer fibers of the bar. If we now introduce the radius of gyration r of the cross-section, this becomes

$$\sigma_{max} = \frac{P}{A}\left[1 + \frac{ec}{r^2}\sec\left(\frac{L}{2r}\sqrt{\frac{P}{AE}}\right)\right]$$

This is the *secant formula* for an eccentrically loaded long column. In it, P/A is the average compressive stress. If the maximum stress is specified to be the yield point of the material, then the corresponding average compressive stress which will first produce yielding may be found from the equation

$$\frac{P_{yp}}{A} = \frac{\sigma_{yp}}{1 + \dfrac{ec}{r^2}\sec\left(\dfrac{L}{2r}\sqrt{\dfrac{P_{yp}}{AE}}\right)}$$

For any designated value of the ratio ec/r^2 this equation may be solved by trial-and-error and a curve of P/A versus L/r plotted to indicate the value of P/A at which yielding first begins in the extreme fibers.

15.15. Discuss column buckling when the average applied stress in the bar exceeds the proportional limit of the material.

In Problem 15.1 the critical load was determined upon the assumption that the entire cross-section of the bar is stressed within the elastic range, hence Hooke's law holds for all fibers and Young's modulus has its nominal (constant) value for the material under consideration. As early as 1889 the German engineer Engesser had suggested that E in (9) of Problem 15.1 be replaced by E_t, the tangent modulus of the material, i.e. the slope of the stress-strain curve at a particular point on the curve. The modified Euler equation thus becomes

$$P_{cr} = \frac{\pi^2 E_t I}{L^2} \tag{a}$$

and the corresponding axial stress in the bar immediately prior to buckling is given by

$$\sigma_{cr} = \frac{\pi^2 E_t}{(L/r)^2} \qquad\qquad (b)$$

This is the so-called tangent-modulus formula and the stress given by (b) and the corresponding axial force $P_{cr} = A\sigma_{cr}$ are termed the Euler-Engesser stress and load, respectively.

This Engesser load represents the maximum load for which the column has only one equilibrium configuration. Up to this load an initially straight column must remain straight. For any greater load it will deflect laterally. Since the tangent modulus decreases rapidly with increasing stress it is evident that the maximum load-carrying capacity of the column is only slightly greater than the Engesser load. This load is usually taken to be the upper limit of column strength.

Figure 15-3 of Problem 15.2 may thus be modified to account for inelastic effects in the region to the left of point A. The stress-strain curve of the material must of course be available and from it a series of values of E_t is determined for various values of σ. It is simplest to write (b) in the form

$$\frac{L}{r} = \pi \sqrt{\frac{E_t}{\sigma_{cr}}} \qquad\qquad (c)$$

and for a given pair of values of σ_{cr} and E_t determine L/r from (c). The result is shown as the region AB in the curve of Fig. 15-13, which corresponds to an aluminum alloy. Since in the elastic range of action the buckling load as predicted by the Engesser theory agrees with that predicted by the Euler theory, the region of the curve to the right of point A is identical in nature to that shown in Problem 15.2, i.e. simply the Euler curve.

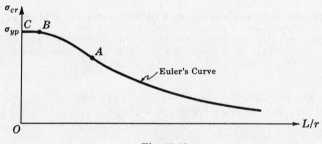

Fig. 15-13

In structural design the effects of inelastic action are also provided for by empirical formulas applicable in the region corresponding to AB above. Linear as well as parabolic empirical representations of this region are discussed in Problems 15.16 and 15.17 respectively.

For values of the slenderness ratio less than that corresponding to point B in Fig. 15-13, the critical stress may be considered to be equal to the yield point of the material. This is represented by the straight line BC in the figure. It is to be noted that no safety factor has been introduced in presenting these values. To obtain working stresses the ordinates to the above diagram must be divided by some number representing a factor of safety. Experimental evidence indicates that both the eccentricity of loading and the initial imperfections always present in the column tend to increase with increasing values of L/r. Hence a variable safety factor is frequently employed, ranging from 2.0 for very short bars to 3.5 for very long slender columns. Formulas for working stresses in columns are presented in Problems 15.16 and 15.17.

15.16. Discuss the various formulas relating to design of columns based upon straight-line relationships between working stress and slenderness ratio.

These linear or straight-line relations assume that the critical stress, when it exceeds the proportional limit of the material, may be represented by an equation of the form

$$\sigma_{cr} = a - b(L/r)$$

where a and b are constants depending upon the physical properties of the material. Such an expression may give the critical stress, or the values of the constants a and b may be adjusted so that a safe working stress is indicated by the above formula.

Usually the latter is the case. One of the most commonly used straight-line formulas is that given by the Chicago Building Code. This expression gives the safe working stress σ_w for structural steel in the following form:

$$\sigma_w = 16{,}000 - 70(L/r)$$

This expression is to be used for $30 < L/r < 120$ for main members and $30 < L/r < 150$ in so-called secondary members, such as lateral bracing in roof and bridge trusses. The same building code specifies a working stress of 14,000 lb/in² for bars having a slenderness ratio less than 30.

It is evident that the purpose of the above formula for the working stress is to reduce the critical compressive stress (based upon very short columns) for increasing values of the slenderness ratio.

15.17. Discuss the various formulas relating to design of columns based upon parabolic relationships between working stress and slenderness ratio.

These parabolic relations assume that the critical stress, when it exceeds the proportional limit of the material, may be represented by an equation of the form

$$\sigma_{cr} = a - b(L/r)^2$$

where again a and b are constants depending upon the physical properties of the material. Usually, the constants a and b are selected so as to make the parabola represented by the above equation tangent to Euler's curve and also to make the critical stress equal to the yield point of the material for very short bars. Again, as in Problem 15.16, such an expression may give the critical stress, or the values of a and b may be adjusted so that a safe working stress is indicated by the formula.

The latter case is exemplified by a formula suggested by the American Institute of Steel Construction (A.I.S.C.) wherein the working stress for structural steel columns is given by

$$\sigma_w = 17{,}000 - 0.485(L/r)^2$$

for the design of main members having L/r less than 120. For secondary members having $120 < L/r < 200$ these same specifications indicate the following formula:

$$\sigma_w = \frac{18{,}000}{1 + \dfrac{(L/r)^2}{18{,}000}}$$

These expressions, as well as those offered in Problem 15.16, assume hinged-end conditions. They may be used for other end conditions by using the modified length concepts mentioned in Problems 15.3 and 15.5.

15.18. What is the maximum safe axial compressive load that an 8 WF 40 section 18 ft long may carry if the bar is pinned at each end? Use the A.I.S.C. formula.

From the table on page 154 we find the minimum moment of inertia of this section to be 49.0 in⁴ and the cross-sectional area 11.76 in². Then the least $r = \sqrt{I/A} = \sqrt{49.0/11.76} = 2.04$ in.

The slenderness ratio is $L/r = 18(12)/2.04 = 106$.

The working stress is $\sigma_w = 17{,}000 - 0.485(L/r)^2 = 11{,}540$ lb/in².

Hence $P = A\sigma_w = 11.76(11{,}540) = 135{,}800$ lb.

15.19. Select a WF section to carry an axial compressive load of 160,000 lb. The bar is 16 ft 8 in. long. Use the A.I.S.C. specifications. The bar is pinned at each end.

The working stress is given by

$$\sigma_w = 17{,}000 - 0.485(L/r)^2 \tag{1}$$

Substituting the given values of P and L $(= 200$ in.) in this expression, we obtain

$$160{,}000/A \;=\; 17{,}000 - 0.485(200/r)^2 \tag{2}$$

The solution of this equation may be obtained by a trial-and-error method. As a first approximation, let us find the minimum area by setting the axial stress equal to 17,000 lb/in², even though this is, of course, greater than the allowable working stress. Doing this we find

$$160{,}000/A \;=\; 17{,}000 \qquad \text{or} \qquad A \;=\; 9.5 \text{ in}^2$$

Hence we need not consider any section having an area less than 9.5 in².

Let us first investigate a 10 WF 37 section. From the table on page 154 the minimum moment of inertia is 42.2 in⁴ and the cross-sectional area is 10.88 in². The least radius of gyration is consequently $r = \sqrt{42.2/10.88} = 1.97$ in. The slenderness ratio is thus $L/r = 200/1.97 = 102$. From (1) the allowable working stress in this bar is

$$\sigma_w \;=\; 17{,}000 - 0.485(102)^2 \;=\; 11{,}950 \text{ lb/in}^2$$

The maximum safe load is thus

$$P \;=\; A\sigma_w \;=\; 10.88(11{,}950) \;=\; 130{,}000 \text{ lb}$$

Since this is less than the design load of 160,000 lb, the section is too light.

Let us next investigate a 10 WF 45 section. According to the table on page 154 this section has a minimum moment of inertia of 53.2 in⁴ and a cross-sectional area of 13.24 in². The least radius of gyration is $r = \sqrt{53.2/13.24} = 2.00$ in. The slenderness ratio is $L/r = 200/2.00 = 100$. The allowable working stress is

$$\sigma_w \;=\; 17{,}000 - 0.485(100)^2 \;=\; 12{,}150 \text{ lb/in}^2$$

The maximum safe load is thus

$$P \;=\; A\sigma_w \;=\; 13.24(12{,}150) \;=\; 161{,}000 \text{ lb}$$

Since this value slightly exceeds the design load, the 10 WF 45 section is adequate.

15.20. Refer to Fig. 15-14. Two bars, AB and BC, are pinned at B as well as at each of the ends A and C. Each bar is initially of length L, and initially point B lies at a distance h above the line AC. The bars are identical, each having cross-sectional area A and Young's modulus E. A vertical force P is applied at B. Determine the relation between P and the displacement Δ under the point of application of P. Point B can only move vertically.

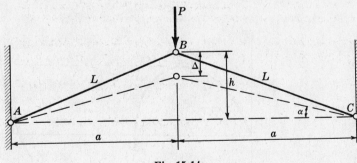

Fig. 15-14

Let us denote by S the axial force in each bar after the displacement Δ has occurred; also, let $L(1-\epsilon)$ denote the length of each bar at that same time. For equilibrium of the joint at B

$$S \;=\; \frac{P}{2 \sin \alpha}$$

Here α is the angle of the displaced system with the horizontal as shown.   From geometry:

$$S = \frac{PL(1-\epsilon)}{2(h-\Delta)} \tag{a}$$

Also from geometry:

$$[L(1-\epsilon)]^2 = a^2 + (h-\Delta)^2 = L^2 - 2h\Delta + \Delta^2 \tag{b}$$

from which

$$\epsilon = 1 - \frac{\sqrt{L^2 - 2h\Delta + \Delta^2}}{L} \tag{c}$$

Use of (a) together with $\sigma = \dfrac{S}{A}$, where σ denotes axial stress in the bar, and Hooke's law, $\sigma = E\epsilon$, now yields

$$P = 2AE\left(\frac{h-\Delta}{L}\right)\left(\frac{\epsilon}{1-\epsilon}\right) \tag{d}$$

which, with the use of (c), becomes

$$P = 2AE\left(\frac{h-\Delta}{L}\right)\left[\frac{L}{\sqrt{L^2 - 2h\Delta + \Delta^2}} - 1\right] \tag{e}$$

Equation (e) is the required relation.   Note that no assumptions have been made regarding the magnitude of the displacement Δ, i.e. it need not be small compared to h.   Equation (e) is plotted in Fig. 15-15 below and from this plot it may be seen that the curve has both a maximum and a minimum.   From (e) it is obvious that the curve intersects the horizontal axis when $\Delta = h$ and also when $\Delta = 2h$.   From the plot it may be seen that if the vertical force is gradually increased from a zero value to that corresponding to point a, a maximum value will be attained. After that the value of P decreases until at point b (when $\Delta = h$) no force is necessary to hold pin B in a horizontal line through A and C.   However, in this configuration, and in fact at all points from a to d on the curve, the state of equilibrium is *unstable* and hence the curve is shown dashed.   If the displacement Δ is further increased past b, then negative (upward) values of the vertical force are required, as represented by the portion bcd of the curve.   When $\Delta = 2h$ (point d) the bars have reached the image of their original configuration and again no force is necessary to hold the system in this position.   In this position, each of the bars is stress-free.   For still further downward vertical displacements the portion of the curve de represents the relationship between downward load P and vertical displacement.   The portions of the curve shown in solid lines correspond to a state of *stable* equilibrium.

From a practical standpoint, if the loading is a dead-weight applied at B, there are usually very small imperfections in the system of bars and possibly some slight initial vibratory motions that cause the system to "jump" from position a on the curve directly to f.   This is termed the "snap-through" phenomenon.

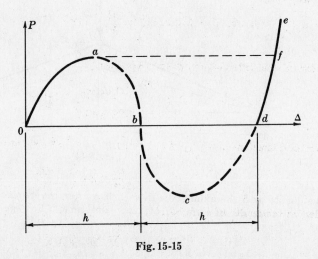

Fig. 15-15

15.21. Determine the deflection curve of a pin-ended bar subject to combined axial compression and a central transverse force.

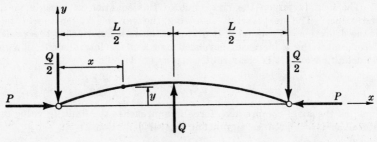

Fig. 15-16

    The free-body diagram of the loaded bar appears in Fig. 15-16. Because of the lateral force Q the bar is always deflected regardless of the magnitude of the axial force. The axial force of course influences the bending of the bar and such a system is termed a *beam-column*. The bar is shown with the unknown deflections in the positive y-direction and the Euler-Bernoulli equation of the deflection curve in the left half of the bar is

$$EI \frac{d^2y}{dx^2} = -Py - \frac{Q}{2}x \qquad 0 < x < L/2$$

If we set $P/EI = k^2$ this equation becomes

$$\frac{d^2y}{dx^2} + k^2y = -\frac{Q}{2P}k^2x$$

The solution of this equation may be found by the methods discussed in Problem 15.4 to be

$$y = A \cos kx + B \sin kx - \frac{Q}{2P}x \qquad (a)$$

from which

$$\frac{dy}{dx} = -Ak \sin kx + Bk \cos kx - \frac{Q}{2P} \qquad (b)$$

As the first boundary condition we have $y = 0$ when $x = 0$. Substituting this in (a) we obtain $A = 0$. For the second boundary condition we have $dy/dx = 0$ when $x = L/2$. Substituting this in (b) we obtain

$$0 = 0 + Bk \cos k\frac{L}{2} - \frac{Q}{2P} \qquad \text{or} \qquad B = \frac{Q}{2Pk \cos k\dfrac{L}{2}}$$

The equation of the deflection curve thus becomes

$$y = \frac{Q \sin kx}{2Pk \cos k\dfrac{L}{2}} - \frac{Q}{2P}x \qquad (c)$$

The maximum deflection $y = y_{max} = \Delta$ occurs at $x = L/2$ and is found to be

$$\Delta = \frac{Q}{2Pk} \tan k\frac{L}{2} - \frac{QL}{4P} \qquad (d)$$

Let us now introduce the new variable

$$u = \frac{kL}{2} = \frac{L}{2}\sqrt{\frac{P}{EI}}$$

We then have from (d)

$$\Delta = \left(\frac{QL^3}{48EI}\right)\frac{3(\tan u - u)}{u^3} \qquad (e)$$

The first factor on the right side of this equation represents the deflection due to the lateral load alone. The next factor, involving the variable u, represents the influence of the axial force on the deflection. From (e) it is evident that as u approaches the value $\pi/2$ the value of the entire factor containing u increases beyond bound and the lateral deflection increases indefinitely. From the definition of u it is clear that when $u = \pi/2$

$$P = \frac{\pi^2 EI}{L^2} \qquad (f)$$

Thus, as the axial compressive force P approaches the limiting value given by (f), even the smallest lateral load will produce significant lateral deflection.

15.22. Obtain the load-deflection relation for a pin-ended column subject to axial compression and undergoing finite lateral displacements.

The treatment presented in Problem 15.1 is restricted to extremely small lateral deflections because this was the assumption made in deriving equation (1), the Euler-Bernoulli equation. To obtain a more general representation let us introduce the angular coordinate θ and arc length s, in addition to the x, y coordinates (see Fig. 15-17).

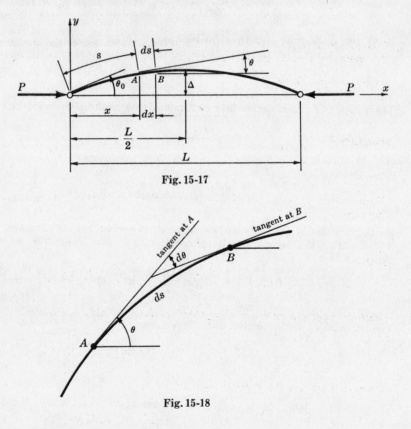

Fig. 15-17

Fig. 15-18

An enlarged view of the deformed bar illustrates the angular coordinates more clearly (Fig. 15-18). Note that $d\theta$ is negative. Let us now examine an element of arc length ds bounded by two adjacent cross-sections of the bar. Prior to loading these cross-sections are parallel to each other but after the bar has deflected laterally they have the appearance shown in Fig. 15-19 in which they subtend a central angle $d\theta$. In a manner similar to that used in Problem 8.1, page 126, we may determine the normal strain of a fiber a distance y from the neutral surface to be

$$\epsilon \;=\; \frac{y\,d\theta}{ds} \;=\; \frac{\sigma}{E}$$

where σ is the longitudinal stress acting on this fiber. But from Problem 8.1 we have $\sigma = My/I$. Thus

$$\frac{y\,d\theta}{ds} \;=\; \frac{My}{EI}$$

or, since $M = -Py$ for the bar,

$$\frac{d\theta}{ds} \;=\; -\frac{Py}{EI} \qquad\qquad (a)$$

If we let $\alpha^2 = P/EI$ then

$$\frac{d\theta}{ds} \;=\; -\alpha^2 y \qquad\qquad (b)$$

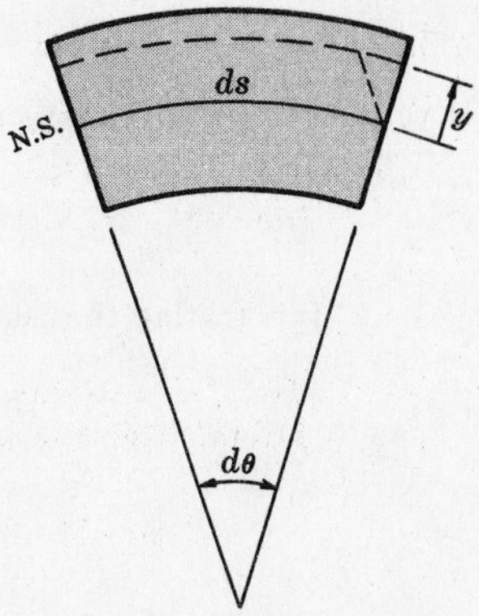

Fig. 15-19

from which

$$\frac{d^2\theta}{ds^2} \;=\; -\alpha^2 \frac{dy}{ds} \;=\; -\alpha^2 \sin\theta \qquad\qquad (c)$$

This equation is valid for large, finite lateral deflections of the bar in contrast to (1) of Problem 15.1 which is limited to very small values of deflection.   To solve (c) let us multiply through by the integrating factor $2\dfrac{d\theta}{ds}$:

$$2\,\frac{d\theta}{ds}\frac{d^2\theta}{ds^2} \;=\; -2\alpha^2(\sin\theta)\frac{d\theta}{ds} \qquad\qquad (d)$$

Integrating:

$$\left(\frac{d\theta}{ds}\right)^2 \;=\; 2\alpha^2\cos\theta \,+\, c_1 \qquad\qquad (e)$$

When $x = 0$, $\theta = \theta_0$ (the initial slope) and at this same point $y = 0$, hence $\dfrac{d\theta}{ds} = 0$ from (b). Thus, from (e),

$$0 \;=\; 2\alpha^2\cos\theta_0 \,+\, c_1$$

so that

$$\frac{d\theta}{ds} \;=\; -\sqrt{2}\,\alpha\,\sqrt{\cos\theta - \cos\theta_0} \qquad\qquad (f)$$

where the negative square root is taken because $d\theta$ is always negative.   This may be transformed to

$$\frac{d\theta}{ds} \;=\; -2\alpha\,\sqrt{\sin^2\frac{\theta_0}{2} - \sin^2\frac{\theta}{2}} \qquad\qquad (g)$$

We next introduce the change of variables

$$\sin\frac{\theta}{2} \;=\; k\sin\phi \qquad\qquad (h)$$

where ϕ is a parameter assuming the value $\pi/2$ when $x = 0$ and the value 0 when $x = L/2$, from which

$$k \;=\; \sin\frac{\theta_0}{2} \qquad\qquad (i)$$

Then $\theta \;=\; 2 \ \text{arc} \ \sin\,(k \ \sin\,\phi)$

and $d\theta \;=\; \dfrac{2k \ \cos\,\phi \ d\phi}{\sqrt{1 - k^2 \ \sin^2\,\phi}}$ (j)

From (g), (h), (i), and (j) we have

$$\frac{d\phi}{\sqrt{1 - k^2 \ \sin^2\,\phi}} \;=\; -\alpha \ ds \tag{k}$$

Integrating the last equation and remembering the definition of ϕ at its endpoint values:

$$\alpha \int_0^{L/2} ds \;=\; -\int_{\pi/2}^0 \frac{d\phi}{\sqrt{1 - k^2 \ \sin^2\,\phi}}$$

or $\alpha\,\dfrac{L}{2} \;=\; \displaystyle\int_0^{\pi/2} \dfrac{d\phi}{\sqrt{1 - k^2 \ \sin^2\,\phi}}$ (l)

The right-hand side of (l) is termed the *complete elliptic integral of the first kind* with modulus k and argument ϕ. Tabulated values of the integral for any specified value of k are readily available; see for example B. O. Peirce, *A Short Table of Integrals*, 4th ed., Ginn, 1957. To employ these tables we must select a value of θ_0 thus fixing k from equation (i). Then (l) may be rewritten in the form

$$P \;=\; \frac{4EI}{L^2}\left[\int_0^{\pi/2} \frac{d\phi}{\sqrt{1 - k^2 \ \sin^2\,\phi}}\right]^2 \tag{m}$$

to determine the axial load P corresponding to this assumed value of θ_0. To find the maximum deflection occurring at $x = L/2$ we have from geometry

$$\frac{dy}{ds} \;=\; \sin\,\theta \;=\; 2 \ \sin\frac{\theta}{2}\,\cos\frac{\theta}{2} \tag{n}$$

From (k) this becomes

$$\frac{dy}{ds} \;=\; -\,\frac{\alpha \ dy \ \sqrt{1 - k^2 \ \sin^2\,\phi}}{d\phi} \tag{o}$$

Equating the right sides of (n) and (o):

$$\frac{-\alpha \ dy \ \sqrt{1 - k^2 \ \sin^2\,\phi}}{d\phi} \;=\; 2k(\sin\,\phi) \ \sqrt{1 - k^2 \ \sin^2\,\phi}$$

or $\alpha \ dy \;=\; -2k \ \sin\,\phi \ d\phi$ (p)

Integrating: $\alpha y \;=\; 2k \ \cos\,\phi \,+\, c_2$

When $y = 0$, $\phi = \dfrac{\pi}{2}$ from which $c_2 = 0$. When $x = L/2$, $\phi = 0$ and $y = y_{\max} = \Delta$. Thus $\alpha\Delta = 2k$ or

$$\Delta \;=\; \frac{2k}{\alpha} \;=\; \frac{2k}{\sqrt{\dfrac{P}{EI}}} \;=\; \frac{kL}{\displaystyle\int_0^{\pi/2} \dfrac{d\phi}{\sqrt{1 - k^2 \ \sin^2\,\phi}}} \tag{q}$$

The procedure is as follows:

(1) Select a value of θ_0 and determine k from equation (i).

(2) Ascertain the value of $\displaystyle\int_0^{\pi/2} \dfrac{d\phi}{\sqrt{1 - k^2 \ \sin^2\,\phi}}$ from tabulated values in, for example, B. O. Peirce, and then calculate the axial force P corresponding to this value of θ_0 from equation (m).

(3) Calculate the central deflection Δ from equation (q).

θ_0	k	$\int_0^{\pi/2} \dfrac{d\phi}{\sqrt{1-k^2\sin^2\phi}}$	PL^2/EI	Δ/L
0°	0	$\pi/2$	9.87 $(=\pi^2)$*	0
40°	0.342	1.6200	10.50	0.211
80°	0.643	1.7868	12.75	0.360
120°	0.866	2.1565	18.56	0.403
160°	0.985	3.1534	39.76	0.313

Fig. 15-20

Results of this computation for selected values appear in the table of Fig. 15-20, in which the starred value 9.87 ($=\pi^2$) indicates that the simple theory of Problem 15.1 actually gives an exact result if it is assumed that the end slopes are zero.

From the above the progressive states of deformation of the bar are as shown in Fig. 15-21.

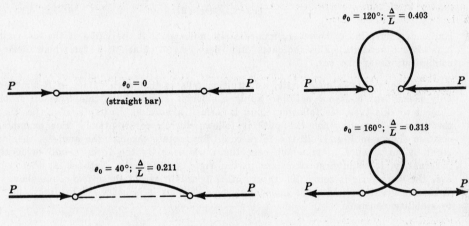

Fig. 15-21

This problem was first investigated by L. Euler in 1744 and the shape of the elastic curve is termed the *elastica*. It is only through use of this more exact finite-deflection theory that the amplitude of the lateral deflection may be determined. The approximate small-deflection treatment of Problem 15.1 does not permit determination of this quantity.

15.23. Determine the critical load for an initially straight bar which is clamped at one end, completely free to translate and rotate at the other, and at this free end is loaded by a single force that always acts tangentially to the deformed bar.

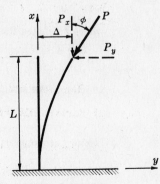

Fig. 15-22

This problem is significantly different from Problem 15.5, in which the force remained at all times directed parallel to the undeflected configuration of the bar. Let us first approach the present problem in a fashion analogous to that employed in Problem 15.1, i.e. we will write the differential equation of the deflected bar using the Euler-Bernoulli relation $M = EI(d^2y/dx^2)$. Noting that for small deflections $P_x \approx P$ and $P_y \approx P\phi$ where ϕ is the angle of rotation of the end, we have

$$EI\frac{d^2y}{dx^2} \;=\; P(\Delta - y) \,-\, P\phi(L - x) \tag{a}$$

If $k^2 = P/EI$, the solution of (a) is

$$y \;=\; c_1 \sin kx \,+\, c_2 \cos kx \,-\, \phi(L - x) \,+\, \Delta \tag{b}$$

For boundary conditions we have

At $x = 0$, $y = 0$: $\qquad\qquad\qquad 0 = 0 + c_2 - \phi L + \Delta \tag{c}$

At $x = 0$, $dy/dx = 0$: $\qquad\qquad 0 = c_1 k + \phi \tag{d}$

At $x = L$, $y = \Delta$: $\qquad\qquad 0 = c_1 \sin kL + c_2 \cos kL \tag{e}$

At $x = L$, $dy/dx = \phi$: $\qquad\quad 0 = c_1 \cos kL - c_2 \sin kL \tag{f}$

This may be regarded as a system of four equations in the four unknowns c_1, c_2, ϕ and Δ. Since the system is homogeneous the condition for existence of a nontrivial solution is the vanishing of the determinant of the coefficients of the unknowns, which is

$$\begin{vmatrix} 0 & 1 & 1 & -L \\ k & 0 & 0 & 1 \\ \sin kL & \cos kL & 0 & 0 \\ \cos kL & -\sin kL & 0 & 0 \end{vmatrix}$$

Expansion by minors, however, immediately indicates that the value of the determinant is -1, i.e. it is always nonzero. This indicates that there are no values of P for which there exist deformed configurations of the bar.

Physically this is unrealistic, as intuition tells us that such values of P must exist. This can only mean that it is *not valid* to apply the Euler approach to this type of problem. The bar subject to this type of "follower" load is termed a *nonconservative* system, i.e. the work done by the force P depends upon the path it follows during deformation. For example, to reach the position shown in Fig. 15-22 the force could first rotate through the angle ϕ, then translate to the right by an amount Δ, in which case the work done by the force would evidently be negative. Alternately, it could first translate to the right through the distance Δ, then rotate, in which case the work done is zero. There are other possibilities but these two alone show that the system is nonconservative. In such nonconservative systems it is *not* valid to apply the Euler approach to stability problems.

Instead it is necessary to turn to a dynamic-type analysis of this problem. Let us assume that the distributed mass of the bar is, for simplicity, replaced by a concentrated mass m at its upper end. Then, the equation governing small lateral motions of the bar may be written directly from the Euler-Bernoulli equation by including in M the contribution of an inertia force on m:

$$EI\frac{\partial^2 y}{\partial x^2} \;=\; P(\Delta - y) \,-\, P\phi(L - x) \,-\, m\frac{\partial^2 \Delta}{\partial t^2}(L - x) \tag{g}$$

To solve this equation, let us take

$$y \;=\; Y(x)e^{i\omega t}; \quad \Delta(t) \;=\; Fe^{i\omega t}; \quad \phi(t) \;=\; \Phi e^{i\omega t} \tag{h}$$

where t denotes time. Substituting these assumed forms into (g), we find

$$\frac{d^2Y}{dx^2} + k^2Y \;=\; k^2F \,-\, k^2\Phi(L - x) \,+\, \frac{m\omega^2 F}{EI}(L - x) \tag{i}$$

which has the solution

$$Y(x) \;=\; c_1 \sin kx \,+\, c_2 \cos kx \,+\, F \,-\, \Phi(L - x) \,+\, \frac{m\omega^2 F}{k^2 EI}(L - x) \tag{j}$$

If one now imposes the boundary conditions (c) through (f), but now expressed in the form

$$Y(0) = 0; \quad Y'(0) = 0; \quad Y(L) = F; \quad Y'(L) = \Phi$$

we again obtain a system of four equations in the four unknowns c_1, c_2, F, and Φ. The determinant of the coefficients of these unknowns must vanish. Thus:

$$
\begin{vmatrix}
0 & 1 & 1 + \dfrac{m\omega^2 L}{k^2 EI} & -L \\[2ex]
k & 0 & -\dfrac{m\omega^2}{k^2 EI} & 1 \\[2ex]
\sin kL & \cos kL & 0 & 0 \\[2ex]
k\cos kL & -k\sin kL & -\dfrac{m\omega^2}{k^2 EI} & 0
\end{vmatrix} = 0
$$

which, upon expansion by minors, leads to

$$
\omega = \sqrt{\frac{P}{mL}\ \frac{1}{\left(\dfrac{\sin kL}{kL} - \cos kL\right)}} \tag{k}
$$

As the force P increases from a value of zero, the natural frequencies ω increase in magnitude. They increase beyond bound when the denominator of (k) vanishes, i.e. when

$$
\tan kL = kL \tag{l}
$$

The smallest root of (l) may be found by trial-and-error to be $kL = 4.493$. Then, since $k^2 = P/EI$ we immediately have the critical value of the "follower" force to be

$$
P_{cr} = \frac{20.19 EI}{L^2} \tag{m}
$$

A "follower"-type force could, for example, be produced by the reaction of a jet attached to the end of a bar or in links in automatic control systems.

Supplementary Problems

15.24. A steel bar of solid circular cross-section is 2 in. in diameter. The bar is pinned at each end and subject to axial compression. If the proportional limit of the material is 36,000 lb/in² and $E = 30 \times 10^6$ lb/in², determine the minimum length for which Euler's formula is valid. Also, determine the value of the Euler buckling load if the column has this minimum length.
Ans. 45.5 in., 112,800 lb

15.25. Determine the slenderness ratio for a steel column of solid circular cross-section 4 in. in diameter and 9 ft long.   *Ans.* 108

15.26. According to the A.I.S.C. specifications, what is the load-carrying capacity of the column described in Problem 15.25? The bar is pin-ended.   *Ans.* 143,000 lb

15.27. Select a WF section to carry an axial compressive load of 120,000 lb. The bar is pin-ended and is 14 ft long. Use the A.I.S.C. specifications. *Ans.* 8 WF 31

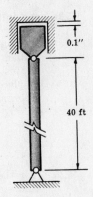

15.28. Select a WF section adequate to support a concentric load of 52,000 lb, together with an eccentric load of 40,000 lb applied 4 in. from the center of the section at a point on the axis of symmetry bisecting the width of the web of the beam. The bar is 18 ft long and has pinned ends. Use the A.I.S.C. specifications.
Ans. 12 WF 40

15.29. The column shown in Fig. 15-23 is pinned at both ends and is free to expand into the opening at the upper end. The bar is steel, 1 in. in diameter and occupies the position shown at 60°F. Determine the highest temperature to which the column may be heated before it will buckle. Take $\alpha = 6 \times 10^{-6}/\text{F}°$ and $E = 30 \times 10^6$ lb/in². Neglect the weight of the column. *Ans.* 95.2°F

Fig. 15-23

15.30. A bar of length L is clamped at its lower end and subject to both vertical and horizontal forces at the upper end, as shown in Fig. 15-24. The vertical force P is equal to one-fourth of the Euler load for this bar. Determine the lateral displacement of the upper end of the bar.

Ans. $16(4 - \pi)RL^3/\pi^3EI$

15.31. Consider a column clamped at one end and pinned at the other, i.e. free to rotate and move axially, but not laterally. Determine an approximate value of the axial force required to cause buckling. Use the energy method, taking as the deflected configuration the shape assumed by a beam with these end conditions and loaded by a uniformly distributed transverse load.

Ans. $P_{cr} = 21EI/L^2$

(*Note.* The solution of the governing differential equation of this problem yields a value $P_{cr} = 20.19EI/L^2$.)

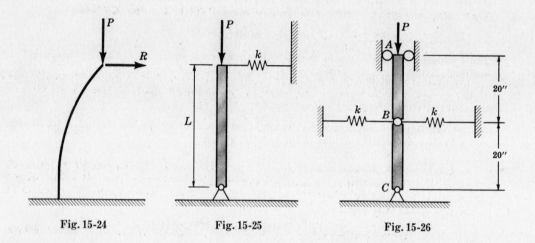

| Fig. 15-24 | Fig. 15-25 | Fig. 15-26 |

15.32. The rigid bar in Fig. 15-25 is pinned at the lower end and partially restrained against lateral motion at the top by a spring having a spring constant k. The upper end of the bar is given a very slight lateral displacement. Determine the critical value of the load P.   *Ans.* $P_{cr} = kL$

15.33. The system of two rigid vertical bars AB and BC shown in Fig. 15-26 is pinned at the base C and restrained against lateral motion at the top A, but is free to rotate there. The bars are also pinned at B. The midpoint B is partially restrained against lateral displacement by the two linear springs, each offering k lb of resistance per inch of lateral movement. The springs are load-free prior to application of P. Determine the buckling load P_{cr}.   *Ans.* $P_{cr} = 12k$

15.34. An initially straight bar is elastically restrained by springs at the upper and lower ends as shown in Fig. 15-27. The torsion spring at A offers a resistance against angular rotation equal to K in-lb per radian of rotation at A. The linear spring at B offers a resistance against horizontal displacement there equal to k lb per unit horizontal displacement. Each spring force is zero when the bar is vertical. Determine the characteristic equation governing buckling of the bar.

Ans. $\alpha L \cot \alpha L + \dfrac{kL}{P - kL} - \dfrac{PL}{K} = 0$ where $\alpha^2 = \dfrac{P}{EI}$

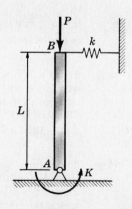

Fig. 15-27

15.35. For the case of the elastica discussed in Problem 15.22 determine the values of PL^2/AE and Δ/L for initial slopes of 20° and also 100°.

Ans. $\dfrac{PL^2}{EI} = 10.02, 15.00;$ $\dfrac{\Delta}{L} = 0.109, 0.395$

15.36.   Consider a round aluminum bar AB supported by two horizontal springs and subject to the vertical compressive load P (see Fig. 15-28).   The spring constants are $k_1 = 22.4$ lb/in and $k_2 = 44.8$ lb/in. The length of the bar is 23.6 inches, its diameter is 0.63 inches, and $E = 10 \times 10^6$ lb/in^2.   Determine the buckling load and the form of the buckled bar.

   Ans. $P_{cr} = L\left(\dfrac{k_1 k_2}{k_1 + k_2}\right) = 352$ lb

   The bar remains straight and rotates about a point $x_0 = 15.8$ in.

15.37.   An initially straight bar AC is pinned at each end and supported at the mid-point B by a spring which resists any lateral movement δ of B with a lateral force $(kEI/L^3)\delta$.   The bar is of length $2L$ and least flexural rigidity EI.   Equal and opposite thrusts P are applied at the end C as well as at the centroid of the bar at B.   In any deflected form the line of action of the thrust applied at B remains parallel to the chord AC.   Determine the minimum buckling load of the system.

   Ans. $P_{cr} = \beta^2 \dfrac{EI}{L^2}$ where β is the smallest positive root of the equation

   $$\frac{\beta}{\tan\beta} = \frac{3k + (9+k)\beta^2 - \beta^4}{3(k - \beta^2)}$$

15.38.   A long thin bar of length L and rigidity EI is supported at each end in an elastic medium which exerts a restoring moment of magnitude λ per radian of angular rotation at the end.   Determine the first buckling load of the bar.

   Ans. $\tan\dfrac{\alpha L}{2} = -\dfrac{P}{\alpha\lambda}$ where $\alpha^2 = \dfrac{P}{EI}$

15.39.   A long thin bar is pinned at each end and is embedded in an elastic packing which exerts a transverse force on the bar when it deflects laterally.   When the transverse deflection at any point is given by y, the packing exerts a transverse force per unit length of the bar equal to ky.   Determine the axial force required to buckle the bar.

   Ans. $P_{cr} = \dfrac{\pi^2 EI}{L^2}\left(n^2 + \dfrac{kL^4}{n^2\pi^4 EI}\right)$ where n is the integer for which P_{cr} is minimum

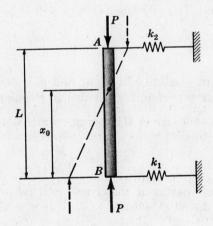

Fig. 15-28

Chapter 16

Strain Energy Methods

Thus far in this book various techniques have been discussed for finding deformations and determining values of indeterminate reactions. These techniques have essentially been based upon geometric considerations. There are, however, many types of problems that can be solved more efficiently through techniques based upon relations between the work done by the external forces and the internal strain energy stored within the body during the deformation process. The present chapter will discuss these techniques which are somewhat more general and more powerful than the various geometric approaches.

INTERNAL STRAIN ENERGY

When an external force acts upon an elastic body and deforms it, the work done by the force is stored within the body in the form of strain energy. The strain energy is always a scalar quantity. For a straight bar subject to a tensile force P, the internal strain energy U is given by

$$U = \frac{P^2 L}{2AE}$$

where L represents the length of the bar, A its cross-sectional area, and E is Young's modulus. This expression is derived in Problem 16.1.

For a circular bar of length L subject to a torque T, the internal strain energy U is given by

$$U = \frac{T^2 L}{2GJ}$$

where G is the modulus of elasticity in shear and J is the polar moment of inertia of the cross-sectional area. This expression is derived in Problem 16.2.

For a bar of length L subject to a bending moment M, the internal strain energy U is given by

$$U = \frac{M^2 L}{2EI}$$

where I is the moment of inertia of the cross-sectional area about the neutral axis. This is derived in Problem 16.3.

Note that in each of these expressions the external load always occurs in the form of a squared magnitude, hence each of these energy expressions is always a positive scalar quantity.

SIGN CONVENTIONS

Strain energy methods are particularly well suited to problems involving several structural members at various angles to one another. The fact that the members may be curved in their planes presents no additional difficulties. One of the great advantages of strain energy methods is that independent coordinate systems may be established for each member without regard for consistency of positive directions of the various coordinate systems. This advantage is essentially due to the fact that the strain energy is always a positive scalar quantity, hence algebraic signs of external forces need be consistent only within each structural member.

CASTIGLIANO'S THEOREM

This theorem is extremely useful for finding displacements of elastic bodies subject to axial loads, torsion, bending, or any combination of these loadings. The theorem states that the partial derivative of the total internal strain energy with respect to any external applied force yields the displacement under the point of application of that force in the direction of that force. Here, the terms force and displacement are used in their generalized sense and could either indicate a usual force and its linear displacement, or a couple and the corresponding angular displacement. In equation form the displacement under the point of application of the force P_n is given according to this theorem by

$$\delta_n = \frac{\partial U}{\partial P_n}$$

This theorem is derived in Problem 16.13.

APPLICATION TO STATICALLY DETERMINATE PROBLEMS

In such problems all external reactions can be found by application of the equations of statics. After this has been done, the deflection under the point of application of any external applied force can be found directly by use of Castigliano's theorem. This is illustrated in Problems 16.14–16.24. If the deflection is desired at some point where there is no applied force, then it is necessary to introduce an auxiliary (i.e., fictitious) force at that point and, treating that force just as one of the real ones, use Castigliano's theorem to determine the deflection at that point. At the end of the problem the auxiliary force is set equal to zero. This is illustrated in Problems 16.17, 16.18, 16.21, 16.23 and 16.24.

APPLICATION TO STATICALLY INDETERMINATE PROBLEMS

Castigliano's theorem is extremely useful for determining the indeterminate reactions in such problems. This is because the theorem can be applied to each reaction, and the displacement corresponding to each reaction is known beforehand and is usually zero. In this manner it is possible to establish as many equations as there are redundant reactions, and these equations together with those found from statics yield the solution for all reactions. After the values of all reactions have been found, the deflection at any desired point can be found by direct use of Castigliano's theorem. This is illustrated in Problems 16.25, 16.27 and 16.28.

ASSUMPTIONS AND LIMITATIONS

Throughout this chapter it is assumed that the material is a linear elastic one obeying Hooke's law. Further, it is necessary that the entire system obey the law of superposition. This implies that certain unusual systems, such as that discussed in Problem 1.14, cannot be treated by the techniques discussed here.

Solved Problems

16.1. Determine the internal strain energy stored within an elastic bar subject to an axial tensile force P.

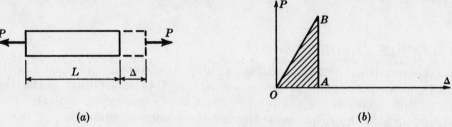

(a)

(b)

Fig. 16-1

For such a bar the elongation Δ has been found in Problem 1.1, page 8, to be $\Delta = PL/AE$, where A represents the cross-sectional area, L the length, and E is Young's modulus. The force-elongation diagram will consequently be linear as shown in Fig. 16-1(b). For any specific value of the force P, such as that corresponding to point B in the force-elongation diagram, the force will have done positive work indicated by the shaded area OBA. This triangular area is given by $\frac{1}{2}P\Delta$. Replacing Δ by the value given above, this becomes $P^2L/2AE$. This is the work done by the external force and the work is stored within the bar in the form of internal strain energy, denoted by U. Hence

$$U = \frac{P^2L}{2AE}$$

Essentially, the elastic bar is acting as a spring to store this energy. The same expression for internal strain energy applies if the load is compressive, since the axial force appears as a squared quantity and hence the final result is the same for either a positive or negative force.

If the axial force P varies along the length of the bar, then in an elemental length dx of the bar the strain energy is

$$dU = \frac{P^2\,dx}{2AE}$$

and the energy in the entire bar is found by integrating over the length:

$$U = \int_0^L \frac{P^2\,dx}{2AE}$$

16.2. Determine the internal strain energy stored within an elastic bar subject to a torque T.

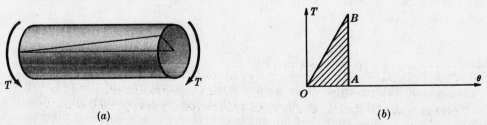

(a)

(b)

Fig. 16-2

In Problem 5.3, page 66, the angle of twist θ has been found to be $\theta = TL/GJ$, where G is the modulus of elasticity in shear, L is the length, and J is the polar moment of inertia of the cross-sectional area. According to this expression, the relation between torque and angle of twist is a linear one, as shown in Fig. 16-2(b). When the torque has reached a specific value such as that indicated by point B, it will have done positive work indicated by the shaded area OBA. This triangular area is given by $\frac{1}{2}T\theta$, or $T^2L/2GJ$. This work done by the external torque is stored within the bar as internal strain energy, denoted by U. Hence

$$U = \frac{T^2L}{2GJ}$$

If the torque T varies along the length of the bar, then in an elemental length dx the strain energy is

$$dU = \frac{T^2\,dx}{2GJ}$$

and in the entire bar it is

$$U = \int_0^L \frac{T^2\,dx}{2GJ}$$

16.3. Determine the internal strain energy stored within an elastic bar subject to a pure bending moment M.

In Problem 8.1, page 126, is shown an initially straight bar subject to the pure bending moment M which deforms it into a circular arc of radius of curvature ρ. In equation (7) of that problem it was shown that $M = EI/\rho$ where I denotes the moment of inertia of the cross-sectional area about the neutral axis. But the length of the bar, L, is equal to the product of the central angle θ subtended by the circular arc and the radius ρ. Thus

$$\frac{M}{EI} = \frac{1}{\rho} = \frac{\theta}{L} \quad\text{or}\quad \theta = \frac{ML}{EI}$$

According to this the relation between moment and angle subtended is a linear one, and this is illustrated in Fig. 16-3. When the moment has reached a specific value M, such as that indicated by point B, it will have done work indicated by the shaded area OAB. This area is given by $\frac{1}{2}M\theta$, or $M^2L/2EI$. This work done by the external moment is stored within the bar as internal strain energy, denoted by U. Hence

$$U = \frac{M^2L}{2EI}$$

If the bending moment M varies along the length of the bar, then in an elemental length dx the strain energy is

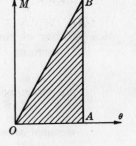

Fig. 16-3

$$dU = \frac{M^2\,dx}{2EI}$$

and in the entire bar it is

$$U = \int_0^L \frac{M^2\,dx}{2EI}$$

16.4. Compare the strain energy stored in each of the three steel bars shown in Fig. 16-4 below, subject to the condition that the axial stress in the lower portion of the second bar is equal to that in the first and third bars, namely, 20,000 lb/in^2.

The strain energy of axial tension is given in Problem 16.1 as $U = P^2L/2AE$. But for axial

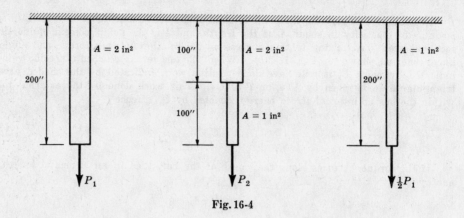

Fig. 16-4

loading, $P = A\sigma$, where σ is axial stress, so let us rewrite this in the form

$$U = \frac{(A\sigma)^2 L}{2AE} = \frac{A\sigma^2 L}{2E}$$

For the first bar:

$$U_1 = \frac{(2)(20,000)^2(200)}{2E} = \frac{16 \times 10^{10}}{2E}$$

For the second bar:

$$U_2 = \frac{(2)(10,000)^2(100)}{2E} + \frac{(1)(20,000)^2(100)}{2E} = \frac{6 \times 10^{10}}{2E}$$

For the third bar:

$$U_3 = \frac{(1)(20,000)^2(200)}{2E} = \frac{8 \times 10^{10}}{2E}$$

The ratio of these strain energies is thus $8 : 3 : 4$.

16.5. A bar of constant cross-section hangs vertically, subjected only to its own weight. Determine the strain energy stored within the bar.

Let us introduce the coordinate system shown in Fig. 16-5 and consider the strain energy stored within the elemental length dx. In general, $U = P^2L/2AE$, and for this element the force acting on it consists of the weight of the material below it. This is given by $Ax\gamma$ where A is the cross-sectional area and γ denotes the weight per unit volume of the material. Thus the strain energy stored in the shaded element of length dx is

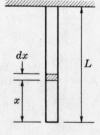

$$dU = \frac{(Ax\gamma)^2\, dx}{2AE}$$

To find the strain energy in the entire bar, we integrate over the length to obtain

Fig. 16-5

$$U = \int dU = \int_0^L \frac{(Ax\gamma)^2\, dx}{2AE} = \frac{A\gamma^2 L^3}{6E}$$

16.6. Consider a compound steel shaft for which $G = 12 \times 10^6$ lb/in². At the lower extremity the shaft is subject to a torque of 50,000 lb-in in the direction shown in Fig. 16-6. At the junction it is subject to a torque of 80,000 lb-in in the direction opposite to that of the first torque. Determine the total internal strain energy stored in the shaft.

For the lower region,

$$J = \frac{\pi}{32}D^4 = \frac{\pi(3)^4}{32} = 7.98 \text{ in}^4$$

For the upper region,

$$J = \frac{\pi}{32}D^4 = \frac{\pi(4)^4}{32} = 25.2 \text{ in}^4$$

The torque acting in the region BC is obviously 50,000 lb-in. In the region AB it is $50,000 - 80,000 = -30,000$ lb-in, i.e. opposite in direction to that in BC.

From Problem 16.2 the internal strain energy is given by

$$U = \frac{(50,000)^2(24)}{2(12 \times 10^6)(7.98)} + \frac{(-30,000)^2(36)}{2(12 \times 10^6)(25.2)} = 354 \text{ lb-in}$$

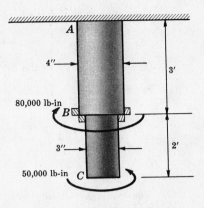

Fig. 16-6

16.7. Consider the two simply supported beams shown in Fig. 16-7. Both are of rectangular cross-section and of equal width. The materials are identical. The first beam has constant height along the length, the second has a small groove in the center which reduces the height by one-fifth. The length of the groove along the axis is negligible. The maximum stress in each bar due to the action of the central force P is the elastic limit of the material. Neglecting the effect of stress concentrations, determine the ratio of internal strain energies in the two bars.

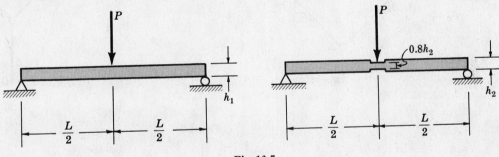

Fig. 16-7

For the first bar, the section modulus is

$$Z = \frac{I}{c} = \frac{\frac{1}{12}bh_1^3}{0.5h_1} = 0.167h_1^2 b$$

For the second bar, in the grooved region the section modulus is

$$Z = \frac{I}{c} = \frac{\frac{1}{12}b(0.8h_2)^3}{0.4h_2} = 0.107h_2^2 b$$

and in the thicker region of depth h_2 the section modulus is

$$Z = \frac{I}{c} = \frac{\frac{1}{12}bh_2^3}{0.5h_2} = 0.167h_2^2 b$$

In general, for bending we have the bending stress at the outer fibers of a bar given by the relation $\sigma = M/Z$. Since the maximum stresses in each bar are equal, we have

$$0.167 h_1^2 b = 0.107 h_2^2 b \quad \text{or} \quad h_2 = 1.25 h_1$$

The strain energy in the first bar is

$$U_1 = \frac{M^2 L}{2EI} = \frac{M^2 L}{2E(\frac{1}{12} b h_1^3)}$$

The strain energy in the second bar, since the groove is of negligible length, is

$$U_2 = \frac{M^2 L}{2E[\frac{1}{12} b (1.25 h_1)^3]}$$

The loadings and lengths are identical, hence we need not calculate $M^2 L$ to obtain the desired ratio, which is

$$U_2 : U_1 = 0.512$$

This indicates that a grooved bar is very ineffective in storing internal strain energy. This is an important consideration in the design of bars to withstand dynamic loadings.

16.8. Two identical bars are pin-connected and support a load Q as shown in Fig. 16-8 (a). Find the vertical displacement of point B by an energy method.

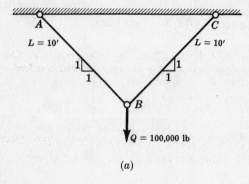
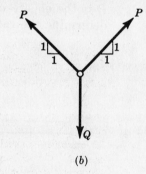

(a) (b)

Fig. 16-8

A free-body diagram of the pin at B is shown in Fig. 16-8(b) where P represents the axial force in each bar. From statics

$$2\left(\frac{1}{\sqrt{2}}\right) P - Q = 0 \quad \text{or} \quad P = \frac{Q\sqrt{2}}{2}$$

The internal strain energy stored in each bar was found in Problem 16.1 to be $U = P^2 L/2AE$. Thus for the two bars we have the internal strain energy

$$U = \frac{P^2 L}{AE} = \frac{(Q\sqrt{2}/2)^2 L}{AE} = \frac{Q^2 L}{2AE}$$

As the applied load gradually increases from 0 to its final value Q it does an amount of work given by its average value times the displacement: $(Q/2)\Delta$. Here Δ is the vertical displacement of point B. We may now equate the work done by Q to the strain energy (since this work is entirely stored within the bars as internal strain energy) to obtain

$$\frac{Q\Delta}{2} = \frac{Q^2 L}{2AE} \quad \text{or} \quad \Delta = \frac{QL}{AE}$$

The reader may compare this solution with the geometric approach to this same problem presented in Problem 1.11.

16.9. Consider a vertical bar of uniform cross-section with a flange at the lower end (Fig. 16-9). A weight W is released from the top of the bar and falls freely along the bar until it strikes the flange. Determine the maximum elongation of the bar and also the maximum stress.

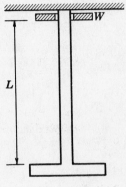

Fig. 16-9

To solve this problem we shall introduce several simplifying assumptions: (a) the weight of the vertical bar is very small compared to W, (b) there are no losses of energy due to friction or local distortion, and (c) the stress-strain diagram of the material of the bar is the same for dynamic loading as for static. Actually, a more sophisticated treatment would take strain wave propagation in the bar into account, but that is beyond the scope of the present study.

The weight W falls through the distance L and after striking the flange extends the bar an unknown amount Δ. At this maximum extension the tension in the bar is maximum and the equation relating work done by W and the internal strain energy of extension at this instant of maximum deformation is

$$W(L+\Delta) \;=\; \frac{P^2 L}{2AE} \qquad\qquad (1)$$

But $\Delta = PL/AE$, and substituting for P in (1), we get

$$W(L+\Delta) \;=\; \frac{AE\Delta^2}{2L} \qquad\qquad (2)$$

The static extension of the bar due to the weight W would be $\Delta_{st} = WL/AE$. If the value of W from this expression is introduced in the above equation and the resulting quadratic equation solved for the unknown extension Δ, we get

$$\Delta \;=\; \Delta_{st} + \sqrt{\Delta_{st}^2 + \frac{\Delta_{st}}{g}v^2} \qquad\qquad (3)$$

where g is the acceleration due to gravity and $v = \sqrt{2gL}$ is the velocity with which W strikes the flange. If the length of the bar, L, is very large compared to Δ_{st}, then the above expression becomes approximately

$$\Delta \;=\; \sqrt{\frac{\Delta_{st}}{g}v^2} \qquad\qquad (4)$$

In this case the axial stress is given by

$$\sigma \;=\; \frac{P}{A} \;=\; \frac{\Delta E}{L} \;=\; \frac{E}{L}\sqrt{\frac{\Delta_{st}}{g}v^2} \;=\; \sqrt{\frac{Wv^2}{2g}\frac{2E}{AL}} \qquad\qquad (5)$$

It is of interest to note that in the dynamic case the stress depends upon the length L as well as the Young's modulus E. The corresponding static stress does not involve either of these factors.

For the special case of a suddenly applied load W acting on the flange, the length L through which the weight falls may be set equal to zero in (3) to obtain

$$\Delta \;=\; 2\Delta_{st} \qquad\qquad (6)$$

Thus, for this particular problem, a suddenly applied load produces a deflection twice as great as would be produced by a gradually applied load.

16.10. A solid brass shaft for which $G = 5 \times 10^6$ lb/in^2 is 1.25 in. in diameter. A 6 ft length projects beyond a bearing and at the end of this 6 ft length a torque of 1000 lb-in is applied. If the bearing suddenly freezes and completely restrains all rotation, determine the peak dynamic torsional shear stress produced.

From Problem 16.2, the internal strain energy is given by

$$U = \frac{T^2 L}{2GJ}$$

But from Problem 5.2, page 65, the torque T is

$$T = \frac{\tau_{max} J}{r}$$

where r is the radius of the shaft and τ_{max} is the peak torsional shearing stress, which of course occurs at the outer fibers. Substituting:

$$U = \frac{(\tau_{max})^2 J^2 L}{2r^2 GJ} = \frac{(\tau_{max})^2 L \pi r^2}{4G}$$

The maximum dynamic shear stress will occur when the load of 1000 lb-in has been absorbed by the shaft. Thus

$$1000 = \frac{(\tau_{max})^2 (72) \pi (0.625)}{4(5 \times 10^6)} \qquad \text{or} \qquad \tau_{max} = 5330 \text{ lb/in}^2$$

16.11. A cantilever beam is struck at its tip by a body of weight W falling freely through a height h above the beam. Neglecting the weight of the beam, determine the total deflection at the tip.

By the time the weight has deflected the tip of the beam to its maximum value, the weight will have done an amount of work given by

$$W(h + \Delta) \qquad\qquad (1)$$

If we let P denote the force exerted by the weight on the beam at the time of peak deflection, then at this moment the strain energy in the beam is given by $P\Delta/2$. Thus, once the work done by the external force is stored within the beam as internal strain energy we have

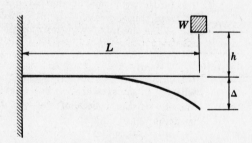

Fig. 16-10

$$W(h + \Delta) = \frac{P\Delta}{2} \qquad\qquad (2)$$

or

$$P = \frac{2W}{\Delta}(h + \Delta) \qquad\qquad (3)$$

But from Problem 9.2, page 158, we know that if this force P acts at the tip of a cantilever beam that the deflection at that point is

$$\Delta = \left[\frac{2W}{\Delta}(h + \Delta) \right] \frac{L^3}{3EI} \qquad\qquad (4)$$

where I is the moment of inertia of the cross-section about the neutral axis through the centroid. However, the deflection due to the weight W, if it were statically applied, is

$$\Delta_{st} = \frac{WL^3}{3EI} \qquad\qquad (5)$$

and hence (4) becomes

$$\Delta^2 - 2\Delta_{st}\Delta - 2h\Delta_{st} = 0 \qquad\qquad (6)$$

Solving:

$$\Delta = \Delta_{st} + \sqrt{\Delta_{st}^2 + 2h\Delta_{st}} \qquad (7)$$

where the positive square root is taken so as to obtain the maximum deflection. For the special case of a suddenly applied load at the tip, $h = 0$, and (7) yields $\Delta = 2\Delta_{st}$. Just as in Problem 16.9, a load suddenly applied produces twice the deflection it would if it were applied gradually.

16.12. A simply supported beam is struck at its midpoint by a weight $W = 250$ lb, falling freely from a height of $h = 5$ in. above the top of the beam. The beam is 14 ft long and of circular cross-section 4 in. in diameter. Take $E = 30 \times 10^6$ lb/in². Determine the maximum deflection of the beam.

The work done by the falling weight in producing the maximum central deflection Δ is

$$W(h + \Delta) \qquad (1)$$

If P denotes the force exerted by the weight on the beam during the moment of maximum deflection, then the strain energy stored in the beam is $P\Delta/2$. Thus

$$\frac{P\Delta}{2} = W(h + \Delta) \qquad (2)$$

or

$$P = \frac{2W(h + \Delta)}{\Delta} \qquad (3)$$

But the central deflection of a centrally loaded, simply supported beam is given in Problem 9.8, page 162, as

$$\Delta = \frac{PL^3}{48EI} \qquad (4)$$

Substituting the above value of P, this becomes

$$\Delta = \frac{2W(h + \Delta)}{\Delta} \frac{L^3}{48EI} \qquad (5)$$

But the static deflection corresponding to W is $\Delta_{st} = \dfrac{WL^3}{48EI}$, and hence (5) can be written in the form

$$\Delta^2 - 2\Delta_{st}\Delta - 2h\Delta_{st} = 0 \qquad (6)$$

Solving:

$$\Delta = \Delta_{st} + \sqrt{\Delta_{st}^2 + 2h\Delta_{st}} \qquad (7)$$

For the beam under consideration,

$$I = \frac{\pi D^4}{64} = 12.6 \text{ in}^4 \qquad \Delta_{st} = \frac{(250)(168)^3}{48(30 \times 10^6)(12.6)} = 0.0655 \text{ in.}$$

The maximum deflection is found from (7) as

$$\Delta = 0.0655 + \sqrt{(0.0655)^2 + 2(5)(0.0655)} = 0.877 \text{ in.}$$

16.13. Derive Castigliano's theorem.

Let us consider a general three-dimensional elastic body loaded by the forces P_1, P_2, etc. (Fig. 16-11 below). These would include forces exerted on the body by the various supports. We shall denote the displacement under P_1 *in the direction* of P_1 by Δ_1, that under P_2 in the direction of P_2 by Δ_2, etc. If we assume that all forces are applied simultaneously and gradually increased from zero to their final values given by P_1, P_2, etc., then the work done by the totality of forces will be

$$U = \frac{P_1}{2}\Delta_1 + \frac{P_2}{2}\Delta_2 + \frac{P_3}{2}\Delta_3 + \cdots \qquad (1)$$

This work is stored within the body as elastic strain energy.

Let us now increase the nth force by an amount dP_n. This changes both the state of deformation and also the internal strain energy slightly. The increase in the latter is given by

$$\frac{\partial U}{\partial P_n}dP_n \qquad (2)$$

Thus, the total strain energy after the increase in the nth force is

$$U + \frac{\partial U}{\partial P_n}dP_n \qquad (3)$$

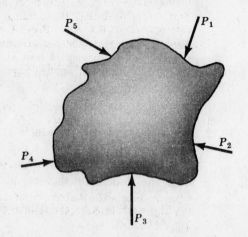

Fig. 16-11

Let us reconsider this problem by first applying a very small force dP_n alone to the elastic body. Then, we apply the same forces as before, namely, P_1, P_2, P_3, etc. Due to the application of dP_n there is a displacement in the direction of dP_n which is infinitesimal and may be denoted by $d\Delta_n$. Now, when P_1, P_2, P_3, etc., are applied their effect on the body will not be changed by the presence of dP_n and the internal strain energy arising from application of P_1, P_2, P_3, etc., will be that indicated in (1). But as these forces are being applied the small force dP_n goes through the additional displacement Δ_n caused by the forces P_1, P_2, P_3, etc. Thus, it gives rise to additional work $(dP_n)\Delta_n$ which is stored as internal strain energy and hence the total strain energy in this case is

$$U + (dP_n)\Delta_n \qquad (4)$$

Since the final strain energy must be independent of the order in which the forces are applied, we may equate (3) and (4):

$$U + \frac{\partial U}{\partial P_n}dP_n = U + (dP_n)\Delta_n$$

or

$$\Delta_n = \frac{\partial U}{\partial P_n}$$

This is *Castigliano's theorem*. In words it states that the displacement of an elastic body under the point of application of any force, in the direction of that force, is given by the partial derivative of the total internal strain energy with respect to that force.

The term force here is used in its most general sense and implies either a true force or a couple. For the case of a couple, Castigliano's theorem gives the angular rotation under the point of application of the couple in the sense of rotation of the couple.

It is important to observe that the above derivation required that we be able to vary the nth force, P_n, independently of the other forces. Thus, P_n must be statically independent of the other external forces, implying that the energy U must always be expressed in terms of the statically independent forces of the system. Obviously, reactions that can be determined by statics cannot be considered as independent forces.

16.14. A cantilever beam of uniform cross-section is loaded by a concentrated force at its tip. Determine the deflection under the point of application of the force by using Castigliano's theorem.

It is simplest to introduce the coordinate system of
Fig. 16-12, such that the bending moment at any section
x is given by $M = Px$. From Problem 16.3, the internal
strain energy stored in the beam is

$$U = \int_0^L \frac{M^2\,dx}{2EI} = \int_0^L \frac{(Px)^2\,dx}{2EI}$$

The desired deflection under the point of application of the
force P is found from Castigliano's theorem as

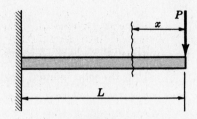

Fig. 16-12

$$\Delta = \frac{\partial U}{\partial P} = \int_0^L \frac{M\frac{\partial M}{\partial P}\,dx}{EI} = \int_0^L \frac{Px(x)\,dx}{EI} = \frac{PL^3}{3EI}$$

which agrees with the results of Problem 9.2. This analysis, just as that of Chapter 9, neglects
the slight additional deflection due to transverse shear effects and obviously considers only bending deflections.

16.15. A cantilever beam of stepwise constant cross-section, as shown in Fig. 16-13, is loaded
by a concentrated force at its tip. Determine the deflection under the point of application of the force by using Castigliano's theorem.

The coordinate system employed in Problem 16.14 is again simplest and the bending moment
is still given at all points along the length of the beam by $M = Px$. The internal strain energy is

$$U = \int_0^L \frac{M^2\,dx}{2EI} = \int_0^{L/2} \frac{(Px)^2\,dx}{2EI_1} + \int_{L/2}^L \frac{(Px)^2\,dx}{2E(2I_1)}$$

According to Castigliano's theorem the deflection at the tip is given by

$$\Delta = \frac{\partial U}{\partial P} = \int_0^L \frac{M\frac{\partial M}{\partial P}\,dx}{EI} \qquad \text{where} \quad M = Px \quad \text{and} \quad \frac{\partial M}{\partial P} = x$$

Thus
$$\Delta = \int_0^{L/2} \frac{(Px)x\,dx}{EI_1} + \int_{L/2}^L \frac{(Px)x\,dx}{E(2I_1)} = \frac{9PL^3}{48EI_1}$$

Note that energy methods are much easier to apply to structures with members of nonconstant
cross-section than either double integration or moment-areas.

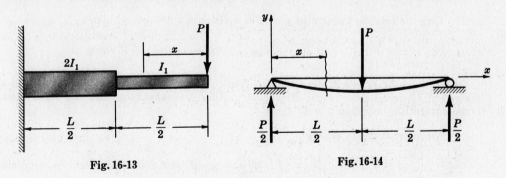

Fig. 16-13   **Fig. 16-14**

16.16. Use Castigliano's theorem to determine the maximum deflection of a simply supported beam subject to a single concentrated force applied at the midpoint.

In the coordinate system shown in Fig. 16-14, the bending moment to the left of the vertical
force P is given by

$$M = \frac{P}{2}x \quad \text{from which} \quad \frac{\partial M}{\partial P} = \frac{x}{2} \quad \text{for} \quad 0 < x < \frac{L}{2}$$

Due to the symmetry about the midpoint, the internal strain energy stored in the half of the beam to the right of the concentrated force is equal to that stored in the half to the left of the force. Thus the total internal strain energy is

$$U = \int_0^L \frac{M^2 \, dx}{2EI} = 2 \int_0^{L/2} \frac{(Px/2)^2 \, dx}{2EI}$$

According to Castigliano's theorem, the deflection under the point of application of the force P is

$$\Delta = \frac{\partial U}{\partial P} = 2 \int_0^{L/2} \frac{M \frac{\partial M}{\partial P} \, dx}{EI} = 2 \int_0^{L/2} \frac{(Px/2)(x/2) \, dx}{EI} = \frac{PL^3}{48EI}$$

which agrees with the result found in Problem 9.8.

16.17. Use Castigliano's theorem to determine the deflection at the tip of a cantilever beam subject to a uniformly distributed load of w lb per unit length.

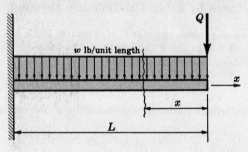

Fig. 16-15

This problem is of a somewhat different nature than Problems 16.14, 16.15, and 16.16 since in each of those we sought the deflection under the point of application of a real concentrated force. Here we seek the deflection at a point where there is no concentrated force, so we must temporarily introduce an auxiliary force Q at the tip. At the end of the problem the value of Q will be taken to be zero.

Thus the bending moment at a section a distance x from the tip is

$$M = Qx + \frac{wx^2}{2} \quad \text{from which} \quad \frac{\partial M}{\partial Q} = x$$

Note that any convenient signs may be introduced for the bending moment (so long as algebraic signs of all terms for any one structural member are consistent) and the signs need not follow the conventions introduced in Chapter 6.

The internal strain energy is

$$U = \int_0^L \frac{M^2 \, dx}{2EI} = \int_0^L \frac{[Qx + (wx^2/2)]^2 \, dx}{2EI}$$

From Castigliano's theorem, the deflection under the point of application of the auxiliary force Q is

$$\Delta = \frac{\partial U}{\partial Q} = \int_0^L \frac{M \frac{\partial M}{\partial Q} \, dx}{EI} = \int_0^L \frac{[Qx + (wx^2/2)]x \, dx}{EI}$$

It is important to note that the auxiliary force Q may be set equal to zero any time *after* the partial derivative has been taken.   Thus

$$\Delta = \int_0^L \frac{(+wx^2/2)(x)\,dx}{EI} = +\frac{wL^4}{8EI}$$

which agrees with the results of Problem 9.5. From this example we can observe the scope of Castigliano's theorem, since we now see that it is not restricted to finding deflections only at those points of a structure where there are actual concentrated forces.

16.18. By the use of Castigliano's theorem, determine the equation of the deflection curve for a simply supported beam loaded by a couple M_1 at the right end, as in Fig. 16-16.

To determine the equation of the deflection curve, we must find the deflection at an arbitrary point a distance x from the left end of the bar.   Since there is no concentrated force acting there, we must temporarily introduce an auxiliary force Q at that point.   Later the value of this force will be set equal to zero.

From statics, the end reactions are found to be

$$R_1 = \frac{M_1}{L} + Q\left(1 - \frac{x}{L}\right)$$

$$R_2 = -\frac{M_1}{L} + \frac{Qx}{L}$$

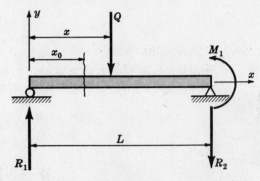

Fig. 16-16

At any section to the left of the auxiliary force Q (this variable section being designated by x_0), the bending moment is given by

$$M = R_1 x_0 = \frac{M_1 x_0}{L} + Q x_0\left(1 - \frac{x}{L}\right) \qquad \text{for } 0 < x_0 < x$$

from which

$$\frac{\partial M}{\partial Q} = x_0\left(1 - \frac{x}{L}\right) \qquad \text{for } 0 < x_0 < x$$

To the right of the auxiliary force Q the bending moment at a section a distance x_0 from the left end is given by

$$M = R_1 x_0 - Q(x_0 - x) = \frac{M_1 x_0}{L} + Q x_0\left(1 - \frac{x}{L}\right) - Q(x_0 - x) \qquad \text{for } x < x_0 < L$$

from which

$$\frac{\partial M}{\partial Q} = x_0\left(1 - \frac{x}{L}\right) - (x_0 - x) \qquad \text{for } x < x_0 < L$$

By Castigliano's theorem, the deflection under the point of application of the auxiliary force Q may now be determined.   Now that the derivatives have been taken, it is legitimate to set the force Q equal to zero, and we have for the deflection, which we may denote by y:

$$\Delta = y = \frac{\partial U}{\partial Q} = \int_0^L \frac{M\left(\frac{\partial M}{\partial Q}\right) dx}{EI}$$

$$= \int_0^x \frac{(M_1 x_0/L) x_0 (1 - x/L)\, dx_0}{EI} + \int_x^L \frac{(M_1 x_0)[x_0(1 - x/L) - (x_0 - x)]\, dx_0}{EI}$$

In the last two integrals x_0 is the variable and x is constant.   Integrating, we find:

$$y = \frac{M_1 L x}{6} - \frac{M_1 x^3}{6L}$$

which agrees with the result found in Problem 9.10. The algebraic signs here are opposite those found by the double-integration method because here we have found the deflection under the auxiliary force Q in the direction of Q, which was taken to be downward. Thus here we have downward deflections as positive, whereas in Problem 9.13 they were positive in the upward direction, i.e. the direction of the positive y-axis.

16.19. Use Castigliano's theorem to determine the slope at the right end of the beam discussed in Problem 16.18.

The applied moment M_1 acts at that point and from Castigliano's theorem the angular rotation under the point of application of M_1 in the direction (or sense) of M_1 is given by

$$Q = \frac{\partial U}{\partial M_1} = \int_0^L \frac{M \frac{\partial M}{\partial M_1} dx}{EI}$$

The auxiliary force Q of Problem 16.18 is no longer needed and thus at any point along the length of the bar the bending moment is simply

$$M = R_1 x = \frac{M_1}{L} x \quad \text{from which} \quad \frac{\partial M}{\partial M_1} = \frac{x}{L}$$

Thus

$$Q = \int_0^L \frac{(M_1 x/L)(x/L) dx}{EI} = \frac{M_1 L}{3EI}$$

16.20. By use of Castigliano's theorem, determine the deflection at the left end of the overhanging beam loaded by the two equal forces P shown in Fig. 16-17(a).

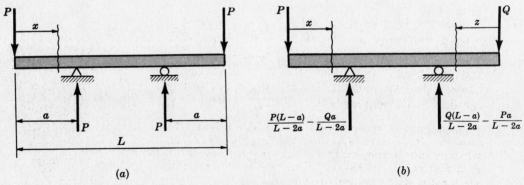

(a)   (b)

Fig. 16-17

Special caution must be exercised in a problem involving two or more forces designated by a common symbol, since direct application of Castigliano's theorem will yield the algebraic sum of the deflections under such forces. Consequently we must temporarily replace one of the forces, in this case say the one at the right end, by a force of different value, designated by Q. Then, after applying the equations of statics, the free-body diagram appears as in Fig. 16-17(b). In the left overhanging region we have

$$M = -Px \quad \text{from which} \quad \frac{\partial M}{\partial P} = -x \quad \text{for } 0 < x < a$$

In the central region

$$M = -Px + \left[\frac{P(L-a)}{L-2a} - \frac{Qa}{L-2a}\right](x-a) \qquad \text{for} \quad a < x < L-a$$

from which

$$\frac{\partial M}{\partial P} = -x + \frac{L-a}{L-2a}(x-a) \qquad \text{for} \quad a < x < L-a$$

In the right overhanging region it is simplest to introduce a new coordinate z having its origin at the right extremity of the beam. In this region

$$M = -Qz \qquad \text{and} \qquad \frac{\partial M}{\partial P} = 0 \qquad \text{for } 0 < z < a$$

The desired deflection at the left end is now found as

$$\Delta = \frac{\partial U}{\partial P} = \int_0^L \frac{M\frac{\partial M}{\partial P}\,dx}{EI}$$

where it must be understood that the integral is essentially symbolic, applying to both variables x and z since the energy in the *entire* bar must be considered. Thus

$$\Delta = \int_0^a \frac{(-Px)(-x)\,dx}{EI} + \int_a^{L-a} \frac{\left\{-Px + \left[\frac{P(L-a)}{L-2a} - \frac{Qa}{L-2a}\right](x-a)\right\}\left[-x + \frac{L-a}{L-2a}(x-a)\right]dx}{EI}$$

Now that the partial derivatives have been taken, it is legitimate to set Q equal to P. Doing this and carrying out the above integration, one obtains

$$\Delta = -\frac{2Pa^3}{3EI} + \frac{Pa^2L}{2EI}$$

Although advantage could have been taken of symmetric deflections in this problem, there are many problems involving no symmetry where this technique is essential.

16.21. A solid conical bar of circular cross-section is suspended vertically as shown in Fig. 16-18. The length of the bar is L, the diameter of the base D, and the weight per unit volume γ. Use Castigliano's theorem to determine the elongation of the bar due to its own weight.

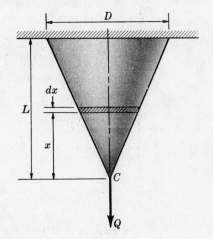

It is desired to determine the vertical deflection of point C. Since there is no external force applied there, it is necessary to temporarily introduce an auxiliary force, Q, at that point. At the end of the problem the value of this force is, of course, set equal to zero.

We consider the shaded element shown. The vertical force acting on this element is due partially to Q and partially to the weight of the material below the element. Consequently, the force tending to elongate the shaded element is

Fig. 16-18

$$P = Q + \frac{1}{3}\frac{\pi}{4}\left(\frac{x}{L}D\right)^2 x\gamma \qquad \text{from which} \qquad \frac{\partial P}{\partial Q} = 1$$

The infinitesimal shaded element may be considered to be of constant cross-section, so that the strain energy in it may be calculated from the expression found in Problem 16.1, or

$$dU = \frac{\left[Q + \frac{1}{3}\frac{\pi}{4}\left(\frac{x}{L}D\right)^2 x\gamma\right]^2 dx}{2\left[\frac{\pi}{4}\left(\frac{x}{L}D\right)^2\right]E}$$

The strain energy in the entire bar is thus

$$U = \int dU = \int_0^L \frac{\left[Q + \frac{1}{3}\frac{\pi}{4}\left(\frac{x}{L}D\right)^2 x\gamma\right]^2 dx}{2\left[\frac{\pi}{4}\left(\frac{x}{L}D\right)^2\right]E}$$

According to Castigliano's theorem, since $U = \int \frac{P^2\,dx}{2AE}$, the desired deflection is given by

$$\Delta = \frac{\partial U}{\partial Q} = \int_0^L \frac{P\frac{\partial P}{\partial Q}dx}{\left[\frac{\pi}{4}\left(\frac{x}{L}D\right)^2\right]E} = \int_0^L \frac{\left[Q + \frac{1}{3}\frac{\pi}{4}\left(\frac{x}{L}D\right)^2 x\gamma\right](1)\,dx}{\left[\frac{\pi}{4}\left(\frac{x}{L}D\right)^2\right]E}$$

Now that the differentiation with respect to Q has been performed, it is permissible to set Q equal to zero for the remainder of the computations. Thus

$$\Delta = \int_0^L \frac{\left[\frac{1}{3}\frac{\pi}{4}\left(\frac{x}{L}D\right)^2 x\gamma\right]dx}{\left[\frac{\pi}{4}\left(\frac{x}{L}D\right)^2\right]E} = \frac{\gamma L^2}{6E}$$

This agrees with the result found in Problem 1.33, page 22.

16.22. A structure is in the form of one quadrant of a thin circular ring of radius R. One end is clamped and the other end is loaded by a vertical force P (see Fig. 16-19). Determine the vertical displacement under the point of application of the force P. Consider only strain energy of bending.

From statics, the reactions at the clamped end consist of a vertical force P and a couple PR. The bending moment at the section in the ring located by the angle θ is given by

$$M = PR - P(R - R\cos\theta) = PR\cos\theta \qquad \text{from which} \qquad \frac{\partial M}{\partial P} = R\cos\theta$$

Castigliano's theorem states that the vertical deflection at A is given by

$$\Delta_v = \frac{\partial U}{\partial P} = \int_0^{\pi/2} \frac{M\frac{\partial M}{\partial P}R\,d\theta}{EI} = \int_0^{\pi/2} \frac{(PR\cos\theta)(R\cos\theta)R\,d\theta}{EI} = \frac{P\pi R^3}{4EI}$$

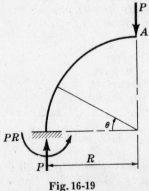

Fig. 16-19

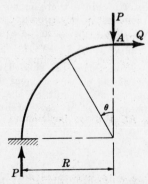

Fig. 16-20

16.23. Determine the horizontal displacement of point A in Problem 16.22.

Since there is no horizontal force applied at A we must temporarily introduce an auxiliary force Q in order to be able to use Castigliano's theorem. This time, let us measure θ from the vertical, making it unnecessary to determine reactions at B. Thus, at the section denoted by θ the bending moment is

$$M = PR \sin \theta + Q(R - R \cos \theta) \qquad \text{from which} \qquad \frac{\partial M}{\partial Q} = R - R \cos \theta$$

The horizontal displacement at A is given by

$$\Delta_h = \frac{\partial U}{\partial Q} = \int_0^{\pi/2} \frac{M \frac{\partial M}{\partial Q} R \, d\theta}{EI}$$

Now that the partial derivative has been taken, Q may be set equal to zero, yielding

$$\Delta_h = \int_0^{\pi/2} \frac{(PR \sin \theta)(R - R \cos \theta) R \, d\theta}{EI} = \frac{PR^3}{2EI}$$

16.24. Consider the frame of Fig. 16-21(a) which is subject to a uniformly distributed load along the horizontal member. Find the horizontal displacement at the roller at D.

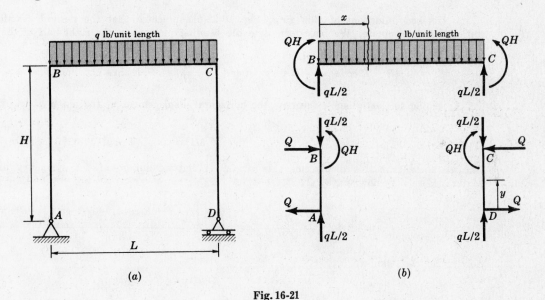

Fig. 16-21

To apply Castigliano's theorem, it is necessary to introduce an auxiliary force Q at point D. Doing this, the free-body diagrams of the various members appear as in Fig. 16-21(b).

In CD, the bending moment is

$$M = Qy; \qquad \frac{\partial M}{\partial Q} = y$$

In BC, the bending moment is

$$M = \frac{qL}{2} x + QH - \frac{qx^2}{2}; \qquad \frac{\partial M}{\partial Q} = H$$

One of course takes advantage of the symmetry of the vertical bars, and the displacement in the horizontal direction at D is given by (after setting $Q = 0$):

$$\Delta = \int \frac{M \frac{\partial M}{\partial Q} dx}{EI} = 2 \int_0^H \frac{(0)y\,dy}{EI} + \int_0^L \frac{[(qLx/2) - (qx^2/2)]H\,dx}{EI} = \frac{qHL^3}{12EI}$$

16.25. A thin semicircular ring is hinged at each end and loaded by a central concentrated force P as shown in Fig. 16-22(a). Determine the horizontal reaction at each hinge.

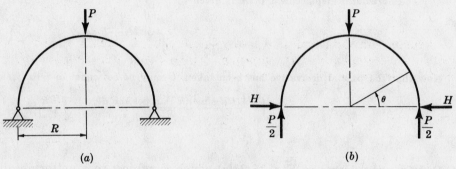

Fig. 16-22

A free-body diagram of this ring, Fig. 16-22(b), indicates that the desired reaction H is statically indeterminate. We may formulate the bending moment in the right half of the ring as follows:

$$M = \frac{P}{2}(R - R\cos\theta) - HR\sin\theta \quad \text{and} \quad \frac{\partial M}{\partial H} = -R\sin\theta \qquad \text{for } 0 < \theta < \frac{\pi}{2}$$

According to Castigliano's theorem, the horizontal displacement at the pin is given by

$$\Delta_H = \frac{\partial U}{\partial H}$$

But we know that this displacement is zero. Taking advantage of the symmetry about the centerline, we may now write

$$0 = \Delta_H = \frac{\partial U}{\partial H} = 2\int_0^{\pi/2} \frac{M\frac{\partial M}{\partial H}R\,d\theta}{EI} = \int_0^{\pi/2} \frac{\left[\frac{P}{2}(R - R\cos\theta) - HR\sin\theta\right](-R\sin\theta)R\,d\theta}{EI}$$

Solving for the unknown H: $H = \dfrac{P}{\pi}$.

16.26. In Problem 16.25, determine the vertical displacement under the point of application of the central force P.

In almost all statically indeterminate problems it is necessary first to determine the redundant reactions before any displacements can be found. For the present ring this has already been done in Problem 16.25.

In the right half of the ring the bending moment is

$$M = \frac{P}{2}(R - R\cos\theta) - \frac{P}{\pi}R\sin\theta \qquad \text{for } 0 < \theta < \frac{\pi}{2}$$

and

$$\frac{\partial M}{\partial P} = \frac{1}{2}(R - R\cos\theta) - \frac{R}{\pi}\sin\theta$$

By Castigliano's theorem, the vertical displacement under the point of application of P is

$$\Delta \;=\; \frac{\partial U}{\partial P} \;=\; 2 \int_0^{\pi/2} \frac{M \dfrac{\partial M}{\partial P} R\, d\theta}{EI}$$

where we have taken advantage of symmetry.   Thus

$$\Delta \;=\; 2 \int_0^{\pi/2} \frac{\left[\dfrac{P}{2}(R - R\cos\theta) - \dfrac{PR}{\pi}\sin\theta\right]\left[\dfrac{1}{2}(R - R\cos\theta) - \dfrac{R}{\pi}\sin\theta\right]R\, d\theta}{EI} \;=\; \frac{PR^3}{EI}\left(\frac{3\pi}{8} - \frac{1}{2\pi} - 1\right)$$

16.27. A structure in the form of a thin semicircular ring lies in a horizontal plane, has both ends clamped, and is subjected to a central vertical force P as shown in Fig. 16-23.   Determine the various reactions.

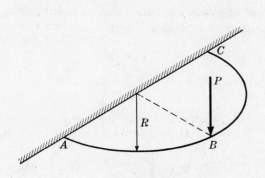

Fig. 16-23

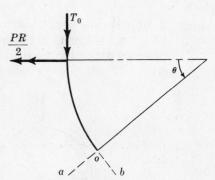

Fig. 16-24

The vertical force reactions at A and C are each $P/2$ and the bending moment exerted by the support on the ring at each of these points is found from statics to be $PR/2$.   There is also another component of reaction exerted by the support on the ring, i.e. a twisting moment T_0 acting at each of the points A and C.   These two types of moment reaction are best illustrated by the vector representation of moment in Fig. 16-24, where a double-headed arrow indicates a moment in the usual sense of the right-hand rule for vector representation of moment.   A segment of the ring to the arbitrary point represented by θ $\left(0 < \theta < \dfrac{\pi}{2}\right)$ is shown and at this cross-section given by θ there is a bending moment about the oa-axis given by

$$M \;=\; \frac{P}{2}R\sin\theta - \frac{PR}{2}\cos\theta - T_0\sin\theta$$

There is a twisting moment about the ob-axis given by

$$T \;=\; \frac{P}{2}(R - R\cos\theta) - \frac{PR}{2}\sin\theta + T_0\cos\theta$$

From these:

$$\frac{\partial M}{\partial T_0} \;=\; -\sin\theta; \qquad \frac{\partial T}{\partial T_0} \;=\; \cos\theta$$

Since the ring is completely restrained at points A and C, we may write (taking advantage of symmetry)

$$0 \;=\; \phi_A \;=\; \phi_C \;=\; 2 \int_0^{\pi/2} \frac{M\dfrac{\partial M}{\partial T_0} R\, d\theta}{EI} \;+\; 2 \int_0^{\pi/2} \frac{T\dfrac{\partial T}{\partial T_0} R\, d\theta}{GJ}$$

where ϕ is used to denote angular rotation of an arbitrary point of the bar, and ϕ_A and ϕ_C are the zero values of this quantity at the points A and C.   Substituting:

$$0 = \int_0^{\pi/2} \frac{\left(\dfrac{PR}{2} \sin\theta - \dfrac{PR}{2}\cos\theta - T_0\sin\theta\right)(-\sin\theta)\,R\,d\theta}{EI}$$

$$+ \int_0^{\pi/2} \frac{\left[\dfrac{P}{2}(R - R\cos\theta) - \dfrac{PR}{2}\sin\theta + T_0\cos\theta\right](\cos\theta)\,R\,d\theta}{GJ}$$

Solving:

$$T_0 = \frac{\dfrac{PR}{2}\left(\dfrac{2-\pi}{EI} + \dfrac{2-\pi}{GJ}\right)}{\dfrac{\pi}{EI} - \dfrac{\pi}{GJ}}$$

16.28. A thin ring is subjected to the equal and opposite diametral forces indicated in Fig. 16-25(a). Determine the bending moment at A and also the increase in diameter of the ring along the diameter CD.

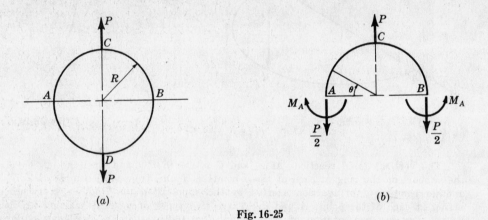

Fig. 16-25

Let us take advantage of symmetry and cut the ring along the horizontal diameter AB in which case it is obvious from statics that there exists a tensile force $P/2$ at each of the points A and B. There must also exist a moment M_A at each of these points but it is indeterminate. However, since by symmetry there can be no angular rotation of the ring at A and B, we may determine this redundant moment by Castigliano's theorem. At a section denoted by θ $\left(0 < \theta < \dfrac{\pi}{2}\right)$ the bending moment is

$$M = +M_A - \frac{P}{2}(R - R\cos\theta) \qquad \text{from which} \qquad \frac{\partial M}{\partial M_A} = 1$$

The condition of zero angular rotation at A is expressed by

$$0 = \int_0^{\pi/2} \frac{M \dfrac{\partial M}{\partial M_A} R\,d\theta}{EI} = \int_0^{\pi/2} \frac{\left[M_A - \dfrac{P}{2}(R - R\cos\theta)\right](1)\,R\,d\theta}{EI}$$

Solving:

$$M_A = \frac{PR}{2}\frac{\pi - 2}{\pi}$$

Thus, the bending moment at the section denoted by θ is

$$M = \frac{PR}{2}\frac{\pi-2}{\pi} - \frac{PR}{2}(1-\cos\theta)$$

from which

$$\frac{\partial M}{\partial P} = \frac{R}{2}\frac{\pi-2}{\pi} - \frac{R}{2}(1-\cos\theta)$$

The increase in the diameter CD is thus

$$\Delta = \frac{\partial U}{\partial P} = 4\int_0^{\pi/2} \frac{M\frac{\partial M}{\partial P}R\,d\theta}{EI}$$

$$= 4\int_0^{\pi/2} \frac{\left[\frac{PR}{2}\frac{\pi-2}{\pi} - \frac{PR}{2}(1-\cos\theta)\right]\left[\frac{R}{2}\frac{\pi-2}{\pi} - \frac{R}{2}(1-\cos\theta)\right]R\,d\theta}{EI}$$

$$= \frac{PR^3}{EI}\frac{\pi^2-8}{4\pi} = 0.149\frac{PR^3}{EI}$$

Supplementary Problems

16.29. A solid conical bar of circular cross-section (Fig. 16-26) hangs vertically, subjected only to its own weight. Determine the strain energy stored within the bar. *Ans.* $U = \pi D^2 L^3 \gamma^2/360E$

16.30. Two circular bars of diameters D_1 and D_2 are made of the same material, have identical lengths, and are subjected to equal twisting moments. Determine the ratio of internal strain energies.
Ans. $U_1 : U_2 = D_2^4 : D_1^4$

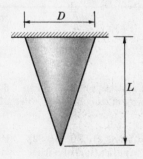

Fig. 16-26

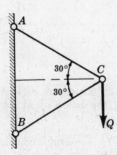

Fig. 16-27

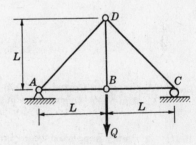

Fig. 16-28

16.31. In Fig. 16-27 two identical pin-connected bars support the vertical load Q. Determine the vertical displacement of point C by equating the work done by Q to the internal strain energy.
Ans. $\Delta = QL/AE$

16.32. The pin-connected truss shown in Fig. 16-28 is composed of five bars, each of area A and modulus of elasticity E. Determine the vertical displacement of point B due to the load Q by equating the work done by Q to the internal strain energy. *Ans.* $\Delta = 2.914QL/AE$

16.33. A rectangular copper bar 2 in. by 3 in. in cross-section is subject to an axial energy input of 120 lb-ft. Determine the minimum length of the bar so as not to exceed an allowable axial stress of 12,000 lb/in². The modulus of elasticity of the bar is 17×10^6 lb/in².
Ans. $L = 56.7$ in.

16.34. Determine the maximum weight W that can be dropped 10 in. onto the flange at the end of the steel bar shown in Fig. 16-29. The bar is 1 in. by 2 in. in cross-section and 6 ft in length. The axial stress is not to exceed 20,000 lb/in². Take $E = 30 \times 10^6$ lb/in².

Ans. $W = 96$ lb

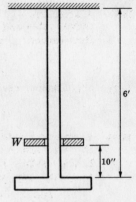

16.35. An 8 WF 19 section is used as a cantilever beam of length 12 ft. A weight W of 300 lb falls freely through a distance of 6 in. before striking the tip of the beam. Determine the maximum bending stress produced. Take $E = 30 \times 10^6$ lb/in². *Ans.* 26,000 lb/in²

16.36. A weight $W = 1500$ lb is dropped 1 in. onto the center of a simply supported 6 WF 15½ beam which is 18 ft long. Determine the maximum deflection of the beam and the maximum bending stress. Take $E = 30 \times 10^6$ lb/in². *Ans.* 1.319 in.; 29,200 lb/in²

Fig. 16-29

16.37. Determine the horizontal and vertical components of displacement of point B in Problem 1.12, by use of Castigliano's theorem.

16.38. Determine the horizontal and vertical components of displacement of point B in Problem 1.36, by use of Castigliano's theorem.

16.39. A cantilever beam is loaded by a moment M_1 applied at the tip (Fig. 16-30). Determine by Castigliano's theorem the deflection of the tip. *Ans.* $M_1L^2/2EI$

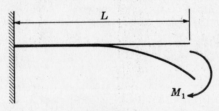

Fig. 16-30

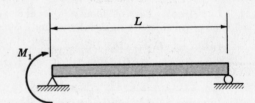

Fig. 16-31

16.40. A simply supported beam is loaded by a moment M_1 at the left end, as shown in Fig. 16-31. Use Castigliano's theorem to determine the deflection at the midpoint of the bar. *Ans.* $M_1L^2/16EI$

16.41. A simply supported beam is loaded by two symmetrically placed loads P (Fig. 16-32). Use Castigliano's theorem to determine the deflection under the point of application of either load.

Ans. $-\dfrac{2Pa^3}{3EI} + \dfrac{Pa^2L}{2EI}$

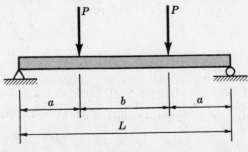

Fig. 16-32

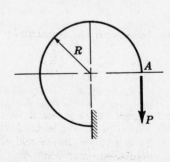

Fig. 16-33

16.42. A structure lies in a vertical plane and is in the form of three quadrants of a thin ring (see Fig. 16-33). One end is clamped, the other is loaded by a vertical force P. Determine the horizontal displacement of point A. Consider only bending energy.   *Ans. $PR^3/2EI$*

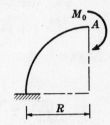

16.43. Find the angular rotation at point A in Problem 16.22.   *Ans. PR^2/EI*

16.44. A structure is in the form of one quadrant of a thin circular ring of radius R. One end is clamped and the other is subject to a couple M_0 as shown in Fig. 16-34. Determine the angular rotation, as well as the vertical and horizontal components of displacement of point A.

Ans. $\dfrac{M_0\pi R}{2EI}$; $\dfrac{M_0R^2}{EI}$; $0.571\dfrac{M_0R^2}{EI}$

Fig. 16-34

16.45. Consider the three-sided framework of Fig. 16-35 wherein all bars are of the same cross-section and material. Determine the amount by which the points A and B approach each other due to the action of the forces P.   *Ans. $PH^2(2H + 3L)/3EI$*

16.46. Solve Problem 16.45 if the forces P are replaced by couples M_0 applied at points A and B.
Ans. $M_0H(H + L)/EI$

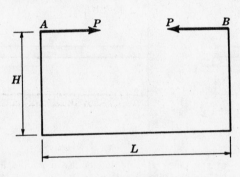

Fig. 16-35

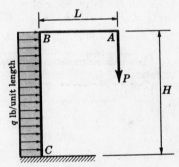

Fig. 16-36

16.47. The two-sided framework of Fig. 16-36 is loaded by a concentrated force P at the tip as well as a uniform load along the vertical member. Determine the vertical displacement of point A.

Ans. $\dfrac{PL^3}{3EI} + \dfrac{PL^2H}{EI} + \dfrac{qLH^3}{6EI}$

16.48. Determine the horizontal displacement of point A in Problem 16.47.

Ans. $\dfrac{PLH^2}{2EI} + \dfrac{qH^4}{8EI}$

16.49. Figure 16-37 shows a thin ring in the form of one quadrant of a circle. One end is fixed, the other is free, and the system is loaded by a moment at the midpoint. Determine the vertical component of displacement of point A.

Ans. $\dfrac{M_0R^2}{\sqrt{2}\,EI}$

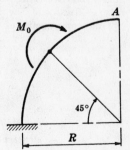

Fig. 16-37

16.50. Determine the angular rotation at A in Problem 16.49.   *Ans.* $M_0\pi R/4EI$

16.51. The beam of Fig. 16-38 is supported at the left end, clamped at the right end and subject to a concentrated load. Determine the reaction at the left support by Castigliano's theorem.
Ans. $Pb^2(2L + a)/2L^3$

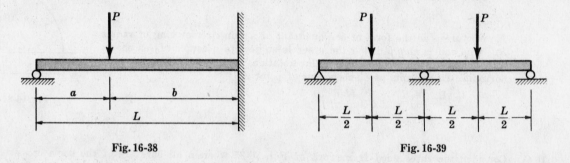

Fig. 16-38   Fig. 16-39

16.52. The two-span continuous beam shown in Fig. 16-39 supports the centrally applied forces. Use Castigliano's theorem to determine the central reaction.   *Ans.* $1.375P$

16.53. A two-span continuous beam is loaded by two centrally applied couples as shown in Fig. 16-40. Use Castigliano's theorem to determine the central reaction.   *Ans.* $2.25M_0/L$

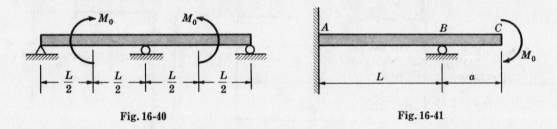

Fig. 16-40   Fig. 16-41

16.54. A beam is clamped at one end, supported at the point B, and loaded by a couple M_0 as shown in Fig. 16-41. Use Castigliano's theorem to determine the reaction at B.   *Ans.* $3M_0/2L$

16.55. The quadrant of a thin circular ring is shown in Fig. 16-42. The right end is free to displace horizontally due to the action of the force P; the left end is clamped. Determine the vertical reaction at the roller at B.   *Ans.* $2P/\pi$

16.56. A thin ring forms one quadrant of a circle and is loaded as shown in Fig. 16-43. One end is fixed and the other is pinned so as to prevent horizontal and vertical displacements. Find the components of reaction at the pin.   *Ans.* $B_v = 0.19M_0/R$, $B_h = 1.12M_0/R$

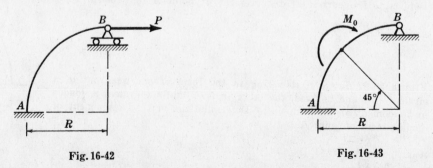

Fig. 16-42   Fig. 16-43

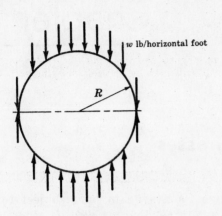

Fig. 16-44

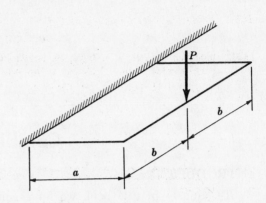

Fig. 16-45

16.57. A thin ring is loaded by forces which are uniformly distributed along the horizontal projection of the ring (see Fig. 16-44). Determine the decrease in the vertical diameter. *Ans.* $wR^4/6EI$

16.58. A structure in the form of a thin three-sided rectangular frame lies in a horizontal plane, has both ends clamped, and is subject to a central vertical force P, as shown in Fig. 16-45. Determine the reactive torque at each support. The frame is of constant cross-section throughout.

Ans. $\dfrac{Pb^2/EI}{\dfrac{b}{EI} + \dfrac{a}{GJ}}$

16.59. A thin structure in the form of one quadrant of a circle (Fig. 16-46) lies in a horizontal plane and is subject to a torque T_0 at the free end. The other end is clamped. Determine the vertical displacement of the free end.

Ans. $T_0R^2\left(\dfrac{\pi}{4EI} + \dfrac{\pi}{4GJ} - \dfrac{1}{GJ}\right)$

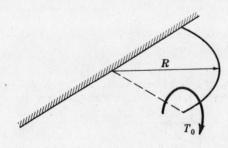

Fig. 16-46

Chapter 17

Combined Stresses

INTRODUCTION

Previously in this book we have considered stresses arising in bars subject to axial loading, shafts subject to torsion, and beams subject to bending, as well as several cases involving thin-walled pressure vessels and riveted joints. It is to be noted that we have considered a bar, for example, to be subject to only *one* loading at a time, such as bending. But frequently such bars are simultaneously subject to several of the previously mentioned loadings, and it is required to determine the state of stress under these conditions. Since normal and shearing stress are vector quantities, considerable care must be exercised in combining the stresses given by the expressions for single loadings as derived in previous chapters. It is the purpose of this chapter to investigate the state of stress on an arbitrary plane through an element in a body subject to several simultaneous loadings.

GENERAL CASE OF TWO-DIMENSIONAL STRESS

In general if a plane element is removed from a body it will be subject to the normal stresses σ_x and σ_y together with the shearing stress τ_{xy} as shown in Fig. 17-1.

SIGN CONVENTION

For normal stresses, tensile stresses are considered to be positive, compressive stress negative. For shearing stresses, the positive sense is that illustrated in Fig. 17-1.

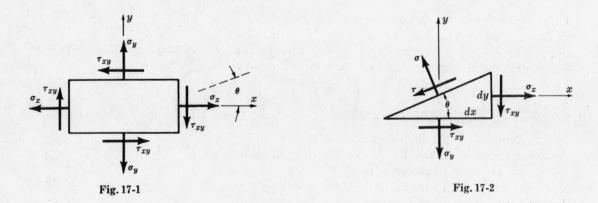

Fig. 17-1   Fig. 17-2

STRESSES ON AN INCLINED PLANE

We shall assume that the stresses σ_x, σ_y, and τ_{xy} are known. (Their determination will be discussed in Chapter 18.) Frequently it is desirable to investigate the state of stress on a plane inclined at an angle θ to the x-axis, as shown in Fig. 17-1. The normal and shearing stresses on such a plane are denoted by σ and τ and appear as in Fig. 17-2.   In Problem 8.17 it is shown that

$$\sigma = \frac{\sigma_x + \sigma_y}{2} - \frac{\sigma_x - \sigma_y}{2}\cos 2\theta + \tau_{xy}\sin 2\theta$$

$$\tau = \frac{\sigma_x - \sigma_y}{2}\sin 2\theta + \tau_{xy}\cos 2\theta$$

Thus, for any value of θ, σ and τ may be obtained from these expressions. For applications see Problems 17.2, 17.9 and 17.11.

PRINCIPAL STRESSES

There are certain values of the angle θ that lead to maximum and minimum values of σ for a given set of stresses σ_x, σ_y, and τ_{xy}. These maximum and minimum values that σ may assume are termed *principal stresses* and are given by

$$\sigma_{\max} = \frac{\sigma_x + \sigma_y}{2} + \sqrt{\left(\frac{\sigma_x - \sigma_y}{2}\right)^2 + (\tau_{xy})^2}$$

$$\sigma_{\min} = \frac{\sigma_x + \sigma_y}{2} - \sqrt{\left(\frac{\sigma_x - \sigma_y}{2}\right)^2 + (\tau_{xy})^2}$$

These expressions are derived in Problem 17.13. For applications see Problems 17.9, 17.11, 17.15, and 17.18.

DIRECTIONS OF PRINCIPAL STRESSES; PRINCIPAL PLANES

The angles designated as θ_p between the x-axis and the planes on which the principal stresses occur are given by the equation

$$\tan 2\theta_p = \frac{-\tau_{xy}}{\left(\dfrac{\sigma_x - \sigma_y}{2}\right)}$$

This expression also is derived in Problem 17.13. For applications see Problems 17.9, 17.11, 17.15, 17.18. As shown there we always have two values of θ_p satisfying this equation. The stress $\sigma_{\max}$ occurs on one of these planes, and the stress $\sigma_{\min}$ occurs on the other. The planes defined by the angles θ_p are known as *principal planes*.

SHEARING STRESSES ON PRINCIPAL PLANES

In Problem 17.13 it is demonstrated that the shearing stresses on the planes on which $\sigma_{\max}$ and $\sigma_{\min}$ occur are always zero, regardless of the values of σ_x, σ_y, and τ_{xy}. Thus, an element oriented along the principal planes and subject to the principal stresses appears as in Fig. 17-3.

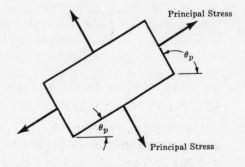

Fig. 17-3

MAXIMUM SHEARING STRESSES

There are certain values of the angle θ that lead to a maximum value of τ for a given set of stresses

σ_x, σ_y, and τ_{xy}. The maximum and minimum values of the shearing stress are given by

$$\tau_{\substack{max \\ min}} = \pm \sqrt{\left(\frac{\sigma_x - \sigma_y}{2}\right)^2 + (\tau_{xy})^2}$$

This expression is derived in Problem 17.13. For applications see Problems 17.3, 17.9, 17.11, 17.15, and 17.18.

DIRECTIONS OF MAXIMUM SHEARING STRESS

The angles θ_s between the x-axis and the planes on which the maximum shearing stresses occur are given by the equation

$$\tan 2\theta_s = \frac{\left(\dfrac{\sigma_x - \sigma_y}{2}\right)}{\tau_{xy}}$$

This expression also is derived in Problem 17.13. For applications see Problems 17.3, 17.9, 17.11, 17.15, and 17.18. There are always two values of θ_s satisfying this equation. The shearing stress corresponding to the positive square root given above occurs on one of the planes designated by θ_s, the shearing stress corresponding to the negative square root occurs on the other plane.

NORMAL STRESSES ON PLANES OF MAXIMUM SHEARING STRESS

In Problem 17.13 it is demonstrated that the normal stress on each of the planes of maximum shearing stress (which are of course 90° apart) is given by

$$\tau' = \frac{\sigma_x + \sigma_y}{2}$$

Thus an element oriented along the planes of maximum shearing stress appears as in Fig. 17-4. This is illustrated in Problems 17.9, 17.11, 17.15, 17.18.

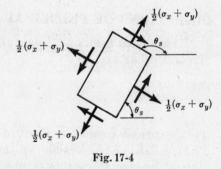

Fig. 17-4

MOHR'S CIRCLE

All of the information contained in the above equations may be presented in a convenient graphical form known as *Mohr's circle*. In this representation normal stresses are plotted along the horizontal axis and shearing stresses along the vertical axis. The stresses σ_x, σ_y, and τ_{xy} are plotted to scale and a circle is drawn through these points having its center on the horizontal axis. Figure 17-5 shows Mohr's circle for an element subject to the general case of plane stress. For applications see Problems 17.4, 17.5, 17.6, 17.8, 17.10, 17.12, 17.14, 17.16, 17.17, and 17.19.

SIGN CONVENTIONS USED WITH MOHR'S CIRCLE

Tensile stresses are considered to be positive and compressive stresses negative. Thus tensile stresses are plotted to the right of the origin in Fig. 17-5 and compressive stresses

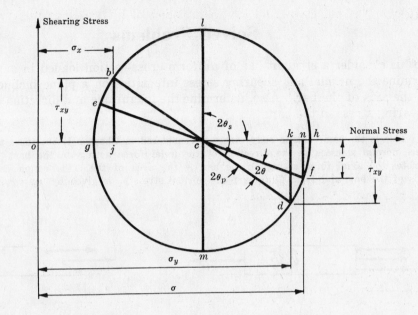

Fig. 17-5

to the left. With regard to shearing stresses it is to be carefully noted that a different sign convention exists than is used
in connection with the above-mentioned equations. We shall
refer to a plane element subject to shearing stresses and
appearing as in the adjoining diagram. We shall say that
shearing stresses are positive if they tend to rotate the element
clockwise, negative if they tend to rotate it counterclockwise.
Thus for the above element the shearing stresses on the vertical faces are positive, those on the horizontal faces are
negative.

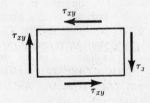

Fig. 17-6

DETERMINATION OF PRINCIPAL STRESSES
BY MEANS OF MOHR'S CIRCLE

When Mohr's circle has been drawn, the principal stresses are represented by the line
segments og and oh. These may either be scaled from the diagram or determined from
the geometry of the figure. This is explained in detail in Problem 17.14. For application
see Problems 17.8, 17.10, 17.12, 17.16, 17.17, 17.19.

DETERMINATION OF STRESSES ON AN ARBITRARY PLANE
BY MEANS OF MOHR'S CIRCLE

To determine the normal and shearing stresses on a plane inclined at a counterclockwise angle θ with the x-axis we measure a counterclockwise angle equal to 2θ from the
diameter bd of Mohr's circle. The endpoints of this diameter bd represent the stress
conditions in the original x-y directions, i.e. they represent the stresses σ_x, σ_y, and τ_{xy}.
The angle 2θ corresponds to the diameter ef. The coordinates of point f represent the
normal and shearing stresses on the plane at an angle θ to the x-axis. That is, the normal
stress σ is represented by the abscissa on and the shearing stress is represented by the
ordinate nf. This is discussed in detail in Problem 17.14. For applications see Problems
17.4, 17.5, 17.6, 17.8, 17.14, 17.17.

Solved Problems

17.1. Let us consider a straight bar of uniform cross-section loaded in axial tension. Determine the normal and shearing stress intensities on a plane inclined at an angle θ to the axis of the bar. Also, determine the magnitude and direction of the maximum shearing stress in the bar.

This is the same elastic body that was considered in Chapter 1, but there the stresses studied were normal stresses in the direction of the axial force acting on the bar. In Fig. 17-7(a) P denotes the axial force acting on the bar, A the area of the cross-section perpendicular to the axis of the bar, and from Chapter 1 the normal stress σ_x is given by $\sigma_x = P/A$.

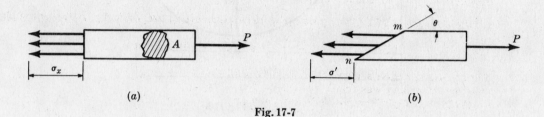

(a)   (b)

Fig. 17-7

Suppose now that instead of using a cutting plane which is perpendicular to the axis of the bar, we pass a plane through the bar at an angle θ with the axis of the bar. Such a plane mn is shown in Fig. 17-7(b). Since we must still have equilibrium of the bar in the horizontal direction there must evidently be distributed horizontal stresses acting over this inclined plane as shown. Let us designate the magnitude of these stresses by σ'. Evidently the area of the inclined cross-section is $A/\sin\theta$ and for equilibrium of forces in the horizontal direction we have

$$\sigma'(A/\sin\theta) = P \qquad \text{or} \qquad \sigma' = (P\sin\theta)/A$$

In Fig. 17-8 we consider only a single stress vector σ' and resolve it into two components, one normal to the inclined plane mn and one tangential to this plane. We shall label the first of these components σ to denote a normal stress, and the second τ to represent a shearing stress.

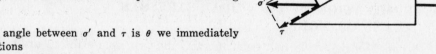

Since the angle between σ' and τ is θ we immediately have the relations

$$\tau = \sigma'\cos\theta \qquad \text{and} \qquad \sigma = \sigma'\sin\theta$$

Fig. 17-8

But $\sigma' = (P\sin\theta)/A$. Substituting this value in the above equations, we obtain

$$\tau = P\sin\theta\cos\theta/A \qquad \text{and} \qquad \sigma = P\sin^2\theta/A$$

But $\sigma_x = P/A$. Hence we may write these in the form

$$\tau = \sigma_x\sin\theta\cos\theta \qquad \text{and} \qquad \sigma = \sigma_x\sin^2\theta$$

Now, employing the trigonometric identities

$$\sin 2\theta = 2\sin\theta\cos\theta \qquad \text{and} \qquad \sin^2\theta = (1-\cos 2\theta)/2$$

we may write

$$\tau = \tfrac{1}{2}\sigma_x\sin 2\theta \tag{1}$$

$$\sigma = \tfrac{1}{2}\sigma_x(1-\cos 2\theta) \tag{2}$$

These expressions give the normal and shearing stresses on a plane inclined at an angle θ to the axis of the bar.

From these equations it is evident that the shearing stress is maximum when $\sin 2\theta$ assumes its maximum value of unity, i.e. when $2\theta = 90°$ or $\theta = 45°$. The value of this maximum shearing stress is evidently $\tau = \frac{1}{2}\sigma_x$. Also, the normal stress is maximum when $\cos 2\theta$ assumes its minimum value of -1, i.e. when $2\theta = 180°$ or $\theta = 90°$. For this value of θ the normal stress has the value $\sigma = \sigma_x$. Consequently the maximum normal stress acts over cross-sections perpendicular to the axis of the bar.

Thus we have the very interesting result that the maximum shearing stress in an axially loaded bar occurs on the planes at 45° to the direction of loading, and, further, that on these planes the value of this maximum shearing stress is $\tau = \frac{1}{2}\sigma_x$, or one-half the maximum normal stress.

17.2. A bar of cross-section 1.3 in² is acted upon by axial tensile forces of 15,000 lb applied at each end of the bar. Determine the normal and shearing stresses on a plane inclined at 30° to the direction of loading.

From Problem 17.1 the normal stress on a cross-section perpendicular to the axis of the bar is

$$\sigma_x = \frac{P}{A} = \frac{15,000}{1.3} = 11,500 \text{ lb/in}^2$$

The normal stress on a plane at an angle θ with the direction of loading was found in Problem 17.1 to be $\sigma = \frac{1}{2}\sigma_x(1 - \cos 2\theta)$. For $\theta = 30°$ this becomes

$$\sigma = \frac{1}{2}(11,500)(1 - \cos 60°) = 2870 \text{ lb/in}^2$$

The shearing stress on a plane at an angle θ with the direction of loading was found in Prob. 17.1 to be $\tau = \frac{1}{2}\sigma_x \sin 2\theta$. For $\theta = 30°$ this becomes

$$\tau = \frac{1}{2}(11,500) \sin 60° = 4940 \text{ lb/in}^2$$

These stresses together with the axial load of 15,000 lb are represented in Fig. 17-9.

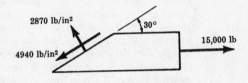

Fig. 17-9

17.3. Determine the maximum shearing stress in the axially loaded bar described in Problem 17.2.

The shearing stress on a plane at an angle θ with the direction of the load was shown in Problem 17.1 to be $\tau = \frac{1}{2}\sigma_x \sin 2\theta$. This is maximum when $2\theta = 90°$, i.e. when $\theta = 45°$. For this loading we have $\sigma_x = 11,500 \text{ lb/in}^2$ and when $\theta = 45°$ the shear stress is

$$\tau = \frac{1}{2}(11,500) \sin 90° = 5750 \text{ lb/in}^2$$

That is, the maximum shearing stress is equal to one-half of the maximum normal stress.

The normal stress on this 45° plane may be found from the expression

$$\sigma = \frac{1}{2}\sigma_x(1 - \cos 2\theta) = \frac{1}{2}(11,500)(1 - \cos 90°) = 5750 \text{ lb/in}^2$$

17.4. Discuss a graphical representation of equations (1) and (2) of Problem 17.1.

According to these equations the normal and shearing stresses on a plane inclined at an angle θ to the direction of loading are given by

$$\sigma = \frac{1}{2}\sigma_x(1 - \cos 2\theta) \quad \text{and} \quad \tau = \frac{1}{2}\sigma_x \sin 2\theta$$

To represent these relations graphically it is customary to introduce a rectangular cartesian coordinate system, plotting normal stresses as abscissas and shearing stresses as ordinates.

Let us proceed by first laying off to some convenient scale the normal stress σ_x (taken to be tensile) along the positive horizontal axis. The midpoint of this line segment, point c in Fig. 17-10, serves as the center of a circle whose diameter is σ_x. The radius of this circle, denoted by $\overline{oc}$, $\overline{ch}$ and $\overline{cd}$, is $\frac{1}{2}\sigma_x$. The angle 2θ is measured positive in a counterclockwise direction from the

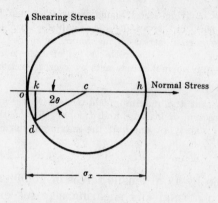

Fig. 17-10

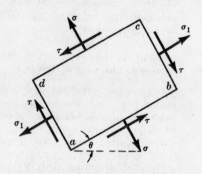

Fig. 17-11

radial line $\overline{oc}$. From the figure we immediately have the relations

$$\overline{kd} = \tau = \tfrac{1}{2}\sigma_x \sin 2\theta \qquad \overline{ok} = \overline{oc} - \overline{kc} = \tfrac{1}{2}\sigma_x - \tfrac{1}{2}\sigma_x \cos 2\theta = \sigma = \tfrac{1}{2}\sigma_x(1 - \cos 2\theta)$$

It is to be noted that the scales used in the horizontal and vertical directions are equal.

Thus the abscissa and ordinate of point d represent respectively the normal stress and the shearing stress acting on a plane at an angle θ with the axis of the bar subject to tension. In plotting this diagram tensile stresses are regarded as positive in algebraic sign and compressive stresses are taken to be negative. Let us return to Problem 17.1 and examine a free-body diagram (Fig. 17-11) of an element taken from the surface of the inclined section on which the stresses σ and τ act. We shall consider shearing stresses to be positive if they tend to rotate the element clockwise, negative if they tend to rotate the element counterclockwise. This sign convention is used only in this graphical representation, not in the analytical treatment of Problem 17.1. Since the shearing stresses found in Problem 17.1 were actually those acting on face dc of the above element, they should be regarded as negative. Hence in the above circular diagram representing normal and shearing stresses the shearing stress on plane dc appears as an ordinate $\overline{kd}$ plotted in the negative sense.

This diagram, termed *Mohr's circle*, was first presented by O. Mohr in 1882. It represents the variation of normal and shearing stresses on all inclined planes passing through a given point in the body. It is a convenient graphical representation of equations (1) and (2) of Problem 17.1.

17.5. Consider again the axially loaded bar discussed in Problem 17.2. Use Mohr's circle to determine the normal and shearing stresses on the 30° plane.

In Fig. 17-12 the normal stress of 11,500 lb/in² is laid off along the horizontal axis to some convenient scale and a circle is drawn with this line as a diameter. The angle $2\theta = 2(30°) = 60°$ is measured counterclockwise from $\overline{oc}$. The coordinates of the point d are

$$\overline{kd} = \tau = -\tfrac{1}{2}(11,500) \sin 60° = -4940 \text{ lb/in}^2$$

$$\overline{ok} = \sigma = \overline{oc} - \overline{kc} = \tfrac{1}{2}(11,500) - \tfrac{1}{2}(11,500) \cos 60°$$
$$= 2870 \text{ lb/in}^2$$

The negative sign accompanying the value of the shearing stress indicates that the shearing stress on this 30° plane tends to rotate an element bounded by this plane in a counterclockwise direction. This is in agreement with the direction of the shearing stress illustrated in Fig. 17-9.

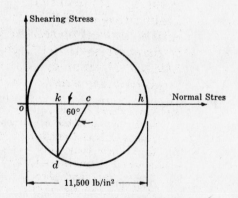

Fig. 17-12

17.6. A bar of cross-section 1.3 in² is acted upon by axial compressive forces of 15,000 lb applied to each end of the bar. Using Mohr's circle, find the normal and shearing stresses on a plane inclined at 30° to the direction of loading. Neglect the possibility of buckling of the bar.

The normal stress on a cross-section perpendicular to the axis of the bar is

$$\sigma_x = P/A = -15,000/1.3 = -11,500 \text{ lb/in}^2$$

We shall first lay off this compressive normal stress to some convenient scale along the negative end of the horizontal axis. The midpoint of this line segment, point c in Fig. 17-13, serves as the center of a circle whose diameter is 11,500 lb/in² to the scale chosen.

The angle $2\theta = 2(30°) = 60°$ with the vertex at c is measured counterclockwise from $\overline{co}$ as shown. The abscissa of point d represents the normal stress and the ordinate the shearing stress on the desired 30° plane. The coordinates of point d are

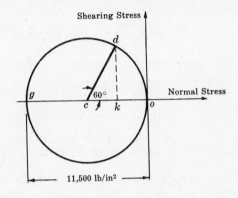

Fig. 17-13

$$\overline{kd} = \tau = \tfrac{1}{2}(11,500) \sin 60° = 4940 \text{ lb/in}^2$$

$$\overline{ok} = \sigma = \overline{oc} - \overline{ck} = \tfrac{1}{2}(11,500) - \tfrac{1}{2}(11,500) \cos 60° = 2870 \text{ lb/in}^2$$

It is to be noted that line segment $\overline{ok}$ lies to the left of the origin of coordinates, hence this normal stress is compressive.

The positive algebraic sign accompanying the shearing stress indicates that the shearing stress on the 30° plane tends to rotate an element (denoted by dashed lines in Fig. 17-14) bounded by this plane in a clockwise direction. The directions of the normal and shearing stresses together with the axial load of 15,000 lb are shown in the figure.

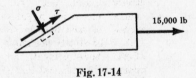

Fig. 17-14

17.7. Consider a plane element removed from a stressed elastic body and subject to the normal and shearing stresses σ_x and τ_{xy} respectively, as shown in Fig. 17-15. (a) Determine the normal and shearing stress intensities on a plane inclined at an angle θ to the normal stress σ_x. (b) Determine the maximum and minimum values of the normal stress that may exist on inclined planes and find the directions of these stresses. (c) Determine the magnitude and direction of the maximum shearing stress that may exist on an inclined plane.

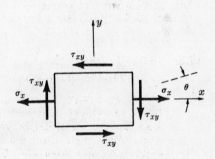

Fig. 17-15

(a) The desired normal and shearing stresses acting on an inclined plane are internal quantities with respect to the element shown in Fig. 17-15. We shall follow the customary procedure of cutting this element with a plane in such a manner as to render the desired stresses external to the new body; that is, we will cut the originally rectangular element along the plane inclined at an angle θ with the x-axis and thus obtain a triangular element as shown in Fig. 17-16. The normal and shearing stresses, designated as σ and τ respectively, represent the effect of the remaining portion of the originally rectangular block that has been removed. Consequently, the problem reduces to finding the unknown stresses σ and τ in terms of the known stresses σ_x and τ_{xy}. It is to be observed that in the free-body diagram of the triangular element, the vectors indicate stresses acting on the various faces of the element and not forces. Each of these stresses is assumed to be uniformly distributed over the area upon which it acts. The thickness of the element perpendicular to the plane of the paper is denoted by t.

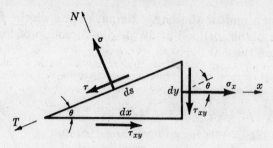

Fig. 17-16

Let us introduce N- and T-axes normal and tangent to the inclined plane as shown in Fig. 17-16. First, we shall sum forces in the N-direction. For equilibrium we have

$$\Sigma F_N \;=\; \sigma t\,ds \;-\; \sigma_x t\,dy\,\sin\theta \;-\; \tau_{xy} t\,dy\,\cos\theta \;-\; \tau_{xy} t\,dx\,\sin\theta \;=\; 0$$

But from trigonometry $dy = ds\,\sin\theta$, $dx = ds\,\cos\theta$. Substituting these relations in the equilibrium equation above, we find

$$\sigma(ds) \;=\; \sigma_x(ds)\,\sin^2\theta \;+\; 2\tau_{xy}(ds)\,\sin\theta\,\cos\theta$$

Next, employing the identities $\sin^2\theta = \tfrac{1}{2}(1 - \cos 2\theta)$ and $\sin 2\theta = 2\sin\theta\cos\theta$, we obtain

$$\sigma \;=\; \tfrac{1}{2}\sigma_x(1 - \cos 2\theta) + \tau_{xy}\sin 2\theta \;=\; \tfrac{1}{2}\sigma_x - \tfrac{1}{2}\sigma_x \cos 2\theta + \tau_{xy}\sin 2\theta \qquad (1)$$

Thus the normal stress σ on any plane inclined at an angle θ with the x-axis is known as a function of σ_x, τ_{xy}, and θ.

Next we shall consider the equilibrium of the forces acting on the triangular element in the T-direction. This leads to the equation

$$\Sigma F_T \;=\; \tau t\,ds \;-\; \sigma_x t\,dy\,\cos\theta \;+\; \tau_{xy} t\,dy\,\sin\theta \;-\; \tau_{xy} t\,dx\,\cos\theta \;=\; 0$$

Substituting $dy = ds\,\sin\theta$ and $dx = ds\,\cos\theta$, we obtain

$$\tau(ds) \;=\; +\sigma_x(ds)\,\sin\theta\,\cos\theta \;-\; \tau_{xy}(ds)\,\sin^2\theta \;+\; \tau_{xy}(ds)\,\cos^2\theta$$

Employing the identities $\cos 2\theta = \cos^2\theta - \sin^2\theta$ and $\sin 2\theta = 2\sin\theta\cos\theta$, this becomes

$$\tau \;=\; \tfrac{1}{2}\sigma_x\sin 2\theta + \tau_{xy}\cos 2\theta \qquad (2)$$

Thus the shearing stress τ on any plane inclined at an angle θ with the x-axis is known as a function of σ_x, τ_{xy}, and θ.

(b) To determine the maximum value that the normal stress σ may assume as the angle θ varies, we shall differentiate equation (1) with respect to θ and set this derivative equal to zero. Thus

$$\frac{d\sigma}{d\theta} \;=\; +\sigma_x\sin 2\theta + 2\tau_{xy}\cos 2\theta \;=\; 0$$

The values of θ leading to maximum and minimum values of the normal stress are consequently

$$\tan 2\theta_p \;=\; -\tau_{xy}/\tfrac{1}{2}\sigma_x \qquad (3)$$

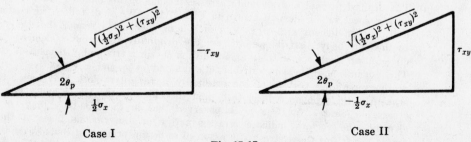

Case I   Case II

Fig. 17-17

The planes defined by the angles θ_p are called *principal planes*. The normal stresses that exist on these planes are designated as *principal stresses*. They are the maximum and minimum values that the normal stress may assume in the element under consideration. The values of the principal stresses may easily be found by interpreting equation (3) graphically, as in Fig. 17-17. Evidently the tangent of either of the angles designated as $2\theta_p$ has the value given in (3). Thus there are two solutions to (3), consequently two values of $2\theta_p$ (differing by 180°) and also two values of θ_p. These values of θ_p differ by 90°. It is to be noted that the triangles of Fig. 17-17 bear no direct relationship to the triangular element whose free-body diagram was considered earlier.

The values of $\sin 2\theta_p$ and $\cos 2\theta_p$ as found from Fig. 17-17 may now be substituted in (1) to yield the maximum and minimum values of the normal stresses. Observing that

$$\sin 2\theta_p = \frac{\mp \tau_{xy}}{\sqrt{(\tfrac{1}{2}\sigma_x)^2 + (\tau_{xy})^2}} \qquad \cos 2\theta_p = \frac{\pm\tfrac{1}{2}\sigma_x}{\sqrt{(\tfrac{1}{2}\sigma_x)^2 + (\tau_{xy})^2}}$$

where the upper signs pertain to Case I and the lower signs to Case II, we obtain from (1):

$$\sigma = \tfrac{1}{2}\sigma_x \mp \tfrac{1}{2}\sigma_x \frac{\tfrac{1}{2}\sigma_x}{\sqrt{(\tfrac{1}{2}\sigma_x)^2 + (\tau_{xy})^2}} \mp \frac{(\tau_{xy})^2}{\sqrt{(\tfrac{1}{2}\sigma_x)^2 + (\tau_{xy})^2}} = \tfrac{1}{2}\sigma_x \pm \sqrt{(\tfrac{1}{2}\sigma_x)^2 + (\tau_{xy})^2} \qquad (4)$$

The maximum normal stress is

$$\sigma_{\max} = \tfrac{1}{2}\sigma_x + \sqrt{(\tfrac{1}{2}\sigma_x)^2 + (\tau_{xy})^2} \qquad (5)$$

The minimum normal stress is

$$\sigma_{\min} = \tfrac{1}{2}\sigma - \sqrt{(\tfrac{1}{2}\sigma_x)^2 + (\tau_{xy})^2} \qquad (6)$$

The stresses given by (5) and (6) are the principal stresses and they occur on the principal planes defined by (3). By substituting one of the values of θ_p from (3) into (1), one may readily determine which of the two principal stresses is acting on that plane. The other principal stress naturally acts on the other principal plane.

By substituting the values of the angles $2\theta_p$ as given by (3) and Fig. 17-17 into (2), it is readily seen that the shearing stresses τ on the principal planes are zero.

(c) To determine the maximum value the shearing stress τ may assume as the angle θ varies, we shall differentiate equation (2) with respect to θ and set this derivative equal to zero. Thus

$$\frac{d\tau}{d\theta} = \sigma_x \cos 2\theta - 2\tau_{xy} \sin 2\theta = 0$$

The values of θ leading to maximum values of the shearing stress are consequently

$$\tan 2\theta_s = \frac{\tfrac{1}{2}\sigma_x}{\tau_{xy}} \qquad (7)$$

The planes defined by the two solutions to this equation are the planes of maximum shearing stress.

Again, a graphical interpretation of (7) is convenient. The two values of the angle $2\theta_s$ satisfying this equation may be represented as in Fig. 17-18. We see that

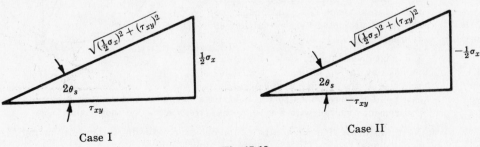

Case I   Case II

Fig. 17-18

$$\sin 2\theta_s = \frac{\pm\frac{1}{2}\sigma_x}{\sqrt{(\frac{1}{2}\sigma_x)^2 + (\tau_{xy})^2}} \qquad \cos 2\theta_s = \frac{\pm\tau_{xy}}{\sqrt{(\frac{1}{2}\sigma_x)^2 + (\tau_{xy})^2}}$$

where the upper (positive) signs pertain to Case I and the lower (negative) signs apply to Case II. Substituting these values in (2) we obtain

$$\tau_{\substack{max \\ min}} = \frac{1}{2}\sigma_x \frac{\pm\frac{1}{2}\sigma_x}{\sqrt{(\frac{1}{2}\sigma_x)^2 + (\tau_{xy})^2}} + (\tau_{xy}) \frac{\pm\tau_{xy}}{\sqrt{(\frac{1}{2}\sigma_x)^2 + (\tau_{xy})^2}} = \pm\sqrt{(\frac{1}{2}\sigma_x)^2 + (\tau_{xy})^2} \qquad (8)$$

Here the positive sign represents the maximum shearing stress, the negative sign the minimum shearing stress.

If we compare (3) and (7) it is evident that the angles $2\theta_p$ and $2\theta_s$ differ by 90°, since the tangents of these angles are the negative reciprocals of one another. Hence the planes defined by the angles θ_p and θ_s differ from one another by 45°; that is, the planes of maximum shearing stress are oriented 45° from the planes of maximum normal stress.

It is also of interest to determine the normal stresses on the planes of maximum shearing stress. These planes are defined by (7). If we now substitute these values of $\sin 2\theta_s$ and $\cos 2\theta_s$ in (1) for the normal stress we find

$$\sigma = \frac{1}{2}\sigma_x - \frac{1}{2}\sigma_x \frac{\pm\tau_{xy}}{\sqrt{(\frac{1}{2}\sigma_x)^2 + (\tau_{xy})^2}} + (\tau_{xy}) \frac{\pm\frac{1}{2}\sigma_x}{\sqrt{(\frac{1}{2}\sigma_x)^2 + (\tau_{xy})^2}} = \frac{1}{2}\sigma_x \qquad (9)$$

Thus on each plane of maximum shearing stress we have a normal stress of magnitude $\frac{1}{2}\sigma_x$.

17.8. Discuss a graphical representation of the analysis presented in Problem 17.7.

For given values of σ_x and τ_{xy} proceed as follows:

1. Introduce a rectangular coordinate system in which normal stresses are represented along the horizontal axis and shearing stresses along the vertical axis. The scales used on there two axes must be equal.

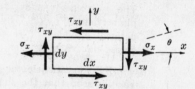

2. With reference to the original rectangular element considered in Problem 17.7 and reproduced in Fig. 17-19, we shall introduce the sign convention that shearing stresses are positive if they tend to rotate the element clockwise, negative if they tend to rotate it counterclockwise. Here the shearing stresses on the vertical faces are positive, those on the horizontal faces are negative. Also, tensile stresses are considered to be positive and compressive stresses negative.

Fig. 17-19

3. We first locate point b by laying out σ_x and τ_{xy} to their given values. The shear stress τ_{xy} on the vertical faces on which σ_x acts is positive, hence this value is plotted as positive in Fig. 17-20. This is drawn on the assumption that σ_x is a tensile stress, although the treatment presented here is valid if σ_x if compressive.

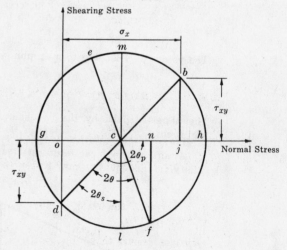

4. We next locate point d in a similar manner by laying off τ_{xy} on the negative side of the vertical axis. Actually, this point d corresponds to the negative shearing stresses τ_{xy} existing on the horizontal faces of the element together with a zero normal stress acting on those same faces.

Fig. 17-20

5. Next, we draw line $\overline{bd}$, locate the midpoint c, and draw a circle having its center at c and radius equal to $\overline{cb}$. This is known as Mohr's circle.

We shall first show that the points g and h along the horizontal diameter of the circle represent the principal stresses. To do this we note that the point c lies at a distance $\frac{1}{2}\sigma_x$ from the origin of the coordinate system. From the right-triangle relationship we have

$$(\overline{cd})^2 = (\overline{oc})^2 + (\overline{od})^2 \quad \text{or} \quad \overline{cd} = \sqrt{(\tfrac{1}{2}\sigma_x)^2 + (\tau_{xy})^2}$$

Also, we have $\overline{cd} = \overline{ch} = \overline{cg}$. Hence, the x-coordinate of point h is $\overline{oc} + \overline{ch}$ or

$$\tfrac{1}{2}\sigma_x + \sqrt{(\tfrac{1}{2}\sigma_x)^2 + (\tau_{xy})^2}$$

But this expression is exactly the maximum principal stress, as found in (5) of Problem 17.7. Likewise the x-coordinate of point g is $\overline{oc} - \overline{cg}$. But this quantity is negative; hence $\overline{og}$ lies to the left of the origin, and point g symbolizes a compressive stress. This stress becomes

$$\tfrac{1}{2}\sigma_x - \sqrt{(\tfrac{1}{2}\sigma_x)^2 + (\tau_{xy})^2}$$

But this expression is exactly the minimum principal stress, as found in (6) of Problem 17.7. Consequently the points g and h represent the principal stresses existing in the original element. We see that the tangent of $\angle ocd$ is $\tau_{xy}/(\frac{1}{2}\sigma_x)$. But from (3) of Problem 17.7, $\tan 2\theta_p = -\tau_{xy}/\frac{1}{2}\sigma_x$; and by comparison of these two relations we see that $\angle hcd = 2\theta_p$, since $\tan(180° - \theta) = -\tan\theta$. Thus a counterclockwise rotation from the diameter $\overline{bd}$ (corresponding to the stresses in the x-y directions) leads us to the diameter $\overline{gh}$, representing the principal planes, on which the principal stresses occur. The principal planes lie at an angle θ_p from the x-direction.

Thus Mohr's circle is a convenient device for finding the principal stresses, since one can merely establish the circle for a given set of stresses σ_x and τ_{xy} then measure $\overline{og}$ and $\overline{oh}$. These abscissas represent the principal stresses to the same scale used in plotting σ_x and τ_{xy}.

It is now apparent that the radius of Mohr's circle, represented by $\overline{cd} = \sqrt{(\frac{1}{2}\sigma_x)^2 + (\tau_{xy})^2}$, corresponds to the maximum shearing stress, as found in (8) of Problem 17.7. Actually, the shearing stress on any plane is represented by the ordinate to Mohr's circle; hence we should consider the radial lines $\overline{cl}$ and $\overline{cm}$ as representing the maximum shearing stresses. The angle dcl is evidently $2\theta_s$ and hence it is apparent that the double angle between the planes of maximum normal stress and the planes of maximum shearing stress ($\angle lch$) is 90°; thus the planes of maximum shearing stress are oriented 45° from the planes of maximum normal stress.

Evidently the endpoints of the diameter $\overline{bd}$ represent the stresses acting in the original x and y directions. We shall now demonstrate that the endpoints of any other diameter, such as $\overline{ef}$ (at any angle 2θ with $\overline{bd}$), represent the stresses on a plane inclined at an angle θ to the x-axis. To do this we note that the abscissa of point f is given by

$$\sigma = \overline{oc} + \overline{cn} = \tfrac{1}{2}\sigma_x + \overline{cf}\cos(2\theta_p - 2\theta)$$

$$= \tfrac{1}{2}\sigma_x + \overline{cf}(\cos 2\theta_p \cos 2\theta + \sin 2\theta_p \sin 2\theta)$$

$$= \tfrac{1}{2}\sigma_x + \sqrt{(\tfrac{1}{2}\sigma_x)^2 + (\tau_{xy})^2}\,(\cos 2\theta_p \cos 2\theta + \sin 2\theta_p \sin 2\theta)$$

But from inspection of triangle cod appearing in Mohr's circle it is evident that

$$\sin 2\theta_p = \frac{\tau_{xy}}{\sqrt{(\tfrac{1}{2}\sigma_x)^2 + (\tau_{xy})^2}} \quad \text{and} \quad \cos 2\theta_p = \frac{-\tfrac{1}{2}\sigma_x}{\sqrt{(\tfrac{1}{2}\sigma_x)^2 + (\tau_{xy})^2}} \tag{1}$$

Substituting the values of τ_{xy} and $\frac{1}{2}\sigma_x$ from these two equations into the previous equation, we find

$$\sigma = \tfrac{1}{2}\sigma_x - \tfrac{1}{2}\sigma_x \cos 2\theta + \tau_{xy}\sin 2\theta$$

But this is exactly the normal stress on a plane inclined at an angle θ to the x-axis as derived in (1) of Problem 17.7.

Next we observe that the ordinate of point f is given by

$$\tau = \overline{nf} = \overline{cf}\sin(2\theta_p - 2\theta)$$

$$= \sqrt{(\tfrac{1}{2}\sigma_x)^2 + (\tau_{xy})^2}\,(\sin 2\theta_p \cos 2\theta - \cos 2\theta_p \sin 2\theta)$$

Again, substituting the values of τ_{xy} and $\frac{1}{2}\sigma_x$ from equations (1) into this equation, we find

$$\tau = \tfrac{1}{2}\sigma_x \sin 2\theta + \tau_{xy}\cos 2\theta$$

But this is exactly the shearing stress on a plane inclined at an angle θ to the x-axis as derived in (2) of Problem 17.7.

Hence the coordinates of point f on Mohr's circle represent the normal and shearing stresses on a plane inclined at an angle θ to the x-axis.

17.9. A plane element in a body is subjected to a normal stress in the x-direction of 12,000 lb/in², as well as a shearing stress of 4000 lb/in², as shown in Fig. 17-21. (a) Determine the normal and shearing stress intensities on a plane inclined at an angle of 30° to the normal stress. (b) Determine the maximum and minimum values of the normal stress that may exist on inclined planes and the directions of these stresses. (c) Determine the magnitude and direction of the maximum shearing stress that may exist on an inclined plane.

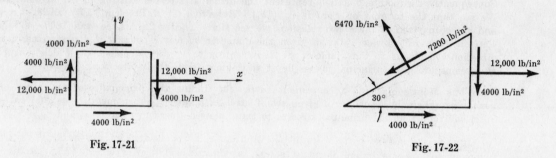

Fig. 17-21   Fig. 17-22

(a) In accordance with the notation of Problem 17.7 we have $\sigma_x = 12,000$ lb/in² and $\tau_{xy} = 4000$ lb/in². From (1) of Problem 17.7, the normal stress on a plane inclined at an angle θ to the x-axis is

$$\sigma = \tfrac{1}{2}\sigma_x - \tfrac{1}{2}\sigma_x \cos 2\theta + \tau_{xy} \sin 2\theta$$

Substituting the above values of σ_x and τ_{xy}, when $\theta = 30°$ this becomes

$$\sigma = \tfrac{1}{2}(12,000) - \tfrac{1}{2}(12,000) \cos 60° + 4000 \sin 60° = 6470 \text{ lb/in}^2$$

From (2) of Problem 17.7 the shearing stress on any plane inclined at an angle θ to the x-axis is

$$\tau = \tfrac{1}{2}\sigma_x \sin 2\theta + \tau_{xy} \cos 2\theta$$

Substituting the above values of σ_x and τ_{xy}, when $\theta = 30°$ this becomes

$$\tau = \tfrac{1}{2}(12,000) \sin 60° + 4000 \cos 60° = 5200 + 2000 = 7200 \text{ lb/in}^2$$

The positive directions of the normal and shearing stresses on an inclined plane were illustrated in Fig. 17-16. In accordance with this sign convention the stresses on the 30° plane appear as in Fig. 17-22.

(b) The values of the principal stresses, that is, the maximum and minimum values of the normal stresses existing in this element, were given by (5) and (6) of Problem 17.7. From (5) for the maximum normal stress we have

$$\sigma_{\max} = \tfrac{1}{2}\sigma_x + \sqrt{(\tfrac{1}{2}\sigma_x)^2 + (\tau_{xy})^2} = 6000 + \sqrt{(6000)^2 + (4000)^2} = 13,220 \text{ lb/in}^2$$

From (6) for the minimum normal stress we have

$$\sigma_{\min} = \tfrac{1}{2}\sigma_x - \sqrt{(\tfrac{1}{2}\sigma_x)^2 + (\tau_{xy})^2} = 6000 - \sqrt{(6000)^2 + (4000)^2} = -1220 \text{ lb/in}^2$$

The directions of the planes on which these principal stresses occur were found in (3) of Problem 17.7 to be

$$\tan 2\theta_p = -\frac{\tau_{xy}}{\tfrac{1}{2}\sigma_x} = -\frac{4000}{6000} = -\frac{2}{3}$$

Since the tangent of the angle $2\theta_p$ is negative, the two values of $2\theta_p$ lie in the second and fourth quadrants.   In the second quadrant, $2\theta_p = 146°20'$; in the fourth quadrant, $2\theta_p' = 326°20'$.   Consequently we have the principal planes defined by $\theta_p = 73°10'$ and $\theta_p' = 163°10'$. If $\theta_p = 73°10'$, together with the given values of σ_x and τ_{xy}, is now substituted in (1) of Problem 17.7, we find

$$\sigma = \tfrac{1}{2}\sigma_x - \tfrac{1}{2}\sigma_x \cos 2\theta + \tau_{xy} \sin 2\theta = 6000 - 6000 \cos 146°20' + 4000 \sin 146°20'$$

$$= 6000 - 6000(-0.833) + 4000(0.554) = 13,220 \text{ lb/in}^2$$

Thus the principal stress of 13,220 lb/in² occurs on the principal plane oriented at 73°10' to the x-axis.   The principal stresses thus appear as in Fig. 17-23.   As stated in Problem 17.7 the shearing stresses on these principal planes are zero.

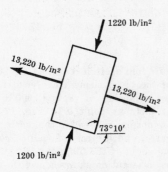

Fig. 17-23

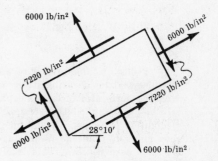

Fig. 17-24

(c)   The values of the maximum and minimum shearing stresses were found in (8) of Problem 17.7 to be

$$\tau_{\substack{max \\ min}} = \pm\sqrt{(\tfrac{1}{2}\sigma_x)^2 + (\tau_{xy})^2} = \pm\sqrt{(6000)^2 + (4000)^2} = \pm 7220 \text{ lb/in}^2$$

The directions of the planes on which these maximum shearing stresses occur were found in (7) of Problem 17.7 to be given by

$$\tan 2\theta_s = \frac{\tfrac{1}{2}\sigma_x}{\tau_{xy}} = \frac{6000}{4000} = \frac{3}{2}$$

The angles $2\theta_s$ are consequently in the first and third quadrants, since the tangent is positive. Thus we have $2\theta_s = 56°20'$ and $2\theta_s' = 236°20'$, or $\theta_s = 28°10'$ and $\theta_s' = 118°10'$.   The shearing stress on any plane inclined at an angle θ with the x-axis was found in (2) of Problem 17.7 to be

$$\tau = \tfrac{1}{2}\sigma_x \sin 2\theta + \tau_{xy} \cos 2\theta$$

Substituting $\sigma_x = 12,000$ lb/in², $\tau_{xy} = 4000$ lb/in² and $\theta = 28°10'$, we find

$$\tau = \tfrac{1}{2}(12,000) \sin 56°20' + 4000 \cos 56°20' = +7220 \text{ lb/in}^2$$

Thus the shearing stress on the 28°10' plane is positive.   The positive sense of shearing stress was shown in Fig. 17-6.

The normal stresses on the planes of maximum shearing stress are found from (9) of Problem 17.7 to be

$$\sigma = \tfrac{1}{2}\sigma_x = \tfrac{1}{2}(12,000) = 6000 \text{ lb/in}^2$$

This normal stress acts on each of the planes of maximum shearing stress, as shown in Fig. 17-24.

17.10. A plane element is subject to the stresses shown in Fig. 17-25 below.   Using Mohr's circle, determine (a) the principal stresses and their directions, (b) the maximum shearing stresses and the directions of the planes on which they occur.

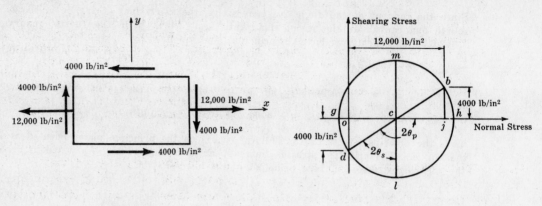

Fig. 17-25   Fig. 17-26

        Following the procedure for the construction of Mohr's circle outlined in Problem 17.8, we realize that the shearing stress on the vertical faces of the given element are positive, whereas those on the horizontal faces are negative. Thus the stress condition of $\sigma_x = 12{,}000$ lb/in^2, $\tau_{xy} = 4000$ lb/in^2 existing on the vertical faces of the element plots as point b in Fig. 17-26. The stress condition of $\tau_{xy} = -4000$ lb/in^2 together with a zero normal stress on the horizontal faces plots as point d. Line $\overline{bd}$ is drawn, its midpoint c is located, and a circle of radius $\overline{cb} = \overline{cd}$ is drawn with c as a center. This is Mohr's circle. The endpoints of the diameter $\overline{bd}$ represent the stress conditions existing in the element if it has the original orientation shown above.

(a)   The principal stresses are represented by points g and h, as shown in Problem 17.8. The principal stresses may be determined either by direct measurement from the above diagram or by realizing that the coordinate of c is 6000, and that $\overline{cd} = \sqrt{(6000)^2 + (4000)^2} = 7220$. Therefore the minimum principal stress is

$$\sigma_{\min} = \overline{og} = \overline{oc} - \overline{cg} = 6000 - 7220 = -1220 \text{ lb/in}^2$$

Also, the maximum principal stress is

$$\sigma_{\max} = \overline{oh} = \overline{oc} + \overline{ch} = 6000 + 7220 = 13{,}220 \text{ lb/in}^2$$

The angle $2\theta_p$ designated above is given by

$$\tan 2\theta_p = -\frac{4000}{6000} = -\frac{2}{3} \quad \text{or} \quad \theta_p = 73°10'$$

This value could also be obtained by measurement of $\angle dch$ in Mohr's circle. From this it is readily seen that the principal stress represented by point h acts on a plane oriented 73°10′ from the original x-axis. The principal stresses thus appear as in Fig. 17-27(a). It is evident from Mohr's circle that the shearing stresses on these planes are zero, since points g and h lie on the horizontal axis of Mohr's circle.

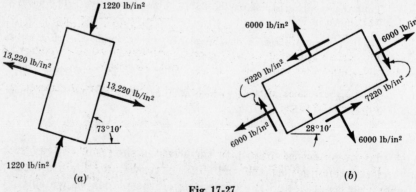

Fig. 17-27

(b) The maximum shearing stress is represented by $\overline{cl}$ in Mohr's circle. This radius has already been found to be equal to 7220 lb/in². The angle $2\theta_s$ may be found either by direct measurement from the above plot or simply by subtracting 90° from the angle $2\theta_p$, which has already been determined. This leads to $2\theta_s = 56°20'$ and $\theta_s = 28°10'$. The shearing stress represented by point l is negative, hence on this 28°10′ plane the shearing stress tends to rotate the element in a counterclockwise direction. Also, from Mohr's circle the abscissa of point l is 6000 lb/in² and this represents the normal stress occurring on the planes of maximum shearing stress. The maximum shearing stresses thus appear as in Fig. 17-27(b).

17.11. A plane element in a body is subject to a normal compressive stress in the x-direction of 12,000 lb/in² as well as a shearing stress of 4000 lb/in², as shown in Fig. 17-28. (a) Determine the normal and shearing stress intensities on a plane inclined at an angle of 30° to the normal stress. (b) Determine the maximum and minimum values of the normal stress that may exist on inclined planes and the direction of these stresses. (c) Find the magnitude and direction of the maximum shearing stress that may exist on an inclined plane.

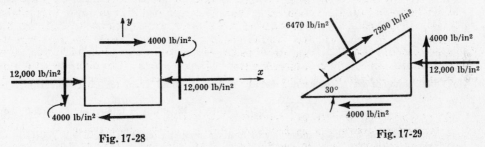

Fig. 17-28   Fig. 17-29

(a) By the sign convention for normal and shearing stresses adopted in Problem 17.7 we have here $\sigma_x = -12,000$ lb/in², $\tau_{xy} = -4000$ lb/in². From (1) of Problem 17.7 the normal stress on the 30° plane is

$$\sigma = -12,000/2 - (-12,000/2)\cos 60° - 4000 \sin 60° = -6470 \text{ lb/in}^2$$

From (2) of Problem 17.7 the shearing stress on the 30° plane is

$$\tau = \tfrac{1}{2}(-12,000)\sin 60° - 4000 \cos 60° = -7200 \text{ lb/in}^2$$

The positive directions of the normal and shearing stresses on an inclined plane were illustrated in Fig. 17-16. By this sign convention the stresses on the 30° plane appear as in Fig. 17-29.

(b) The values of the principal stresses were given by (5) and (6) of Problem 17.7. From (5),

$$\sigma_{\max} = -12,000/2 + \sqrt{(-12,000/2)^2 + (-4000)^2} = 1220 \text{ lb/in}^2$$

From (6),

$$\sigma_{\min} = -12,000/2 - \sqrt{(-12,000/2)^2 + (-4000)^2} = -13,220 \text{ lb/in}^2$$

The tensile principal stress is usually referred to as the maximum, even though its absolute value is smaller than that of the compressive stress.

The directions of the planes on which these principal stresses occur are given by (3) of Problem 17.7 to be

$$\tan 2\theta_p = -\frac{\tau_{xy}}{\tfrac{1}{2}\sigma_x} = -\frac{-4000}{-12,000/2} = -2/3$$

The angles defined by $2\theta_p$ lie in the second and fourth quadrants since the tangent is negative. Hence $2\theta_p = 146°20'$ and $2\theta_p' = 326°20'$. Thus the principal planes are defined by $\theta_p = 73°10'$ and $\theta_p' = 163°10'$ If $\theta_p = 73°10'$, together with the given values of σ_x and τ_{xy}, is now substituted in (1) of Problem 17.7 we find

$$\sigma = \tfrac{1}{2}\sigma_x - \tfrac{1}{2}\sigma_x \cos 2\theta + \tau_{xy} \sin 2\theta$$

$$= -12{,}000/2 - (-12{,}000/2) \cos 146°20' - 4000 \sin 146°20' = -13{,}220 \text{ lb/in}^2$$

Thus the principal stress of $-13{,}220$ lb/in² occurs on the principal plane oriented at $73°10'$ to the x-axis. The principal stresses are shown in Fig. 17-30. The shearing stresses on these principal planes are zero.

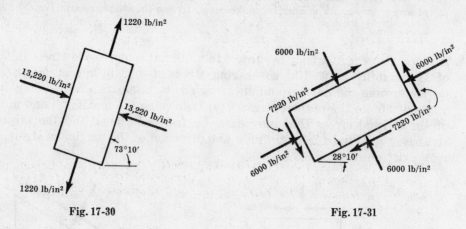

Fig. 17-30   Fig. 17-31

(c) The value of the maximum shearing stress is found from (8) of Problem 17.7 to be

$$\tau_{\substack{\max \\ \min}} = \pm\sqrt{(\tfrac{1}{2}\sigma_x)^2 + (\tau_{xy})^2} = \pm\sqrt{(-12{,}000/2)^2 + (-4000)^2} = \pm 7220 \text{ lb/in}^2$$

The directions of the planes on which these shearing stresses occur was found in (7) of Problem 17.7 to be

$$\tan 2\theta_s = \frac{\tfrac{1}{2}\sigma_x}{\tau_{xy}} = \frac{-12{,}000/2}{-4000} = \frac{3}{2}$$

Thus $2\theta_s = 56°20'$ and $2\theta_s' = 236°20'$; or $\theta_s = 28°10'$ and $\theta_s' = 118°10'$. From (2) of Problem 17.7, the shearing stress on any plane inclined at an angle θ with the x-axis is

$$\tau = \tfrac{1}{2}\sigma_x \sin 2\theta + \tau_{xy} \cos 2\theta = \tfrac{1}{2}(-12{,}000) \sin 56°20' - 4000 \cos 56°20' = -7220 \text{ lb/in}^2$$

Thus the shearing stress on the $28°10'$ plane is negative. The positive sense of shearing stress was shown in Fig. 17-16.

The normal stresses on the planes of maximum shearing stress were found in (9) of Problem 17.7 to be

$$\sigma = \tfrac{1}{2}\sigma_x = -12{,}000/2 = -6000 \text{ lb/in}^2$$

This normal stress acts on each of the planes of maximum shearing stress, as shown in Fig. 17-31.

17.12. A plane element is subject to the stresses shown in Fig. 17-32 below. Using Mohr's circle, determine (a) the principal stresses and their directions, (b) the maximum shearing stresses and the directions of the planes on which they occur.

The procedure for the construction of Mohr's circle was outlined in Problem 17.8. Following the instructions there, the shearing stresses on the vertical faces of the above element are negative, those on the horizontal faces are positive. Thus the stress condition of $\sigma_x = -12{,}000$ lb/in², $\tau_{xy} = -4000$ lb/in² existing on the vertical faces of the element plots as point b in Fig. 17-33. The stress condition of $\tau_{xy} = 4000$ lb/in², together with a zero normal stress on the horizontal faces, plots as point d. Line $\overline{bd}$ is drawn, its midpoint c is located, and a circle of radius $\overline{cb} = \overline{cd}$ is drawn with c as a center. This is Mohr's circle. The endpoints of the diameter $\overline{bd}$ represent the stress conditions existing in the element if it has the original orientation shown in Fig. 17-32.

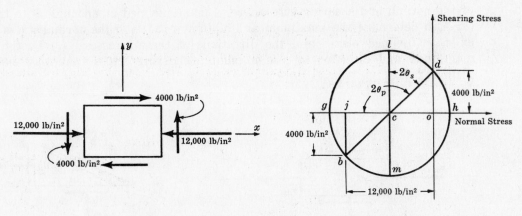

Fig. 17-32   Fig. 17-33

(a) The principal stresses are represented by points g and h, as demonstrated in Problem 17.8. They may be determined either by direct measurement from the above diagram or by realizing that the coordinate of c is -6000, and that $\overline{cd} = \sqrt{(6000)^2 + (4000)^2} = 7220$. Thus the minimum principal stress is

$$\sigma_{\min} = \overline{og} = +(\overline{oc} + \overline{cg}) = -6000 - 7220 = -13,220 \text{ lb/in}^2$$

The maximum principal stress is

$$\sigma_{\max} = \overline{oh} = \overline{ch} - \overline{co} = 7220 - 6000 = 1220 \text{ lb/in}^2$$

The angle $2\theta_p$ designated above is given by $\tan 2\theta_p = -4000/6000 = -2/3$ since $\tan(180° - \theta) = -\tan\theta$. Hence $2\theta_p = 146°20'$, $\theta_p = 73°10'$. This value could of course have been obtained by direct measurement of angle dcg in Mohr's circle. Thus the principal stress of $-13,220$ lb/in² represented by point g acts on a plane oriented $73°10'$ from the original x-axis. The principal stresses thus appear as in Fig. 17.34. It is evident from Mohr's circle that the shearing stresses on these planes are zero, since points g and h lie on the horizontal axis of Mohr's circle.

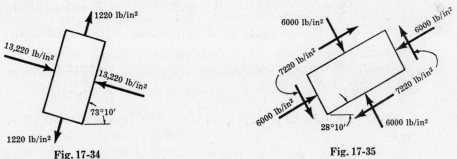

Fig. 17-34   Fig. 17-35

(b) The maximum shearing stress is represented by $\overline{cl}$ in Mohr's circle. This radius has already been found to be equal to 7220 lb/in². The angle $2\theta_s$ may be found either by direct measurement from Mohr's circle or simply by subtracting 90° from the above value of $2\theta_p$. This leads to $\theta_s = 28°10'$. The shearing stress represented by point l is positive; hence on this $28°10'$ plane the shearing stress tends to rotate the element in a clockwise direction. Also, from Mohr's circle the abscissa of point l is -6000 lb/in² and this represents the normal stress occurring on the planes of maximum shearing stresses, as shown in Fig. 17-35.

17.13. Consider a plane element removed from a stressed elastic member. In general such an element will be subject to normal stresses in each of two perpendicular directions, as well as shearing stresses. Let these stresses be denoted by σ_x, σ_y, and τ_{xy} and have the positive directions shown in Fig. 17-36 below. (a) Determine the magnitudes of the normal and shearing stresses on a plane inclined at an angle θ to the x-axis. (*b*) Also determine the maximum and minimum values of the normal stress that may exist on inclined planes and the directions of these stresses. (*c*) Lastly, find the magnitude and direction of the maximum shearing stress that may exist on an inclined plane.

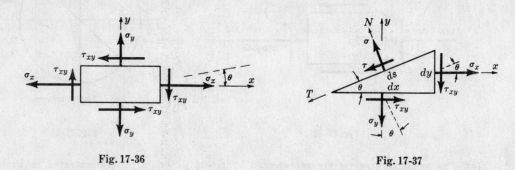

Fig. 17-36   Fig. 17-37

(*a*) Evidently the desired stresses acting on the inclined planes are internal quantities with respect to the element shown in Fig. 17-36. Following the usual procedure of introducing a cutting plane so as to render the desired quantities external to the new section, we cut the originally rectangular element along the plane inclined at the angle θ to the x-axis and thus obtain the triangular element shown in Fig. 17-37. Since we have removed half of the material in the rectangular element, we must replace it by the effect that it exerted upon the remaining lower triangle shown and this effect in general consists of both normal and shearing forces acting along the inclined plane. We shall designate the magnitudes of the normal and shearing stresses corresponding to these forces by σ and τ respectively. Thus our problem reduces to finding the unknown stresses σ and τ in terms of the known stresses σ_x, σ_y, and τ_{xy}. Chapter 18 illustrates the manner of determination of the stresses σ_x, σ_y, and τ_{xy}. It is to be carefully noted that the free-body diagram, Fig. 17-37, indicates stresses acting on the various faces of the element, and not forces. Each of these stresses is assumed to be uniformly distributed over the area on which it acts.

We shall introduce the N- and T-axes normal and tangential to the inclined plane as shown. Let t denote the thickness of the element perpendicular to the plane of the page. Let us begin, by summing forces in the N-direction. For equilibrium we have

$$\Sigma F_N = \sigma t\, ds - \sigma_x t\, dy \sin\theta - \tau_{xy} t\, dy \cos\theta - \sigma_y t\, dx \cos\theta - \tau_{xy} t\, dx \sin\theta = 0$$

Substituting $dy = ds \sin\theta$, $dx = ds \cos\theta$ in the equilibrium equation,

$$\sigma\, ds = \sigma_x\, ds \sin^2\theta + \sigma_y\, ds \cos^2\theta + 2\tau_{xy}\, ds \sin\theta \cos\theta$$

Introducing the identities $\sin^2\theta = \tfrac{1}{2}(1 - \cos 2\theta)$, $\cos^2\theta = \tfrac{1}{2}(1 + \cos 2\theta)$, $\sin 2\theta = 2\sin\theta \cos\theta$, we find

$$\sigma = \tfrac{1}{2}\sigma_x(1 - \cos 2\theta) + \tfrac{1}{2}\sigma_y(1 + \cos 2\theta) + \tau_{xy} \sin 2\theta$$

or

$$\sigma = \tfrac{1}{2}(\sigma_x + \sigma_y) - \tfrac{1}{2}(\sigma_x - \sigma_y) \cos 2\theta + \tau_{xy} \sin 2\theta \tag{1}$$

Thus the normal stress σ on any plane inclined at an angle θ with the x-axis is known as a function of σ_x, σ_y, τ_{xy}, and θ.

Next, summing forces acting on the element in the T-direction, we find

$$\Sigma F_T = \tau t\, ds - \sigma_x t\, dy \cos\theta + \tau_{xy} t\, dy \sin\theta - \tau_{xy} t\, dx \cos\theta + \sigma_y t\, dx \sin\theta = 0$$

Substituting for dx and dy as before we get

$$\tau\, ds = \sigma_x\, ds \sin\theta \cos\theta - \tau_{xy}\, ds \sin^2\theta + \tau_{xy}\, ds \cos^2\theta - \sigma_y\, ds \sin\theta \cos\theta$$

Introducing the previous identities and the relation $\cos 2\theta = \cos^2\theta - \sin^2\theta$, this last equation becomes

$$\tau = \tfrac{1}{2}(\sigma_x - \sigma_y) \sin 2\theta + \tau_{xy} \cos 2\theta \tag{2}$$

Thus the shearing stress τ on any plane inclined at an angle θ with the x-axis is known as a function of σ_x, σ_y, τ_{xy}, and θ.

(*b*) To determine the maximum value that the normal stress σ may assume as the angle θ varies, we shall differentiate equation (*1*) with respect to θ and set this derivative equal to zero. Thus

$$\frac{d\sigma}{d\theta} = (\sigma_x - \sigma_y)\sin 2\theta + 2\tau_{xy}\cos 2\theta = 0$$

Hence the values of θ leading to maximum and minimum values of the normal stress are given by

$$\tan 2\theta_p = -\frac{\tau_{xy}}{\frac{1}{2}(\sigma_x - \sigma_y)} \tag{3}$$

The planes defined by the angles θ_p are called *principal planes*. The normal stresses that exist on these planes are designated as *principal stresses*. They are the maximum and minimum values that the normal stress may assume in the element under consideration. The values of the principal stresses may easily be found by considering the graphical interpretation of (3) given in Fig. 17-38.

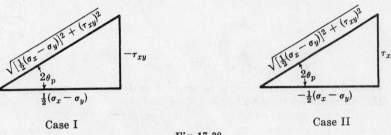

Case I   Case II

Fig. 17-38

Evidently the tangent of either of the angles designated as $2\theta_p$ has the value given in (3). Thus there are two solutions of (3), consequently two values of $2\theta_p$ (differing by 180°) and also two values of θ_p (differing by 90°). It is to be noted that Fig. 17-38 bears no direct relationship to the triangular element whose free-body diagram was given in Fig. 17-37.

The values of $\sin 2\theta_p$ and $\cos 2\theta_p$ as found from the above two diagrams may now be substituted in (1) to yield the maximum and minimum values of the normal stresses. Observing that

$$\sin 2\theta_p = \frac{\mp\tau_{xy}}{\sqrt{[\frac{1}{2}(\sigma_x - \sigma_y)]^2 + (\tau_{xy})^2}} \qquad \cos 2\theta_p = \frac{\pm\frac{1}{2}(\sigma_x - \sigma_y)}{\sqrt{[\frac{1}{2}(\sigma_x - \sigma_y)]^2 + (\tau_{xy})^2}}$$

where the upper signs pertain to Case I and the lower to Case II, we obtain from (1)

$$\sigma = \frac{1}{2}(\sigma_x + \sigma_y) \pm \sqrt{[\frac{1}{2}(\sigma_x - \sigma_y)]^2 + (\tau_{xy})^2} \tag{4}$$

The maximum normal stress is

$$\sigma_{max} = \frac{1}{2}(\sigma_x + \sigma_y) + \sqrt{[\frac{1}{2}(\sigma_x - \sigma_y)]^2 + (\tau_{xy})^2} \tag{5}$$

The minimum normal stress is

$$\sigma_{min} = \frac{1}{2}(\sigma_x + \sigma_y) - \sqrt{[\frac{1}{2}(\sigma_x - \sigma_y)]^2 + (\tau_{xy})^2} \tag{6}$$

The stresses given by (5) and (6) are the principal stresses and they occur on the principal planes defined by (3). By substituting one of the values of θ_p from (3) into equation (1), one may readily determine which of the two principal stresses is acting on that plane. The other principal stress naturally acts on the other principal plane.

By substituting the values of the angle $2\theta_p$ as given by (3) or by Fig. 17-38 into (2), it is readily seen that the shearing stresses τ on the principal planes are zero.

(c) To determine the maximum value that the shearing stress τ may assume as the angle θ varies, we shall differentiate equation (2) with respect to θ and set this derivative equal to zero.

Thus
$$\frac{d\tau}{d\theta} = (\sigma_x - \sigma_y)\cos 2\theta - 2\tau_{xy}\sin 2\theta = 0$$

The values of θ leading to the maximum values of the shearing stress are thus

$$\tan 2\theta_s = \frac{1}{2}(\sigma_x - \sigma_y)/\tau_{xy} \tag{7}$$

The planes defined by the two solutions to this equation are the planes of maximum shearing stress.

Again, a graphical interpretation of (7) is convenient. The two values of the angle $2\theta_s$ satisfying this equation may be represented as in Fig. 17-39.

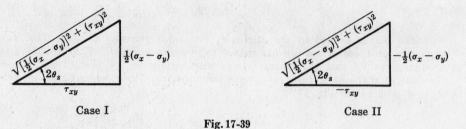

Case I   Case II

Fig. 17-39

From these diagrams we have

$$\sin 2\theta_s = \frac{\pm\frac{1}{2}(\sigma_x - \sigma_y)}{\sqrt{[\frac{1}{2}(\sigma_x - \sigma_y)]^2 + (\tau_{xy})^2}} \qquad \cos 2\theta_s = \frac{\pm\tau_{xy}}{\sqrt{[\frac{1}{2}(\sigma_x - \sigma_y)]^2 + (\tau_{xy})^2}}$$

where the upper (positive) sign refers to Case I and the lower (negative) sign applies to Case II. Substituting these values in (2) we find

$$\tau_{\substack{max \\ min}} = \pm\sqrt{[\frac{1}{2}(\sigma_x - \sigma_y)]^2 + (\tau_{xy})^2} \qquad (8)$$

Here the positive sign represents the maximum shearing stress, the negative sign the minimum shearing stress.

If we compare (3) and (7), it is evident that the angles $2\theta_p$ and $2\theta_s$ differ by 90°, since the tangents of these angles are the negative reciprocals of one another. Hence the planes defined by the angles θ_p and θ_s differ by 45°, i.e. the planes of maximum shearing stress are oriented 45° from the planes of maximum normal stress.

It is also of interest to determine the normal stresses on the planes of maximum shearing stress. These planes are defined by (7). If we now substitute the values of $\sin 2\theta_s$ and $\cos 2\theta_s$ in equation (1) for normal stress we find

$$\sigma = \frac{1}{2}(\sigma_x + \sigma_y) \qquad (9)$$

Thus on each of the planes of maximum shearing stress is a normal stress of magnitude $\frac{1}{2}(\sigma_x + \sigma_y)$.

17.14. Discuss a graphical representation of the analysis presented in Problem 17.13.

For given values of σ_x, σ_y, and τ_{xy} we proceed thus:

1. Introduce a rectangular coordinate system in which normal stresses are represented along the horizontal axis and shearing stresses along the vertical axis. The scales used on these two axes must be equal.

2. With reference to the original rectangular element considered in Problem 17.13 and reproduced in Fig. 17-40, we shall introduce the sign convention that shearing stresses are positive if they tend to rotate the element clockwise, negative if they tend to rotate it counterclockwise. Here the shearing stresses on the vertical faces are positive, those on the horizontal faces are negative. Also, tensile normal stresses are considered to be positive, compressive stresses negative.

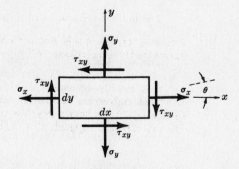

Fig. 17-40

3. We first locate point b by laying out σ_x and τ_{xy} to their given values. The shear stress τ_{xy} on the vertical faces on which σ_x acts is positive, hence this value is plotted as positive in Fig. 17-41.

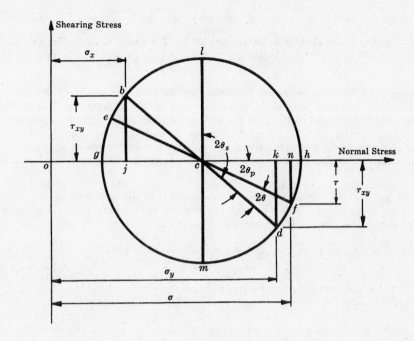

Fig. 17-41

4. We next locate point d in a similar manner by laying off σ_y and τ_{xy} to their given values. Figure 17-41 is drawn on the assumption that $\sigma_y > \sigma_x$ although the treatment presented here holds if $\sigma_y < \sigma_x$. The shear stress τ_{xy} on the horizontal faces on which σ_y acts is negative, hence this value is plotted below the reference axis.

5. Next, we draw line $\overline{bd}$, locate midpoint c, and draw a circle having its center at c and radius equal to $\overline{cb}$. This is known as Mohr's circle.

We shall first show that the points g and h along the horizontal diameter of the circle represent the principal stresses. To do this we note that the point c lies at a distance $\frac{1}{2}(\sigma_x + \sigma_y)$ from the origin of the coordinate system. Also, the line segment $\overline{jk}$ is of length $\sigma_y - \sigma_x$, hence $\overline{ck}$ is of length $\frac{1}{2}(\sigma_y - \sigma_x)$. From the right-triangle relationship we have

$$(\overline{cd})^2 = (\overline{ck})^2 + (\overline{kd})^2 \quad \text{or} \quad \overline{cd} = \sqrt{[\tfrac{1}{2}(\sigma_x - \sigma_y)]^2 + (\tau_{xy})^2}$$

Also, $\overline{cg} = \overline{ch} = \overline{cd}$. Hence the x-coordinate of point h is $\overline{oc} + \overline{ch}$ or

$$\tfrac{1}{2}(\sigma_x + \sigma_y) + \sqrt{[\tfrac{1}{2}(\sigma_x - \sigma_y)]^2 + (\tau_{xy})^2}$$

But this expression is exactly the maximum principal stress, as found in (5) of Problem 17.13. Likewise the x-coordinate of point g is $\overline{oc} - \overline{gc}$ or

$$\tfrac{1}{2}(\sigma_x + \sigma_y) - \sqrt{[\tfrac{1}{2}(\sigma_x - \sigma_y)]^2 + (\tau_{xy})^2}$$

and this expression is exactly the minimum principal stress, as found in (6) of Problem 17.13. Consequently the points g and h represent the principal stresses existing in the original element. We see that the tangent of $\angle kcd = \overline{dk}/\overline{ck} = \tau_{xy}/\frac{1}{2}(\sigma_y - \sigma_x)$. But from (3) of Problem 17.13 we had

$$\tan 2\theta_p = -\frac{\tau_{xy}}{\frac{1}{2}(\sigma_x - \sigma_y)}$$

and by comparison of these two relations we see that $\angle kcd = 2\theta_p$, i.e. a counterclockwise rotation from the diameter $\overline{bd}$ (corresponding to the stresses in the x-y directions) leads us to the diameter $\overline{gh}$, representing the principal planes, on which the principal stresses occur. The principal planes lie at an angle θ_p from the x-direction.

Thus Mohr's circle is a convenient device for finding the principal stresses, since one can merely establish the circle for a given set of stresses σ_x, σ_y, τ_{xy}, then measure $\overline{og}$ and $\overline{oh}$. These abscissas represent the principal stresses to the same scale used in plotting σ_x, σ_y, τ_{xy}.

It is now apparent that the radius of Mohr's circle,

$$\overline{cd} = \sqrt{[\tfrac{1}{2}(\sigma_x - \sigma_y)]^2 + (\tau_{xy})^2}$$

corresponds to the maximum shearing stress as found in (8) of Problem 17.13. Actually, the shearing stress on any plane is represented by the ordinate to Mohr's circle; hence we should consider the radial lines $\overline{cl}$ and $\overline{cm}$ as representing the maximum shearing stresses. The angle dcl is evidently $2\theta_s$ and hence it is apparent that the double angle between the planes of maximum normal stress and the planes of maximum shearing stress ($\angle kcl$) is 90°; hence the planes of maximum shearing stress are oriented 45° from the planes of maximum normal stress.

Evidently the endpoints of the diameter $\overline{bd}$ represent the stresses acting in the original x- and y-directions. We shall now demonstrate that the endpoints of any other diameter such as $\overline{ef}$ (at an angle 2θ with $\overline{bd}$) represent the stresses on a plane inclined at an angle θ to the x-axis. To do this we note that the abscissa of point f is given by

$$\begin{aligned}
\sigma = \overline{oc} + \overline{cn} &= \tfrac{1}{2}(\sigma_x + \sigma_y) + \overline{cf}\cos(2\theta_p - 2\theta) \\
&= \tfrac{1}{2}(\sigma_x + \sigma_y) + \overline{cf}(\cos 2\theta_p \cos 2\theta + \sin 2\theta_p \sin 2\theta) \\
&= \tfrac{1}{2}(\sigma_x + \sigma_y) + \sqrt{[\tfrac{1}{2}(\sigma_x - \sigma_y)]^2 + (\tau_{xy})^2}\,(\cos 2\theta_p \cos 2\theta + \sin 2\theta_p \sin 2\theta)
\end{aligned}$$

But from an inspection of triangle ckd in Mohr's circle it is evident that

$$\sin 2\theta_p = \frac{\tau_{xy}}{\sqrt{[\tfrac{1}{2}(\sigma_x - \sigma_y)]^2 + (\tau_{xy})^2}} \qquad \cos 2\theta_p = \frac{\tfrac{1}{2}(\sigma_y - \sigma_x)}{\sqrt{[\tfrac{1}{2}(\sigma_x - \sigma_y)]^2 + (\tau_{xy})^2}} \qquad (1)$$

Substituting the values of τ_{xy} and $\tfrac{1}{2}(\sigma_y - \sigma_x)$ from these last two equations into the previous equation, we find

$$\sigma = \tfrac{1}{2}(\sigma_x + \sigma_y) - \tfrac{1}{2}(\sigma_x - \sigma_y)\cos 2\theta + \tau_{xy}\sin 2\theta$$

But this is exactly the normal stress on a plane inclined at an angle θ to the x-axis as derived in (1) of Problem 17.13.

Next we observe that the ordinate of point f is given by

$$\tau = \overline{nf} = \overline{cf}\sin(2\theta_p - 2\theta) = \overline{cf}(\sin 2\theta_p \cos 2\theta - \cos 2\theta_p \sin 2\theta)$$

$$= \sqrt{[\tfrac{1}{2}(\sigma_x - \sigma_y)]^2 + (\tau_{xy})^2}\,(\sin 2\theta_p \cos 2\theta - \cos 2\theta_p \sin 2\theta)$$

Again, substituting the values of τ_{xy} and $\tfrac{1}{2}(\sigma_y - \sigma_x)$ from (1) into this equation we find

$$\tau = \tau_{xy}\cos 2\theta + \tfrac{1}{2}(\sigma_x - \sigma_y)\sin 2\theta$$

But this is exactly the shearing stress on a plane inclined at an angle θ to the x-axis as derived in (2) of Problem 17.13.

Hence the coordinates of point f on Mohr's circle represent the normal and shearing stresses on a plane inclined at an angle θ to the x-axis.

17.15. A plane element is subject to the stresses shown in Fig. 17-42. Determine (a) the principal stresses and their directions, (b) the maximum shearing stresses and the directions of the planes on which they occur.

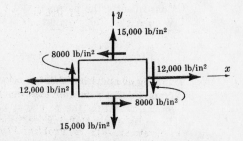

Fig. 17-42

(a) In accordance with the notation of Problem 17.13 we have $\sigma_x = 12{,}000$ lb/in², $\sigma_y = 15{,}000$ lb/in², and $\sigma_{xy} = 8000$ lb/in². The maximum normal stress is, by (5) of Problem 17.13,

$$\sigma_{\max} = \tfrac{1}{2}(\sigma_x + \sigma_y) + \sqrt{[\tfrac{1}{2}(\sigma_x - \sigma_y)]^2 + (\tau_{xy})^2}$$

$$= \tfrac{1}{2}(12{,}000 + 15{,}000) + \sqrt{[\tfrac{1}{2}(12{,}000 - 15{,}000)]^2 + (8000)^2}$$

$$= 13{,}500 + 8150 = 21{,}650 \text{ lb/in}^2$$

The minimum normal stress is given by (6) of Problem 17.13 to be

$$\sigma_{\min} = \tfrac{1}{2}(\sigma_x + \sigma_y) - \sqrt{[\tfrac{1}{2}(\sigma_x - \sigma_y)]^2 + (\tau_{xy})^2} = 13{,}500 - 8150 = 5350 \text{ lb/in}^2$$

From (3) of Problem 17.13 the directions of the principal planes on which these stresses of 21,650 lb/in² and 5350 lb/in² occur are given by

$$\tan 2\theta_p = -\frac{\tau_{xy}}{\tfrac{1}{2}(\sigma_x - \sigma_y)} = -\frac{8000}{\tfrac{1}{2}(12{,}000 - 15{,}000)} = 5.33$$

Then $2\theta_p = 79°24', 259°24'$ and $\theta_p = 39°42', 129°42'$.

To determine which of the above principal stresses occurs on each of these planes we return to (1) of Problem 17.13, namely

$$\sigma = \tfrac{1}{2}(\sigma_x + \sigma_y) - \tfrac{1}{2}(\sigma_x - \sigma_y)\cos 2\theta + \tau_{xy}\sin 2\theta$$

and substitute $\theta = 39°42'$ together with the given values of σ_x, σ_y, and τ_{xy} to obtain

$$\sigma = \tfrac{1}{2}(12{,}000 + 15{,}000) - \tfrac{1}{2}(12{,}000 - 15{,}000)\cos 79°24' + 8000 \sin 79°24' = 21{,}650 \text{ lb/in}^2$$

Thus an element oriented along the principal planes and subject to the above principal stresses appears as in Fig. 17-43. The shearing stresses on these planes are zero.

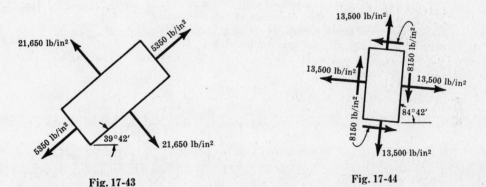

Fig. 17-43   Fig. 17-44

(b) The maximum and minimum shearing stresses were found in (8) of Problem 17.13 to be

$$\tau_{\substack{\max\\\min}} = \pm\sqrt{[\tfrac{1}{2}(\sigma_x - \sigma_y)]^2 + (\tau_{xy})^2}$$

$$= \pm\sqrt{[\tfrac{1}{2}(12{,}000 - 15{,}000)]^2 + (8000)^2} = \pm 8150 \text{ lb/in}^2$$

From (7) of Problem 17.13 the planes on which these maximum shearing stresses occur are defined by the equation

$$\tan 2\theta_s = \frac{\tfrac{1}{2}(\sigma_x - \sigma_y)}{\tau_{xy}} = -0.188$$

Then $2\theta_s = 169°24', 349°24'$ and $\theta_s = 84°42', 174°42'$. Evidently these planes are located 45° from the planes of maximum and minimum normal stress.

To determine whether the shearing stress is positive or negative on the 84°42' plane, we return to (2) of Problem 17.13, namely

$$\tau = \tfrac{1}{2}(\sigma_x - \sigma_y)\sin 2\theta + \tau_{xy}\cos 2\theta$$

and substitute $\theta = 84°42'$ together with the given values of σ_x, σ_y, and τ_{xy} to obtain

$$\tau = \tfrac{1}{2}(12{,}000 - 15{,}000) \sin 169°24' + 8000 \cos 169°24' = -8150 \text{ lb/in}^2$$

The negative sign indicates that the shearing stress is directed oppositely to the assumed positive direction shown in Fig. 17-36. Lastly, the normal stresses on these planes of maximum shearing stress are found from (9) of Problem 17.13 to be

$$\sigma = \tfrac{1}{2}(\sigma_x + \sigma_y) = \tfrac{1}{2}(12{,}000 + 15{,}000) = 13{,}500 \text{ lb/in}^2$$

The orientation of the element for which the shearing stresses are maximum is as in Fig. 17-44.

17.16. A plane element is subject to the stresses shown in Fig. 17-45. Using Mohr's circle, determine (a) the principal stresses and their directions, (b) the maximum shearing stresses and the directions of the planes on which they occur.

The procedure for the construction of Mohr's circle was outlined in Problem 17.14. Following the instructions there, we realized that the shearing stresses on the vertical faces of the given element are positive, whereas those on the horizontal faces are negative. Thus the stress condition of $\sigma_x = 12{,}000$ lb/in^2, $\tau_{xy} = 8000$ lb/in^2 existing on the vertical faces of the element plots as point b in Fig. 17-46. The stress condition of $\sigma_y = 15{,}000$ lb/in^2, $\tau_{xy} = -8000$ lb/in^2 existing on the horizontal faces plots as point d. Line $\overline{bd}$ is drawn, its midpoint c is located, and a circle of radius $\overline{cb} = \overline{cd}$ is drawn with c as a center. This is Mohr's circle. The endpoints of the diameter $\overline{bd}$ represent the stress conditions existing in the element if it has the original orientation of Fig. 17-45.

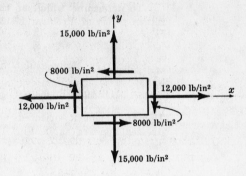

Fig. 17-45

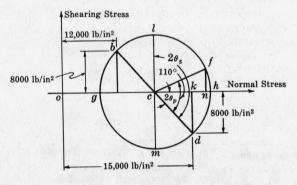

Fig. 17-46

(a) The principal stresses are represented by points g and h as demonstrated in Problem 17.14. The principal stress may be determined either by direct measurement from Fig. 17-46 or by realizing that the coordinate of c is 13,500, that $\overline{ck} = 1500$ and that $\overline{cd} = \sqrt{(1500)^2 + (8000)^2} = 8150$. Thus the minimum principal stress is

$$\sigma_{\min} = \overline{og} = \overline{oc} - \overline{cg} = 13{,}500 - 8150 = 5350 \text{ lb/in}^2$$

Also, the maximum principal stress is

$$\sigma_{\max} = \overline{oh} = \overline{oc} + \overline{ch} = 13{,}500 + 8150 = 21{,}650 \text{ lb/in}^2$$

The angle $2\theta_p$ is given by $\tan 2\theta_p = 8000/1500 = 5.33$ from which $\theta_p = 39°42'$. This value could also be obtained by measurement of $\angle dck$ in Mohr's circle. From this it is readily seen that the principal stress represented by point h acts on a plane oriented $39°42'$ from the original *x*-axis. The principal stresses thus appear as in Fig. 17-47. It is evident that the shearing stresses on these planes are zero, since points *g* and *h* lie on the horizontal axis of Mohr's circle.

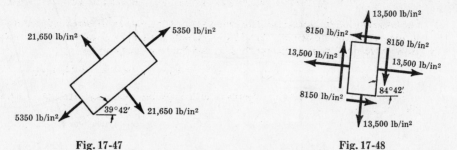

| Fig. 17-47 | Fig. 17-48 |

(*b*) The maximum shearing stress is represented by $\overline{cl}$ in Mohr's circle. This radius has already been found to represent 8150 lb/in². The angle $2\theta_s$ may be found either by direct measurement from the above plot or simply by adding 90° to the angle $2\theta_p$, which has already been determined. This leads to $2\theta_s = 169°24'$ and $\theta_s = 84°42'$. The shearing stress represented by point *l* is positive, hence on this 84°42′ plane the shearing stress tends to rotate the element in a clockwise direction.

Also, from Mohr's circle the abscissa of point *l* is 13,500 lb/in² and this represents the normal stress occurring on the planes of maximum shearing stress. The maximum shearing stresses thus appear as in Fig. 17-48.

17.17. For the element discussed in Problem 17.16, determine the normal and shearing stresses on a plane making an angle of 55° measured counterclockwise from the positive end of the *x*-axis.

According to the properties of Mohr's circle discussed in Problem 17.14, we realize that the endpoints of the diameter $\overline{bd}$ represent the stress conditions occurring on the original *x-y* planes. On any plane inclined at an angle θ to the *x*-axis the stress conditions are represented by the coordinates of a point *f*, where the radius $\overline{cf}$ makes an angle of 2θ with the original diameter $\overline{bd}$. This angle 2θ appearing in Mohr's circle is measured in the same direction as the angle representing the inclined plane, namely counterclockwise.

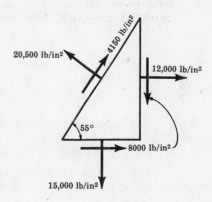

Fig. 17-49

Hence in the Mohr's circle appearing in Problem 17.16, we merely measure a counterclockwise angle of $2(55°) = 110°$ from line $\overline{cd}$. This locates point *f*. The abscissa of point *f* represents the normal stress on the desired 55° plane and may be found either by direct measurement or by realizing that

$$\overline{on} = \overline{oc} + \overline{cn} = 13,500 + 8150 \cos(110° - 79°24') = 20,500 \text{ lb/in}^2$$

The ordinate of point *f* represents the shearing stress on the desired 55° plane and may be found from the relation

$$\overline{fn} = 8150 \sin(110° - 79°24') = 4150 \text{ lb/in}^2$$

The stresses acting on the 55° plane may thus be represented as in Fig. 17-49.

17.18. A plane element is subject to the stresses shown in Fig. 17-50 below. Determine (*a*) the principal stresses and their directions, (*b*) the maximum shearing stresses and the directions of the planes on which they occur.

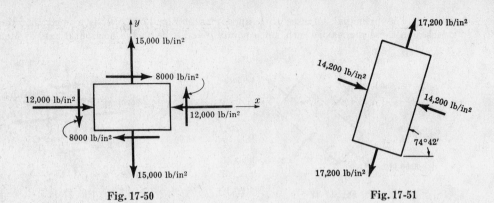

Fig. 17-50   Fig. 17-51

(a) In accordance with the notation of Problem 17.13, $\sigma_x = -12{,}000$ lb/in², $\sigma_y = 15{,}000$ lb/in², and $\tau_{xy} = -8000$ lb/in². The maximum normal stress is given by (5) of Problem 17.13 to be

$$\sigma_{max} = \tfrac{1}{2}(\sigma_x + \sigma_y) + \sqrt{[\tfrac{1}{2}(\sigma_x - \sigma_y)]^2 + (\tau_{xy})^2}$$

$$= \tfrac{1}{2}(-12{,}000 + 15{,}000) + \sqrt{[\tfrac{1}{2}(-12{,}000 - 15{,}000)]^2 + (-8000)^2}$$

$$= 1500 + 15{,}700 = 17{,}200 \text{ lb/in}^2$$

The minimum normal stress is given by (6) of Problem 17.13 to be

$$\sigma_{min} = \tfrac{1}{2}(\sigma_x + \sigma_y) - \sqrt{[\tfrac{1}{2}(\sigma_x - \sigma_y)]^2 + (\tau_{xy})^2} = 1500 - 15{,}700 = -14{,}200 \text{ lb/in}^2$$

From (3) of Problem 17.13 the directions of the principal planes on which these stresses of 17,200 lb/in² and −14,200 lb/in² occur are given by

$$\tan 2\theta_p = -\frac{\tau_{xy}}{\tfrac{1}{2}(\sigma_x - \sigma_y)} = -\frac{-8000}{\tfrac{1}{2}(-12{,}000 - 15{,}000)} = -0.592$$

Then $2\theta_p = 149°24', 329°24'$ and $\theta_p = 74°42', 164°42'$.

To determine which of the above principal stresses occurs on each of these planes we return to (1) of Problem 17.13, namely

$$\sigma = \tfrac{1}{2}(\sigma_x + \sigma_y) - \tfrac{1}{2}(\sigma_x - \sigma_y)\cos 2\theta + \tau_{xy}\sin 2\theta$$

and substitute $\theta = 74°42'$ together with the given values of σ_x, σ_y, and τ_{xy} to obtain

$$\sigma = \tfrac{1}{2}(-12{,}000 + 15{,}000) - \tfrac{1}{2}(-12{,}000 - 15{,}000)\cos 149°24' - 8000\sin 149°24' = -14{,}200 \text{ lb/in}^2$$

Consequently an element oriented along the principal planes and subject to the above principal stresses appears as in Fig. 17-51. The shearing stresses on these planes are zero.

(b) The maximum and minimum shearing stresses were found in (8) of Problem 17.13 to be

$$\tau_{\substack{max \\ min}} = \pm\sqrt{[\tfrac{1}{2}(\sigma_x - \sigma_y)]^2 + (\tau_{xy})^2} = \pm\sqrt{[\tfrac{1}{2}(-12{,}000 - 15{,}000)]^2 + (-8000)^2} = \pm 15{,}700 \text{ lb/in}^2$$

From (7) of Problem 17.13 the planes on which these maximum shearing stresses occur are defined by

$$\tan 2\theta_s = \tfrac{1}{2}(\sigma_x - \sigma_y)/\tau_{xy} = 1.69$$

Then $2\theta_s = 59°24', 239°24'$ and $\theta_s = 29°42', 119°42'$. It is apparent that these planes are located 45° from the planes of maximum and minimum normal stress.

To determine whether the shearing stress is positive or negative on the 29°42' plane we return to (2) of Problem 17.13, namely

$$\tau = \tfrac{1}{2}(\sigma_x - \sigma_y)\sin 2\theta + \tau_{xy}\cos 2\theta$$

and substitute $\theta = 29°42'$ together with the given values of σ_x, σ_y, and τ_{xy} to obtain

$$\tau = \tfrac{1}{2}(-12{,}000 - 15{,}000)\sin 59°24' - 8000\cos 59°24' = -15{,}700 \text{ lb/in}^2$$

The negative sign indicates that the shearing stress on the 29°42' plane is directed oppositely to the assumed positive direction shown in Fig. 17-36. The normal stresses on these planes of maximum shearing stress were found in (9) of Problem 17.13 to be

$$\sigma = \tfrac{1}{2}(\sigma_x + \sigma_y)$$

$$= \tfrac{1}{2}(-12{,}000 + 15{,}000) = 1500 \text{ lb/in}^2$$

Consequently the orientation of the element for which the shearing stresses are a maximum appears as in Fig. 17-52.

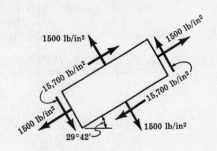

Fig. 17-52

17.19. A plane element is subject to the stresses shown in Fig. 17-53. Using Mohr's circle determine (a) the principal stresses and their directions, (b) the maximum shearing stresses and the directions of the planes on which they occur.

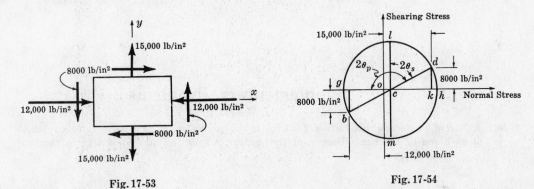

Fig. 17-53   Fig. 17-54

Again we refer to Problem 17.14 for the procedure for constructing Mohr's circle. In accordance with the sign convention outlined there the shearing stresses on the vertical faces of the element are negative, those on the horizontal faces positive. Thus the stress condition of $\sigma_x = -12{,}000$ lb/in², $\tau_{xy} = -8000$ lb/in² existing on the vertical faces of the element plots as point b in Fig. 17-54. The stress condition of $\sigma_y = 15{,}000$ lb/in², $\tau_{xy} = 8000$ lb/in² existing on the horizontal faces plots as point d. Line $\overline{bd}$ is drawn, its midpoint c is located, and a circle of radius $\overline{cb} = \overline{cd}$ is drawn with c as a center. This is Mohr's circle. The endpoints of the diameter $\overline{bd}$ represent the stress conditions existing in the element if it has the original orientation shown above.

(a) The principal stresses are represented by points g and h as shown in Problem 17.14. They may be found either by direct measurement from the above diagram or by realizing that the coordinate of c is 1500, that $\overline{ck} = 13{,}500$, and that $\overline{cd} = \sqrt{(13{,}500)^2 + (8000)^2} = 15{,}700$. Thus the minimum principal stress is

$$\sigma_{\min} = \overline{og} = \overline{oc} - \overline{cg} = 1500 - 15{,}700 = -14{,}200 \text{ lb/in}^2$$

Also, the maximum principal stress is

$$\sigma_{\max} = \overline{oh} = \overline{oc} + \overline{ch} = 1500 + 15{,}700 = 17{,}200 \text{ lb/in}^2$$

The angle $2\theta_p$ is given by $\tan 2\theta_p = -8000/13{,}500 = -0.592$, from which $\theta_p = 74°42'$. This value could also be obtained by measurement of $\angle dcg$ in Mohr's circle. From this it is readily seen that the principal stress represented by point g acts on a plane oriented 74°42' from the original x-axis. The principal stresses thus appear as in Fig. 17-55 below. Since the ordinates of points g and h are each zero the shearing stresses on these planes are zero.

(b) The maximum shearing stress is represented by $\overline{cl}$ in Mohr's circle. This radius has already been found to represent 15,700 lb/in². The angle $2\theta_s$ may be found either by direct measurement from the above plot or simply by subtracting 90° from the angle $2\theta_p$ which has already been determined. This leads to $2\theta_s = 59°24'$ and $\theta_s = 29°42'$. The shearing stress represented by point l is positive, hence on this 29°42' plane the shearing stress tends to rotate the element in a clockwise direction.

Also, from Mohr's circle the abscissa of point l is 1500 lb/in² and this represents the normal stress occurring on the planes of maximum shearing stress. The maximum shearing stresses thus appear as in Fig. 17-56.

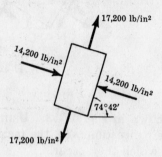

Fig. 17-55

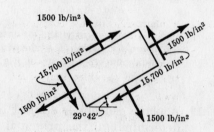

Fig. 17-56

Supplementary Problems

17.20. A bar of uniform cross-section 2 × 3 in. is subject to an axial tensile force of 108,000 lb applied at each end of the bar. Determine the maximum shearing stress existing in the bar.
Ans. 9000 lb/in²

17.21. In Problem 17.20, determine the normal and shearing stresses acting on a plane inclined at 20° to the line of action of the axial loads.   *Ans.* $\sigma = 2100$ lb/in², $\tau = 5800$ lb/in²

17.22. A square steel bar 1 in. on a side is subject to an axial compressive load of 8000 lb. Determine the normal and shearing stresses acting on a plane inclined at 30° to the line of action of the axial loads. The bar is so short that the possibility of buckling as a column may be neglected.
Ans. $\sigma = -2000$ lb/in², $\tau = -3460$ lb/in²

17.23. Rework Problem 17.22 by use of Mohr's circle.
Ans. See Fig. 17-57 below. $\sigma = \overline{ko} = -2000$ lb/in², $\tau = \overline{dk} = 3460$ lb/in²

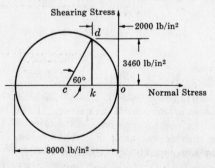

Fig. 17-57

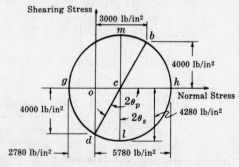

Fig. 17-58

17.24. A plane element in a body is subject to the stresses $\sigma_x = 3000$ lb/in², $\sigma_y = 0$ and $\tau_{xy} = 4000$ lb/in². Determine analytically the normal and shearing stresses existing on a plane inclined at 45° to the x-axis.   *Ans.* $\sigma = 5500$ lb/in², $\tau = 1500$ lb/in²

17.25. For the element of Problem 17.24, determine analytically the principal stresses and their directions as well as the maximum shearing stresses and the directions of the planes on which they occur.
Ans. $\sigma_{max} = 5780$ lb/in² at 55°15′, $\sigma_{min} = -2780$ lb/in² at 145°15′, $\tau_{max} = 4280$ lb/in² at 10°15′

17.26. Rework Problem 17.25 by the use of Mohr's circle.   *Ans.* See Fig. 17-58 above.

17.27. A plane element in a body is subject to the stresses shown in Fig. 17-59. Determine analytically (*a*) the principal stresses and their directions, (*b*) the maximum shearing stresses and the directions of the planes on which they occur.

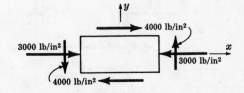

Ans. $\sigma_{max} =$ 2780 lb/in² at 145°15′

$\sigma_{min} = -5780$ lb/in² at   55°15′

$\tau_{max} =$ 4280 lb/in² at   10°15′

Fig. 17-59

17.28. For the element treated in Problem 17.27 determine the normal and shearing stresses acting on a plane inclined at 30° to the x-axis.   *Ans.* $\sigma = -5710$ lb/in², $\tau = -3300$ lb/in²

17.29. A plane element is subject to the stresses $\sigma_x = 8000$ lb/in² and $\sigma_y = 8000$ lb/in². Determine analytically the maximum shearing stress existing in the element.   *Ans.* 0

17.30. What form does Mohr's circle assume for the loading described in Problem 17.29?
Ans. A point on the horizontal axis, located a distance of 8000 lb/in² (to scale) from the origin.

17.31. A plane element is subject to the stresses $\sigma_x = 8000$ lb/in², and $\sigma_y = -8000$ lb/in². Determine analytically the maximum shearing stress existing in the element. What is the direction of the planes on which the maximum shearing stresses occur?   *Ans.* 8000 lb/in² at 45°

17.32. For the element described in Problem 17.31, determine analytically the normal and shearing stresses acting on a plane inclined at 30° to the x-axis.   *Ans.* $\sigma = -4000$ lb/in², $\tau = 6920$ lb/in²

17.33. Draw Mohr's circle for a plane element subject to the stresses $\sigma_x = 8000$ lb/in² and $\sigma_y = -8000$ lb/in². From Mohr's circle determine the stresses acting on a plane inclined at 20° to the x-axis.
Ans. See Fig. 17-60. $\sigma = \overline{on} = -6130$ lb/in², $\tau = \overline{nf} = -5130$ lb/in²

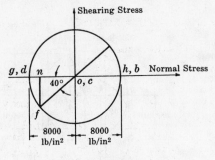

Fig. 17-60

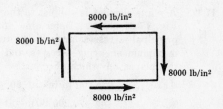

Fig. 17-61

17.34. A plane element removed from a thin-walled cylindrical shell loaded in torsion is subject to the shearing stresses shown in Fig. 17-61. Determine the principal stresses existing in this element and the directions of the planes on which they occur.   *Ans.* 8000 lb/in² at 45°

17.35. A plane element is subject to the stresses shown in Fig. 17-62. Determine analytically (a) the principal stresses and their directions, (b) the maximum shearing stresses and the directions of the planes on which they act.

Ans. $\sigma_{max} = 24{,}940$ lb/in² at 121°45′, $\sigma_{min} = 7060$ lb/in² at 31°45′, $\tau_{max} = 8940$ lb/in² at 76°45′

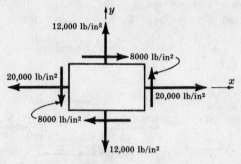

Fig. 17-62

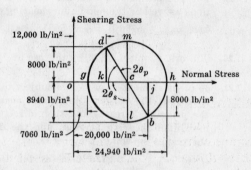

Fig. 17-63

17.36. Rework Problem 17.35 by the use of Mohr's circle.   *Ans.* See Fig. 17-63.

17.37. Consider again the element shown in Fig. 17-62. Determine analytically the normal and shearing stresses on a plane inclined at an angle of 20° to the x-axis.

Ans. $\sigma = 7790$ lb/in², $\tau = -3530$ lb/in²

17.38. Rework Problem 17.37 by the use of Mohr's circle.

Ans. See Fig. 17-64. $\sigma = \overline{on} = 7790$ lb/in², $\tau = \overline{nf} = 3530$ lb/in²

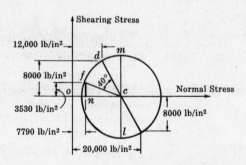

Fig. 17-64

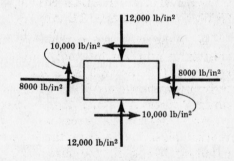

Fig. 17-65

17.39. A plane element is subject to the stresses shown in Fig. 17-65 above. Determine analytically (a) the principal stresses and their directions, (b) the maximum shearing stresses and the directions of the planes on which they act.

Ans. $\sigma_{max} = 200$ lb/in² at 50°40′

   $\sigma_{min} = -20{,}200$ lb/in² at 140°40′

   $\tau_{max} = 10{,}200$ lb/in² at 5°40′

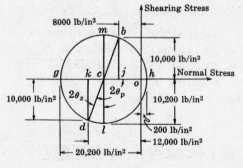

Fig. 17-66

17.40. Rework Problem 17.39 by the use of Mohr's circle.
    Ans. See Fig. 17-66.

Chapter 18

Members Subject to Combined Loadings; Theories of Failure

AXIALLY LOADED MEMBERS SUBJECT TO ECCENTRIC LOADS

In Chapters 1 and 2, where we considered straight bars subject to either tensile or compressive loads, it was always required that the action line of the applied force pass through the centroid of the cross-section of the member. In the present chapter we shall consider those cases where the action line of the applied force acting on a bar in either tension or compression does *not* pass through the centroid of the cross-section. A typical example of such an eccentric loading is shown in Fig. 18-1. For those cross-sections of the bar that are perpendicular to the direction of the load, the resultant stress at any point is the sum of the direct stress due to a concentric load of equal magnitude P plus a bending stress due to a couple of moment Pe. This first stress is found from the expression derived in Chapter 1, namely $\sigma = P/A$. The second stress is found from the formula for bending stress presented in Chapter 8, namely $\sigma = My/I$. An application may be found in Problem 18.1.

CYLINDRICAL SHELLS SUBJECT TO COMBINED INTERNAL PRESSURE AND AXIAL TENSION

In Chapter 3 we considered the stresses arising in a thin-walled cylindrical shell subject to uniform internal pressure. There it was shown that a longitudinal stress given by $\sigma = pr/2t$, as well as a circumferential stress given by $\sigma = pr/t$, exists because of the internal pressure p. If in addition an axial tension P is acting simultaneously with the internal pressure, then there arises an additional longitudinal stress given by $\sigma = P/A$ where A denotes the cross-sectional area of the shell. The resultant stress in the longitudinal direction is thus the algebraic sum of these two longitudinal stresses, and the resultant stress in the circumferential direction is equal to that due to the internal pressure.

CYLINDRICAL SHELLS SUBJECT TO COMBINED TORSION AND AXIAL TENSION

In Chapter 5 we considered the stresses arising in a thin-walled cylindrical shell subject to torsion. There it was shown that a shearing stress given by $\tau_{xy} = T\rho/J$ exists on cross-sections perpendicular to the axis of the cylinder. If in addition an axial tension P is acting simultaneously with the torque, then there arises a longitudinal stress given by $\sigma = P/A$. This loading is illustrated in Fig. 18-2. In this case the stresses due to

Fig. 18-1

Fig. 18-2

these two loadings are acting in different directions and use must be made of the results obtained in Chapter 17. In this manner it will be possible to obtain the principal stresses due to these two loads acting simultaneously. For an application see Problem 18.2.

CIRCULAR SHAFT SUBJECT TO COMBINED AXIAL TENSION AND TORSION

This loading is illustrated in Fig. 18-3. Due to the axial tensile force P there exists a uniform longitudinal tensile stress given by $\sigma = P/A$, where A denotes the cross-sectional area of the bar. From Chapter 5 we know that there exists a torsional shearing stress over any cross-section perpendicular to the axis given by $\tau_{xy} = T\rho/J$. Again, the stresses due to these two loadings are acting in different directions and the results of Chapter 17 must be employed to obtain the values of the principal stresses at any point or to obtain the state of stress on any plane inclined at some angle to a generator of the shaft. For an application see Problem 18.3.

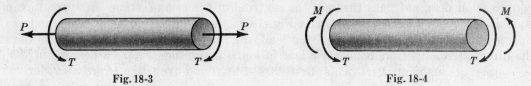

Fig. 18-3   Fig. 18-4

CIRCULAR SHAFT SUBJECT TO COMBINED BENDING AND TORSION

This loading is illustrated in Fig. 18-4. Again from Chapter 5 we know that there exists a torsional shearing stress over any cross-section perpendicular to the axis given by $\tau_{xy} = T\rho/J$. From Chapter 8 we know that there also exists a bending stress perpendicular to this cross-section, i.e. in the direction of the axis of the shaft, given by $\sigma = My/I$. Since these stresses are acting in different directions the results of Chapter 17 must be employed to obtain the values of the principal stresses at any point in the shaft or to obtain the state of stress on any plane inclined to a generator of the shaft. For applications see Problems 18.4 and 18.5.

DESIGN OF MEMBERS SUBJECT TO COMBINED LOADINGS

So far we have discussed only *analysis*, i.e. determination of principal stresses in a member subject to combined loadings. The inverse problem, i.e. *design* of a member to withstand combined loads, is somewhat more complex and must necessarily be related to experimentally determined mechanical properties of the materials. Because such properties cannot be determined for all possible combinations of loadings, the mechanical characteristics are usually determined in very simple tensile, compressive, or shear tests. The problem then arises as to how to relate the strength of an elastic body subject to combined loadings to these known strength characteristics under the simpler loading conditions. Relations between strength under various combined loads and simple mechanical properties of the material are termed *theories of failure*. Many such theories are available but we shall discuss only the three most commonly used, one applicable to brittle materials and two suitable for use in design of ductile members.

MAXIMUM NORMAL STRESS THEORY

This theory states that failure of the material subject to biaxial or triaxial stresses occurs when the maximum normal stress reaches the value at which failure occurs in a simple tension test on the same material. Failure is usually defined as either yielding or fracture — whichever occurs first. This theory is in good agreement with experimental evidence on brittle materials. For applications, see Problems 18.6, 18.9, and 18.10.

MAXIMUM SHEARING STRESS THEORY

This theory states that failure of the material subject to biaxial or triaxial stresses occurs when the maximum shearing stress reaches the value of the shearing stress at failure in a simple tension or compression test on the same material. The theory is widely used for design of ductile materials. For applications see Problems 18.7 and 18.11.

HUBER–VON MISES–HENCKY (MAXIMUM ENERGY OF DISTORTION) THEORY

For an element subject to the principal stresses σ_1, σ_2, σ_3 this theory states that yielding begins when

$$(\sigma_1 - \sigma_2)^2 + (\sigma_2 - \sigma_3)^2 + (\sigma_1 - \sigma_3)^2 = 2(\sigma_{yp})^2$$

where σ_{yp} is the yield point of the material. This theory is in excellent agreement with experiments on ductile materials. For applications see Problems 18.8, 18.9, and 18.12.

Solved Problems

18.1. A short block is loaded by a compressive force of 100,000 lb acting 2 in. from one axis and 3 in. from another axis of an 8×8 in. cross-section as shown in Fig. 18-5(a). Determine the maximum tensile and compressive stresses in the cross-section.

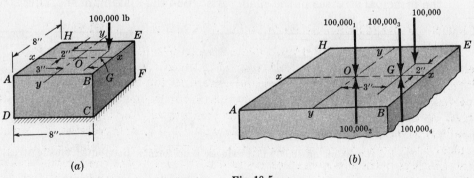

(a) (b)

Fig. 18-5

Let us consider point O, the geometric center of the cross-section, and point G located on one axis of symmetry and lying 3 in. from O. Through each of these two points let us introduce a pair of equal and opposite forces, each of magnitude 100,000 lb. The upper surface then has the appearance of Fig. 18-5(b).

These four forces that have been added are designated as $100,000_1$, $100,000_2$, etc., and constitute a self-equilibrating system. Thus they do not change the original stressed state of the body but merely provide a simplified medium of calculation.

The force $100{,}000_1$ lb produces a uniformly distributed compressive stress over any horizontal cross-section.   The forces $100{,}000_4$ lb and $100{,}000$ lb constitute a couple giving rise to bending about the x-x axis.   The forces $100{,}000_2$ lb and $100{,}000_3$ lb constitute a couple giving rise to bending about the y-y axis.

Due to the force $100{,}000_1$ lb we have a uniform compressive stress

$$\sigma_1 \;=\; \frac{100{,}000}{8(8)} \;=\; 1560 \text{ lb/in}^2$$

The couple consisting of the forces $100{,}000_4$ lb and $100{,}000$ lb gives rise to maximum tension along the line AB and maximum compression along the line HE.   The values of these extreme fiber stresses are

$$\sigma_2 \;=\; \frac{M_x c}{I_x} \;=\; \frac{100{,}000(2)(4)}{8(8)^3/12} \;=\; 2350 \text{ lb/in}^2$$

The couple consisting of the forces $100{,}000_2$ lb and $100{,}000_3$ lb gives rise to maximum tension along the line AH and maximum compression along the line BE.   The values of these extreme fiber stresses are

$$\sigma_3 \;=\; \frac{M_y c}{I_y} \;=\; \frac{100{,}000(3)(4)}{8(8)^3/12} \;=\; 3520 \text{ lb/in}^2$$

The maximum compressive stress is thus along line EF and is given by

$$\sigma_4 \;=\; -1560 - 2350 - 3520 \;=\; -7430 \text{ lb/in}^2$$

The maximum tensile stress occurs along line AD and is equal to

$$\sigma_5 \;=\; -1560 + 2350 + 3520 \;=\; +4310 \text{ lb/in}^2$$

The stresses σ_4 and σ_5 are directed vertically.

It is to be observed that this method of analysis is valid only for those cases where the x and y axes are axes of symmetry of the cross-section.

18.2.   Consider a thin-walled cylindrical shell subject to combined axial tension and torsion. The shell is of diameter 16 in., and the wall thickness is 0.10 in.   The shell is subject to an axial tension of 40,000 lb together with a torque of 400,000 lb-in.   Determine the principal stresses in the shell.   Also, find the maximum shearing stress.

$$(a) \qquad\qquad\qquad\qquad\qquad\qquad (b)$$

Fig. 18-6

The axial tension of 40,000 lb produces a uniformly distributed tensile stress given by

$$\sigma_x \;=\; \frac{P}{A} \;=\; \frac{40{,}000}{\pi(16)(0.10)} \;=\; 7950 \text{ lb/in}^2$$

acting over every cross-section.   This stress appears as in Fig. 18-6(a).

The shearing stresses due to the torque of 400,000 lb-in were determined in Problem 5.2 The shearing stress in the wall of the shell was found to be $\tau_{xy} = T\rho/J$.   For a thin-walled tube such as we have here, the polar moment of inertia is seen from Problem 5.8 to be

$$J \;=\; 2\pi R^3 t \;=\; 2\pi(8^3)(0.10) \;=\; 321 \text{ in}^4$$

The shearing stress in the shell is thus

$$\tau_{xy} = \frac{T\rho}{J} = \frac{400,000(8)}{321} = 10,000 \text{ lb/in}^2$$

These stresses appear as in Fig. 18-6(b).

Since both loadings are acting simultaneously it is necessary to combine these stresses. Because the stress directions are different, the vector methods described in Chapter 17 must be employed. The case of one normal stress together with a shearing stress acting on an element was treated in Problem 17.7. Using the notation of that problem, we have here

$$\sigma_x = 7950 \text{ lb/in}^2 \qquad \tau_{xy} = 10,000 \text{ lb/in}^2$$

From Problem 17.7 the principal stresses are

$$\sigma_{max} = \tfrac{1}{2}\sigma_x + \sqrt{(\tfrac{1}{2}\sigma_x)^2 + (\tau_{xy})^2} = 7950/2 + \sqrt{(7950/2)^2 + (10,000)^2} = 14,800 \text{ lb/in}^2$$

$$\sigma_{min} = \tfrac{1}{2}\sigma_x - \sqrt{(\tfrac{1}{2}\sigma_x)^2 + (\tau_{xy})^2} = 7950/2 - \sqrt{(7950/2)^2 + (10,000)^2} = -6800 \text{ lb/in}^2$$

These stresses occur on planes defined by (3) of Problem 17.7:

$$\tan 2\theta_p = -\frac{\tau_{xy}}{\tfrac{1}{2}\sigma_x} = -\frac{10,000}{7950/2} = -2.50 \quad \text{or} \quad \theta_p = 55°50', 145°50'$$

Substituting in (1) of Problem 17.7, letting $\theta = 55°50'$ we have

$$\sigma = 7950/2 - (7950/2) \cos 111°40' + 10,000 \sin 111°40' = 14,800 \text{ lb/in}^2$$

The maximum principal stress of 14,800 lb/in² thus occurs on a plane oriented 55°50' to the longitudinal axis of the shell.

From (8) of Problem 17.7, the maximum shearing stresses are

$$\tau = \pm\sqrt{(\tfrac{1}{2}\sigma_x)^2 + (\tau_{xy})^2} = \pm\sqrt{(7950/2)^2 + (10,000)^2} = \pm10,800 \text{ lb/in}^2$$

These stresses occur on planes oriented at 45° to the planes on which the maximum normal stresses occur.

18.3. A shaft 2 in. in diameter is loaded by an axial compressive force of 50,000 lb together with a twisting moment of 30,000 lb-in. Determine the principal stresses and also the maximum shearing stress in the shaft.

The axial force gives rise to a uniform compressive stress given by

$$\sigma_x = \frac{P}{A} = \frac{50,000}{\tfrac{1}{4}\pi(2)^2} = 15,900 \text{ lb/in}^2$$

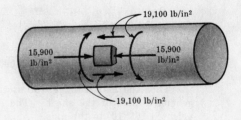

The shearing stress due to the applied twisting moment was shown in Problem 5.2, to be $\tau_{xy} = T\rho/J$. This is maximum at the outer fibers of the shaft and becomes

$$\tau_{xy} = \frac{T\rho}{J} = \frac{30,000(1)}{\pi(2)^4/32} = 19,100 \text{ lb/in}^2$$

Fig. 18-7

An element at the outer surface of the shaft is thus subject to the stresses shown in Fig. 18-7. The principal stresses for such a loading on an element were derived in Problem 17.7. They are

$$\sigma_{max} = \tfrac{1}{2}\sigma_x + \sqrt{(\tfrac{1}{2}\sigma_x)^2 + (\tau_{xy})^2} = -15,900/2 + \sqrt{(-15,900/2)^2 + (19,100)^2} = 12,750 \text{ lb/in}^2$$

$$\sigma_{min} = \tfrac{1}{2}\sigma_x - \sqrt{(\tfrac{1}{2}\sigma_x)^2 + (\tau_{xy})^2} = -15,900/2 - \sqrt{(-15,900/2)^2 + (19,100)^2} = -28,650 \text{ lb/in}^2$$

The maximum shearing stress is found from (8) of Problem 17.7 to be

$$\tau = \pm\sqrt{(\tfrac{1}{2}\sigma_x)^2 + (\tau_{xy})^2} = \pm\sqrt{(-15,900/2)^2 + (19,100)^2} = \pm20,700 \text{ lb/in}^2$$

18.4. Consider a hollow circular shaft whose outside diameter is 3 in. and whose inside diameter is equal to one-half the outside diameter. The shaft is subject to a twisting moment of 20,000 lb-in as well as a bending moment of 30,000 lb-in. Determine the principal stresses in the body. Also, determine the maximum shearing stress.

The twisting moment gives rise to shearing stresses that attain their peak values in the outer fibers of the shaft. From Problem 5.2 these shearing stresses are given by $\tau_{xy} = T\rho/J$. From Problem 5.1 it is seen that for the hollow circular area

$$J = \frac{\pi}{32}(D_o^4 - D_i^4) = \frac{\pi}{32}[3^4 - (1.5)^4] = 7.46 \text{ in}^4$$

where D_o denotes the outer diameter of the section and D_i represents the inner diameter. At the outer fibers the torsional shearing stresses are thus

$$\tau_{xy} = \frac{T\rho}{J} = \frac{20,000(1.5)}{7.46} = 4000 \text{ lb/in}^2$$

Let the bending moments lie in a vertical plane. Then the upper and lower fibers of the beam are subject to the peak bending stresses. These are found from the expression $\sigma_x = My/I$. The moment of inertia I for the hollow circular cross-section may be seen from Problem 7.9 to be

$$I = \frac{\pi}{64}(D_o^4 - D_i^4) = \frac{\pi}{64}[3^4 - (1.5)^4] = 3.73 \text{ in}^4$$

Substituting,

$$\sigma_x = \frac{My}{I} = \frac{30,000(1.5)}{3.73} = 12,000 \text{ lb/in}^2$$

Thus an element located at the lower extremity of the shaft is subject to the stresses shown in Fig. 18-8.

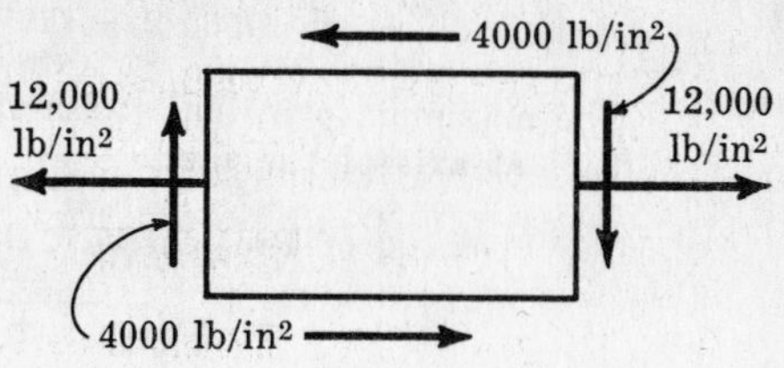

Fig. 18-8

From Problem 17.7 the principal stresses for this element are

$$\sigma_{max} = \tfrac{1}{2}\sigma_x + \sqrt{(\tfrac{1}{2}\sigma_x)^2 + (\tau_{xy})^2} = 12,000/2 + \sqrt{(12,000/2)^2 + (4000)^2} = 13,200 \text{ lb/in}^2$$

$$\sigma_{min} = \tfrac{1}{2}\sigma_x - \sqrt{(\tfrac{1}{2}\sigma_x)^2 + (\tau_{xy})^2} = 12,000/2 - \sqrt{(12,000/2)^2 + (4000)^2} = -1200 \text{ lb/in}^2$$

These stresses occur on planes defined by (3) of Problem 17.7:

$$\tan 2\theta_p = -\frac{\tau_{xy}}{\tfrac{1}{2}\sigma_x} = -\frac{4000}{12,000/2} = -\frac{2}{3} \quad \text{or} \quad \theta_p = 73°10', 163°10'$$

Substituting in (1) of Problem 17.7, letting $\theta = 73°10'$ we have

$$\sigma = 12,000/2 - (12,000/2)\cos 146°20' + 4000 \sin 146°20' = 13,200 \text{ lb/in}^2$$

Thus the maximum tensile stress is 13,200 lb/in², occurring on a plane oriented 73°10' to the geometric axis of the shaft. The other principal stress, $\sigma_{min} = -1200$ lb/in², occurs on a plane oriented 163°10' to the axis.

The maximum shearing stress is given by (8) of Problem 17.7. It is

$$\tau = \pm\sqrt{(\tfrac{1}{2}\sigma_x)^2 + (\tau_{xy})^2} = \pm\sqrt{(12,000/2)^2 + (4000)^2} = \pm7200 \text{ lb/in}^2$$

and occurs on planes oriented at 45° to the planes found above on which the principal stresses act.

18.5. The shaft shown in Fig. 18-9(a) rotates with constant angular velocity. The belt pulls create a state of combined bending and torsion. Neglect the weights of the shaft and pulleys and assume that the bearings can exert only concentrated force reactions. The diameter of the shaft is 1.25 in. Determine the principal stresses in the shaft.

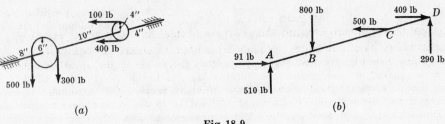

(a)   (b)

Fig. 18-9

The transverse forces acting on the shaft are not parallel and the bending moments caused by them must be added vectorially to obtain the resultant bending moment. This vector addition need be carried out at only a few apparently critical points along the length of the shaft. The loads causing bending, together with the reactions they produce, are shown above in Fig. 18-9(b); they are considered as passing through the axis of the shaft. The upper and lower shaded portions of Fig. 18-10 respectively represent the bending moment diagrams for a vertical and for a horizontal plane.

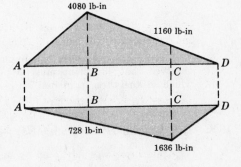

Fig. 18-10

The resultant bending moments at B and C are

$$M_B = \sqrt{(4080)^2 + (728)^2} = 4140 \text{ lb-in}$$

$$M_C = \sqrt{(1160)^2 + (1636)^2} = 2000 \text{ lb-in}$$

The twisting moment between the two pulleys is constant and equal to

$$T = (400 - 100)(4) = 1200 \text{ lb-in}$$

Since the torque is the same at B and C, the critical element lies at the outer fibers of the shaft at point B. The maximum bending stress is given by

$$\sigma_x = \frac{My}{I} = \frac{(4140)(1.25/2)}{\pi(1.25)^4/64} = 21,500 \text{ lb/in}^2$$

The maximum shearing stress, occurring at the outer fibers of the shaft, is given by

$$\tau_{xy} = \frac{T\rho}{J} = \frac{1200(1.25/2)}{\pi(1.25)^4/32} = 3100 \text{ lb/in}^2$$

The principal stresses were found in Problem 17.7 to be

$$\sigma_{max} = \tfrac{1}{2}\sigma_x + \sqrt{(\tfrac{1}{2}\sigma_x)^2 + (\tau_{xy})^2} = 21,500/2 + \sqrt{(21,500/2)^2 + (3100)^2} = 22,000 \text{ lb/in}^2$$

$$\sigma_{min} = \tfrac{1}{2}\sigma_x - \sqrt{(\tfrac{1}{2}\sigma_x)^2 + (\tau_{xy})^2} = 21,500/2 - \sqrt{(21,500/2)^2 + (3100)^2} = -400 \text{ lb/in}^2$$

18.6. Discuss a failure criterion for *brittle* materials.

The criterion which is in best agreement with experimental evidence was advanced by the English engineer W. J. M. Rankine and is termed the *maximum normal stress theory*. It states that failure of the material (i.e. either yielding or fracture – whichever occurs first) occurs when the maximum normal stress reaches the value at which failure occurs in a simple tension test on the same material. Alternatively, if the loading is compressive, failure occurs when the minimum normal stress reaches the value at which failure occurs in a simple compression test. Evidently this criterion considers only the greatest (or smallest) of the principal stresses and disregards the influence of the other principal stresses.

18.7. Discuss the *maximum shearing stress* failure criterion for *ductile* materials.

This criterion is in good agreement with experimental evidence, provided the yield point of the material in tension is equal to that in compression. It was advanced first by C. A. Coulomb in 1773 and later by H. Tresca in 1864; in fact, it is often called the *Tresca criterion*. The criterion states that failure of the material subject to biaxial or triaxial stress occurs when the maximum shearing stress at any point reaches the value of the shearing stress at failure in a simple tension or compression test on the same material. In Problem 17.13 it was shown that the maximum shear stress is one-half the difference between the maximum and minimum principal stresses and always occurs on a plane inclined at 45 degrees to the principal planes. Thus, if σ_{yp} denotes the yield point of the material in simple tension or compression, then the corresponding maximum shear stress is $\sigma_{yp}/2$. Accordingly, the maximum shearing stress criterion may be formulated as

$$(\sigma_{max} - \sigma_{min})/2 = \sigma_{yp}/2$$

or

$$\sigma_{max} - \sigma_{min} = \sigma_{yp} \tag{1}$$

where σ_{max} and σ_{min} are maximum and minimum principal stresses respectively. It is to be observed that judgment must be used in analysis of three-dimensional situations to determine which of the three principal stresses lead to the greatest difference on the left-hand side of (1).

18.8.   Discuss the *Huber–von Mises–Hencky* failure criterion for *ductile* materials.

This theory was advanced by M. T. Huber in Poland in 1904 and independently by R. von Mises in Germany in 1913 and H. Hencky in 1925. It is in even better agreement with experimental evidence concerning failure of ductile materials subject to biaxial or triaxial stresses than the maximum shearing stress theory discussed in Problem 18.7.

Development of this widely accepted criterion first necessitates determination of the strain energy per unit volume in a simple tension specimen. If the axial tensile stress arising in this test is σ_1 and the corresponding axial strain is ϵ_1, then the work done on a unit volume of the test specimen is the product of the mean value of force per unit area, i.e. $\sigma_1/2$, times the displacement in the direction of the force, or ϵ_1. The work is thus $U = \sigma_1\epsilon_1/2$ and this work is stored as internal strain energy.

The strain energy per unit volume in an element subject to triaxial *principal stresses* σ_1, σ_2, σ_3 is readily found by superposition (since energy is a scalar quantity) to be

$$U = \tfrac{1}{2}\sigma_1\epsilon_1 + \tfrac{1}{2}\sigma_2\epsilon_2 + \tfrac{1}{2}\sigma_3\epsilon_3 \tag{a}$$

where $\epsilon_1, \epsilon_2, \epsilon_3$ are the normal strains in the directions of the principal stresses respectively. If the strains are expressed in terms of the stresses according to the relations given in Problem 1.20, equation (a) becomes

$$U = \frac{1}{2E}[(\sigma_1^2 + \sigma_2^2 + \sigma_3^2) - 2\mu(\sigma_1\sigma_2 + \sigma_1\sigma_3 + \sigma_2\sigma_3)] \tag{b}$$

The triaxial principal stresses may be represented as in Fig. 18-11(a). Alternatively, this general state of stress may be represented as the sum of the two triaxial states shown in Fig. 18-11(b) and (c).

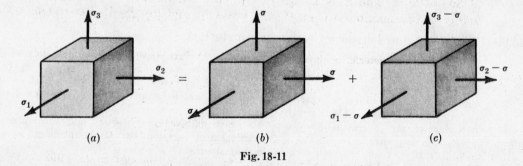

$$(a) \qquad\qquad\qquad (b) \qquad\qquad\qquad (c)$$

Fig. 18-11

The strain energy U given by equation (b) may be resolved into two components, one portion U_v corresponding to a change of volume with no distortion of the element, the other, U_d, corresponding to distortion of the element with no change of volume. The stresses indicated in Fig. 18-11(c) represent *distortion only* with no change of volume, provided the expression for *dilatation* given in Problem 1.20 is set equal to zero. Thus

$$\epsilon_1 + \epsilon_2 + \epsilon_3 = \frac{1}{E} [(\sigma_1 - \sigma) - \mu(\sigma_2 + \sigma_3 - 2\sigma)$$

$$+ (\sigma_2 - \sigma) - \mu(\sigma_1 + \sigma_3 - 2\sigma)$$

$$+ (\sigma_3 - \sigma) - \mu(\sigma_1 + \sigma_2 - 2\sigma)] = 0 \qquad (c)$$

Solving (c) we find

$$\sigma = (\sigma_1 + \sigma_2 + \sigma_3)/3 \qquad (d)$$

for the uniform stresses in Fig. 18-11(b) which correspond to change of volume with no distortion. The normal strains corresponding to the stresses given in (d) are readily found from the three-dimensional form of Hooke's law given in Problem 1.20 to be

$$\epsilon = (1 - 2\mu)\sigma/E \qquad (e)$$

Thus, the internal strain energy corresponding to the unit volume indicated in Fig. 18-11(b) is found by substituting the expressions (d) and (e) in (a), with $\sigma_1 = \sigma_2 = \sigma_3 = \sigma$ and $\epsilon_1 = \epsilon_2 = \epsilon_3 = \epsilon$, to obtain

$$U_v = 3\left(\frac{\sigma\epsilon}{2}\right) = \frac{1 - 2\mu}{6E}(\sigma_1 + \sigma_2 + \sigma_3)^2 \qquad (f)$$

The strain energy corresponding to *distortion only*, with no change of volume, is now found to be

$$U_d = U - U_v = \frac{1 + \mu}{6E}[(\sigma_1 - \sigma_2)^2 + (\sigma_2 - \sigma_3)^2 + (\sigma_1 - \sigma_3)^2] \qquad (g)$$

The Huber–von Mises–Hencky theory assumes that failure takes place when the internal strain energy of distortion given by (g) is equal to that at which failure occurs in a simple tension test. In such a test $\sigma_2 = \sigma_3 = 0$, $\sigma_1 = \sigma_{yp}$ and the right side of (g) becomes

$$\frac{1 + \mu}{6E}[2\sigma_{yp}^2] \qquad (h)$$

Equating the right side of (g) to (h) we find

$$(\sigma_1 - \sigma_2)^2 + (\sigma_2 - \sigma_3)^2 + (\sigma_1 - \sigma_3)^2 = 2\sigma_{yp}^2 \qquad (i)$$

as the criterion for failure. This is sometimes called the *maximum energy of distortion* theory. It assumes that U_v is ineffective in causing failure.

18.9. A thin-walled cylindrical pressure vessel is subject to an internal pressure of 600 lb/in². The mean radius of the cylinder is 15 in. If the material has a yield point of 39,000 lb/in² and a safety of 3 is employed, determine the required wall thickness using (a) the maximum normal stress theory, and (b) the Huber–von Mises–Hencky theory.

The stresses determined in Problem 3.1 are principal stresses. Thus we have

$$\sigma_1 = \sigma_c = \frac{pr}{h} = \frac{600(15)}{h} = \frac{9000}{h}$$

$$\sigma_2 = \sigma_l = \frac{pr}{2h} = \frac{600(15)}{2h} = \frac{4500}{h}$$

The third principal stress varies from zero at the outside of the shell to the value $-p$ at the inside. It is customary to neglect this third component in thin-shell design, so we shall assume that $\sigma_3 = 0$.

(a) Using the maximum normal stress theory we have

$$\frac{9000}{h} = \frac{39,000}{3} \qquad \text{from which} \qquad h = 0.69 \text{ in.}$$

(b) Using the Huber–von Mises–Hencky theory we have from (i) of Problem 18.8

$$\left(\frac{9000}{h} - \frac{4500}{h}\right)^2 + \left(\frac{4500}{h} - 0\right)^2 + \left(\frac{9000}{h} - 0\right)^2 = 2\left(\frac{39,000}{3}\right)^2$$

whence $h = 0.60$ in.

18.10. The solid circular shaft in Fig. 18-12(a) is subject to belt pulls at each end and is simply supported at the two bearings. The material has a yield point of 36,000 lb/in². Determine the required diameter of the shaft using the maximum normal stress theory together with a safety factor of 3.

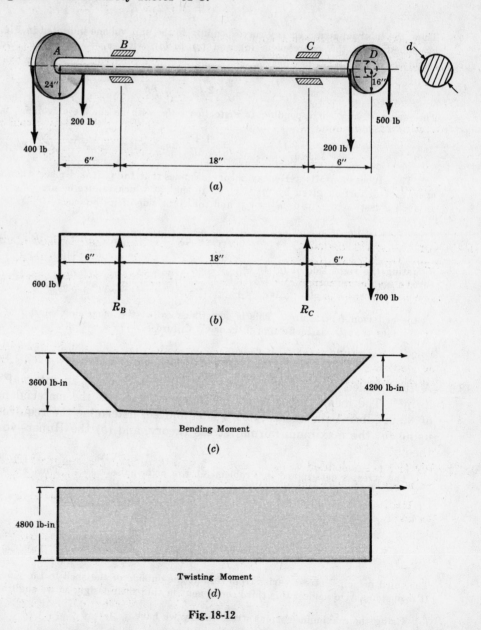

Fig. 18-12

The bearing reactions, which are in a vertical plane, are denoted by R_B and R_C in the free-body diagram, Fig. 18-12(b). From statics it is found that $R_B = 567$ lb and $R_C = 733$ lb. The variation of bending moment along the length of the shaft is shown in Fig. 18-12(c). Similarly, the twisting moment along the length of the shaft may be depicted as a constant, as in Fig. 18-12(d).

Evidently the shaft is most critically stressed at its outer fibers at point C where a top view of the uppermost element indicates the stresses σ_x and τ_{xy} shown in Fig. 18-13. The normal stress σ_x arises because of bending action and is found from Problem 8.1 to be

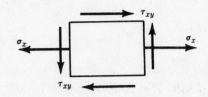

$$\sigma_x = \frac{Mc}{I} = \frac{4200(d/2)}{\pi d^4/64} = \frac{42,800}{d^3} \qquad (a)$$

The other normal stresses, σ_y and σ_z, are zero. The shearing stresses τ_{xy} arise from the torsion due to the unequal belt pulls and are found from Problem 5.2 to be

Fig. 18-13

$$\tau_{xy} = \frac{Tr}{J} = \frac{4800(d/2)}{\pi d^4/32} = \frac{24,480}{d^3} \qquad (b)$$

According to the maximum normal stress theory, yielding of the shaft occurs when the maximum normal stress reaches the value at which yielding occurs in a simple tensile test. The maximum normal stress is found as the maximum principal stress of Problem 17.13 to be

$$\sigma_{max} = \frac{\sigma_x + \sigma_y}{2} + \sqrt{\left(\frac{\sigma_x - \sigma_y}{2}\right)^2 + (\tau_{xy})^2} \qquad (c)$$

Substituting the results of (a) and (b) into (c) and introducing the safety factor of 3 yields

$$\frac{36,000}{3} = \frac{42,800 + 0}{2d^3} + \sqrt{\left(\frac{42,800 - 0}{2d^3}\right)^2 + \left(\frac{24,480}{d^3}\right)^2}$$

from which

$$d = 1.66 \text{ in.}$$

18.11. For the shaft loaded as in Problem 18.10 determine the required diameter using the maximum shearing stress theory together with a safety factor of 3.

The maximum normal stress is given in (c) of Problem 18.10. From Problem 17.13 the minimum normal stress is given by

$$\sigma_{min} = \frac{\sigma_x + \sigma_y}{2} - \sqrt{\left(\frac{\sigma_x - \sigma_y}{2}\right)^2 + (\tau_{xy})^2} \qquad (a)$$

It is to be carefully noted that the difference between the σ_{max} and σ_{min} indicated above leads to the *greatest* possible difference, since the third principal stress is zero and σ_{min} is evidently negative. Substituting in (1) of Problem 18.7 we have

$$2\sqrt{\left(\frac{42,800 - 0}{2d^3}\right)^2 + \left(\frac{24,480}{d^3}\right)^2} = \frac{36,000}{3} \qquad \text{or} \qquad d = 1.76 \text{ in.}$$

18.12. For the shaft loaded as in Problem 18.10 determine the required diameter using the Huber–von Mises–Hencky theory together with a safety factor of 3.

The criterion is expressed by (i) of Problem 18.8 where σ_1, σ_2, and σ_3 are principal stresses. We take these principal stresses to be

$$\sigma_1 = \sigma_{max} = \left(\frac{42,800 + 0}{2d^3}\right) + \sqrt{\left(\frac{42,800 - 0}{2d^3}\right)^2 + \left(\frac{24,480}{d^3}\right)^2}$$

$$\sigma_2 = 0$$

$$\sigma_3 = \sigma_{min} = \left(\frac{42,800 + 0}{2d^3}\right) - \sqrt{\left(\frac{42,800 - 0}{2d^3}\right)^2 + \left(\frac{24,480}{d^3}\right)^2}$$

Substituting in (*i*) of Problem 18.8 we have

$$\left[\left\{\left(\frac{42,800+0}{2d^3}\right) + \sqrt{\left(\frac{42,800-0}{2d^3}\right)^2 + \left(\frac{24,480}{d^3}\right)^2}\right\} - 0\right]^2$$

$$+ \left[0 - \left\{\left(\frac{42,800+0}{2d^3}\right) - \sqrt{\left(\frac{42,800-0}{2d^3}\right)^2 + \left(\frac{24,480}{d^3}\right)^2}\right\}\right]^2$$

$$+ \left[\left\{\left(\frac{42,800+0}{2d^3}\right) + \sqrt{\left(\frac{42,800-0}{2d^3}\right)^2 + \left(\frac{24,480}{d^3}\right)^2}\right\}\right.$$

$$\left. - \left\{\left(\frac{42,800+0}{2d^3}\right) - \sqrt{\left(\frac{42,800-0}{2d^3}\right)^2 + \left(\frac{24,480}{d^3}\right)^2}\right\}\right]^2$$

$$= \quad 2\left(\frac{36,000}{3}\right)^2$$

Solving: $\qquad\qquad\qquad\qquad d = 1.54$ in.

Supplementary Problems

18.13. A short block is loaded by a compressive force of 300,000 lb. The force is applied with an eccentricity of 1.5 in. as shown in Fig. 18-14. The block is 12 in. by 12 in. in cross-section. Determine the stresses at the outer fibers m and n.   *Ans.* $\sigma_m = -520$ lb/in^2, $\sigma_n = -3640$ lb/in^2

18.14. In Problem 18.13 how large an eccentricity must exist if the resultant stress at fiber m is to be zero?   *Ans.* 2.00 in.

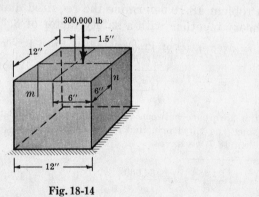

Fig. 18-14

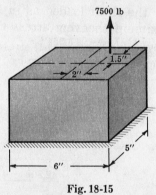

Fig. 18-15

18.15. A block is loaded by the eccentric tensile force shown in Fig. 18-15. Determine the maximum tensile stress.   *Ans.* 1200 lb/in^2

18.16. A thin-walled cylinder is 10 in. in diameter and of wall thickness 0.10 in. The cylinder is subject to a uniform internal pressure of 100 lb/in^2. What additional axial tension may act simultaneously without the maximum tensile stress exceeding 20,000 lb/in^2?   *Ans.* 55,000 lb

18.17. A thin-walled cylindrical shell is subject to an axial compression of 50,000 lb together with a torsional moment of 30,000 lb-in. The diameter of the cylinder is 12 in. and the wall thickness 0.125 in. Determine the principal stresses in the shell. Also determine the maximum shearing stress. Neglect the possibility of buckling of the shell.

Ans. $\sigma_{max} = 120$ lb/in^2, $\sigma_{min} = -10,680$ lb/in^2, $\tau = 5400$ lb/in^2

18.18. A shaft 2.50 in. in diameter is subject to an axial tension of 40,000 lb together with a twisting moment of 35,000 lb-in. Determine the principal stresses in the shaft. Also determine the maximum shearing stress. *Ans.* $\sigma_{max} = 16,180$ lb/in², $\sigma_{min} = -8020$ lb/in², $\tau = 12,100$ lb/in²

18.19. Consider a solid circular shaft subject to a twisting moment of 20,000 lb-in together with a bending moment of 30,000 lb-in. The diameter of the shaft is 3 in. Determine the principal stresses, as well as the maximum shearing stress in the shaft.

Ans. $\sigma_{max} = 12,450$ lb/in², $\sigma_{min} = -1150$ lb/in², $\tau = 6800$ lb/in²

18.20. The shaft shown in Fig. 18-16 rotates with constant angular velocity and is subject to combined bending and torsion due to the indicated belt pulls. The weights of the shaft and pulleys may be neglected and the bearings can exert only concentrated force reactions. The diameter of the shaft is 1.75 in. Determine the principal stresses in the shaft.

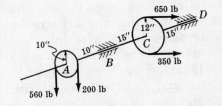

Ans. $\sigma_{max} = 16,600$ lb/in², $\sigma_{min} = -750$ lb/in²

Fig. 18-16

18.21. Consider the thin-walled pressure vessel mentioned in Problem 18.9. Use the maximum shearing stress theory to determine the required wall thickness. *Ans.* 0.69 in.

18.22. Consider a thin-walled cylindrical pressure vessel with mean diameter 6 in. subject to a twisting moment of 10,000 lb-in together with an internal pressure of 400 lb/in². If the allowable working stress in tension is 20,000 lb/in², determine the wall thickness as required by the maximum normal stress theory. *Ans.* 0.0625 in.

18.23. For Problem 18.22 determine the wall thickness as required by the maximum shearing stress theory. *Ans.* 0.0625 in.

18.24. For Problem 18.22 determine the wall thickness as required by the Huber–von Mises–Hencky theory. *Ans.* 0.054 in.

Chapter 19

Theory of Elasticity

The preceding chapters have dealt with various types of analysis of bars, beams, and shafts subject to either concentrated or distributed forces. The general approach has essentially been based upon a consideration of (a) the geometry of deformation, (b) equilibrium relations, and (c) stress-strain relations. In each type of analysis certain simplifying assumptions were introduced. The results presented are of course limited to those situations where the assumptions are satisfied.

In many problems of practical concern the types of analysis examined so far are inadequate, i.e. the physical situation under investigation does not correspond to the assumptions introduced into the analyses. In such cases it is necessary to turn to a more comprehensive and more sophisticated type of analysis which does not involve as many simplifying assumptions. Such an analysis is to be found in the science termed *theory of elasticity*. Analyses based upon theory of elasticity give much more detailed and more precise information about the state of stress, strain, and deformation at any point within the body than the more simplified type of study presented up to now and termed *strength of materials*. However, in general it is necessary to deal with much more complex mathematical situations in the theory of elasticity than are encountered in strength of materials.

STRESS CONCENTRATION

In particular, the theory of elasticity is excellent for investigating the state of stress and deformation in the immediate vicinity of small holes, notches, and cuts in an elastic body. This is illustrated in Problem 19.20. Obviously such effects cannot be treated by the strength of materials approach. As we shall see, the presence of a small hole can significantly increase the value of localized stress over that predicted by a strength of materials approach. The study of stress concentrations due to holes and other local irregularities is an important aspect of the theory of elasticity.

BOUNDARY CONDITIONS

The theory of elasticity permits a much more rigorous and more detailed treatment of boundary conditions than can be carried out in strength of materials. For example, in the theory of elasticity we can examine the boundary conditions at every point throughout the depth of a bar, as in Problems 19.8 and 19.9, whereas in strength of materials we can specify only the resultant shear force or bending moment acting over the end section. Moreover, we can specify the deflection or slope, which in strength of materials is assumed to be the same at every point along the depth of the bar.

In general, the external applied forces may be regarded as continuations of the internal stresses as determined by elasticity theory. That is, on surface elements of the body, the stresses must be in equilibrium with the applied external forces. Occasionally, as we shall see in Problem 19.9, we will not be able to satisfy precisely all of the boundary conditions prescribed and in some problems the solution found will indicate the presence of small additional forces applied at the boundaries in addition to the true loading. However, these supplemental forces can usually be shown to have a zero resultant and, further, to lead to a zero resultant moment. Thus, their presence is not serious except in the immediate vicinity of their points of application. This is intuitively evident and is expressed more generally as:

SAINT-VENANT'S PRINCIPLE

Let us consider a number of statically equivalent force systems acting over a specified small portion of the surface of an elastic body. By statically equivalent it is implied that the systems all have the same force and moment resultants. Saint-Venant's principle states that although these various statically equivalent systems may have considerably different localized effects, all have essentially the same effect on stresses at any distance which is large compared to the dimensions of the part of the surface on which these forces are applied.

The solution of any problem is exact only if the surface forces actually have the distribution indicated by the theory of elasticity solution. If the true forces are not distributed in such a manner, the solution is still of value if one remembers Saint-Venant's principle and does not employ the solution in the immediate vicinity of the points of application of the surface forces.

NOMENCLATURE

With reference to a rectangular Cartesian coordinate system (x, y, z) we shall designate the normal stresses parallel to these axes by σ_x, σ_y, and σ_z, respectively. A double subscript notation will be employed for shearing stresses, the first indicating the direction of the normal to the plane under consideration and the second the direction of the particular stress component. Thus, for shearing stresses we have the symbols τ_{xy}, τ_{xz}, and τ_{yz}. This is discussed in Problem 19.1. With reference to a plane polar coordinate system the radial and tangential normal stresses are designated as σ_r and σ_θ, respectively, and the shearing stress by $\tau_{r\theta}$. This is illustrated in Problem 19.11.

Displacement components parallel to the x-, y-, and z-axes are denoted by u, v, and w, respectively, and in the plane polar coordinate system the radial and tangential components of displacement are represented as u and v, respectively. See Problems 19.3 and 9.13.

Normal strains in the rectangular Cartesian coordinate system are denoted by ϵ_x, ϵ_y, and ϵ_z in the directions of the x-, y-, and z-axes, respectively. The shear strains are represented by γ_{xy}, γ_{xz}, and γ_{yz}, as discussed in Problem 9.3. In the plane polar coordinate system the normal strains in the radial and tangential directions are given by ϵ_r and ϵ_θ and the shearing strain by $\gamma_{r\theta}$, as mentioned in Problem 9.13.

BODY FORCES

Forces distributed continuously throughout the volume of a body are called *body forces*. Examples of body forces are gravitational forces, magnetic forces, and inertia forces.

SURFACE FORCES

Forces distributed over the surface of a body are called *surface forces*. The mechanical action of one body upon another is an example of a surface force.

EQUATIONS OF EQUILIBRIUM

The theory of elasticity is based upon consideration of the equilibrium of an infinitesimal element. One of the significant features of the theory of elasticity is that account is taken of the increment of stress between two faces of the element a distance dx apart. This is in contrast to the strength of materials approach, which does not consider stress increments between two closely adjacent surfaces. In Problem 19.1 it is shown that equilibrium of an element in the x-direction leads to the equation

$$\frac{\partial \sigma_x}{\partial x} + \frac{\partial \tau_{yx}}{\partial y} + \frac{\partial \tau_{zx}}{\partial z} + X = 0$$

together with two more analogous equations corresponding to the y- and z-directions.

EQUATIONS OF COMPATIBILITY

The six strain components $\epsilon_x, \epsilon_y, \epsilon_z, \gamma_{xy}, \gamma_{xz}, \gamma_{yz}$ are expressed in terms of the displacement components $u, v,$ and w as shown in Problem 19.3. Thus, the strains cannot be prescribed arbitrarily but instead must satisfy a set of six partial differential equations which are obtained in Problem 19.4. The first of these is

$$\frac{\partial^2 \epsilon_x}{\partial y^2} + \frac{\partial^2 \epsilon_y}{\partial x^2} = \frac{\partial^2 \gamma_{xy}}{\partial x \, \partial y}$$

Satisfaction of these six equations, termed the *equations of compatibility*, ensures the existence of a single-valued continuous strain field.

FORMULATION OF AN ELASTICITY PROBLEM

A solution to any problem in the theory of elasticity consists of determining stress components satisfying the differential equations of equilibrium, strain components satisfying compatibility conditions, and stress and displacement components satisfying boundary conditions. Obviously this is a matter of considerable mathematical complexity.

AIRY STRESS FUNCTION

The problem of simultaneously satisfying equilibrium and compatibility can be simplified by introducing a function defined in such a manner that equilibrium is automatically satisfied, leaving only the compatibility conditions to be satisfied. For the two-dimensional case this is discussed in Problem 19.6. There we obtain a function $\Phi(x, y)$ which when appropriately differentiated leads to the stresses $\sigma_x, \sigma_y,$ and τ_{xy}. This function is termed the *Airy stress function* in honor of the English mathematician G. B. Airy, who introduced it in 1862. Introduction of this function makes the problem more compact, since we must now satisfy only the compatibility equations together with boundary conditions.

Solved Problems

19.1. Derive the differential equations of equilibrium.

Let us consider the equilibrium of a small rectangular parallelepiped with edges of lengths dx, dy and dz. The element, together with the stresses acting on the various faces, is shown in Fig. 19-1. The stresses are indicated there in their positive directions. One of the essential distinctions of the theory of elasticity is that the small changes of stress occurring in a distance of dx, or dy, or dz are considered, whereas in the more elementary treatment of strength of materials such refinements are not introduced. Because of this variation of stress from one face to another an infinitesimal distance away, the normal stress on the back face of the element (normal to the x-axis) is denoted by σ_x, whereas that on the forward face is denoted by

$$\sigma_x + \frac{\partial \sigma_x}{\partial x} dx$$

i.e., the *increment* of stress over that existing on the back face is given by the rate of change of this stress component with respect to the x-coordinate, times the change (dx) of this coordinate. Further, it now becomes necessary to use a subscript to distinguish the three components of normal stress. For example, for the sides of the element perpendicular to the x-axis the normal stress is denoted by σ_x. Similarly, the subscript y indicates that the normal stress is acting on a plane normal to the y-axis. The normal stress is considered to be positive for tension and negative for compression.

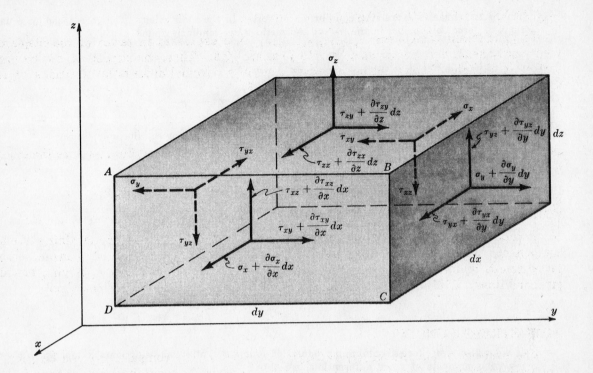

Fig. 19-1

The shearing stresses on the various faces are resolved into two components parallel to the coordinate axes. A double subscript is used for shearing stresses, the first indicating the direction of the normal to the plane under consideration and the second the direction of the particular stress component. For example, on the face $ABCD$ the component in the y-direction is denoted by τ_{xy} and that in the z-direction by τ_{xz}. The positive directions of shearing stresses on any face are taken as the positive directions of the corresponding coordinate axes if a tensile stress on that face is in the positive direction of the corresponding coordinate axis. If a tensile stress is in the negative direction of the coordinate axis, then the positive direction of the shearing stress component is reversed.

To write the equilibrium equations we must of course consider the *forces* acting on the element and to do this we multiply the stress on any face by the area of that face. The element is presumed to be sufficiently small that the normal and shearing stresses can be considered to be uniformly applied over each face, i.e., the resultant normal force, for example, acts at the center of each face. We must also consider a body force through the centroid of the element and having components X, Y, Z per unit volume of the element. Thus, the equation expressing equilibrium of forces in the x-direction is

$$-\sigma_x \, dy \, dz + \left(\sigma_x + \frac{\partial \sigma_x}{\partial x} dx \right) dy \, dz$$

$$- \tau_{yx} \, dx \, dz + \left(\tau_{yx} + \frac{\partial \tau_{yx}}{\partial y} dy \right) dx \, dz$$

$$- \tau_{zx} \, dx \, dy + \left(\tau_{zx} + \frac{\partial \tau_{zx}}{\partial z} dz \right) dx \, dy + X \, dx \, dy \, dz = 0$$

Since the volume of the element, $dx \, dy \, dz$, is not zero we may divide out this expression to obtain

$$\frac{\partial \sigma_x}{\partial x} + \frac{\partial \tau_{yx}}{\partial y} + \frac{\partial \tau_{zx}}{\partial z} + X = 0 \qquad (1)$$

This is the equation expressing equilibrium of forces in the x-direction. The corresponding equations of force equilibrium in the y- and z-directions are similarly found to be

$$\frac{\partial \sigma_y}{\partial y} + \frac{\partial \tau_{xy}}{\partial x} + \frac{\partial \tau_{zy}}{\partial z} + Y = 0 \tag{2}$$

$$\frac{\partial \sigma_z}{\partial z} + \frac{\partial \tau_{yz}}{\partial y} + \frac{\partial \tau_{xz}}{\partial x} + Z = 0 \tag{3}$$

These three equations must be satisfied at all points in the body for equilibrium to exist.

19.2. Obtain relations between the various components of shearing stress.

From the element shown on page 365, it is apparent that we have described the shearing stresses by the six quantities $\tau_{xy}, \tau_{yx}, \tau_{xz}, \tau_{zx}, \tau_{yz},$ and τ_{zy}. Let us write an equation of moment equilibrium about an axis parallel to the x-axis and passing through the center of the face $ABCD$. We get

$$-\tau_{yz}\, dx\, dz\, \frac{dy}{2} - \left(\tau_{yz} + \frac{\partial \tau_{yz}}{\partial y}\, dy\right) dx\, dz\, \frac{dy}{2}$$

$$+ \tau_{zy}\, dx\, dy\, \frac{dz}{2} + \left(\tau_{zy} + \frac{\partial \tau_{zy}}{\partial z}\, dz\right) dx\, dy\, \frac{dz}{2} = 0$$

If we neglect those terms containing products of four differentials as being small compared to those involving products of three differentials, we obtain

$$\tau_{yz} = \tau_{zy}$$

Two other relations

$$\tau_{xy} = \tau_{yx} \qquad \text{and} \qquad \tau_{xz} = \tau_{zx}$$

can be obtained in an analogous fashion. Thus, the number of necessary symbols for shearing stresses is reduced from six to three.

Thus, the six quantities $\sigma_x, \sigma_y, \sigma_z, \tau_{xy}, \tau_{yz},$ and τ_{xz} are adequate to describe the stresses acting on the coordinate planes through a point.

19.3. Derive expressions for the strains in terms of displacements.

Let us consider within a continuously deforming elastic solid a line element designated as PQ which joins the points P and Q. With reference to a Cartesian coordinate system such as is shown on page 365 we designate the coordinates of P and Q in the unstrained state prior to any deformation by (x_P, y_P, z_P) and (x_Q, y_Q, z_Q). During deformation, point P will undergo displacements parallel to the x, y, z axes designated by $u_P, v_P,$ and w_P, respectively, and correspondingly the point Q will have displacement components $u_Q, v_Q,$ and w_Q. Thus, after deformation the ends of the line element will have coordinates

$$(x_P + u_P, y_P + v_P, z_P + w_P) \qquad \text{and} \qquad (x_Q + u_Q, y_Q + v_Q, z_Q + w_Q)$$

The length L of PQ in the unstrained state is given by

$$L^2 = (x_Q - x_P)^2 + (y_Q - y_P)^2 + (z_Q - z_P)^2 \tag{1}$$

and after deformation the line element has a length L' given by

$$(L')^2 = (x_Q + u_Q - x_P - u_P)^2 + (y_Q + v_Q - y_P - v_P)^2 + (z_Q + w_Q - z_P - w_P)^2 \tag{2}$$

We may denote the extension of the line element PQ by Δ in which case

$$L + \Delta = L' \tag{3}$$

and we have from (2):

$$L^2 + 2L\Delta + \Delta^2 = (x_Q - x_P)^2 + (y_Q - y_P)^2 + (z_Q - z_P)^2$$
$$+ 2[(x_Q - x_P)(u_Q - u_P) + (y_Q - y_P)(v_Q - v_P) + (z_Q - z_P)(w_Q - w_P)]$$
$$+ (u_Q - y_P)^2 + (v_Q - v_P)^2 + (w_Q - w_P)^2 \qquad (4)$$

Note that the first three terms on the right-hand side of this equation cancel with L^2 on the left-hand side by virtue of (1). Also, we shall assume that the displacements u, v, w are so small that their squares may be neglected in comparison with the quantities themselves. The same is true of the extension Δ, whose square we shall neglect in comparison with the first power. Thus (4) becomes

$$L\Delta = (x_Q - x_P)(u_Q - u_P) + (y_Q - y_P)(v_Q - v_P) + (z_Q - z_P)(w_Q - w_P) \qquad (5)$$

and the elongation per unit length, denoted by ϵ, is

$$\epsilon = \frac{\Delta}{L} = \frac{1}{L^2}[(x_Q - x_P)(u_Q - u_P) + (y_Q - y_P)(v_Q - v_P) + (z_Q - z_P)(w_Q - w_P)] \qquad (6)$$

If now we let l, m, and n represent the direction cosines of the line element in its undeformed state, then

$$l = \frac{x_Q - x_P}{L} \qquad m = \frac{y_Q - y_P}{L} \qquad n = \frac{z_Q - z_P}{L} \qquad (7)$$

Thus (6) becomes

$$\epsilon = \frac{l}{L}(u_Q - u_P) + \frac{m}{L}(v_Q - v_P) + \frac{n}{L}(w_Q - w_P) \qquad (8)$$

Next, we may express u_Q as a Taylor series expansion in the form

$$u_Q = x_Q\left(\frac{\partial u}{\partial x}\right) + y_Q\left(\frac{\partial u}{\partial y}\right) + z_Q\left(\frac{\partial u}{\partial z}\right) \qquad (9)$$

where only first-order terms have been retained for small displacements. A similar expansion may be written for u_P and the difference formed to yield

$$u_Q - u_P = (x_Q - x_P)\frac{\partial u}{\partial x} + (y_Q - y_P)\frac{\partial u}{\partial y} + (z_Q - z_P)\frac{\partial u}{\partial z} \qquad (10)$$

or
$$u_Q - u_P = Ll\frac{\partial u}{\partial x} + Lm\frac{\partial u}{\partial y} + Ln\frac{\partial u}{\partial z} \qquad (11)$$

Similarly we can form the expressions

$$v_Q - v_P = Ll\frac{\partial v}{\partial x} + Lm\frac{\partial v}{\partial y} + Ln\frac{\partial v}{\partial z} \qquad (12)$$

$$w_Q - w_P = Ll\frac{\partial w}{\partial x} + Lm\frac{\partial w}{\partial y} + Ln\frac{\partial w}{\partial z} \qquad (13)$$

If these last three results are substituted in (8), we obtain

$$\epsilon = l^2\frac{\partial u}{\partial x} + m^2\frac{\partial u}{\partial y} + n^2\frac{\partial u}{\partial z} + mn\left(\frac{\partial v}{\partial z} + \frac{\partial w}{\partial y}\right) + nl\left(\frac{\partial w}{\partial x} + \frac{\partial u}{\partial z}\right) + lm\left(\frac{\partial u}{\partial y} + \frac{\partial v}{\partial x}\right) \qquad (14)$$

This expression gives at any point the strain ϵ in the direction having direction cosines l, m, and n, in terms of the displacements u, v, and w existing at that point.

If the element PQ originally lies along the x-axis, then $l = 1$ and $m = n = 0$ and (14) becomes

$$\epsilon = \epsilon_x = \frac{\partial u}{\partial x} \qquad (15)$$

where we have denoted the strain in the direction of the x-axis by ϵ_x. Similarly, the strains in the directions of the y- and z-axes are found to be

$$\epsilon_y = \frac{\partial v}{\partial y} \tag{16}$$

$$\epsilon_z = \frac{\partial w}{\partial z} \tag{17}$$

The quantities ϵ_x, ϵ_y, and ϵ_z given above are called *normal strains*.

We may seek an interpretation of the last three terms in (14) in the following fashion. Let us consider two line elements PQ and PR, both of which lie in the x-y plane. Prior to deformation these line elements lay along the x- and y-axes, respectively, but after deformation they may have the positions shown in Fig. 19-2, where the point P has moved to P', Q has displaced to Q' and R has displaced to R'.

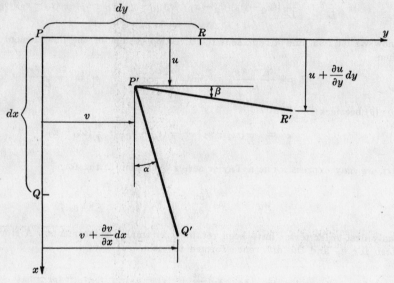

Fig. 19-2

The displacement components of the point P are u and v, while the y-component of displacement of Q is $v + (\partial v/\partial x)\, dx$. Consequently,

$$\tan \alpha = \frac{\left(v + \dfrac{\partial v}{\partial x}\, dx \right) - v}{dx}$$

But since we are concerned only with small displacements the angular rotations of these line elements will also be small so that we may replace the tangent of α by α itself in radian measure (as an approximation). Thus,

$$\alpha = \frac{\partial v}{\partial x}$$

Similarly, the x-component of displacement of R is $u + (\partial u/\partial y)\, \partial y$, and

$$\beta = \frac{\partial u}{\partial y}$$

Since PQ and PR were originally at right angles, the sum $\alpha + \beta$ denotes the deviation from a right angle after deformation. This deviation from a right angle is termed *shear strain* and in the x-y plane we shall denote it by γ_{xy}. Thus,

$$\gamma_{xy} = \frac{\partial u}{\partial y} + \frac{\partial v}{\partial x} \tag{18}$$

Similarly, in the x-z and y-z planes, the shear strain is given by

$$\gamma_{xz} = \frac{\partial u}{\partial z} + \frac{\partial w}{\partial x} \tag{19}$$

$$\gamma_{yz} = \frac{\partial v}{\partial z} + \frac{\partial w}{\partial y} \tag{20}$$

19.4. Derive the equations of compatibility.

Let us consider the strain-displacement relations derived in Problem 19.3. They are

$$\epsilon_x = \frac{\partial u}{\partial x} \tag{1}$$

$$\epsilon_y = \frac{\partial v}{\partial y} \tag{2}$$

$$\epsilon_z = \frac{\partial w}{\partial z} \tag{3}$$

$$\gamma_{xy} = \frac{\partial u}{\partial y} + \frac{\partial v}{\partial x} \tag{4}$$

$$\gamma_{xz} = \frac{\partial u}{\partial z} + \frac{\partial w}{\partial x} \tag{5}$$

$$\gamma_{yz} = \frac{\partial v}{\partial z} + \frac{\partial w}{\partial y} \tag{6}$$

We seek to eliminate the displacements from these equations. To do this we differentiate (1) twice with respect to y and (2) twice with respect to x. Adding:

$$\frac{\partial^2 \epsilon_x}{\partial y^2} + \frac{\partial^2 \epsilon_y}{\partial x^2} = \frac{\partial^3 u}{\partial x\, \partial y^2} + \frac{\partial^3 v}{\partial y\, \partial x^2} \tag{7}$$

Next, we shall differentiate (4) once with respect to x and then once with respect to y. Thus

$$\frac{\partial^2 \gamma_{xy}}{\partial x\, \partial y} = \frac{\partial^3 u}{\partial x\, \partial y^2} + \frac{\partial^3 v}{\partial x^2\, \partial y} \tag{8}$$

Since the right-hand sides of (7) and (8) are equal, we may equate the left-hand sides to obtain

$$\frac{\partial^2 \epsilon_x}{\partial y^2} + \frac{\partial^2 \epsilon_y}{\partial x^2} = \frac{\partial^2 \gamma_{xy}}{\partial x\, \partial y} \tag{9}$$

Two additional equations analogous to (8) may be formed in a similar fashion:

$$\frac{\partial^2 \epsilon_y}{\partial z^2} + \frac{\partial^2 \epsilon_z}{\partial y^2} = \frac{\partial^2 \gamma_{yz}}{\partial y\, \partial z} \tag{10}$$

$$\frac{\partial^2 \epsilon_z}{\partial x^2} + \frac{\partial^2 \epsilon_x}{\partial z^2} = \frac{\partial^2 \gamma_{xz}}{\partial x\, \partial z} \tag{11}$$

Equations (9), (10) and (11) are three of the six compatibility equations.

Another compatibility equation may be formed by differentiating (1) once with respect to y and then once with respect to z. This gives

$$\frac{\partial^2 \epsilon_x}{\partial y\, \partial z} = \frac{\partial^3 u}{\partial x\, \partial y\, \partial z} \tag{12}$$

Another expression for $\dfrac{\partial^3 u}{\partial x \, \partial y \, \partial z}$ may be found as follows. First, differentiate (4) once with respect to x and then once with respect to z to yield

$$\frac{\partial^2 \gamma_{xy}}{\partial x \, \partial z} \;=\; \frac{\partial^3 u}{\partial x \, \partial y \, \partial z} + \frac{\partial^3 v}{\partial x^2 \, \partial z} \tag{13}$$

Second, differentiate (5) once with respect to x and then once with respect to y:

$$\frac{\partial^2 \gamma_{xz}}{\partial x \, \partial y} \;=\; \frac{\partial^3 u}{\partial x \, \partial y \, \partial z} + \frac{\partial^3 w}{\partial x^2 \, \partial y} \tag{14}$$

Third, differentiate (6) twice with respect to x to get

$$\frac{\partial^2 \gamma_{yz}}{\partial x^2} \;=\; \frac{\partial^3 v}{\partial x^2 \, \partial z} + \frac{\partial^3 w}{\partial x^2 \, \partial y} \tag{15}$$

Finally, add (13) and (14) then subtract (15), to get

$$\frac{\partial^2 \gamma_{xy}}{\partial x \, \partial z} + \frac{\partial^2 \gamma_{xz}}{\partial x \, \partial y} - \frac{\partial^2 \gamma_{yz}}{\partial x^2} \;=\; 2 \, \frac{\partial^3 u}{\partial x \, \partial y \, \partial z} \tag{16}$$

From (12) and (16) we immediately have

$$2 \, \frac{\partial^2 \epsilon_x}{\partial y \, \partial z} \;=\; \frac{\partial}{\partial x} \left(-\frac{\partial \gamma_{yz}}{\partial x} + \frac{\partial \gamma_{xz}}{\partial y} + \frac{\partial \gamma_{xy}}{\partial z} \right) \tag{17}$$

Analogously, two more equations may be formed:

$$2 \, \frac{\partial^2 \epsilon_y}{\partial x \, \partial z} \;=\; \frac{\partial}{\partial y} \left(-\frac{\partial \gamma_{xz}}{\partial y} + \frac{\partial \gamma_{xy}}{\partial z} + \frac{\partial \gamma_{yz}}{\partial x} \right) \tag{18}$$

$$2 \, \frac{\partial^2 \epsilon_z}{\partial x \, \partial y} \;=\; \frac{\partial}{\partial z} \left(-\frac{\partial \gamma_{xy}}{\partial z} + \frac{\partial \gamma_{yz}}{\partial x} + \frac{\partial \gamma_{xz}}{\partial y} \right) \tag{19}$$

Equations (17), (18), and (19) constitute the other three compatibility equations. The complete set of six equations, i.e. (9), (10), (11), (17), (18), and (19), are termed the conditions of compatibility for strain. The strains in any problem must satisfy all six of these equations. This is essentially because the six strain components given in equations (1) through (6) are expressed in terms of three displacement components, hence the strains cannot be independent of one another. Satisfaction of these six compatibility equations ensures the existence of a single-valued continuous strain field.

19.5. **Discuss the effect of a nonuniform temperature field in an elastic body. Which basic elastic relations change and which ones remain unchanged?**

The equilibrium equations (1), (2), and (3) of Problem 19.1 are based upon essentially mechanical considerations and hence are not influenced by a variable temperature field. The same is true of the shearing stress relations found in Problem 19.2. The strain-displacement relations (15), (16), (17), (18), (19), and (20) of Problem 19.3 are based entirely on geometric considerations, hence they remain unchanged. The same is true of the six compatibility equations derived in Problem 19.4.

However, the form of Hooke's law given in the introduction of Chapter 1 is no longer adequate. We shall assume that the temperature variation is not sufficiently large to cause any change in the value of the modulus of elasticity or Poisson's ratio. It is to be realized that an infinitesimal unconstrained isotropic element subject to a change of temperature dilates uniformly in all directions at any point. Thus there will be equal normal strains in all directions due to the temperature change, but no shearing strains will be induced by the thermal effects.

To determine the normal strain corresponding to the change of temperature ΔT we recall from Chapter 1 that the coefficient of linear expansion α (which is defined as the change of length per unit length per degree change of temperature) need merely be multiplied by ΔT to find the strain.

It is now possible to superpose the strains arising from thermal changes upon those strains discussed in Chapter 1 to get the following:

$$\epsilon_x = \frac{1}{E}\left[\sigma_x - \mu(\sigma_y + \sigma_z)\right] + \alpha\,\Delta T \tag{1}$$

$$\epsilon_y = \frac{1}{E}\left[\sigma_y - \mu(\sigma_x + \sigma_z)\right] + \alpha\,\Delta T \tag{2}$$

$$\epsilon_z = \frac{1}{E}\left[\sigma_z - \mu(\sigma_x + \sigma_y)\right] + \alpha\,\Delta T \tag{3}$$

$$\gamma_{xy} = \frac{1}{G}\tau_{xy} \tag{4}$$

$$\gamma_{xz} = \frac{1}{G}\tau_{xz} \tag{5}$$

$$\gamma_{yz} = \frac{1}{G}\tau_{yz} \tag{6}$$

19.6. Discuss the special forms of the equations of equilibrium and compatibility for the two-dimensional elastic medium.

Two possible special cases are of extreme practical importance. The first is called

PLANE STRESS

By definition, this is the case for which

$$\sigma_z = \tau_{xz} = \tau_{yz} = 0 \tag{1}$$

An example of this type of stress distribution is a body one of whose dimensions is very small compared to the other two and which is loaded by forces lying in the plane of symmetry of the body, i.e. a thin plate with loads in its middle plane. For simplicity we assume that the other three components of stress do not vary in magnitude through the thickness of the body. The equilibrium equations of Problem 19.1 become

$$\frac{\partial \sigma_x}{\partial x} + \frac{\partial \tau_{xy}}{\partial y} + X = 0 \tag{2}$$

$$\frac{\partial \sigma_y}{\partial y} + \frac{\partial \tau_{xy}}{\partial x} + Y = 0 \tag{3}$$

the other terms and the third equation vanishing because the stresses are functions of x and y only.

In most cases of practical interest, the body forces X and Y are derivable from a so-called potential function $\Omega(x,y)$ defined by the following equations

$$X = -\frac{\partial \Omega}{\partial x} \tag{4}$$

$$Y = -\frac{\partial \Omega}{\partial y} \tag{5}$$

Hooke's law, assuming no changes of temperature, is

$$\epsilon_x = \frac{1}{E}\left[\sigma_x - \mu(\sigma_y + \sigma_z)\right] = \frac{1}{E}(\sigma_x - \mu\sigma_y) \tag{6}$$

$$\epsilon_y = \frac{1}{E}\left[\sigma_y - \mu(\sigma_x + \sigma_z)\right] = \frac{1}{E}(\sigma_y - \mu\sigma_x) \tag{7}$$

$$\epsilon_z = \frac{1}{E}\left[\sigma_z - \mu(\sigma_x + \sigma_y)\right] = -\frac{\mu}{E}(\sigma_x + \sigma_y) \qquad (8)$$

$$\gamma_{xy} = \frac{1}{G}\tau_{xy}; \quad \gamma_{xz} = \gamma_{yz} = 0 \qquad (9)$$

From this it may be seen that for the plane stress case the strains depend only on x and y and are independent of z.

We shall now introduce a function $\Phi(x, y)$ known as *Airy's stress function*, defined by the following three equations:

$$\sigma_x = \frac{\partial^2 \Phi}{\partial y^2} + \Omega \qquad (10)$$

$$\sigma_y = \frac{\partial^2 \Phi}{\partial x^2} + \Omega \qquad (11)$$

$$\tau_{xy} = -\frac{\partial^2 \Phi}{\partial x\, \partial y} \qquad (12)$$

It is readily seen that these stresses satisfy the equilibrium equations (2) and (3).

If relations (10), (11), and (12) are introduced into Equations (6), (7), (8), and (9), and the resulting strain expressions substituted into the compatibility equation (9) of Problem 19.4, one obtains

$$\frac{\partial^4 \Phi}{\partial x^4} + 2\frac{\partial^4 \Phi}{\partial x^2\, \partial y^2} + \frac{\partial^4 \Phi}{\partial y^4} = -(1 - \mu)\left(\frac{\partial^2 \Omega}{\partial x^2} + \frac{\partial^2 \Omega}{\partial y^2}\right) \qquad (13)$$

The quantity ∇^4, called the biharmonic operator, is defined as

$$\nabla^4 = \frac{\partial^4}{\partial x^4} + 2\frac{\partial^4}{\partial x^2\, \partial y^2} + \frac{\partial^4}{\partial y^4}$$

and the quantity ∇^2, the Laplacian operator, is defined as

$$\nabla^2 = \frac{\partial^2}{\partial x^2} + \frac{\partial^2}{\partial y^2}$$

Thus, (13) becomes

$$\nabla^4 \Phi = -(1 - \mu)\nabla^2 \Omega \qquad (14)$$

Observe that no use has been made of the remaining five compatibility equations. Two of these vanish because the stress field under consideration here is independent of z, but in general the other three will not be satisfied. However, the stresses indicated by (10), (11), and (12) are good approximations to the exact stress field provided the body under consideration is very thin.

For the case of zero body forces, (14) reduces to the so-called biharmonic equation

$$\nabla^4 \Phi = 0 \qquad (15)$$

The second special case is termed

PLANE STRAIN

For this case, by definition

$$\epsilon_z = \gamma_{xz} = \gamma_{yz} = 0 \qquad (16)$$

and also the body force Z must be zero.

An example of this type of strain distribution is an extremely long body subject to lateral loads. In the regions some distance from each of the two ends strain at any section in the direction of the axis (z-axis) is prevented by the action of adjacent material.

The equations of equilibrium are the same as given in (2) and (3). Again, we shall consider that the body forces are derivable from a potential function as represented in (4) and (5). Hooke's law becomes

$$\epsilon_x = \frac{1}{E}\left[\sigma_x - \mu(\sigma_y + \sigma_z)\right] \qquad (17)$$

$$\epsilon_y = \frac{1}{E}\left[\sigma_y - \mu(\sigma_x + \sigma_z)\right] \qquad (18)$$

$$0 = \frac{1}{E}\left[\sigma_z - \mu(\sigma_x + \sigma_y)\right] \qquad (19)$$

from which

$$\sigma_z = \mu(\sigma_x + \sigma_y) \qquad (20)$$

We shall employ the same Airy stress function $\Phi(x, y)$ as for the plane stress case and as given in (10), (11), and (12). For the plane strain case it is readily seen that the compatibility equations (10), (11), (17), (18), and (19) of Problem 19.4 are identically satisfied, leaving only (9) of that problem to be considered. If the strains given by (17), (18), and (19) are expressed in terms of the Airy function and the results substituted into this remaining compatibility equation, one obtains

$$\nabla^4\Phi = -\frac{1}{1-\mu}\nabla^2\Omega \qquad (21)$$

Note that if there are no body forces, this governing equation reduces to

$$\nabla^4\Phi = 0 \qquad (22)$$

the same as was obtained for the plane stress case.

19.7. Can the stress function $\Phi = Bx^3y^2$ describe a state of plane stress for the case of zero body forces?

To investigate this, we shall check to see if this function satisfies (15) of Problem 19.6. The biharmonic operator here becomes

$$\nabla^4(Bx^3y^2) = \frac{\partial^4}{\partial x^4}(Bx^3y^2) + 2\frac{\partial^4}{\partial x^2\,\partial y^2}(Bx^3y^2) + \frac{\partial^4}{\partial y^4}(Bx^3y^2) = 0 + 24Bx + 0$$

Because this does not vanish for all values of x and y, equation (15) is not satisfied and $\Phi = Bx^3y^2$ cannot describe a state of plane stress.

19.8. Given the function $\Phi = Axy^2$. Can this describe a state of plane stress? If so, describe the stress state. Assume that all body forces vanish.

First, we must check to ascertain if this function satisfies (15) of Problem 19.6. The biharmonic operator becomes

$$\nabla^4(Axy^2) = \frac{\partial^4}{\partial x^4}(Axy^2) + 2\frac{\partial^4}{\partial x^2\,\partial y^2}(Axy^2) + \frac{\partial^4}{\partial y^4}(Axy^2) = 0$$

Thus the biharmonic equation is satisfied and $\Phi = Axy^2$ can be regarded as an Airy stress function. The stresses corresponding to it are found from (10), (11), and (12) of Problem 19.6 to be

$$\sigma_x = \frac{\partial^2\Phi}{\partial y^2} = 2Ax \qquad \sigma_y = \frac{\partial^2\Phi}{\partial x^2} = 0 \qquad \tau_{xy} = -\frac{\partial^2\Phi}{\partial x\,\partial y} = -2Ay$$

Evidently this could describe the state of stress in a plane rectangular block subject to the loadings applied at the boundaries as shown below in Fig. 19-3(a).

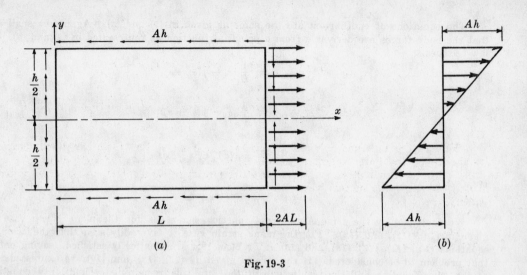

Fig. 19-3

Note that the shearing stresses of magnitude Ah on the upper and lower faces are both directed to the left, in accordance with the rule for positive directions of shearing stresses as given in Problem 19.1. In particular, on the lower face where $y = -h/2$, the shearing stress is $\tau_{xy} = -2A(-h/2) = Ah$. Thus, since a tensile stress on that face is in the negative y-direction, the positive direction of shear is in the negative x-direction as indicated. On the end faces, $x = 0$ and $x = L$, the shear stress varies linearly from zero at $y = 0$ to the value Ah at the upper and lower faces. The directions of these shearing stresses are found from the above-mentioned rule given in Problem 19.1. Their linear variation is illustrated in Fig. 19-3(b).

19.9. Consider a plane prismatic bar subject to the uniformly varying normal load indicated in Fig. 19-4. Investigate the state of stress within the body. Neglect body forces.

The intensity of load p at any arbitrary value of the axial coordinate x is $p = (p_B/L)x$. The total load on the bar is $\frac{1}{2}p_B L$ and thus the resultant reactions are $\frac{1}{6}p_B L$ at the left end and $\frac{1}{3}p_B L$ at the right end, respectively.

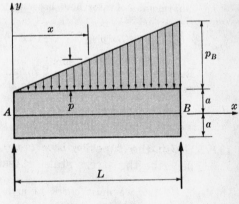

Fig. 19-4

As boundary conditions we have

$$\tau_{xy} = 0 \quad \text{for} \quad y = \pm a \qquad (1)$$

$$\sigma_y = 0 \quad \text{for} \quad y = -a \qquad (2)$$

$$\sigma_y = -p \quad \text{for} \quad y = a \qquad (3)$$

Also, at the ends, $x = 0$ and $x = L$, the resultant of the distributed vertical shearing forces must equal the reactions for each of these ends. In addition, it is desirable to have zero normal stresses σ_x on the ends $x = 0$ and $x = L$.

We must satisfy (15) of Problem 19.6, $\nabla^4\Phi = 0$. Let us investigate an Airy stress function of the form

$$\Phi = \frac{p_B}{4a^3 L}\left[\frac{1}{6}x^3 y^3 - \frac{1}{10}xy^5 - \frac{1}{2}a^2 x^3 y + \left(\frac{a^2}{5} - \frac{L^2}{6}\right)xy^3 + \left(-\frac{a^4}{10} + \frac{a^2 L^2}{2}\right)xy - \frac{a^3}{3}x^3\right] \qquad (4)$$

which can be shown to satisfy $\nabla^4\Phi = 0$.

This function implies the following stress field:

$$\sigma_x = \frac{\partial^2 \Phi}{\partial y^2} = \frac{p_B}{4a^3 L}[x^3 y - 2xy^3 + (\tfrac{6}{5}a^2 - L^2)xy] \tag{5}$$

$$\sigma_y = \frac{\partial^2 \Phi}{\partial x^2} = \frac{p_B}{4a^3 L}[xy^3 - 3a^2 xy - 2a^3 x] \tag{6}$$

$$\tau_{xy} = \frac{-\partial^2 \Phi}{\partial x\, \partial y} = -\frac{p_B}{4a^3 L}\left[\tfrac{3}{2}x^2 y^2 - \tfrac{1}{2}y^4 - \tfrac{3}{2}a^2 x^2 + \left(\tfrac{3}{5}a^2 - \frac{L^2}{2}\right)y^2 - \frac{a^4}{10} + \frac{a^2 L^2}{2}\right] \tag{7}$$

With regard to boundary conditions, it is to be observed that the condition on shearing stress at the upper and lower edges, given as (1) above, is satisfied by (7) for all values of x. Further, from (6) the normal stress σ_y vanishes along the lower edge for all values of x, satisfying (2). Lastly, from (5) the normal stress along the upper edge is equal to $-p$ for all values of x, thus satisfying (3).

Let us now examine the boundary condition pertaining to resultant vertical shear due to the reaction at $x = 0$. The resultant vertical shear according to (7) is

$$\int_{-a}^{a} [\tau_{xy}]_{x=0}\, dy = -\frac{p_B}{4a^3 L}\left[-\frac{a^5}{5} + \left(\frac{3}{5}a^2 - \frac{L^2}{2}\right)\frac{2a^3}{3} + \left(\frac{a^2 L^2}{2} - \frac{a^4}{10}\right)(2a)\right] = -\frac{p_B L}{6} \tag{8}$$

which of course must be correct from statics. At the right end, $x = L$, the resultant vertical shear is

$$\int_{-a}^{a} [\tau_{xy}]_{x=L}\, dy = -\frac{p_B}{4a^3 L}\left[L^2 a^3 - \frac{a^5}{5} + \left(\frac{3}{5}a^2 - \frac{L^2}{2}\right)\left(\frac{2a^3}{3}\right)\right.$$

$$\left. - \left(\frac{3}{2}a^2 L^2 + \frac{a^4}{10} - \frac{a^2 L^2}{2}\right)(2a)\right] = \frac{p_B L}{3} \tag{9}$$

which again agrees with the value given by statics.

Everywhere over the left end, $x = 0$, the normal stress σ_x vanishes from (5). However, at the right end this normal stress does not vanish but instead is given by

$$[\sigma_x]_{x=L} = \frac{p_B}{4a^3} y(\tfrac{6}{5}a^2 - 2y^2) \tag{10}$$

Thus the above solution does imply the existence of normal stresses given by (10) over the end $x = L$. However, it is readily verified that these stresses have a zero resultant in the x-direction, since

$$\int_{-a}^{a} [\sigma_x]_{x=L}\, dy = 0 \tag{11}$$

and further, that they have a zero resultant moment, since

$$\int_{-a}^{a} [\sigma_x]_{x=L}\, y\, dy = 0 \tag{12}$$

Thus, the assumed stress function (4) gives a solution that is completely satisfactory except for the presence of these normal stresses over the right end. Numerical evaluation would indicate that these stresses given by (10) are small compared to the other normal stresses in the body, and in addition (11) and (12) indicate that they have a zero force and zero moment resultant. Thus, invoking St. Venant's principle, the stress distribution found in (5), (6), and (7) is satisfactory everywhere except in the region immediately adjacent to the right end of the bar.

19.10. Transform the governing equation $\nabla^4 \Phi = 0$ of Problem 19.6 to polar coordinates.

From Fig. 19-5,

$$x = r \cos \theta$$

$$y = r \sin \theta$$

$$r = \sqrt{x^2 + y^2}$$

$$\theta = \arctan \frac{y}{x}$$

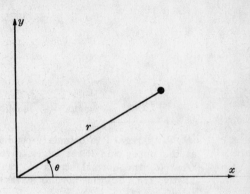

Fig. 19-5

from which

$$\frac{\partial r}{\partial x} = \cos \theta \qquad \frac{\partial r}{\partial y} = \sin \theta$$

$$\frac{\partial \theta}{\partial x} = -\frac{\sin \theta}{r} \qquad \frac{\partial \theta}{\partial y} = \frac{\cos \theta}{r}$$

We seek $\Phi = \Phi(r, \theta)$. Thus, we must calculate

$$\frac{\partial \Phi}{\partial x} = \frac{\partial \Phi}{\partial r}\frac{\partial r}{\partial x} + \frac{\partial \Phi}{\partial \theta}\frac{\partial \theta}{\partial x} = \frac{\partial \Phi}{\partial r}\left(\frac{x}{r}\right) + \frac{\partial \Phi}{\partial \theta}\left(-\frac{y}{r^2}\right) = \frac{\partial \Phi}{\partial r}(\cos \theta) - \frac{\partial \Phi}{\partial \theta}\left(\frac{\sin \theta}{r}\right)$$

$$\frac{\partial^2 \Phi}{\partial x^2} = \frac{\partial}{\partial r}\left(\frac{\partial \Phi}{\partial x}\right)\frac{\partial r}{\partial x} + \frac{\partial}{\partial \theta}\left(\frac{\partial \Phi}{\partial x}\right)\frac{\partial \theta}{\partial x}$$

$$= \left[\frac{\partial^2 \Phi}{\partial r^2}(\cos \theta) - \frac{\partial \Phi}{\partial \theta}\left(-\frac{\sin \theta}{r^2}\right) - \frac{\partial^2 \Phi}{\partial r \, \partial \theta}\left(\frac{\sin \theta}{r}\right)\right]\cos \theta$$

$$+ \left[\frac{\partial \Phi}{\partial r}(-\sin \theta) + \frac{\partial^2 \Phi}{\partial r \, \partial \theta}(\cos \theta) - \frac{\partial \Phi}{\partial \theta}\left(\frac{\cos \theta}{r}\right) - \frac{\partial^2 \Phi}{\partial \theta^2}\left(\frac{\sin \theta}{r}\right)\right]\left(-\frac{\sin \theta}{r}\right)$$

$$= \frac{\partial^2 \Phi}{\partial r^2}\cos^2 \theta - 2\frac{\partial^2 \Phi}{\partial r \, \partial \theta}\frac{\sin \theta \cos \theta}{r} + 2\frac{\partial \Phi}{\partial \theta}\frac{\sin \theta \cos \theta}{r^2} + \frac{\partial \Phi}{\partial r}\frac{\sin^2 \theta}{r} + \frac{\partial^2 \Phi}{\partial \theta^2}\frac{\sin^2 \theta}{r^2} \qquad (1)$$

Similarly, we get

$$\frac{\partial^2 \Phi}{\partial y^2} = \frac{\partial^2 \Phi}{\partial r^2}\sin^2 \theta + 2\frac{\partial^2 \Phi}{\partial r \, \partial \theta}\frac{\sin \theta \cos \theta}{r} - 2\frac{\partial \Phi}{\partial \theta}\frac{\sin \theta \cos \theta}{r^2} + \frac{\partial \Phi}{\partial r}\frac{\cos^2 \theta}{r} + \frac{\partial^2 \Phi}{\partial \theta^2}\frac{\cos^2 \theta}{r^2} \qquad (2)$$

Adding (1) and (2), we get

$$\frac{\partial^2 \Phi}{\partial x^2} + \frac{\partial^2 \Phi}{\partial y^2} = \frac{\partial^2 \Phi}{\partial r^2} + \frac{1}{r}\frac{\partial \Phi}{\partial r} + \frac{1}{r^2}\frac{\partial^2 \Phi}{\partial \theta^2} \qquad (3)$$

The biharmonic equation in polar coordinates is thus

$$\nabla^4 \Phi = \nabla^2(\nabla^2 \Phi) = \left(\frac{\partial^2}{\partial r^2} + \frac{1}{r}\frac{\partial}{\partial r} + \frac{1}{r^2}\frac{\partial^2}{\partial \theta^2}\right)\left(\frac{\partial^2 \Phi}{\partial r^2} + \frac{1}{r}\frac{\partial \Phi}{\partial r} + \frac{1}{r^2}\frac{\partial^2 \Phi}{\partial \theta^2}\right) = 0 \qquad (4)$$

19.11. Derive the differential equations of equilibrium in polar coordinates. Consider only a two-dimensional case.

One possible approach would be to employ the results from Problem 19.6, together with the transformation relations of Problem 17.13. However, it is perhaps simpler to start anew by consideration of the equilibrium of an element in polar coordinates (see Fig. 19-6).

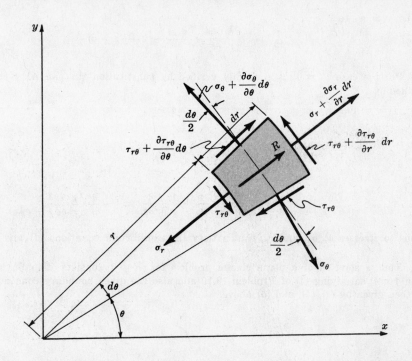

Fig. 19-6

The radial and tangential stresses are denoted by σ_r and σ_θ, respectively, and the shearing stress in the plane by $\tau_{r\theta}$. From Problem 19.2, $\tau_{r\theta} = \tau_{\theta r}$. We shall consider only a radial body force, denoted by R per unit volume.

For equilibrium in the radial direction:

$$-\sigma_r r\, d\theta + \left(\sigma_r + \frac{\partial \sigma_r}{\partial r}\, dr\right)(r + dr)\, d\theta - \sigma_\theta\, dr \sin\frac{d\theta}{2}$$

$$- \left(\sigma_\theta + \frac{\partial \sigma_\theta}{\partial \theta}\, d\theta\right) dr \sin\frac{d\theta}{2} - \tau_{r\theta}\, dr \cos\frac{d\theta}{2}$$

$$+ \left(\tau_{r\theta} + \frac{\partial \tau_{r\theta}}{\partial \theta}\, d\theta\right) dr \cos\frac{d\theta}{2} + Rr\, dr\, d\theta \;=\; 0$$

If we neglect higher order differentials, and take $\sin\dfrac{d\theta}{2} \approx \dfrac{d\theta}{2}$ and $\cos\dfrac{d\theta}{2} \approx 1$, then we get

$$\frac{\partial \sigma_r}{\partial r} + \frac{1}{r}\frac{\partial \tau_{r\theta}}{\partial \theta} + \frac{\sigma_r - \sigma_\theta}{r} + R \;=\; 0 \qquad\qquad (1)$$

For equilibrium in the tangential direction:

$$-\tau_{r\theta} r\, d\theta + \left(\tau_{r\theta} + \frac{\partial \tau_{r\theta}}{\partial r}\, dr\right)(r + dr)\, d\theta - \sigma_\theta\, dr \cos\frac{d\theta}{2}$$

$$+ \left(\sigma_\theta + \frac{\partial \sigma_\theta}{\partial \theta}\, d\theta\right) dr \cos\frac{d\theta}{2} + \tau_{r\theta}\, dr \sin\frac{d\theta}{2}$$

$$+ \left(\tau_{r\theta} + \frac{\partial \tau_{r\theta}}{\partial \theta}\, d\theta\right) dr \sin\frac{d\theta}{2} \;=\; 0$$

which reduces to

$$\frac{1}{r}\frac{\partial \sigma_\theta}{\partial \theta} + \frac{\partial \tau_{r\theta}}{\partial r} + 2\frac{\tau_{r\theta}}{r} = 0 \tag{2}$$

For the case $R = 0$ it is readily verified by substitution that an Airy stress function $\Phi(r, \theta)$ defined by

$$\sigma_r = \frac{1}{r}\frac{\partial \Phi}{\partial r} + \frac{1}{r^2}\frac{\partial^2 \Phi}{\partial \theta^2} \tag{3}$$

$$\sigma_\theta = \frac{\partial^2 \Phi}{\partial r^2} \tag{4}$$

$$\tau_{r\theta} = \frac{1}{r^2}\frac{\partial \Phi}{\partial \theta} - \frac{1}{r}\frac{\partial^2 \Phi}{\partial r\,\partial \theta} = -\frac{\partial}{\partial r}\left(\frac{1}{r}\frac{\partial \Phi}{\partial \theta}\right) \tag{5}$$

leads to stresses σ_r, σ_θ, and $\tau_{r\theta}$ that satisfy the equilibrium equations (1) and (2).

Thus a solution of a plane elastic problem for $R = 0$ consists of finding an Airy stress function $\Phi(r, \theta)$ satisfying (4) of Problem 19.10 and also the given boundary conditions. The stress field is then given by (3), (4), and (5) above.

19.12. Determine the Airy stress function for the case of stress distribution symmetric about an axis. Consider all body forces to be zero.

For this case all derivatives with respect to θ vanish and the biharmonic equation (4) of Problem 19.10 becomes

$$\nabla^4\Phi = \left(\frac{\partial^2}{\partial r^2} + \frac{1}{r}\frac{\partial}{\partial r}\right)\left(\frac{\partial^2 \Phi}{\partial r^2} + \frac{1}{r}\frac{\partial \Phi}{\partial r}\right) = 0 \tag{1}$$

or

$$\nabla^4\Phi = \frac{d^4\Phi}{dr^4} + \frac{2}{r}\frac{d^3\Phi}{dr^3} - \frac{1}{r^2}\frac{d^2\Phi}{dr^2} + \frac{1}{r^3}\frac{d\Phi}{dr} = 0 \tag{2}$$

Note that we may now use ordinary rather than partial derivatives.

This type of differential equation may be transformed to a simpler equation with constant coefficients by the substitution $r = e^t$. Let us first examine the term $d\Phi/dr$:

$$\frac{d\Phi}{dr} = \frac{d\Phi}{dt}\frac{dt}{dr} = \frac{d\Phi}{dt}e^{-t} \tag{3}$$

Continuing:

$$\frac{d^2\Phi}{dr^2} = \frac{d}{dr}\left(\frac{d\Phi}{dr}\right) = \frac{d}{dt}\left(\frac{d\Phi}{dt}e^{-t}\right)\frac{dt}{dr} = \left[\frac{d^2\Phi}{dt^2}e^{-t} + \frac{d\Phi}{dt}(-e^{-t})\right]e^{-t} = e^{-2t}\left(\frac{d^2\Phi}{dt^2} - \frac{d\Phi}{dt}\right) \tag{4}$$

$$\frac{d^3\Phi}{dr^3} = \frac{d}{dr}\left(\frac{d^2\Phi}{dr^2}\right) = \frac{d}{dt}\left[\left(\frac{d^2\Phi}{dt^2} - \frac{d\Phi}{dt}\right)e^{-2t}\right]\frac{dt}{dr}$$

$$= \left[\frac{d^2\Phi}{dt^2}(-2e^{-2t}) + \frac{d^3\Phi}{dt^3}e^{-2t} - \frac{d\Phi}{dt}(-2e^{-2t}) - \frac{d^2\Phi}{dt^2}e^{-2t}\right]e^{-t}$$

$$= e^{-3t}\left(\frac{d^3\Phi}{dt^3} - 3\frac{d^2\Phi}{dt^2} + 2\frac{d\Phi}{dt}\right) \tag{5}$$

Another analogous differentiation would yield

$$\frac{d^4\Phi}{dr^4} = e^{-4t}\left(\frac{d^4\Phi}{dt^4} - 6\frac{d^3\Phi}{dt^3} + 11\frac{d^2\Phi}{dt^2} - 6\frac{d\Phi}{dt}\right) \tag{6}$$

If these results (3), (4), (5), and (6) are substituted in the compatibility equation (2) we obtain the much simpler linear differential equation with constant coefficients:

$$\frac{d^4\Phi}{dt^4} - 4\frac{d^3\Phi}{dt^3} + 4\frac{d^2\Phi}{dt^2} = 0 \tag{7}$$

To solve, let $\Phi = e^{mt}$ and substitute to obtain the auxiliary equation $m^4 - 4m^3 + 4m^2 = 0$, which has roots $m = 0, 0, 2, 2$.　Thus, the general solution of (7) is

$$\Phi = Ae^{0t} + Bte^{0t} + Ce^{2t} + Dte^{2t} = A + B\ln r + Cr^2 + D(\ln r)r^2 \tag{8}$$

where ln stands for the logarithm to the base e.

From equations (3), (4), and (5) of Problem 19.11, the stresses corresponding to this Airy function are

$$\sigma_r = \frac{1}{r}\frac{\partial\Phi}{\partial r} = \frac{B}{r^2} + 2C + D(1 + 2\ln r) \tag{9}$$

$$\sigma_\theta = \frac{\partial^2\Phi}{\partial r^2} = -\frac{B}{r^2} + 2C + D(3 + 2\ln r) \tag{10}$$

$$\tau_{r\theta} = 0 \tag{11}$$

These stresses represent a stress distribution symmetric about an axis.　The constants B, C, and D must of course be determined from boundary conditions.

19.13. Determine the strain-displacement relations in plane polar coordinates.

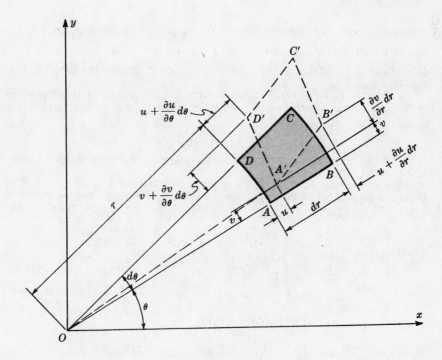

Fig. 19-7

Let u and v represent the radial and tangential components of displacement, respectively. The element originally at $ABCD$ lies after deformation at $A'B'C'D'$ (see Fig. 19-7, page 379). For the side AB the radial strain, denoted by ϵ_r, is given by

$$\epsilon_r = \frac{\left(dr + u + \frac{\partial u}{\partial r}dr - u\right) - dr}{dr} = \frac{\partial u}{\partial r} \tag{1}$$

The tangential strain, ϵ_θ, is somewhat more intricate. Due to the radial displacement u, one portion of the tangential strain is given by

$$\frac{(r+u)\,d\theta - r\,d\theta}{r\,d\theta} = \frac{u}{r}$$

There is another portion of the tangential strain arising from the incremental tangential displacement $\frac{\partial v}{\partial \theta}\,d\theta$ of point D as it moves to D'. This portion of the strain is given by

$$\frac{\frac{\partial v}{\partial \theta}\,d\theta}{r\,d\theta} = \frac{1}{r}\frac{\partial v}{\partial \theta}$$

The total tangential strain is thus

$$\epsilon_\theta = \frac{u}{r} + \frac{1}{r}\frac{\partial v}{\partial \theta} \tag{2}$$

The shearing strain, i.e. the change in the right angle originally at A, is found from a consideration of Fig. 19-7. First, we examine the contribution due to radial displacement and along the side AD this gives rise to a small change of angle given by

$$\frac{\frac{\partial u}{\partial \theta}\,d\theta}{r\,d\theta} = \frac{1}{r}\frac{\partial u}{\partial \theta}$$

Similarly, along AB there is apparently another small change of angle

$$\frac{\frac{\partial v}{\partial r}\,dr}{dr} = \frac{\partial v}{\partial r}$$

However, it must be carefully observed that a portion of this last angle is due to rigid body rotation of the element about point O. This rigid body rotation angle is clearly given by $\frac{v}{r}$ and this term does not contribute to the elastic deformation. Hence, the total change of angle, giving the shearing strain, is

$$\gamma_{r\theta} = \frac{1}{r}\frac{\partial u}{\partial \theta} + \frac{\partial v}{\partial r} - \frac{v}{r} \tag{3}$$

19.14. Find the displacement field for the case of axisymmetric stress distribution.

The stresses for this situation are given as equations (9), (10), and (11) of Problem 19.12. If these are substituted in the two-dimensional form of Hooke's law we obtain

$$\epsilon_r = \frac{1}{E}[\sigma_r - \mu\sigma_\theta] \tag{1}$$

whence

$$\frac{\partial u}{\partial r} = \frac{1}{E}\left[\frac{(1+\mu)B}{r^2} + 2(1-\mu)C + (1-3\mu)D + 2(1-\mu)D\ln r\right] \tag{2}$$

$$\epsilon_\theta = \frac{1}{E}(\sigma_\theta - \mu\sigma_r) \tag{3}$$

$$\frac{u}{r} + \frac{1}{r}\frac{\partial v}{\partial \theta} = \frac{1}{E}\left[-\frac{(1+\mu)B}{r^2} + 2(1-\mu)C + (3-\mu)D + 2(1-\mu)D\ln r\right] \tag{4}$$

Integrating (2) we obtain

$$u = \frac{1}{E}\left[-\frac{(1+\mu)B}{r} + 2(1-\mu)Cr + 2(1-\mu)Dr\ln r - D(1+\mu)r\right] + f(\theta) \qquad (5)$$

where $f(\theta)$ denotes a function of integration, independent of r and depending only on θ. Now, if this value of u be substituted in (4), we obtain

$$\frac{\partial v}{\partial \theta} = \frac{4Dr}{E} - f(\theta) \qquad (6)$$

Let us now integrate (6) to obtain

$$v = \frac{4Dr\theta}{E} - \int f(\theta)\,d\theta + g(r) \qquad (7)$$

where $g(r)$ is another function of integration, independent of θ and depending only on r. If now (5) and (7) are substituted in the expression for shearing strain $\gamma_{r\theta}$ as given by (3) of Problem 19.13, we obtain (since the shear strain vanishes for axisymmetric stress distributions):

$$\frac{1}{r}\frac{df(\theta)}{d\theta} + \frac{4D\theta}{E} + \frac{dg(r)}{dr} - \frac{4D\theta}{E} + \frac{1}{r}\int f(\theta)\,d\theta - \frac{1}{r}g(r) = 0 \qquad (8)$$

After multiplying through by r it is evident that this equation contains certain terms that are functions of r only, and others that are functions of θ only. The only way this can be satisfied is for the sum of the r-terms to vanish, and likewise for the θ-terms. Thus

$$r\frac{dg(r)}{dr} - g(r) = 0$$

Solving:
$$g(r) = Gr \qquad (9)$$

where G is a constant. Likewise, equating θ-terms to zero:

$$\frac{df(\theta)}{d\theta} + \int f(\theta)\,d\theta = 0$$

which is readily solved to yield

$$f(\theta) = H\sin\theta + K\cos\theta \qquad (10)$$

where H and K are constants. The displacements are found by substituting (9) and (10) into (5) and (7) to yield

$$u = \frac{1}{E}\left[-\frac{(1+\mu)B}{r} + 2(1-\mu)Cr + 2(1-\mu)Dr\ln r - D(1+\mu)r\right] + H\sin\theta + K\cos\theta \qquad (11)$$

$$v = \frac{4Dr\theta}{E} + H\cos\theta - K\sin\theta + Gr \qquad (12)$$

It is to be observed that for the case of a continuous body in the form of a complete 360° circle that it is necessary to set $D = 0$; otherwise (12) would lead to multiple-valued components of tangential displacement v. That is, at any point in the body represented by the coordinates r and θ a nonzero value of D would imply a different displacement of this same point if θ were to be replaced by, say, $\theta + 2\pi$ which could represent the same point.

19.15. Determine equations for the stresses in a thick-walled cylinder subject to both internal and external pressure. Consider the cylinder to be very long and take all body forces to be zero.

Since body forces are to be considered to be zero, the axisymmetric stress distribution determined in Problem 19.12 is suitable. Thus

$$\sigma_r = \frac{B}{r^2} + 2C + D(1 + 2\ln r) \tag{1}$$

$$\sigma_\theta = -\frac{B}{r^2} + 2C + D(3 + 2\ln r) \tag{2}$$

$$\tau_{r\theta} = 0 \tag{3}$$

From Problem 19.14 we must take $D = 0$ to avoid multiple-valued tangential displacements. Then, B and C must be determined from boundary conditions at the inner and outer surfaces of the cylinder. We shall take the internal pressure to be p_i and it acts over the inner radius r_i. The external pressure is p_o acting on the outer radius r_o, as shown in Fig. 19-8. Thus we have as boundary conditions:

$$[\sigma_r]_{r=r_i} = -p_i \tag{4}$$

$$[\sigma_r]_{r=r_o} = -p_o \tag{5}$$

Substituting (1) into (4) and into (5), we obtain

$$\frac{B}{r_i^2} + 2C = -p_i \tag{6}$$

$$\frac{B}{r_o^2} + 2C = -p_o \tag{7}$$

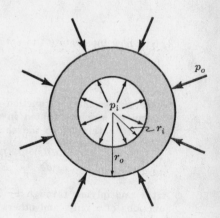

Fig. 19-8

Solving these equations for B and C and substituting the solutions in (1) and (2) we obtain the general solution for the thick-walled cylinder subject to both external and internal pressure:

$$\sigma_r = \frac{r_i^2 r_o^2 (p_o - p_i)}{r^2(r_o^2 - r_i^2)} + \frac{p_i r_i^2 - p_o r_o^2}{r_o^2 - r_i^2} \tag{8}$$

$$\sigma_\theta = -\frac{r_i^2 r_o^2 (p_o - p_i)}{r^2(r_o^2 - r_i^2)} + \frac{p_i r_i^2 - p_o r_o^2}{r_o^2 - r_i^2} \tag{9}$$

These equations indicate that for the special case $p_o = 0$ the radial stress σ_r is everywhere compressive and the tangential stress σ_θ everywhere tensile. The tangential stress from (9) clearly attains its maximum value at $r = r_i$. The maximum tangential stress is always greater (in absolute value) than the maximum radial stress.

19.16. A thick-walled steel cylinder of 3 in. inside diameter is subjected to an internal pressure of 6000 lb/in². There is no external pressure. The working stress of the material is 18,000 lb/in². Determine the required outside diameter of the cylinder.

Since the critical stress is the tangential stress at the inner fibers, we employ (9) of Problem 19.15 with $r_i = 1.5$ inches, $p_o = 0$, $p_i = 6000$ lb/in², to get

$$18,000 = -\frac{(1.5)^2(r_o^2)(0 - 6000)}{(1.5)^2[r_o^2 - (1.5)^2]} + \frac{6000(1.5)^2 - 0}{r_o^2 - (1.5)^2}$$

Solving:

$$r_o = 2.12 \text{ in.} \qquad \text{outer diameter} = 2r_o = 4.24 \text{ in.}$$

19.17. Determine the radial displacement at any point for the thick-walled cylinder treated in Problem 19.15.

In Problem 19.14 we found the equation (11) for the radial displacement u for axisymmetric stress distribution. As mentioned there we must set $D = 0$ for the case of a complete 360° body. Also, for the thick-walled cylinder u is clearly independent of θ; hence we must set $H = K = 0$ in (11) leaving

$$u = \frac{1}{E}\left[-\frac{(1+\mu)B}{r} + 2Cr(1-\mu)\right] \tag{1}$$

But for the thick-walled cylinder B and $2C$ are found from (6) and (7) of Problem 19.15 to be

$$B = \frac{r_i^2 r_o^2}{(r_o^2 - r_i^2)}(p_o - p_i) \tag{2}$$

$$2C = \frac{p_i r_i^2 - p_o r_o^2}{r_o^2 - r_i^2} \tag{3}$$

Substituting:

$$u = \frac{r^2(p_i r_i^2 - p_o r_o^2)(1-\mu) + (p_i - p_o)r_i^2 r_o^2(1+\mu)}{Er(r_o^2 - r_i^2)} \tag{4}$$

19.18. A steel cylinder has an inside diameter 5 in. and outside diameter 8 in. Another cylinder of inside diameter 7.995 in. and outside diameter 10 in. is heated and shrunk over the inner cylinder. Determine the stresses in the composite cylinder due to an internal pressure of 25,000 lb/in² acting in the inner cylinder.

We must first determine the interfacial pressure acting at the boundary of the two cylinders where the radius is approximately 4 in. Let us for the moment neglect the internal pressure of 25,000 lb/in² and treat the inner cylinder ($r_i = 2.5$ in., $r_o = 4$ in.) as subject only to the unknown interfacial pressure which we shall designate as p. The outer boundary of the inner cylinder according to (4) of Problem 19.17 moves radially an amount

$$u_{\text{inner}} = \frac{(4)^2[0 - p(4)^2](1-0.3) + (0-p)(2.5)^2(4)^2(1+0.3)}{(30\times10^6)(4)[(4)^2 - (2.5)^2]} = -7.92\frac{p}{30\times10^6}$$

due to the interfacial pressure. The negative sign indicates that the displacement is toward the center of the cylinder. The inner boundary of the outer cylinder ($r_i = 4$ in., $r_o = 5$ in.) moves radially an amount

$$u_{\text{outer}} = \frac{(4)^2[p(4)^2 - 0](1-0.3) + (p-0)(4)^2(5)^2(1+0.3)}{(30\times10^6)(4)[(5)^2 - (4)^2]} = 19.4\frac{p}{30\times10^6}$$

Since the initial interference of diameters is 0.005 in., we have

$$7.92\frac{p}{30\times10^6} + 19.4\frac{p}{30\times10^6} = 0.0025 \quad \text{or} \quad p = 2750 \text{ lb/in}^2$$

This interfacial pressure acts as an external pressure on the inner cylinder and produces tangential stresses given by (9) of Problem 19.15 as

$$[\sigma_\theta]_{r=2.5} = -\frac{(2.5)^2(4)^2(2750-0)}{(2.5)^2[(4)^2-(2.5)^2]} + \frac{0 - 2750(4)^2}{[(4)^2-(2.5)^2]} = -9000 \text{ lb/in}^2$$

$$[\sigma_\theta]_{r=4} = -\frac{(2.5)^2(4)^2(2750-0)}{(4)^2[(4)^2-(2.5)^2]} + \frac{0 - 2750(4)^2}{[(4)^2-(2.5)^2]} = -6280 \text{ lb/in}^2$$

It also acts as an internal pressure on the outer cylinder and gives rise to the following tangential stresses in that cylinder:

$$[\sigma_\theta]_{r=4} = -\frac{(4)^2(5)^2(0-2750)}{(4)^2[(5)^2-(4)^2]} + \frac{2750(4)^2-0}{[(5)^2-(4)^2]} = 12,500 \text{ lb/in}^2$$

$$[\sigma_\theta]_{r=5} = -\frac{(4)^2(5)^2(0-2750)}{(5)^2[(5)^2-(4)^2]} + \frac{2750(4)^2-0}{[(5)^2-(4)^2]} = 9800 \text{ lb/in}^2$$

Next, we consider the entire assembly as one thick-walled cylinder ($r_i = 2.5$ in., $r_o = 5$ in.) subject to an applied internal pressure of 25,000 lb/in^2. The tangential stresses at $r = 2.5$ in., $r = 4$ in., and $r = 5$ in. are again found from (9) of Problem 19.15 to be

$$[\sigma_\theta]_{r=2.5} = -\frac{(2.5)^2(5)^2(0-25,000)}{(2.5)^2[(5)^2-(2.5)^2]} + \frac{25,000(2.5)^2-0}{[(5)^2-(2.5)^2]} = 41,800 \text{ lb/in}^2$$

$$[\sigma_\theta]_{r=4} = -\frac{(2.5)^2(5)^2(0-25,000)}{(4)^2[(5)^2-(2.5)^2]} + \frac{25,000(2.5)^2-0}{[(5)^2-(2.5)^2]} = 21,300 \text{ lb/in}^2$$

$$[\sigma_\theta]_{r=5} = -\frac{(2.5)^2(5)^2(0-25,000)}{(5)^2[(5)^2-(2.5)^2]} + \frac{25,000(2.5)^2-0}{[(5)^2-(2.5)^2]} = 16,700 \text{ lb/in}^2$$

The true stress field is now found by superposition of the above results. In the inner cylinder we have

$$[\sigma_\theta]_{r=2.5} = -9000 + 41,800 = 32,800 \text{ lb/in}^2$$

$$[\sigma_\theta]_{r=4} = -6280 + 21,300 = 15,020 \text{ lb/in}^2$$

and in the outer cylinder

$$[\sigma_\theta]_{r=4} = 12,500 + 21,300 = 33,800 \text{ lb/in}^2$$

$$[\sigma_\theta]_{r=5} = 9,800 + 16,700 = 26,500 \text{ lb/in}^2$$

Since the maximum tangential stress is always greater (in absolute value) than the maximum radial stress (which in this case is 25,000 lb/in^2, at $r_i = 2.5$ in.) it is seen from the above that the shrink fit has the effect of reducing the peak stress from a value of 41,800 lb/in^2 in a cylinder with no shrink fit to a value of 33,800 lb/in^2 in the shrink-fit assembly.

19.19. Determine the stress distribution due to a concentrated normal force acting on a straight boundary of a semi-infinite plate.

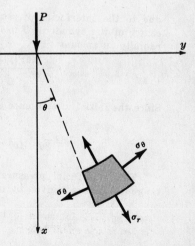

We shall consider the plate to be of unit thickness and take the distribution of force across the thickness to be uniform. The resultant force acting on this unit thickness we shall designate by P.

Let us investigate a stress function of the form $\Phi = Kr\theta \sin \theta$ where K is a constant to be determined so as to satisfy equilibrium. From the general equations for stresses in polar coordinates as discussed in Problem 19.11, this stress function leads to the stresses

$$\sigma_r = \frac{1}{r}\frac{\partial\Phi}{\partial r} + \frac{1}{r^2}\frac{\partial^2\Phi}{\partial\theta^2} = +2K\frac{\cos\theta}{r}$$

$$\sigma_\theta = \frac{\partial^2\Phi}{\partial r^2} = 0$$

$$\tau_{r\theta} = -\frac{\partial}{\partial r}\left(\frac{1}{r}\frac{\partial\Phi}{\partial\theta}\right) = 0$$

Fig. 19-9

Thus, on the shaded element shown in Fig. 19-9, the tangential stress σ_θ and shearing stress $\tau_{r\theta}$ are everywhere zero. The boundary conditions are satisfied because σ_θ and $\tau_{r\theta}$ vanish everywhere along the straight edge of the plate, which is necessary since that surface is unloaded except directly under the point of application of the force P. The constant K may be determined by considering the equilibrium of forces acting on any cylindrical surface of radius r as shown in Fig. 19-10.

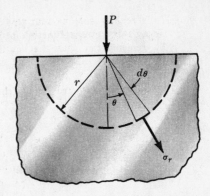

$$-P - 2 \int_0^{\pi/2} \sigma_r (r\,d\theta) \cos\theta = 0$$

or

$$P = -2 \int_0^{\pi/2} \left(2K \frac{\cos\theta}{r} \right) (\cos\theta) r\,d\theta$$

Fig. 19-10

Hence

$$K = -\frac{P}{\pi} \quad \text{and} \quad \sigma_r = -\frac{2P}{\pi} \frac{\cos\theta}{r}$$

It is seen that the stress becomes infinite under the point of application $(r = 0)$ of the concentrated force. Actually there is some yielding or flow of the material which distributes the concentrated force P over a finite area.

Thus the stress function investigated led to the satisfaction of boundary conditions corresponding to a load-free boundary at all points except directly under the point of application of the concentrated force where the stress was theoretically infinitely large. This problem was solved by Boussinesq in 1892.

19.20. Consider the case of an infinitely large, thin flat plate with a small circular hole. The plate is subject to a uniaxial tension T in its plane. Determine the state of stress in the vicinity of the hole. Neglect body forces.

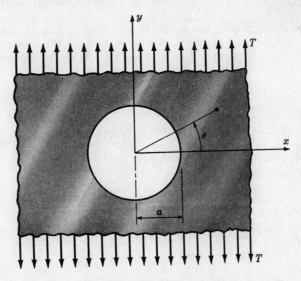

Fig. 19-11

The plate is shown in Fig. 19-11, with the polar coordinate system having its origin at the center of the small hole of radius a. We shall consider an Airy stress function of the form

$$\Phi = \frac{T}{4}\left[r^2 - 2a^2 \ln r - \frac{(r^2 - a^2)^2}{r^2} \cos 2\theta \right] \tag{1}$$

It is readily verified that this function indeed satisfies (4) of Problem 19.10. Thus, it may serve as an Airy stress function provided it satisfies the boundary conditions of the problem. It is to be remembered that the equilibrium equations are automatically satisfied by the stresses implied by (1), since the biharmonic equation is satisfied by Φ.

The stresses are found from Problem 19.11 to be

$$\sigma_r = \frac{1}{r}\frac{\partial \Phi}{\partial r} + \frac{1}{r^2}\frac{\partial^2 \Phi}{\partial \theta^2} = \frac{T}{2}\left[1 - \frac{a^2}{r^2} - \left(1 - \frac{4a^2}{r^2} + \frac{3a^4}{r^4}\right)\cos 2\theta\right] \qquad (2)$$

$$\sigma_\theta = \frac{\partial^2 \Phi}{\partial r^2} = \frac{T}{2}\left[1 + \frac{a^2}{r^2} + \left(1 + 3\frac{a^4}{r^4}\right)\cos 2\theta\right] \qquad (3)$$

$$\tau_{r\theta} = -\frac{\partial}{\partial r}\left(\frac{1}{r}\frac{\partial \Phi}{\partial \theta}\right) = \frac{T}{2}\left(+1 + \frac{2a^2}{r^2} - \frac{3a^4}{r^4}\right)\sin 2\theta \qquad (4)$$

With regard to boundary conditions, we must examine the stresses both at the boundary of the hole and at a location far remote from the hole. At the hole ($r = a$) equations (1) and (4) indicate that both σ_r and $\tau_{r\theta}$ vanish, as indeed they must since the boundary of the hole is free of any surface loads. At a point remote from the hole, i.e. as $r \to \infty$, the above equations indicate the following state of stress:

$$\sigma_r = \frac{T}{2}(1 - \cos 2\theta) \qquad (5)$$

$$\sigma_\theta = \frac{T}{2}(1 + \cos 2\theta) \qquad (6)$$

$$\tau_{r\theta} = +\frac{T}{2}\sin 2\theta \qquad (7)$$

But comparison with the results of Problem 17.1 indicates that (5), (6), and (7) correspond to the case of uniaxial tension. It is obvious that this should be the case at any point a considerable distance from the small hole. Thus, all boundary conditions are satisfied by the stress function.

The most interesting feature of this problem lies in the evaluation of the tangential stress at $\theta = 0$ and $\theta = \pi$, at $r = a$, i.e. an element at the edge of the hole. The tangential stress on such an element is of course in the direction of the applied tensile loading. From (2) this stress is

$$[\sigma_\theta]_{\substack{r=a \\ \theta=0,\pi}} = \frac{T}{2}[1 + 1 + (1+3)(1)] = 3T \qquad (8)$$

Thus, a small hole in a plate subject to uniaxial tension gives rise to a maximum normal stress three times the value that would exist if there were no hole in the plate. This is an illustration of the stress concentration that can arise because of holes, cuts, notches, and imperfections in an elastic body.

19.21. Determine the relations governing the torsion of prismatic bars.

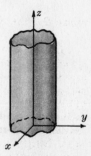

Fig. 19-12

Let us consider a prismatic bar of arbitrary cross-section subject to torsional moments M applied at each of the ends. We shall take the z-axis to coincide with the geometric axis of the bar and the x- and y-axes to lie in a plane normal to the geometric axis, as in Fig. 19-12. It is evident that, in general, deformations consist of rotations of the cross-sections together with displacements out of the plane of the cross-section, termed warping. For the coordinate system shown, we have

$$u = -\theta yz \tag{1}$$

$$v = \theta xz \tag{2}$$

for the displacements parallel to the x- and y-axes, respectively. Here θ is the rotation of the cross-section per unit length of the z-axis. We shall assume that we can represent the warping by a function of the form

$$w = \theta\Omega(x, y) \tag{3}$$

From the strain-displacement relations of Problem 19.3 we now have

$$\epsilon_x = \frac{\partial u}{\partial x} = 0; \quad \epsilon_y = \frac{\partial v}{\partial y} = 0; \quad \epsilon_z = \frac{\partial w}{\partial z} = 0 \tag{4}$$

$$\gamma_{xy} = \frac{\partial u}{\partial y} + \frac{\partial v}{\partial x} = 0; \quad \gamma_{xz} = \frac{\partial u}{\partial z} + \frac{\partial w}{\partial x} = -\theta y + \theta\frac{\partial\Omega}{\partial x}; \quad \gamma_{yz} = \frac{\partial v}{\partial z} + \frac{\partial w}{\partial y} = +\theta x + \theta\frac{\partial\Omega}{\partial y} \tag{5}$$

The three-dimensional form of Hooke's law thus indicates the following stress field:

$$\sigma_x = \sigma_y = \sigma_z = \tau_{xy} = 0 \tag{6}$$

$$\tau_{xz} = G\theta\left(\frac{\partial\Omega}{\partial x} - y\right) \tag{7}$$

$$\tau_{yz} = G\theta\left(\frac{\partial\Omega}{\partial y} + x\right) \tag{8}$$

As is to be expected, there are no longitudinal normal stresses.

We must now determine the warping function Ω so that the equilibrium equations of Problem 19.1 are satisfied. If (6), (7), and (8) are substituted into (3) of Problem 19.1, we obtain

$$\frac{\partial^2\Omega}{\partial x^2} + \frac{\partial^2\Omega}{\partial y^2} = 0 \tag{9}$$

Equations (1) and (2) of Problem 19.1 are satisfied identically.

We may also use (3) of Problem 19.1 in the form (since $\sigma_z = 0$)

$$\frac{\partial\tau_{yz}}{\partial y} + \frac{\partial\tau_{xz}}{\partial x} = 0 \tag{10}$$

Let us now introduce a stress function $\phi(x, y)$ defined by the equations

$$\tau_{xz} = \frac{\partial\phi}{\partial y}; \quad \tau_{yz} = -\frac{\partial\phi}{\partial x} \tag{11}$$

If these last relations be substituted in (7) and (8) and the variable Ω eliminated, we obtain

$$\frac{\partial^2\phi}{\partial x^2} + \frac{\partial^2\phi}{\partial y^2} = -2G\theta \tag{12}$$

where G is the modulus of rigidity in shear.

Let us now examine the condition at the cylindrical boundary of the bar. In Fig. 19-13 are shown the forces acting on a surface element of an arbitrary cross-section. According to (6), (7), and (8) there are no other forces acting on the element. We must also realize that the cylindrical boundary is free of external loads; hence the resultant shearing stress for the element shown must be tangent to the boundary with no component normal to the boundary.

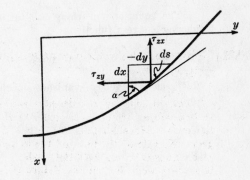

Fig. 19-13

The presence of a normal component on the cross-section should, by Problem 19.2, necessitate an applied shearing force on the cylindrical boundary, which is contradictory to the load-free stipulation. Thus, for a zero normal component we must have

$$\tau_{zx} \sin \alpha + \tau_{zy} \cos \alpha = 0 \tag{13}$$

$$\tau_{zx}\left(\frac{-dy}{ds}\right) + \tau_{zy}\left(\frac{dx}{ds}\right) = 0 \tag{14}$$

Substituting from (11):

$$\frac{\partial \phi}{\partial y}\left(\frac{-dy}{ds}\right) - \frac{\partial \phi}{\partial x}\frac{dx}{ds} = 0 \tag{15}$$

The negative sign accompanying dy/ds is required because as s increases in a counterclockwise direction, a positive increment of s implies a negative increment of y.

But from (15) we have

$$\frac{d\phi}{ds} = 0 \tag{16}$$

i.e., the stress function $\phi(x, y)$ must be constant along the boundary. We shall take this constant to be zero and thus the solution of the elastic torsion problem reduces to solving (12) for a stress function ϕ which vanishes on the boundary.

It is also necessary to determine the relation between the stress function ϕ and the applied torsional moments. From Fig. 19-13,

$$M = \int (\tau_{zx}y - \tau_{zy}x)\, dx\, dy = \int \left(y\frac{\partial \phi}{\partial y} + x\frac{\partial \phi}{\partial x}\right) dx\, dy$$

This is readily integrated by parts to yield

$$M = 2 \iint \phi\, dx\, dy \tag{17}$$

since ϕ vanishes at the boundary. The integration is of course extended over the cross-section of the bar.

19.22. Investigate the torsion of a bar of elliptical cross-section.

For an ellipse with semi-major axis a and semi-minor axis b the boundary of the cross-section is given by

$$\frac{x^2}{a^2} + \frac{y^2}{b^2} = 1 \tag{1}$$

The equation governing the torsion problem, namely (12) of Problem 19.19, is satisfied by considering a stress function

$$\phi = C\left(\frac{x^2}{a^2} + \frac{y^2}{b^2} - 1\right) \tag{2}$$

which, as is required, vanishes on the boundary. If (2) is substituted in (12) of Problem 19.21, we find

$$C = \frac{a^2b^2}{2(a^2+b^2)}(-2G\theta) \tag{3}$$

Thus

$$\phi = \frac{-a^2b^2G\theta}{a^2+b^2}\left(\frac{x^2}{a^2} + \frac{y^2}{b^2} - 1\right) \tag{4}$$

If this value of ϕ is now substituted in (17) of Problem 19.21,

$$M = \frac{\pi a^3 b^3 G \theta}{a^2 + b^2} \qquad (5)$$

which is the relation between applied torsional moment and angle of twist θ per unit length.

If now (5) is substituted in (4), we obtain

$$\phi = -\frac{M}{\pi ab}\left(\frac{x^2}{a^2} + \frac{y^2}{b^2} - 1\right) \qquad (6)$$

The shearing stresses are readily found from (11) of Problem 19.21 to be

$$\tau_{xz} = -\frac{2My}{\pi ab^3} \qquad \tau_{yz} = \frac{2Mx}{\pi a^3 b} \qquad (7)$$

From these equations it is evident that the maximum shearing stress occurs at the ends of the minor axis of the elliptical cross-section and is given numerically by

$$\tau_{max} = \frac{2M}{\pi ab^2} \qquad (8)$$

If $a = b$ the results agree with those found in Chapter 5 for a bar of circular cross-section.

With the values of stresses known from (7), one may use (5) to find θ then substitute these values in equations (3), (7), and (8) of Problem 19.21 to obtain the displacement

$$w = \frac{M(b^2 - a^2)xy}{\pi a^3 b^3 G} \qquad (9)$$

due to warping of the cross-section.

Supplementary Problems

19.23.   Consider the set of orthogonal displacements $u = Ax$, $v = Ay$, $w = Az$ in the rectangular Cartesian coordinate system.   What type of deformation does this describe?

Ans.   uniform dilatation

19.24.   Consider the functions $\epsilon_x = axy$; $\epsilon_y = by^3$; $\gamma_{xy} = c - dy^2$ where a, b, c, and d are arbitrary constants.   Can this describe a state of plane stress in a continuous elastic medium?   *Ans.* yes

19.25.   Consider the functions $\epsilon_x = axy^2$; $\epsilon_y = ax^2y$; $\epsilon_z = axy$; $\gamma_{xy} = 0$; $\gamma_{yz} = az^2 + by$; $\gamma_{xz} = ax^2 + by^2$. Can this describe the strains in a continuous elastic medium?   The parameters a and b are nonzero.

Ans.   no

19.26.   Given the functions $\Phi = Axy^4 + Bx^3y^2$.   Determine the relation between A and B so that Φ can be an Airy stress function for the two-dimensional state of stress with zero body forces.

Ans. $A = -B$

19.27.   Given the stress function $\Phi = Ay^3$.   Describe the state of two-dimensional stress represented for the case of zero body forces.

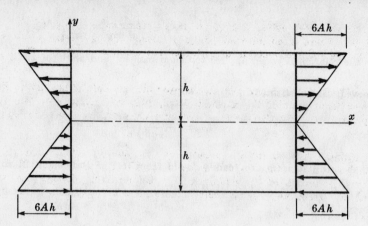

Fig. 19-14

19.28. The plane-stress state of the semicircular ring shown in Fig. 19-15 is described by $\sigma_r = (-A/r)\sin\theta$, $\sigma_\theta = (B/r)\sin\theta$, $\tau_{r\theta} = (B/r)\cos\theta$. Determine the loadings responsible for this state of stress.

Ans. Relative to the Oy axis there is a symmetric radial and shear loading. The radial (normal) load increases from both the right and left edges toward the axis of symmetry. It acts as a compressive edge load at $r = a$ and also at $r = b$. The shear force decreases and changes algebraic sign at the axis of symmetry, having its numerical maximum values at $\theta = 0$ and $\theta = \pi$. The radial faces $\theta = 0$ and $\theta = \pi$ are free of normal stress.

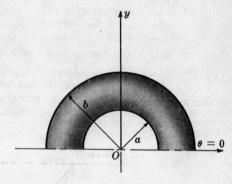

Fig. 19-15

19.29. Consider a function of the form $\Phi = f(r)\theta$. Determine $f(r)$ so that Φ is an Airy stress function for the plane polar coordinate system.   *Ans.* $f(r) = A + B\ln r + Cr^2 + Dr^2\ln r$

19.30. Consider a function in polar coordinates of the form $\Phi = f(r)\cos n\theta$, where n is an integer. Determine $f(r)$ so that Φ is an Airy stress function.

Ans. $f(r) = Ar^n + Br^{-n} + Cr^{2+n} + Dr^{2-n}$ $(n > 1)$

$f(r) = Ar + Br^{-1} + Cr^3 + Dr\ln r$ $(n = 1)$

19.31. Determine f_1 and f_2 so that the function $\Phi = f_1(r)\sin\theta + f_2(r)\theta\cos\theta$ is an Airy stress function for the plane polar coordinate case.

Ans. $f_1 = Ar^3 + Br\ln r + Cr + \dfrac{D}{r}$, $f_2 = A_1 r$

19.32. Determine the stress field corresponding to the stress function discussed in Problem 19.31.

Ans. $\sigma_r = \left(2Ar + \dfrac{B}{r} - \dfrac{2D}{r^3}\right)\sin\theta - \dfrac{2A_1}{r}\sin\theta$

$\sigma_\theta = \left(6Ar + \dfrac{B}{r} + \dfrac{2D}{r^3}\right)\sin\theta$

$\tau_{r\theta} = -\left(2Ar + \dfrac{B}{r} - \dfrac{2D}{r^3}\right)\sin\theta$

19.33. A thick-walled steel cylinder of 3 in. inside diameter is subjected to an external pressure of 6000 lb/in². There is no internal pressure. The working stress of the material is 18,000 lb/in². Determine the required outside diameter of the cylinder.   *Ans.* 5.20 in.

19.34. A thick-walled steel cylinder has an external diameter of 7.5 in. and an inside diameter of 3.5 in. The proportional limit of the material is 52,000 lb/in². Determine the maximum permissible internal pressure so that the action is still entirely elastic.   *Ans.* 33,500 lb/in²

19.35. A steel cylinder has an inside diameter of 4 in. and an outside diameter of approximately 5 in. Over it is shrunk another steel cylinder of outside diameter 7 in. The assembly is subjected to an internal pressure of 22,000 lb/in². If the maximum permissible stress is not to exceed 38,000 lb/in², determine the necessary interference of diameters prior to the shrink fit.

Ans. 0.00129 in.

19.36. Consider the two-dimensional elastic medium with no body forces present. For a function of the form $\Phi = f(y) \sin \alpha x$ to represent an Airy stress function, what must the function $f(y)$ be? Here $f(y)$ is a function of y only, $\alpha = n\pi/L$, and n and L are integers. This type of function is frequently useful for solving problems involving varying loads acting along the boundaries of a rectangular block.   *Ans.* $f(y) = c_1 \cosh \alpha y + c_2 \sinh \alpha y + c_3 y \cosh \alpha y + c_4 y \sinh \alpha y$

19.37. Determine the stress distribution due to a concentrated tangential force P applied along the straight boundary of a semi-infinite plate (Fig. 19-16).

Ans. $\sigma_r = -\dfrac{2P}{\pi}\dfrac{\sin \theta}{r}, \quad \sigma_\theta = \tau_{r\theta} = 0$

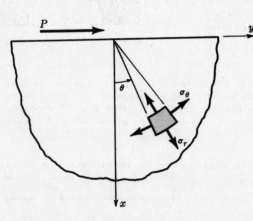

Fig. 19-16

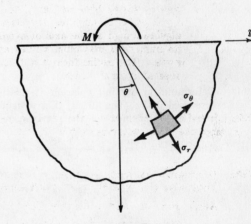

Fig. 19-17

19.38. Determine the stress distribution due to a concentrated moment applied at a point on the straight boundary of a semi-infinite plate (Fig. 19-17).

Ans. $\sigma_r = -\dfrac{4M}{\pi r^2} \sin \theta \cos \theta, \quad \sigma_\theta = 0,$

$\tau_{r\theta} = -\dfrac{2M}{\pi r^2} \cos^2 \theta$

19.39. Using a stress function of the type employed in Problem 19.19, determine the stress distribution in an elastic wedge loaded by a concentrated force at its apex as shown in Fig. 19-18.

Ans. $\sigma_r = -\dfrac{P \cos \theta}{r(\alpha + \frac{1}{2} \sin 2\alpha)}, \quad \sigma_\theta = \tau_{r\theta} = 0$

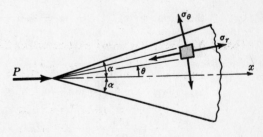

Fig. 19-18

19.40. Consider the infinite wedge of Fig. 19-19, subject to a uniform normal loading p distributed along one edge. Using a stress function of the form $\Phi = r^2(c_1 \cos 2\theta + c_2 \sin 2\theta + c_3 \theta + c_4)$, determine the stress distribution in the wedge so that all boundary conditions are satisfied. The lower surface is load free.

Ans. $\sigma_r = -\dfrac{p}{2}(1 + 2b\theta + c \sin 2\theta)$

         $\sigma_\theta = -\dfrac{p}{2}(1 + 2b\theta - c \sin 2\theta)$

         $\tau_{r\theta} = -\dfrac{p}{2}(-b + c \cos 2\theta)$

where

$b = \dfrac{\cos 2\beta}{2\beta \cos 2\beta - \sin 2\beta}$

$c = \dfrac{1}{2\beta \cos 2\beta - \sin 2\beta}$

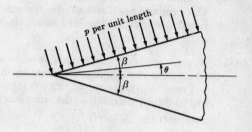

Fig. 19-19

19.41. In Problem 19.11 we considered the equilibrium conditions in two-dimensional plane polar coordinates. Extend this to the three-dimensional case by introducing a z-axis through the origin of coordinates. Neglect body forces.

Ans. $\dfrac{\partial \sigma_r}{\partial r} + \dfrac{1}{r}\dfrac{\partial \tau_{r\theta}}{\partial \theta} + \dfrac{\partial \tau_{rz}}{\partial z} + \dfrac{\sigma_r - \sigma_\theta}{r} = 0$

         $\dfrac{\partial \tau_{r\theta}}{\partial r} + \dfrac{1}{r}\dfrac{\partial \sigma_\theta}{\partial \theta} + \dfrac{\partial \tau_{\theta z}}{\partial z} + \dfrac{2\tau_{r\theta}}{r} = 0$

         $\dfrac{\partial \tau_{rz}}{\partial r} + \dfrac{1}{r}\dfrac{\partial \tau_{\theta z}}{\partial \theta} + \dfrac{\partial \sigma_z}{\partial z} + \dfrac{\tau_{rz}}{r} = 0$

19.42. Determine the strain-displacement relations for the three-dimensional cylindrical polar coordinates discussed in Problem 19.41. Let w denote the z-component of displacement. Compare the results with those obtained for the plane system in Problem 19.13.

Ans. $\epsilon_r = \dfrac{\partial u}{\partial r}, \quad \epsilon_\theta = \dfrac{1}{r}\dfrac{\partial v}{\partial \theta} + \dfrac{u}{r}, \quad \epsilon_z = \dfrac{\partial w}{\partial z}$

         $\gamma_{r\theta} = \dfrac{\partial v}{\partial r} + \dfrac{1}{r}\dfrac{\partial u}{\partial \theta} - \dfrac{v}{r}, \quad \gamma_{\theta z} = \dfrac{1}{r}\dfrac{\partial w}{\partial \theta} + \dfrac{\partial v}{\partial z}$

         $\gamma_{rz} = \dfrac{\partial u}{\partial z} + \dfrac{\partial w}{\partial r}$

19.43. Investigate the torsion of a bar of equilateral triangular cross-section. Use the coordinate system shown in Fig. 19-20 and consider a stress function of the form $\Phi = C\left(x^3 - hx^2 - 3y^2 x - hy^2 + \dfrac{4h^3}{27}\right)$.

Ans. $M = \dfrac{G\theta a h^3}{30}$

         $\tau_{xz} = -\dfrac{30M}{ah^4} y (3x + h)$

         $\tau_{yz} = \dfrac{15M}{ah^4}(3y^2 - 3x^2 + 2hx)$

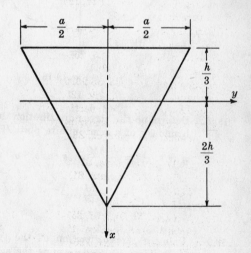

Fig. 19-20

INDEX